Applying Physical Ergonomics to Modern Ship Design

Alexander Arnfinn Olsen

Applying Physical Ergonomics to Modern Ship Design

Springer

Alexander Arnfinn Olsen
Southampton, UK

ISBN 978-3-031-57973-8 ISBN 978-3-031-57974-5 (eBook)
https://doi.org/10.1007/978-3-031-57974-5

This Springer imprint is published by the registered company Springer Nature Switzerland AG
The registered company address is: Gewerbestrasse 11, 6330 Cham, Switzerland

Paper in this product is recyclable.

Preface

The focus of ergonomics is the study of the role of humans in the safe and efficient operation of complex industrial systems and the application of ergonomic principles and data to the design of equipment and systems. The importance of the "human element" and ergonomics in maritime safety is increasingly recognised and embraced by the maritime community (O'Neil 2001). This book intends to inform vessel designers of the ergonomic principles and data appropriate to the design of navigation bridges on ships. This book draws on the principal international statutory and discretionary guidance addressing the design of bridge systems, including the International Association of Classification Societies Standard for Bridge Design, Equipment, and Arrangement; the International Safety of Life at Sea Convention (SOLAS), as amended, and the International Maritime Organisation's Guidelines on Ergonomic Criteria for Bridge Equipment and Layout. This book is appropriate to seagoing vessels of various types and categories, such as container vessels, oil carriers, bulk carriers, workboats, ferries and surface-type utility vessels and passenger vessels. This being the case, this book provides:

1. General ergonomic design guidance (design principles) for navigation bridges (Chap. 2).
2. Specific bridge design guidance is gleaned from international sources, such as the IMO and the IACS (see Chaps. 3 through 11).
3. A process to identify individual vessel bridge requirements to guide the application of ergonomic design principles (see Appendix A).

This book is not intended to provide more generally applicable detailed ergonomic design guidance or data, such as control dimensions, environmental limits or human anthropometry as this data can be readily found in other authoritative references. Rather, this book can be used to support the identification or refinement of requirements based on ergonomic analysis techniques and the application of accepted guiding principles of ergonomic design as related to ship's navigation bridges.

Southampton, UK
February 2024

Alexander Arnfinn Olsen

Reference

O'Neil, WA (2001) the human element in shipping. Keynote Address, Biennial Symposium of the Seafarers International Research Centre, Cardiff, Wales, 29 June 2001

Contents

Part I
Introducing the Human Element and Ergonomics to the Maritime Environment

Chapter 1
Introduction

The importance of the human element in maritime safety is increasingly being recognised by the shipping and offshore communities and is receiving increased levels of attention due to the efforts of organisations such as the United States Coast Guard (USCG), the United Kingdom's Health and Safety Executive (HSE), and the International Maritime Organisation (IMO). The IMO's primary efforts have concentrated on human element issues relating to management, training, and personnel, as reflected by the International Management Code for the Safe Operation of Ships and for Pollution Prevention (the ISM Code) in 1993 and the update of the International Convention on Standards of Training, Certification, and Watchkeeping for Seafarers (STCW) in 2012. While organisations such as the IMO have issued other documents, circulars, and guidelines related to aspects of ergonomics, systematic application of ergonomics in the maritime industry remains limited. This lack of systematic application occurs even though ergonomics has been recognised to be central to improving safety and productivity. This book is intended to promote the application and understanding of ergonomic knowledge to maritime and offshore systems. To that end, ergonomics, or human factors engineering as it is also referred to, can be defined as follows: The importance of the human element in maritime safety is increasingly being recognised by the shipping and offshore communities and is receiving increased levels of attention due to the efforts of organisations such as the United States Coast Guard (USCG), the United Kingdom's Health and Safety Executive (HSE), and the International Maritime Organisation (IMO). The IMO's primary efforts have concentrated on human element issues relating to management, training, and personnel, as reflected by the International Management Code for the Safe Operation of Ships and for Pollution Prevention (the ISM Code) in 1993 and the update of the International Convention on Standards of Training, Certification, and Watchkeeping for Seafarers (STCW) in 2012. While organisations such as the IMO have issued other documents, circulars, and guidelines related to aspects of ergonomics, systematic application of ergonomics in the maritime industry remains limited. This lack of systematic application occurs even though ergonomics has been

A. A. Olsen, *Applying Physical Ergonomics to Modern Ship Design*,
https://doi.org/10.1007/978-3-031-57974-5_1

recognised to be central to improving safety and productivity. This book is intended to promote the application and understanding of ergonomic knowledge to maritime and offshore systems. To that end, ergonomics, or human factors engineering as it is also referred to, can be defined as follows:

> Ergonomics is the scientific discipline concerned with the understanding of interactions among humans and other elements of a system, and the profession that applies theoretical principles, data, and methods to design in order to optimise human well-being and overall system performance.
>
> ***International Ergonomics Association, 2012***

The focus of ergonomics is the design of the human-system interface. This includes interfaces between personnel and the hardware, software, and physical environments associated with systems. It also involves the interfaces between personnel, individual tasks, and the overall work system (e.g., its structure, management, policies, and procedures). Throughout this book we will introduce important ergonomic principles and criterion that should be considered during the design of a vessel or offshore installation. A development effort without ergonomic considerations is likely to result in designs that encourage human error. Applying the ergonomic principles and criteria contained in this book within the context of engineering design processes will allow for improved safety, productivity, and efficiency. Although we fully support and recommend adherence to the ergonomic guidance provided throughout this book, compliance is voluntary and not mandatory.

The Human Elements that Enhance Safety at Sea

The human factors model shown in Fig. 1.1, Human factors model, encapsulates important elements that influence safety and efficiency in job performance: vessel or offshore installation design and layout considerations, workplace ambient environmental elements, management and organisational issues related to operations, and the personnel who operate the vessel or offshore installation. Insufficient attention to any of these elements may adversely affect safety, productivity, and efficiency. This book primarily addresses ergonomic design and layout concerns, and to a lesser extent the ambient environment.

Design and layout considerations. It is the design and layout aspect of the human factors model that forms the basis for the guidance and recommendations made in this book. Importantly, design and layout consider the integration of personnel with equipment, systems, and interfaces. Examples of interfaces include controls, displays, alarms, video display units, computer workstations, labels, ladders, stairs, and overall workspace arrangements. Ambient environmental considerations. The ambient environmental aspect of the human factors model addresses the habitability and occupational health characteristics related to human whole-body vibration, noise, indoor climate and lighting. Considerations related to personnel. Personnel readiness and fitness-for-duty are essential for vessel or offshore installation safety. This is

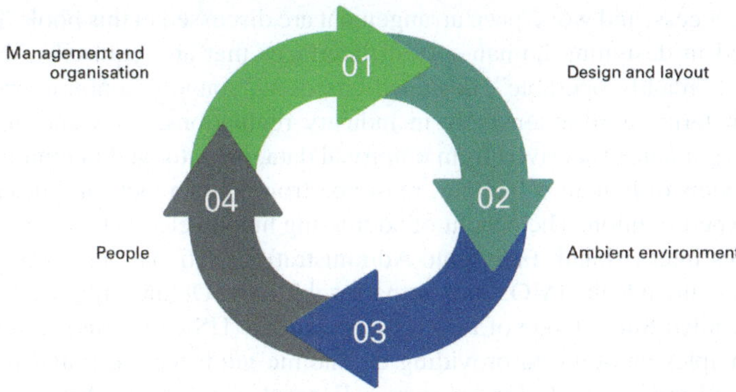

Fig. 1.1 Human elements model

particularly so as tasks and equipment increase in complexity, requiring ever greater vigilance, skills, and experience. The following factors should be considered when selecting personnel for a task:

- Knowledge, skills, and attitudes (KSA) that stem from an individual's basic knowledge, general training, and experience,
- Maritime-specific or craft-specific training and abilities (certifications and licenses), and vessel or offshore installation-specific skills and abilities,
- Bodily dimensions and characteristics of personnel such as stature, shoulder breadth, eye height, functional reach, overhead reach, weight, and strength,
- Physical stamina; capabilities, and limitations, such as resistance to and freedom from fatigue; visual acuity; physical fitness and endurance; acute or chronic illness; and substance dependency, and
- Psychological characteristics, such as individual tendencies for risk taking, risk tolerance, and resistance to psychological stress.

Management and organisational considerations. This aspect of the human factors model considers management and organisational considerations that impact safety throughout a system lifecycle. The effective implementation of a well-designed safety policy that includes ergonomics creates an environment that minimises risks and reflects a good corporate safety culture. Commitment of top management is essential if a safety policy is to succeed.

Scope of the Book

This book has been written to provide readers with design and layout guidance related to ergonomics at sea, including the design of human-system interfaces at the individual task and workstation levels. Physical and perceptual issues for the design of the personnel interface with controls, displays, alarms, video display units, labelling, workspace access, and workspace arrangement are discussed in this book. This book can be used in designing human-system interfaces that are required to be "easily accessible", "readily operable", designed based on "sound ergonomic principles", and similar terms used extensively in industry regulations, rules and documents. Ergonomic guidance is derived from empirical data, operator and maintainer expectations, models of human behaviour, reasoned from known facts and research, and user and expert opinion. The benefit of addressing human element concerns is noted in numerous international, Flag State Administration, national, and industry documents. These include the IMO, the International Labour Organisation (ILO), the UK HSE, the United States Code of Federal Regulations (US CFR), and so forth. Most of the principles and criteria providing ergonomic guidance are found in industry standards and ergonomic design references. Examples include the American Society for Testing and Materials (ASTM), the International Organisation for Standardisation (ISO), and British Standards (BS). The source of all guidelines, standards, and recommendations discussed in this book are listed in the References.

Structure of the Book

The book is divided into six parts with each part relating to a specific area of concern in the application of human factors to ship design and onboard operations. Accordingly:

- *Part 1* examines the application of human factors within the maritime environment,
- *Part 2* discusses the ergonomics associated with the design of the navigational bridge, equipment and systems,
- *Part 3* discusses the ergonomics associated with crew habitability on ships with a focus on the crew accommodation, vibration, noise, the indoor climate, and lighting and illumination,
- *Part 4* discusses the ergonomics associated with crew habitability on workboats with a focus on the crew accommodation, vibration, noise, the indoor climate, and lighting and illumination,
- *Part 5* discusses the ergonomics associated with personnel habitability on offshore installations with a focus on personnel accommodation, vibration, noise, the indoor climate, and lighting and illumination, and
- *Part 6* examines the human factors associated with living and working in low temperature marine environments.

Chapter 2
Controls

General

This chapter examines and discusses the design and usability of controls on a vessel's navigational bridge. The guidance contained in this chapter applies to devices used for control of equipment, consoles and panels.

Principles

Labelling. Controls should be labelled to convey the information needed for proper identification and utilisation of the control. The principles of control labelling are described in Chap. 8, Labelling, signs, graphics, and symbols. *Feedback.* Personnel should be provided with a positive indication of control activation. This can be provided via feel (e.g., snap action), an audible clicking noise, or a display *Integration with displays and alarms.* Displays and alarms can be used to provide feedback of control action. For principles and criteria covering the integration of controls with their associated displays and alarms, refer to Chap. 5, Integration of controls, displays, and alarms. *Coding.* Any coding used to enhance the transfer of information to personnel should be consistent and meaningful. The coding system should also be appropriate for the industry, organisation, and culture of the personnel aboard the vessel or offshore installation. *Simultaneous operation of controls.* Controls should be placed so that simultaneous operation of two controls will not require a crossing or interchanging of hands. Controls required to be used by two operators should be duplicated, where practical, or otherwise centred between the two operators, or positioned nearest to the operator having the greatest need. If any control is required to be operated with the operator's preferred hand, such as operating a keyboard or fine setting a continuous control knob, duplicate controls should be provided. *Controls for maintenance.* Controls used solely for maintenance or adjustment should be covered

A. A. Olsen, *Applying Physical Ergonomics to Modern Ship Design*,
https://doi.org/10.1007/978-3-031-57974-5_2

or otherwise protected during normal operations, but within arm's reach and visible when needed for maintenance. *Prevention of accidental activation.* Controls should be designed and located so they are not susceptible to accidental activation. Methods to reduce the likelihood of accidental activation include:

- Locating and orienting the control so that bumping is unlikely to cause an activation.
- Providing sufficient control resistance to prevent unintentional movements.
- Requiring complex motions for control activation, such as an interlock or rotary motion.
- Restricting access to controls by isolating them or by providing a cover guard or physical barrier.

Compatibility with clothing. It should be possible to operate controls wearing both normal clothing and personnel protective equipment (e.g., gloves, boots).

General Guidance

Control selection. With respect to control selection, the following guidelines should be applied in selecting controls:

- Controls requiring rapid or precise setting should be assigned to the hands.
- Controls requiring large or continuous forward applications of force generally should be assigned to the feet. Although a considerable number and variety of controls can be assigned to the hands, each foot should not have more than two controls assigned to it, and these should require only fore-aft or ankle flexion movement.
- Controls should be distributed so that no one limb is overburdened.
- Select, locate, and orient controls so that their motion is compatible with the movement of the associated display element, equipment, component, vessel or offshore installation.
- Select multi-rotation controls (e.g., cranks) when precise settings are required over a wide range of adjustments. Because the range of movement of a linear control is limited, it does not permit high precision over a wide range of adjustment. With this type of control, desired precision can be obtained by appropriate gearing, although consideration will need to be given to its effect on operating time.
- Select discrete-adjustment (detent) controls or push button arrays rather than continuous-adjustment controls when the controlled object is to be adjusted for discrete positions or values only. Discrete-adjustment controls are preferred when a limited number of settings are required, or when precision requirements are such that a limited number of settings can represent the entire continuum.
- Discrete-adjustment controls (i.e., controls that snap into place) can be positioned with one movement. Continuous adjustment controls require a slowing movement and a fine adjustment movement and thus more time and attention.

- Continuous-adjustment controls should be selected when precise adjustment along a continuum is needed.
- Select controls that can be easily identified. All controls should be made identifiable, in part by standardising their locations. All critical and emergency controls should be identifiable both visually and by touch (e.g., shape coding). Identification information should not hinder the manipulation of the control nor increase the likelihood of accidental activation.
- Combine functionally related controls to reduce reaching movements, to aid in sequential or simultaneous operations, or to economise panel space.
- Multi-axis continuous controllers (e.g., joysticks, track balls) should be used to provide continuous control within a two-dimensional space (e.g., positioning a pointer on a computer screen) or a three-dimensional space (e.g., operating a crane).

Control movement. Direction of control movement should be consistent among the same function and application and should operate according to the cultural expectations of the intended operators. Widely accepted movement expectations are indicated in Fig. 2.1, Control movement expectations and Table 2.1, Control movement expectations. This figure demonstrates that control movement expectations change depending on the location of the control in reference to an operator or maintainer's body position. Note that a toggle switch located above and behind the head operates in a different orientation to a toggle switch located directly in front of the body. For this reason, this figure should be closely examined when assigning control settings. Control spacing. Adequate spacing should be provided between controls and controls and obstructions so that manipulation of one will not inadvertently affect the others. Minimum spacing between controls is shown in Table 2.2, Minimum spacing between two controls. Spacing between a control and an obstruction or two different types of controls should be at least the comparable distance in Table 2.2. Unless otherwise specified, dimensions for hand-operated controls assume bare-handed operation, and should be increased as necessary if the personnel may be wearing gloves.

Control Spacing. Adequate spacing should be provided between controls and controls and obstructions so that manipulation of one will not inadvertently affect the others. Minimum spacing between controls is shown in Table 2.2, Minimum spacing between two controls. Spacing between a control and an obstruction or two different types of controls should be at least the comparable distance in Table 2.2. Unless otherwise specified, dimensions for hand-operated controls assume bare-handed operation, and should be increased as necessary if the personnel may be wearing gloves.

Mounting. (1) *Mounting heights on panels.* Frequently used controls should be located within a radius of 460 mm (18 in.) from the operator's centreline while less frequently used controls should be located within a radius of 800 mm (31.5 in.) from the operator's centreline. (2) *Mounting heights for kneeling and squatting.* While not preferred for normal operations or maintenance, if necessary, kneeling or squatting positions may be adopted by personnel. For kneeling, see Table 2.3,

Fig. 2.1 Control movement expectations

Table 2.1 Control movement expectations

Direction of movement	Function
Up, right, forward, clockwise, pull (push–pull type switch)	On
Down, left, rearward, counter-clockwise, push	Off
Clockwise, right	Right
Counter-clockwise, left	Left
Up, back	Raise
Down, forward	lower
Up, rearward, pull	Retract
Down, forward, push	Extend
Forward, up, right, clockwise	Increase
Rearward, down, left, counter-clockwise	Decrease
Counter-clockwise (valve)	Open valve
Clockwise (valve)	Close valve

[a] Reprinted with permission from the Annual Book of ASTM Standards, copyright American Society for Testing and Materials, 100 Barr Harbor Drive, West Conshohocken, PA 19428–2959

Controls and kneeling dimensions, and for squatting, see Table 2.4, Controls and squatting dimensions.

Coding. Coding refers to the use of size, shape, colour, texture, location, labelling, or other scheme to assist with the identification of states and control functions. Several

Table 2.2 Minimum spacing between two controls[a]

	Toggle switches	Push buttons	Continuous rotary controls	Rotary selector switches	Discrete thumbwheel controls
Toggle switches	19 mm (0.75 in.)	13 mm (0.5 in.)	19 mm (0.75 in.)	19 mm (0.75 in.)	13 mm (0.5 in.)
Push button	13 mm (0.5 in.)	13 mm (0.5 in.)	13 mm (0.5 in.)	13 mm (0.5 in.)	13 mm (0.5 in.)
Continuous rotary controls	19 mm (0.75 in.)	13 mm (0.5 in.)	25 mm (1.0 in.)	25 mm (1.0 in.)	19 mm (0.75 in.)
Rotary selector switches	19 mm (0.75 in.)	13 mm (0.5 in.)	25 mm (1.0 in.)	25 mm (1.0 in.)	19 mm (0.75 in.)
Discrete thumbwheel controls	13 mm (0.5 in.)	13 mm (0.5 in.)	19 mm (0.75 in.)	19 mm (0.75 in.)	10 mm (0.40 in.)

[a] For push buttons not separated by barriers

Table 2.3 Controls and kneeling dimensions[a]

Dimension	Size
Maximum height	1450 mm (57 in.)
Preferred maximum height	1080 mm (42.5 in.)
Preferred minimum height	540 mm (21 in.)
Minimum height	460 mm (18 in.)
Minimum depth for kneeling	1070 mm (42 in.)

[a] Dimensions are based on North American males

Table 2.4 Controls and squatting dimensions[a]

Dimension	Size
Maximum height	1250 mm (49 in.)
Preferred maximum height	800 mm (32 in.)
Preferred minimum height	400 mm (16 in.)
Minimum height	360 mm (14 in.)
Minimum depth for kneeling	915 mm (36 in.)

[a] Dimensions are based on North American males

coding methods may be combined to achieve maximum identification and differentiation. *Colour.* Colour coding should be used as a redundant form of information with other coding techniques. Colour coding is most effective when a specific meaning can be attached to a colour, and the colour is used consistently with the associated meaning. Cultural expectations[1] should be considered when colour coding. Only the

[1] A study of differences in colour perception between populations of the US and China revealed that of the 784 Chinese participants, comprising both sexes from various occupational groups, less than

following colours should be used for colour coding of controls: red, green, orange-yellow, and white. Blue may be used only if an additional colour is deemed absolutely necessary. If red ambient lighting could be used at any time in which the colour-coded control would be operated (e.g., night-time on the navigation bridge), controls that would otherwise be colour coded red should be coded orange-yellow with black striping. Controls that are not colour coded should be black or grey. *Shape.* Coding by shape is useful when controls should be identified without the use of vision. Using different shapes provides for tactile identification of controls and aids in visual identification. Where feasible, shapes that suggest the purpose of the control should be used. *Size.* Coding by size is useful when controls are to be identified without the use of vision. Controls can be coded by size, but if personnel are to rely on touch alone, at most three sizes of controls should be used. Size coding, as with other coding methods, should be used consistently. In general, knobs should differ by at least 13 mm (0.5 in.) in diameter or by 10 mm (0.4 in.) in thickness to be identified accurately by touch. *Precautions with shape and size coding.* The shape or size coding should not interfere with operation of the control and should be identifiable by feel regardless of the position or orientation of the control. Knobs and handles, which are shape or size coded, should be attached to their shafts so to preclude removal and incorrect replacement. If individuals wearing gloves will use the control, the shape or size should be identifiable by feel through the gloves. *Texture.* Use of texture differences can provide effective coding when knobs are to be identified by feel. Three surface characteristics can be used together to provide reasonably accurate discrimination: smooth, fluted, and knurled. Different degrees of fluting or knurling should not be used together to provide discrimination.

50% associated the colour red with stop and green with go. The ratio for their American counterparts was almost 100% (Courtney, 1986).

Chapter 3
Displays

General

The principles and guidance in this chapter apply to instruments or devices that present visual or audible information.

Principles

Labelling. Visual displays should be labelled to convey the basic information needed to identify and interpret displayed information. Principles of display labelling are described in Chap. 8, Labelling, signs, graphics, and symbols. *Content.* Displays are intended to convey information to operators or maintainers. Displayed information should be clear, concise, consistent, and follow appropriate standards and conventions. For principles covering the integration of displays with their associated alarms and controls, refer to Chap. 5, Integration of controls, displays, and alarms. *Coding.* Any coding used to enhance the transfer of information to personnel should be consistent and meaningful. The coding system should also be appropriate for the industry, organisation, and culture of the personnel aboard the vessel or offshore installation. *Status information.* Visual displays should provide a positive indication of the state of the equipment such as: ready, running, not running, or "out-of-tolerance". The absence, or non-activated state, of a visual display should not be relied upon to convey status information. Where equipment status must always be available (e.g., stop/start, on/off, etc.), it is appropriate to provide status indication of each state. *Operations versus maintenance.* Separate and distinct information required for operations and maintenance should not be combined in a single display. Displays used solely for maintenance should be covered or non-visible during normal operations but should be immediately available to the maintainer when required for use. *Avoidance of clutter.* Markings (e.g., trademarks, company names, or other similar markings)

A. A. Olsen, *Applying Physical Ergonomics to Modern Ship Design*,
https://doi.org/10.1007/978-3-031-57974-5_3

should not interfere with the reading of the visual display. *Detectability of audible displays.* When information is conveyed to personnel through sound signals, the signals should be distinct and detectable under all expected conditions.

General Visual Display

Line-of-sight (LOS) and orientation. A display should be perpendicular to the normal line-of-sight (LOS) but not less than 45°. The normal LOS is typically 15° below the horizontal LOS. *Readability. Accuracy.* Visual displays should be located to allow reading to the required degree of accuracy from the normal operating position. The normal reading distance for displays is 330 mm (13 in.) to 710 mm (28 in.). *Local displays.* Individual local displays, such as sight gauges on tanks and thermometers or pressure gauges attached to pipes, should be mounted so they are read upright without requiring personnel to assume awkward, uncomfortable, or unsafe postures. In addition, personnel should not have to use or stand on structures not meant to be walking or standing surfaces to view displays. *Lighting levels.* Lighting levels should be sufficient to permit visual display viewing under all conditions of operation without the use of a flashlight or other temporary light source. *Location for operations.* In general, visual displays used for operations should be readily visible at all times. Display locations which require the removal of a cover or deck plate should only be used for non-critical operations and infrequently used displays. These locations should be provided with a clearly marked quick-opening access door requiring no tools to open or close it. *Mounting. Field-of-vision: primary displays.* Primary displays include those used frequently, those used for obtaining precise readings or those used in emergencies. These should be located within personnel's immediately readable field-of-vision. *Field-of-vision: secondary displays.* Secondary displays include those used for normal operations and those not requiring accurate readings. These should be located within personnel's easily readable field-of-vision. Reference displays that are used infrequently may be located in an area outside personnel's easily readable field-of-vision. *Height on panels.* Preferred mounting heights for vertical displays on a flat surface such as a panel or bulkhead are shown in Fig. 3.1, Display mounting height for standing personnel and Displays located immediately adjacent to a walkway should not be located within 460 mm (18 in.) of the standing surface (e.g., deck) in order to prevent them from being easily damaged. *Heights for kneeling and squatting.* While not preferred for normal operations or maintenance, if necessary, kneeling or squatting positions may be adopted by personnel. For kneeling, see Table 3.1, Displays and kneeling dimensions, and for squatting, see Table 3.2, Displays and squatting dimensions.

Size coding. When size coding is used with displays such as simple indicator lights, no more than three distinct sizes should be used, and two are preferred. Each size progression should be a minimum of twice (or half) the size (in square area of the next larger [or smaller] size). Size-coded displays should be in close proximity to

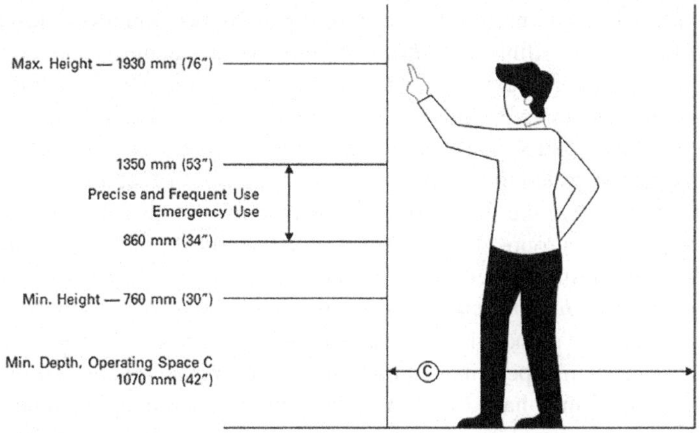

Fig. 3.1 Display mounting height for standing personnel

Table 3.1 Displays and kneeling dimensions

Dimension	Size
Maximum height	1450 mm (57 in.)
Preferred maximum height	1320 mm (52 in.)
Preferred minimum height	950 mm (37 in.)
Minimum height	700 mm (28 in.)
Minimum depth for kneeling	1150 mm (45 in.)

Table 3.2 Displays and squatting dimensions

Dimension	Size
Maximum height	1250 mm (49 in.)
Preferred maximum height	1020 mm (40 in.)
Preferred minimum height	750 mm (30 in.)
Minimum height	520 mm (20 in.)
Minimum depth for kneeling	1050 mm (41 in.)

one another. Most displays are designed for a viewing distance of up to 710 mm (28 in.). Where the viewing distance (y) is greater, the size (x) of the display markings and numbers must be increased, using the following formula:

$$\text{Dimension at y (mm)} = [\text{dimension at 710 mm}] \times \left[\text{y mm}\right]/710$$

or

$$\text{Dimension at y (in)} = [\text{dimension at 28 in}] \times \left[\text{y in}\right]/28$$

Colour coding. The interpretation of colour codes by personnel is contingent on a number of factors, including (1) *culture.* Where colour meanings may be particular to a geographic area. For example, in parts of Europe, the colours white-on-green are used on emergency exit signs, while in the US, the colours white and red are used for emergency exit signs. Another example of the effect of culture on colour meanings is demonstrated in a study between populations of the US and China. The results revealed that of the 784 Chinese participants, comprising both sexes from various occupational groups, less than 50% associated the colour red with stop and green with go. The ratio for their American counterparts was almost 100% (Courtney, 1986); (2) *personnel history and expectations. Personnel's* history and expectations influence the learned associations of colour meanings, and it is based on an individual's experience in operating equipment that employs specific colour coding. Different personnel may have different expectations. For example, some personnel with commercial electrical-power generation experience associate an open/flowing state with red and a closed/stop state with green, while some US Navy personnel have the reverse associations (EPRI NP-3659). In other instances, an individual may have been exposed to many different, often conflicting colour codes; and (3) *context.* The context of colour-coded information varies according to the context of presentation, and personnel quickly learn to associate coded meanings within those contexts. For example, a yellow-orange alarm may convey an out-of-tolerance condition, while yellow-orange on a pipe marker label may indicate the pipe contains gas. In most industrial contexts, there are many more demands to code information than can be provided using eleven or so different colours. For example, in North American culture, red may be used to denote: danger, immediate operator attention is required, a pump is running, a valve is open, a circuit breaker is closed or a pipe is conveying firefighting media, among others. It is context that allows personnel to quickly and reliably identify the colour's meaning. For displays, two basic contexts exist: display of equipment status and display of alarm/annunciation conditions. When an operator is overlooking a sea of red and green indicator lights associated with, for example, a ballast control panel, there is no immediate call for operator action as long as the sea of lights presents the normal status of valves, pumps, and so on. When that same operator notes red indicators within a panel dedicated to alarms (an entirely different context), red immediately assumes an entirely different meaning, and a different operator response. Given this, there are several general principles associated with the use of colour coding:

- Devise and use colour codes that are consistent with the expectations of the majority of personnel, whether derived from native culture or operating experience.
- Employ those colour codes consistency throughout a vessel, fleet, offshore installation, or a company.
- Use colour coding consistently and strictly within each context. For example, when designing a control station or computer program to align and control ballasting systems, avoid physically locating an alarm such that it could be hidden amongst the alignment indicator status lights. It is better to dedicate specific alarm

Table 3.3 Typical display and alarm colour codes for US industry

Colour	Typical display/alarm colour meanings
Red	Danger, hot, unsafe condition, critical parameter value is out-of-tolerance Immediate operator action is required
Amber	Caution, test, hazard, potentially unsafe condition Amber can be used to indicate that a condition exists that is marginal or beyond operation action, or where caution, recheck, or unexpected delay is necessary Amber may also be used to indicate that equipment is moving from one condition to another (e.g., valve is opening)
Green	Parameters normal, condition is clear or safe, no operator action required, parameter values are within tolerance Green can also indicate that monitored equipment is operating satisfactorily or that a condition is satisfactory, and it is safe to proceed
White	White should be used to indicate system conditions that do not have right or wrong implications, such as alternative functions (e.g., Pump No. 1 selected), or transitory conditions (e.g., action or test in progress, function available), provided such indication does not imply success or failure of operations
Blue	Blue may be used for an advisory light, but use of blue should be avoided whenever possible

panels that allow quick scanning and that resolve all contextual ambiguity where colour coding of alarms is concerned.

Table 3.3, Typical display and alarm colour codes for US industry, presents commonly used colour-coding schemes in the US. This table does not provide specific recommendations for colour coding of displays and alarms for all applications or geographical regions, as there can be variance across industries and cultures. Designers should identify regional and industry colour meaning expectations, as well as any statutory requirements, when defining colour-coding requirements. *Scale coding.* Where operating values fall within a range, the ranges may be identified by a colour or shape coding applied to or adjacent to the face of the display as indicated in Fig. 3.2, Colour and shape coding of ranges on an analogue display.

Auditory Display

There are three basic forms of information displays using sound: (1) voice communications (aided and unaided); (2) alarm and warning signals (denoting off-normal or emergency situations); and (3) general auditory display of information (to present simple status information such as to indicate completion of a time interval or to orient operator attention). Further information and guidance relating to the design of the audible portion of alarm systems is given in Chap. 4, Alarms. Guidance regarding best practice for design of general auditory displays is provided here. *Use of auditory signals.* Auditory signals can be used where messages or information are simple and

Fig. 3.2 Colour and shape coding of ranges on an analogue display

short (e.g., on/off, cycle complete, watertight doors closed); when information will not need to be referred to at a later time; to orient an operator to a location (e.g., to direct visual attention to a radar display); when an immediate action is required (e.g., alarms and warnings); when vision is poor (e.g., smoke filled spaces, in holds and tanks); and to present information independent of visual orientation. *Design characteristics of auditory display.* Design characteristics of auditory displays should be consistent with best practice. This means each signal has a single specific meaning or required operator action, and all signals are clearly distinct from each other. Ideally, the number of coded signals should be limited to five. Where more than five coded signals are needed, it is advisable to use a two-phase signal instead. When using a two-phase signal, use the same signal characteristics for the alerting portion of the signal. For the second phase, use intelligible spoken messages to indicate the condition that initiated the alarm and/or any actions to be taken. Each signal should have sufficient attention-demanding properties (refer to Chap. 4, Table 4.2, General recommendations for sound loudness and frequency and Chap. 4, Table 4.3, Guidelines for selecting audible signal devices). This means signals are clearly audible in the expected normal or worst-case ambient noise environment. Furthermore, the signal sound source should propagate from the immediate vicinity of where operator or maintainer action (e.g., examine a display or operate a control) is performed. For improved effectiveness, signals should be of a fixed duration with automatic termination (e.g., operator or maintainers do not have to acknowledge or manually silence an audible signal, except under alarm conditions). Finally, audible signals should be clearly detectable but should not be so loud as to invoke a startle response.

Chapter 4
Alarms

General

The IMO Resolution A.1021(26), Code on Alerts and Indicators, provides us with the "general design guidance to promote uniformity of type, location, and priority…" for alerts and indicators. Where applicable, that IMO resolution should be used for determining design and implementation of alarms, alerts, and associated indicators. The document also presents tables of required alarms and indicators for spaces of different functionality, machinery, and ship or structure type. The principles and guidelines of this chapter apply to visual and auditory alarms/alerts and associated controls, local and remote, which require the attention of personnel. Alarms can be categorised as visual alarms, visual alarms accompanied by audible tones, and audible alarms. These principles apply to visual and audible alarms located on local panels or central control area panels or consoles (e.g., control rooms, navigation bridges). These principles do not necessarily apply to general emergency alarms.

Alarm Profiles

Alarms are used to indicate an out-of-tolerance condition to personnel. Visual alarms can be coded for priority as system conditions escalate or change. Visual alarms may be used in high noise areas to attract attention and inform personnel of out-of-tolerance conditions (e.g., a hydrogen-sulphide beacon). Audible alarms can also be used to indicate an out-of-tolerance condition to personnel. These alarms are used in conjunction with visual alarms to call attention to or emphasise the information. Audible alarms with verbal messages may include the general emergency alarm, abandonment alarm, and those alarms demanding immediate personnel response to attend to conditions such as a fire, collision, or flooding. These allow specific instructions to be conveyed to personnel. *Alarm set points.* The limits or set points

A. A. Olsen, *Applying Physical Ergonomics to Modern Ship Design*,
https://doi.org/10.1007/978-3-031-57974-5_4

for initiating the alarm signals should not occur so frequently as to be considered a nuisance by personnel; and provide personnel adequate time to respond to the condition. *False alarms.* A means should be provided to prevent normal operating conditions from causing false alarms. These methods may include the provision of time-delays or allowance for transient signals. *Alarm priorities.* When simultaneous alarms are active, the alarms should be prioritised so that personnel can first respond to those conditions that are more urgent. Prioritisation should be accomplished using two to four levels. Priority can be conveyed with either visual or auditory coding methods. Prioritisation should be based on a combination of relative severity of the consequences of not responding to the condition or situation; and time constraints imposed on personnel to respond. If alarms are presented on a dedicated printer, those of higher priority should be presented first, and the others presented in descending order of priority. If alarms are presented on a video display, active alarms should be viewable as sorted by priority, onset sequence or by system function. *Alarm integration.* In the event of a complete system failure, a single summary alarm (e.g., "Diesel Generator Set B Failure") should indicate the failure rather than requiring personnel to integrate the information presented by numerous component level alarms (e.g., "Low Bus Voltage", "Stator Trouble", or "Lube Oil Pressure Low"). *Alarm response.* Each alarm should prompt personnel action. Where multiple alarm conditions prompt the same action, there should only be one alarm. If there is no action required, then that function should not be conveyed via an alarm, but through a status indication. *Alarm acknowledgement.* Alarms should require personnel acknowledgment after alarm onset. An acknowledge control should be provided. The control should terminate the flashing of a visual alarm and allow the alarm to remain illuminated until the alarm condition is cleared. The control can also mute the audible signal, or a separate silence control can be provided. An acknowledge control should be provided near the controls and displays associated with the alarm. *Master silence control.* The master silence control should only silence active audible signals. It should not block audible signals at the onset of subsequent alarms. The master silence control should not affect the visual portion of the alarm. *Subsequent alarms.* Each alarm onset should activate visual and audible signals such as a flashing visual indicator and audible alerting signal. This should occur regardless of the condition of any other active alarms (e.g., acknowledged, not acknowledged, cleared, uncleared, or reset). If a single alarm has multiple inputs, any new alarm condition should reactivate that alarm. *Repetitive alarms/controls.* Repetitive groups of alarms should have the same arrangement and relative location on different panels and consoles. Any alarm controls (e.g., silence, acknowledge, reset, clear) should be consistent between panels and consoles. *Test function.* For control consoles or panels, a means should be provided to test the flashing and auditory signals associated with alarms without disrupting the normal operation of the alarm system. *Temporary disconnection of alarms.* Alarm circuits may be temporarily disabled (e.g., for maintenance) if such action is coordinated with appropriate personnel (e.g., operations centres, the bridge, engine control room) and is clearly indicated at all locations where such information may be required. These locations include the specific piece of equipment, the local control panel or console, the central control room, and work permits. Permanent alarms (e.g., fixed lights or tiles, as opposed to computer driven displays) should be provided with a means to indicate their status (e.g., by tag out or sticker indicating that the alarm is disabled).

Visual alarms. ***Note****: Additional general guidance related to alarms can be found in* Chap. 3, *Displays. Alarm onset.* Visual alarms should flash at onset. Flash rates should be three to five flashes per second with approximately equal on and off time. The flashing display should change to a steady state upon alarm acknowledgment. The steady state should remain activated, either individually or in a summarised fashion, until the alarm condition is corrected. *Alarms and normal operations.* Under normal operating conditions, no visual alarms should be illuminated. Exceptions include provisions for alarms associated with equipment under repair or receiving servicing. When this is the case, alarms should be tagged or otherwise marked to indicate the status of the display and equipment. Alarms that are illuminated as a general case should undergo review to determine if that alarm should be reclassified as a status indicator, and if so, removed and relocated. *Priority coding.* A method for coding the visual alarms for the various priority levels should be employed. Acceptable methods for priority coding include colour, position, shape, size, flash rate, or symbols. *Colour coding.* Where applicable, visual alarms should be colour coded in conformance with the IMO Assembly Resolution A.830(19), Code on Alarms and Indicators. In other cases, Table 4.1, Guidelines for colour coding should be followed. This colour-coding scheme should be used throughout the vessel or offshore installation. *Flasher failure.* If a panel or console alarm flasher fails, then at the onset of an alarm condition the alarm light should illuminate and burn steadily, rather than not illuminate. *Contrast detection.* There should be sufficient contrast between flashing and fully illuminated alarms, and between illuminated and non-illuminated alarms, so that personnel can reliably discriminate each state. *Text visibility and legibility.* The text on visual alarms should be easily readable whether or not the alarm is illuminated. Wording criteria. If text is used for alarms, wording should be in accordance with the following criteria: wording should be concise, specific, and unambiguous; wording should address a specific condition (e.g., do not use one alarm for **HIGH-LOW, TEMPERATURE–PRESSURE**); abbreviations and acronyms should have consistent usage throughout the vessel or offshore installation; words should be legible in worst-case conditions; and personnel should be able to read all the text associated with visual alarms from the position at the panel or console where the alarms acknowledgment control is located.

Audible Alarms

Significant benefits of audible alarms, or of an audible component of a visual alarm, include providing indication to personnel about conditions without requiring constant monitoring of visual displays; a general reduction in personnel workload associated with equipment and safety monitoring; a general increase in the reliability of equipment and status monitoring, enhancing safety, production, and efficiency; the

Table 4.1 Guidelines for colour coding

Colour	Meaning	Explanation	Typical applications
Red	Danger or alarm	Warning of danger or a situation that requires immediate action	Stop of essential machinery, (e.g., lubricating oil pump for the propulsion engine or motor, failure of pressure in the lubricating system for the propulsion machine, etc.) Temperature and pressure values (water, oil, etc.) at critical levels Activation of a safety system
Yellow	Caution	Change or impending change of conditions	Temperature and pressure values which differ from the normal levels, but not critical
Blue	Instruction/ information (specific meaning assigned according to the case considered)	Blue may be given specific meaning that is not covered by the colours: red, yellow, and green	Motor ready to start No-load generator ready to be connected Electrical heating circuit for idle electrical machines
White	No specific meaning assigned (neutral)	Any meaning; may be used whenever doubt exists about the application of the colours red, yellow, and green, often used for confirmation	Earth insulation indicators Synchronising lamps Telephone calls Appliances automatically controlled

ability to announce the status and conditions of equipment to a wide operating area, immediately alerting those conditions to a potentially wide audience; the ability to immediately compel personnel to attend to off-normal or emergency conditions; and the ability to quickly and effectively influence safety behaviour in the form of general emergency alarms. *Content.* Audible alarms may be tones or tones used with verbal messages. Audible alarms with verbal messages should consist of an initial alerting (non-speech) signal to attract attention and a brief, standardised verbal message to identify the specific condition and/or suggest an appropriate action. *Sound characteristics.* The selection of loudness, frequency, and use of sound waveforms should be in accordance with the guidelines given in Table 4.2, General recommendations for sound volume and frequency. *Number of distinct audible alarms.* The number of distinct audible alarms (in terms of the auditory characteristics of each) should not exceed seven, and it is recommended that the number not exceed four to prevent confusion as to the meaning of different alarms. *Selection of audible signal devices.* Devices such as a whistle, siren, bell, klaxon, or horn can be used to provide audible signals. General guidance for the selection of devices is given in Table 4.3, Guidelines for selecting audible signal devices. For specific vessel or offshore installation applications, the IMO documents referenced in the subsection 'Alarms required' by the IMO should be used. *Sound volume.* Recommended sound volume qualities are as follows:

Table 4.2 General recommendations for sound volume and frequency

Conditions	Recommendations
If the distance to the listener is great	Use increased loudness and low frequencies
If the sound is to be heard around obstacles and pass through partitions	Use low frequencies
If background noise is present	Select frequencies lower than those of the back- ground noise Use several tones differing by an octave
To demand personnel attention	Modulate the signal to give intermittent "beeps" or modulate the frequency to make pitch rise and fall at rate of 1–3 Hz
To distinguish between different alarms	Use different sound waveforms, with pulse frequency between 0.5 and 2.0 Hz Consistently use the same waveform for the same function or purpose Use no more than three or four different forms in the same space

Table 4.3 Guidelines for selecting audible signal devices

Device	Loudness	Frequency	Attention demanding	Noise-penetration ability
Foghorn	Very high	Very low	Good	Poor with low frequency background noise Good with high frequency background noise
Horn	High	Low to high	Good	Good
Whistle	High	Low to high	Good if intermittent	Good if frequency is properly chosen
Siren	High	Low to high	Very good with frequent pitch rising and falling	Very good with rising and falling frequency
Bell	Medium	Medium to high	Good	Good in low frequency noise
Buzzer	Low to medium	Low to medium	Good	Fair if spectrum is suited to background noise

- Audible signals should be at least 10 dB(A), preferably, 20 dB(A), above ambient noise levels during normal operations,
- Audible alarm sound loudness should be at least 75 dB(A) at a distance of 1000 mm (40 in) from the source and at sleeping positions in cabins,
- A signal-to-noise ratio of at least 20 dB(A) should be provided in at least one octave band between 200 and 5000 Hz at the operating position of the intended receiver,
- Audible signal sound levels in a space should not exceed 120 dB(A),
- For large spaces, more than one audible alarm may be required for the sound level to be heard throughout the space, and

- For spaces of unusually high noise levels, a beacon, strobe light, or similar device, installed in a location visible from all parts of the space should supplement the audible signal. Red light beacons should only be used for fire alarms, unless this conflicts with appropriate statutory, regulatory, or class requirements.

Sound frequency. With the exception of bells, audible signals should have a signal frequency between 200 and 2500 Hz, with the preferable range between 500 and 1500 Hz. *Control of loudness.* The loudness of an audible signal should be designed to be adjustable. The control mechanism should be restricted and administratively controlled to prevent reducing the volume to an inaudible level or increasing it to an unacceptably high level. *Detection level.* Each audible signal should be detectable in all locations where personnel may be when their response to the condition is required. For example, signals to be detected by control room personnel should be readily detectable anywhere in the control room. General emergency alarms (e.g., emergency or evacuation) should be readily detectable throughout the vessel or offshore installation (e.g., in hull, topsides, accommodations, etc.). *Automatic reset.* An automatic reset function should be provided whether the audible signal is designed to terminate automatically, manually, or both. It should be controlled by a sensing mechanism that recycles the alarm system to a specified condition (e.g., alarm active or inactive) as a function of time or the state of the signalling system so that the alarm can sound again if the condition reoccurs. *Manual reset.* If an automatic reset function is not provided, a control should be provided to reset the system manually after alarms have been cleared. *Cleared alarms.* Cleared alarms should be accompanied by a dedicated, distinctive audible signal of finite duration. *Audible coding of priority levels.* Coding methods for audible signals include pulse, duration, and frequency. Intensity should not be used as a coding method. If pulse coding is used, the number of levels should not exceed three. If duration is used, the number of levels should not exceed three. If frequency coding is used, the number of levels should not exceed five. *Caution signal controls.* Audible caution signals should be provided with manual reset and volume controls. *Location of equipment generating audible signals.* Equipment for generating audible signals should be positioned in a location and with an orientation that will maximise perception of the signal. The equipment should not be located where personnel may be immediately nearby unless precautions have been taken so that the activation of audible signals will not startle them. The equipment generating audible signals should be located and arranged to remain operable during incidents such as flooding or fire. *Verbal messages.* (1) *Content.* Verbal messages should indicate the required response from personnel on the vessel or offshore installation (e.g., abandon ship or offshore installation, or report to emergency stations) rather than simply the nature of the emergency. In selecting words to be used in verbal messages, priority should be given to intelligibility, appropriateness, and conciseness, in that order; (2) *type of voice and delivery style.* The voice used in recording verbal messages should be distinctive and mature. Verbal messages should be presented in a formal, impersonal manner; (3) *speech processing.* Verbal messages should be electronically processed only if necessary to increase or preserve intelligibility (e.g., by increasing the strength of consonant sounds relative to vowel strength); and (4) *verbal message intensity.* In respect of verbal message intensity, the verbal message intensity for critical functions should be at least 20 dB(A) above ambient noise levels at the typical operating positions of the intended receivers, but should not exceed 85 dB(A). The message should capture the personnel's attention but should not cause irritation or a startled reaction. The verbal message intensity, where adjustable, should be controlled by administrative procedures; and changes in the intensity of verbal message broadcasts should be followed by testing and verification of speech intelligibility.

General emergency alarms. General emergency alarms intended to alert a general audience (e.g., all personnel on a vessel or offshore installation) should be indicated primarily by audible alarms, including a unique tone and preferably a verbal message. These include the general emergency alarm, abandonment alarm, and those alarms warning of immediate personnel hazards such as the release of certain fire extinguishing media. *Guidance by exception.* This subsection modifies some of the alarm guidance provided above to address specific issues related to the design of general emergency alarms. *Acknowledgement of alarms.* General emergency alarms should not be acknowledged but cleared when the condition that has triggered the alarm has been resolved or a general alarm of greater priority is engaged. *Master silence control.* A master silence control associated with any system or equipment alarms should not silence any general emergency alarm. A separate control should be provided for general emergency alarms. *Alarm integration.* General emergency alarms should not be integrated with other alarms or alerting devices (e.g., equipment or system alarms). Temporary disconnection of alarms. General emergency alarm circuits are of such importance to safety that they should not be disabled without the provision of another system for alarm notification. *Repetitive controls.* Controls that activate the general emergency alarm system should be the same in appearance and operation at each location at which they exist throughout the vessel or offshore installation. *Alarm content.* General emergency alarms may be distinct tones or tones with a verbal message added at regular intervals. *Caution signal controls.* General emergency alarms should not be used as caution signals. *Duration of alarm flashing.* Visual alarms associated with general emergencies should flash until the emergency is cleared. *Colour and priority coding.* General emergency alarms should use a coding scheme that indicates their high level of priority for the audible portion of the signal. Any visual signal should make use of the appropriate colour code by referring to appropriate statutory, regulatory, or class requirements. The use of a prominent position for alarm indication should be considered as well as the use of flashing mechanisms for demanding attention; and the general emergency colour-coding scheme should be adhered to throughout the vessel or offshore installation. *Navigational bridge alarms.* Alarms on the navigation bridge should be limited to those that are critical to safety of the vessel or offshore installation. Visual alarms and indicators should not interfere with night vision.

Alarm Requirements

Requirements stipulated by the IMO. Alarm requirements stipulated by the IMO should follow the guidelines contained in IMO Assembly Resolution A.1021(26), Code on Alerts and Indicators (2009), and further the IMO resolution MSC.128(75) requiring a Bridge Navigational Watch Alarm System (BNWAS). *Requirements imposed by other regulatory bodies.* Where requirements for alarms specified by other regulatory bodies (e.g., Flag State Administrations and Port State Authorities) are inconsistent with the guidance contained herein, those requirements must take precedence over those incorporated into this book.

Chapter 5
Integration of Controls, Displays and Alarms

General

The principles and guidance provided in this chapter apply to the design, layout, and placement of individual controls, displays, and alarms within the context of operations and maintenance activities performed in association with consoles and panels. The figures in this chapter are provided as examples to demonstrate control, display, and alarm integration concepts. These figures are not intended to serve as functional models or designs for any system or equipment. Actual designs must adhere to appropriate Statutory, Class, or other applicable requirements from regulatory bodies.

Principles

Control/display relationship. The relationship of a control with its associated display should be immediately apparent and unambiguous to personnel. Controls should be located adjacent to (e.g., normally under or to the right of) their associated displays and positioned so that neither the control nor the hand normally used for operating the control will obscure the display. Control/display/equipment relationship. Where controls and displays are provided for local operation of a specific piece of equipment, they should be located above or immediately adjacent to it. For example, a board of gauges associated with a pump or filter would be located next to the pump or filter, so that it is visually obvious that all of the components are functionally related. Where other design considerations dictate, or where a deliberate choice is made to place the controls and displays remote from the equipment, the controls and displays should be mounted in such a manner that the relationship of the location of the controls and displays matches the arrangement of the associated equipment. Control/display/alarm relationship. On consoles or panels, controls associated with displays

A. A. Olsen, *Applying Physical Ergonomics to Modern Ship Design*,
https://doi.org/10.1007/978-3-031-57974-5_5

or alarms should be located below or to the right of the display or alarm. Display/alarm relationship. The relationship between a display and the associated alarm indication should be immediately apparent and unambiguous to personnel.

Displays and alarms should be located at the top of a console or panel, with the associated controls below or to the right. Alarms should be located adjacent to (e.g., normally above or to the left of) associated displays and controls. Alarms should be positioned so that neither the control nor the hand normally used for operating the control will obscure the display or alarm. Control and display movement ratio. The response of a display to a control movement should be consistent, predictable, and based on personnel's movement expectations. For principles of control movement expectations, refer to the section 'Control and display movement relationships'. Grouping relationships. There are two general types of ergonomic principles for arranging controls and displays on a console or panel. The first set of principles applies to determining the general location of the controls or displays and the second set of principles is for their specific arrangement within a given panel area. General location. The more important or frequently used controls and displays should be located in the centre of a console or panel. Control and display arrangement and grouping. Controls and displays should be arranged according to task requirements (e.g., by sequence of use, precision, frequency of use) and by common function (e.g., locating two sets of auxiliary power generator unit controls and displays in two adjacent areas on a panel or console). Segregation of groups of controls and displays. Groups of controls and displays should be physically and visually segregated (e.g., a group of controls and displays associated with Generator Set A are adjacent to, but physically and visibly separated from, Generator Set B). Labelling of controls, displays, and alarms. Console and panel labels should be used to identify individual or groups of controls, displays, or alarms. Each control, display, and alarm should be labelled as described in Chap. 8, Labelling, signs, graphics, and symbols. Colour coding. Colour used as a coding technique should be consistent with that of displays and should follow the guidance in Chap. 3, Colour coding. Feedback. Feedback on equipment/ system response to a control action should be provided as rapidly as possible. Feedback can be provided through direct visual observation of equipment or by an associated display.

Position Relationship of Displays and Alarms

Display arrangement, whenever possible, should provide viewing from left-to-right or top-to-bottom. This arrangement should take into consideration the cultural expectations of the intended operators and maintainers. Displays that are functionally or otherwise related to each other should be positioned in a consistent manner to each other. As an example, dual displays providing **IN** and **OUT** type information (e.g., suction and discharge pressure), should be arranged as shown below unless the **IN** and **OUT** displays are a part of a mimic display that does not permit these locations:

IN On the left (preferred) or top
OUT On the right (preferred) or bottom

Lighted annunciators used as alarms to identify two critical positions of a measured parameter (e.g., high and low temperature, high and low pressure, high and low flow rate), should be located so that the high alarm annunciator is directly above or to the right of the low alarm annunciator. Lighted annunciators or backlit push buttons for **ON** and **OFF**, **START** and **STOP**, or **OPEN** and **CLOSE** should be located as follows. The arrangement of these types of displays should also consider cultural expectations:

ON On the right or top
OFF On the left or bottom
START On the right or top
STOP On the left or bottom
OPEN On the right or top
CLOSE On the left or bottom

Position Relationship of Controls to Associated Displays and Alarms

Positioning. The positioning of controls to displays and visual alarms should be such that neither the control itself nor the hand normally used for operating the control blocks the view of the display or visual alarm. For multiple rows of displays or visual alarms where corresponding controls cannot be located directly under each display or visual alarm or adjacent to the displays or visual alarms, then they should be placed in a consistent and uncluttered manner. *Design.* Control/display relationships should be apparent through proximity, similarity of groupings, coding, boundary lines, borders, labelling, or similar techniques. *Multiple controls, displays, and alarms.* For controls that cannot be located beneath or to the right of their respective displays or visual alarms, the controls should be grouped in the same row arrangement as are the displays.

Location of Alarms and Alarm Response Controls

Alarm panel location. Matrices of visual alarms should be located on local or control room panels, above the related controls and displays which are required for corrective or diagnostic action in response to the alarm. Alarms that refer personnel to another, or to a more detailed annunciator panel outside the primary operating area, should be minimised. *Positioning of alarms and alarm response controls.* Repetitive groups of alarms should have the same arrangement and relative location at different workstations. Any alarm response controls (e.g., silence, acknowledge, reset, clear) should be consistent from control panel to control panel.

Control and Display Movement Relationships

The response of a display to a control movement should be consistent, predictable and based on the expectations of personnel who will use the equipment. For principles of control movement expectations, refer to the section on 'Control movement' in Chap. 2. *Control/display movement consistency.* Display indicators should clearly and unambiguously direct and guide the appropriate control response. The direction of movement of a display should be the same as the direction of movement of the control. Clockwise movement of a rotary control or movement of a linear control forward, up, or to the right should produce a clockwise movement of the pointer in a circular display or a movement up or to the right for horizontal and vertical displays causing an increase in the magnitude of the display reading. *Control/display response characteristics.* For continuous-adjustment controls, such as knobs, and if speed of adjustment is more important than accuracy, the control/display ratio should be low (high gain). If accuracy is more important that speed, a high control/display ratio (low gain) is preferred. If both speed and accuracy are important, the ratio may be optimised for the specific task.

Grouping Relationships (Principles of Arrangement)

Functionally related controls and displays should be located close to one another. *Determining the general location.* A combination of the following principles should be used to determine the item's location:

- *Importance.* More important or critical controls or displays should be centrally located on a console or panel.
- *Frequency of use.* More frequently used items should be centrally located.
- *Function.* Functionally related controls and displays should be located together.

Determining the specific arrangement within a general location. Items should be arranged within a given panel area using the most appropriate of the following:

- *Common function.* Group items having the same or similar function together, such as similar generator displays,
- *Common equipment.* Group common equipment, such as all controls, displays, and/or visual alarms associated with a specific piece of equipment,
- *Common systems.* Group all controls, displays, and/or visual alarms associated with a complete system, and

- *Sequence of use.* Group items in accordance with the sequence in which they are used. For example, controls, displays, and/or visual alarms used in a specific sequence, such as startup or shut down of equipment or a system.

Consistency. Controls and displays used for similar equipment or systems should be oriented the same way on different panels or consoles. An exception to this guidance would be situations where the spatial relationships of sets of equipment or subsystems differ from location-to-location and where the personnel are performing operations within sight of the equipment or subsystems. For additional guidance, refer to the section on 'Spatial relationship of controls and displays to equipment'. *Design.* Control/display relationships should be apparent through proximity, coding, the use of boundary lines and borders, labelling, or other similar techniques. Functional group consistency. Location of recurring functional groups and individual items should be similar from panel to panel. Generally, ***mirror-imaged arrangements should be avoided*** where the two groups of equipment have the same function and where spatial orientation is not important to the use of the equipment, thus the mirror-imaged arrangement is likely to confuse personnel and result in human errors. Mirror-imaged arrangements are acceptable when the corresponding arrangement of the equipment or systems is also mirror-imaged and where personnel are mindful of their location at a console or panel in reference to the layout of the equipment or system itself. *Feedback.* Feedback on control response (displays or equipment) should be located close to its control. *Emergency use.* Emergency controls and displays that require immediate action to prevent injury to personnel or damage to the equipment or the environment should be located where they can be seen and reached with minimum delay (Fig. 5.1).

Fig. 5.1 Control arrangements (engine control room)

Grouping Segregation

The general principle of grouping segregation is to provide a natural relationship among controls, displays, and equipment. Once a grouping is established based on a chosen relationship, this grouping should be segregated from other groupings. Refer to the section on 'Grouping relationships—principles of arrangement' for guidance about how to establish groupings. One or more of the following methods can be used to accomplish segregation of groups:

- Use of physical separation between groups of controls; horizontal separations are preferred over vertical separations.
- Use of boundary lines and borders around particular groups.
- Use of coloured or shaded areas around a particular group to create a different background colour from the rest of the panel or console.
- Use of separate sub panels mounted to a background plate.

Spatial Relationship of Controls and Displays to Equipment

Local position. Where local control of equipment is provided, controls or displays associated with a specific piece of equipment (e.g., controls and pressure gauges for a pump) should be mounted directly above, or adjacent to, the equipment so that it is visually obvious that all of the components are related. Controls and displays for redundant or identical systems, separated within a space, or in different spaces, should be positioned with the same spatial relationship to the equipment at each location. Where other design considerations dictate, or where a deliberate choice is made to place the controls and displays remote from the equipment, then the controls and displays should be mounted in such a manner that the relationship of the location of the controls and display matches the arrangement of the equipment to which they belong. Where the relationship of controls, displays, and equipment is according to a process (for example, a flow arrangement of pumps, tanks, and valves), consider the use of mimic lines and demarcations to make the relationships apparent. *Position on local console.* Controls and displays on a local console (or panel) should be arranged such that their physical location on the console has a direct relationship, called *spatial relationship*, with the physical location of the associated equipment. This is particularly important where personnel working at the console are able to view the equipment from the console location. As an example, controls and displays on the left side of the console should be associated with equipment physically located to the left side of personnel, and controls and displays on the right side of the console should be associated with equipment physically located to personnel's right while facing the console. When a console is arranged athwartships on a vessel to control or monitor equipment or systems which are physically arranged fore and aft on the vessel (e.g., piping and pumps on a tanker), the preferred control and display arrangement on the console is should be positioned such that:

- Controls and displays positioned on the right side of the console should be for equipment located forward. Correspondingly, those on the left side of the console are for equipment located aft.
- Controls and displays positioned at the top of the console should be for equipment on the port side. Correspondingly, those at the bottom of the console are for equipment located on the starboard side.
- This arrangement may also be used for panels and consoles oriented where personnel face port.
- The same rules apply to vertical control panels.

Panel orientation for equipment packages. For equipment packages, including skid mounted equipment, control panels should be located and oriented so personnel can directly view the equipment being controlled and/or monitored. The panel(s) should be oriented such that personnel face the equipment and panel. *Panel layout and spatial relationships.* Where a panel is arranged so that the control and display arrangement correlate directly to equipment, the controls and displays closest to the right side of the panel should control or monitor the equipment closest on the right. The controls and displays on the left side of the panel should be related to equipment on the left.

General Console Arrangements

Console dimensions. Console types include those designed for standing operations, seated operations, or for a combination of standing and seated operations. Console dimensions for these variations with and without vision over the top of the console are shown on Fig. 5.2, Console dimensions. Dimensions for consoles with video-display units are shown in Fig. 5.3, Seated video-display-unit console, and Fig. 5.4, Sit/stand video-display-unit console. Controls and displays should be grouped by importance and frequency of use. Frequently used controls and displays and those requiring precise reading or settings should be centred in the primary view and reach area. Controls and displays that are used less frequently should be located nearer to the extreme ends of the console. Additional information regarding video-display units which are not embedded into consoles is contained in Chap. 6, Computer workstation design.

Extra-width consoles. Seated consoles should be 1120 mm (44 in.) or less in width and standing consoles should be 1525 mm (60 in.) or less in width. There may, however, be conditions where it is preferred to have wider consoles. Acceptable alternatives include:

- *Wrap-around.* A console may be configured as a wrap-around console. A seated console is shown in Fig. 5.5, Wrap-around seated console, with an 1120 mm (44 in.) centre span and 610 mm (24 in.) wings. A standing console would be similar

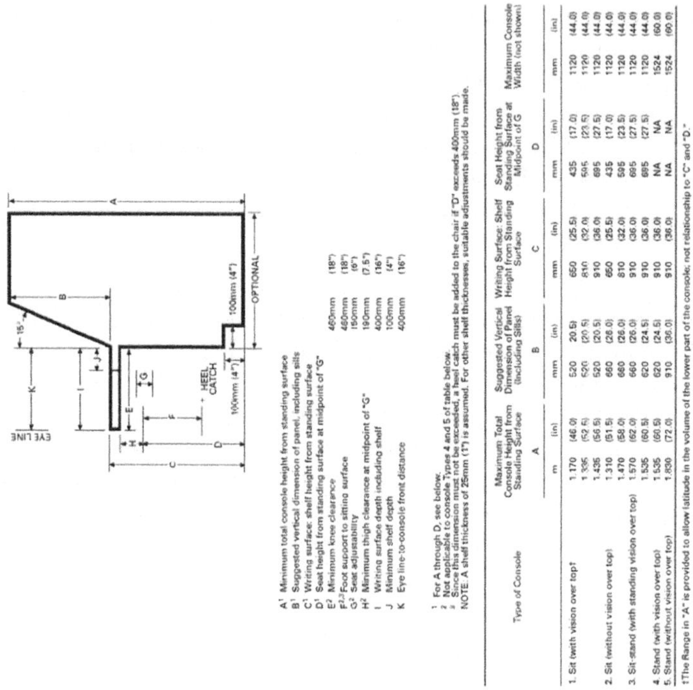

Type of Console	Maximum Total Console Height from Standing Surface A		Suggested Vertical Dimension of Panel (Including Sills) B		Writing Surface: Shelf Height from Standing Surface C		Seat Height from Standing Surface at Midpoint of G D		Maximum Console Width (not shown)	
	m	(in)	mm	(in)	mm	(in)	mm	(in)	mm	(in)
1. Sit (with vision over top†)	1.170	(46.0)	520	20.5)	650	(25.5)	435	(17.0)	1120	(44.0)
	1.335	(52.5)	520	(20.5)	810	(32.0)	595	(23.5)	1120	(44.0)
	1.435	(56.5)	520	(20.5)	910	(36.0)	695	(27.5)	1120	(44.0)
2. Sit (without vision over top)	1.310	(51.5)	660	(26.0)	650	(25.5)	435	(17.0)	1120	(44.0)
	1.470	(58.0)	660	(26.0)	810	(32.0)	595	(23.5)	1120	(44.0)
3. Sit-stand (with standing vision over top)	1.570	(62.0)	660	(26.0)	910	(36.0)	695	(27.5)	1120	(44.0)
	1.535	(60.5)	620	(24.5)	910	(36.0)	695	(27.5)	1120	(44.0)
4. Stand (with vision over top)	1.535	(60.5)	620	(24.5)	910	(36.0)	NA	NA	1524	(60.0)
5. Stand (without vision over top)	1.830	(72.0)	910	(36.0)	910	(36.0)	NA	NA	1524	(60.0)

†The Range in "A" is provided to allow latitude in the volume of the lower part of the console; not relationship to "C" and "D."

Fig. 5.2 Console dimensions

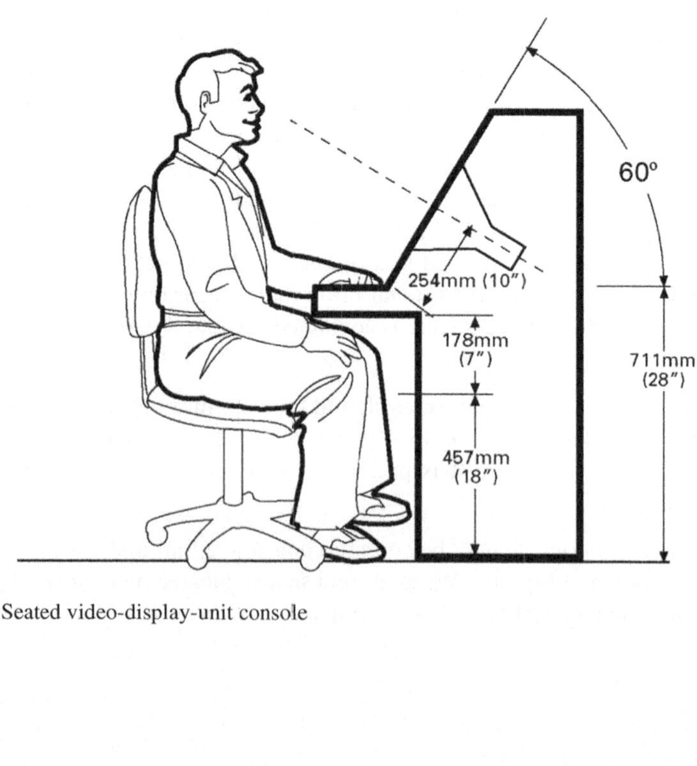

Fig. 5.3 Seated video-display-unit console

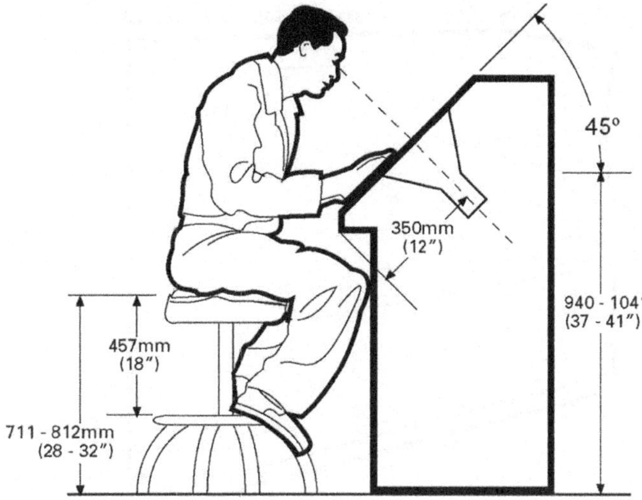

Fig. 5.4 Sit/stand video-display-unit console

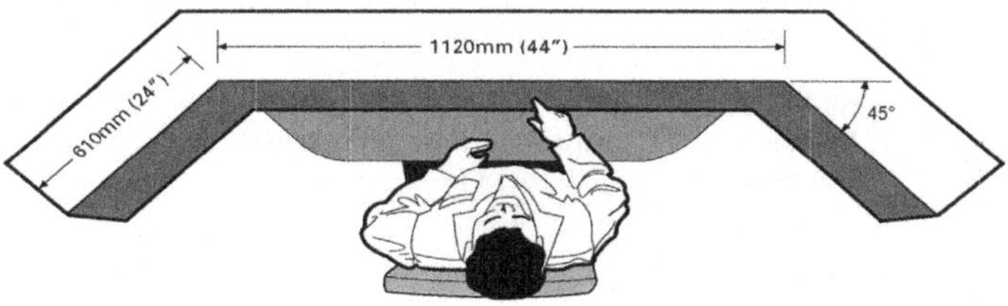

Fig. 5.5 Wrap-around seated console

but with a 1525 mm (60 in.) centre span and 915 mm (36 in.) wings. Additional dimensions should be as shown in Fig. 5.2, Console dimensions.

- *Special-width.* A console may be configured as an extra-width seated console as shown in Fig. 5.6, Special width console. Frequently used controls should be within a distance of 460 mm (18 in.) from the consoles centreline and the less frequently used controls may be located within a distance of 800 mm (31.5 in.) from the consoles centreline. These dimensions are based on the assumption of a single-person workstation. All other dimensions should be as shown in Fig. 5.2, Console dimensions.
- *Additional information* regarding video displays which are not embedded into consoles is contained in Chap. 6, Computer workstation design.

Extra-height, multi-tiered console. For standing and seated consoles where vision over the top is not required, the console height may be increased provided the angled upper tier is used only for displays and the displays remain within the field of view. *Desktop consoles for standing personnel.* Dimensions for desktop consoles

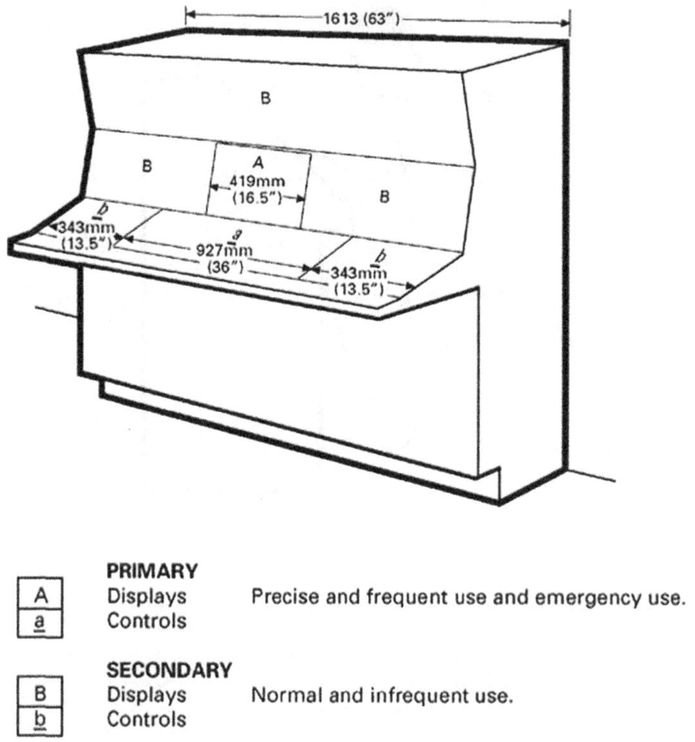

Fig. 5.6 Special-width console

for standing personnel are shown in Fig. 5.9, Dimensions for desktop standing console. Although it is preferred that desktop standing consoles be 1220 mm (48 in.) or less in width, consoles of greater width may be allowed for specific applications, such as a navigation bridge or propulsion control consoles. In those instances, the console layouts should separate control and display groups into segments no greater than 1220 mm (48 in.) in width on the console face. Additional information regarding video displays which are not embedded into consoles is contained in Chap. 6, Computer workstation design (Figs. 5.7, 5.8, and 5.9).

Specific Console Applications

Cargo and ballast transfer consoles. Cargo and ballast transfer controls and displays on the console should be arranged so their orientation is spatially related to the actual tanks, valves, pumps, etc., located on the vessel or offshore installation. In the case of vessels, control panels and consoles should be installed so that personnel face forward or port. On multi-tier consoles which use a combination of mimics, controls, and their

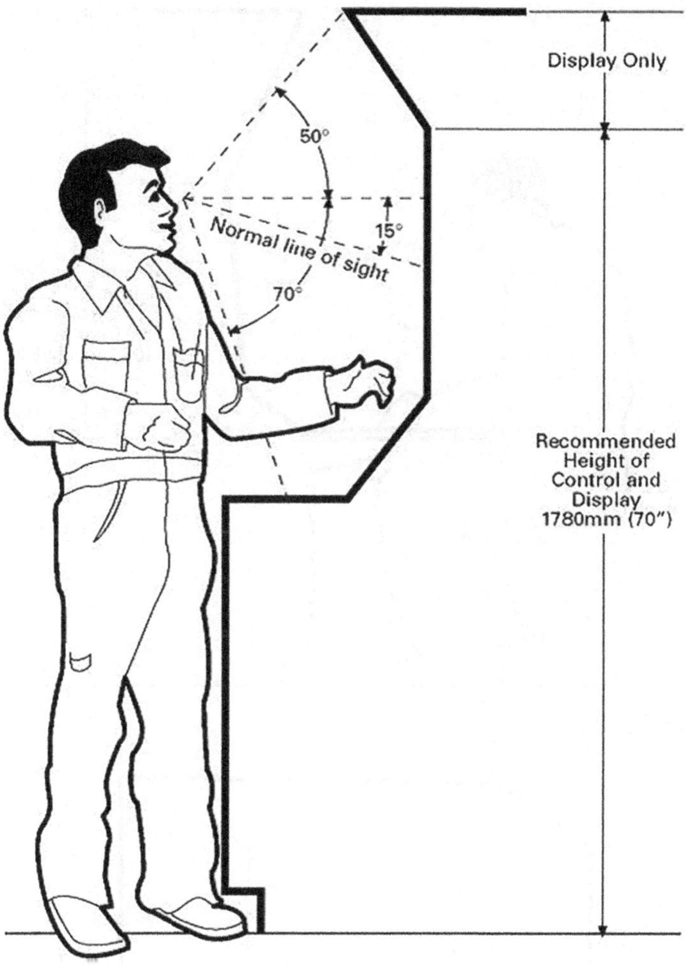

Fig. 5.7 Multi-tiered standing console

associated displays and/or visual alarms (e.g., cargo transfer console on tankers), the layout of the bottom portion of the console and the upper tier should maintain the same relationship as shown in Fig. 5.10, Cargo and ballast transfer console. The top or upper area of the console face should represent the port side of the vessel, and the bottom or lower area of the console face should represent the starboard side of the vessel. *Propulsion, navigation, and steering consoles.* Consoles associated with the control of propulsion engines; navigation, including chart tables, radar, electronic chart displays, and steering should be installed so that personnel face forward in the direction in which the vessel travels. This should be done whether or not personnel have visual contact with the outside of the vessel. *Auxiliary machinery consoles.* Consoles associated with auxiliary machinery, such as vessel service generators,

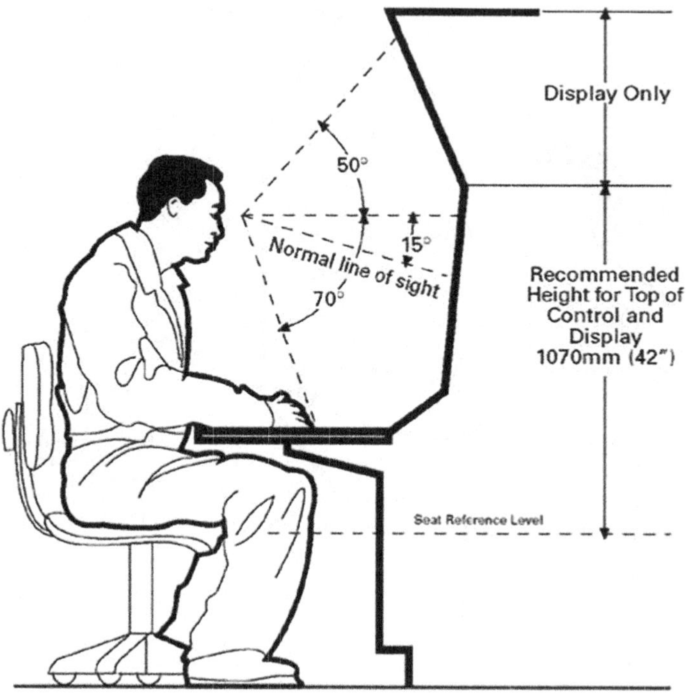

Fig. 5.8 Multi-tiered seated console

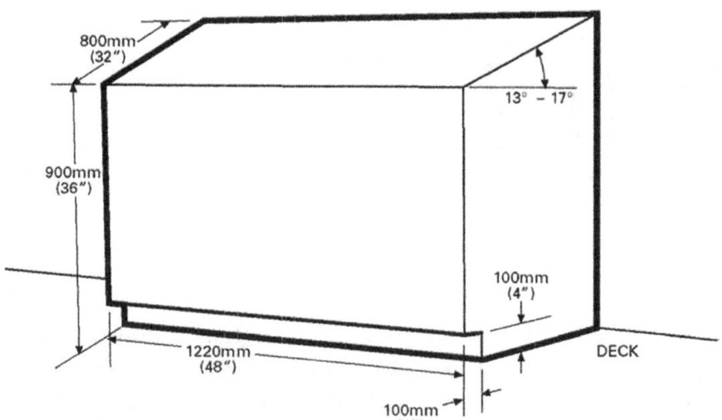

Fig. 5.9 Dimensions for desktop standing console

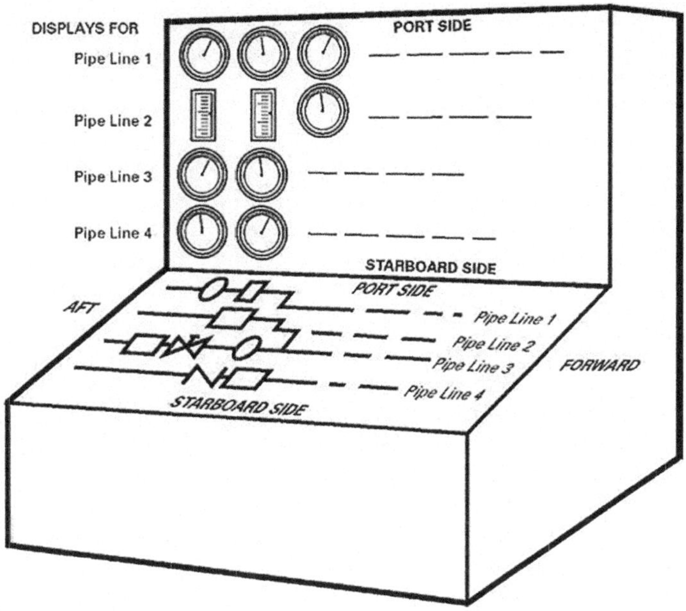

Fig. 5.10 Cargo and ballast transfer console

pumps, mooring winches, or thrusters, should be oriented to maintain a spatial relationship between the console controls, displays, and the actual equipment they control or monitor. If there is more than one piece of identical machinery located in separate spaces, such as identical thrusters located in a forward or aft thruster room, the console for each piece of equipment should be located and oriented the same in each space.

Chapter 6
Computer Workstation Design

General

Computer workstations should comply with the following design recommendations to facilitate task performance and minimise fatigue in personnel. Computer workstations should be designed to accommodate the expected range of personnel. Figure 6.1, Display heights for standing personnel and Fig. 6.2, Seated workstation illustration, present general computer workstation arrangements. These figures' dimensions are based on North American males. All designs based on anthropometric data should consider differences in size and stature of personnel from different ethnic and geographic groups. Given the international nature of the marine and offshore industries, consideration should be made for the widest range of possible personnel populations. Chap. 13, Anthropometric design principles and dimensions, contains detailed anthropometric data for many international populations.

Computer Workstation Design Principles

Usability and workstation adjustability. In order for a workstation to be usable, it may be necessary for an operator or maintainer to configure that workstation to meet individual requirements. In particular, seated workstations should be provided with adjustable seating, sufficient arm support and workspace for the intended task. *Maintainability.* Workstations should be designed to provide access to necessary hardware and equipment for maintenance purposes. *Workstation ambient environment.* Whole-body vibration, noise, indoor climate, and lighting characteristics of computer workstation areas should follow the guidance presented in Chap. 12, Crew habitability. *Consoles and desks.* A console is a single piece of furniture that accommodates the video display and associated hardware. Desks are more general-purpose workspaces and typically do not provide storage for computers or video displays. The console or

A. A. Olsen, *Applying Physical Ergonomics to Modern Ship Design*,
https://doi.org/10.1007/978-3-031-57974-5_6

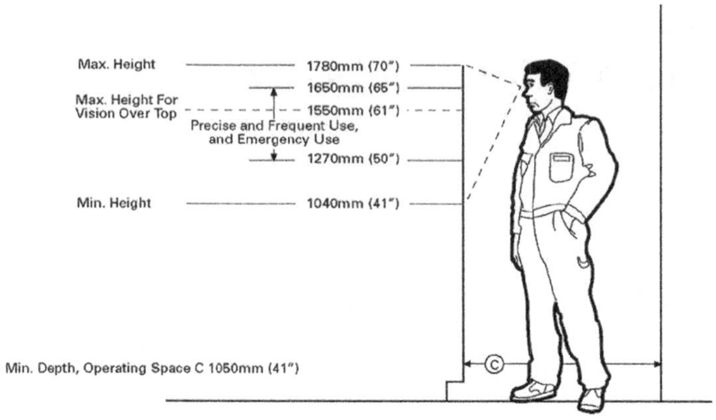

Fig. 6.1 Display heights for standing personnel

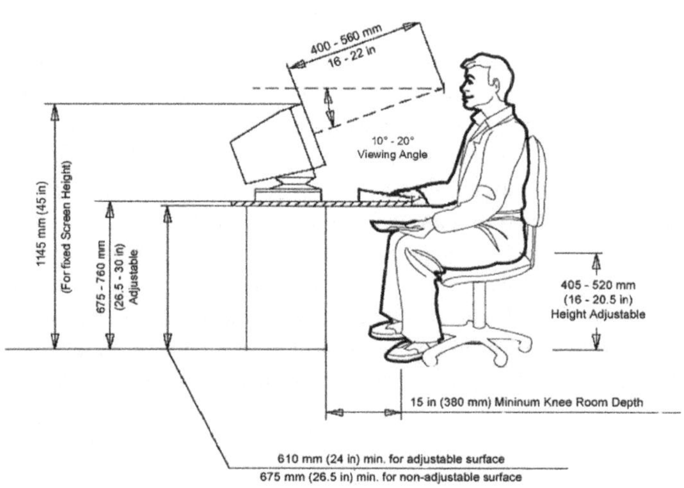

Fig. 6.2 Seated workstation illustration

desk supplied should provide a suitable work surface height, working area, and knee room. Refer to Fig. 6.2, Seated workstation illustration, for design recommendations. All workstation designs based on anthropometric data should consider differences in size and stature of persons from different ethnic and geographic groups. Chapter 13, Anthropometric design principles and dimensions, contains detailed anthropometric data for many international populations. Also refer to the section 'General console arrangements' in Chap. 5 for examples of video-display units embedded in consoles.

Writing surfaces. A writing surface should be at least 400 mm (16 in.) deep and 610 mm (24 in.) wide. For surfaces where a keyboard and mouse may be used, the width should be at least 760 mm (30 in.). It is suggested that writing and working surface heights be adjustable in order to prevent overexertion or strain of the back or appendages. *Working surface height.* The working surface should be between 710 mm (28 in.) and 760 mm (30 in.) above the floor surface on which the chair sits. Where practical, writing and working surface heights should be adjustable. Seating. For most tasks, office chair seating is acceptable. The height of the seat pan should be adjustable in order to accommodate a variety of personnel. This height should be adjustable from 380 mm (15 in.) to 520 mm (20.5 in.) when measured from the surface on which the chair sits to the top of the seat pan. Seating should also have adjustable seat pan angle, seat back height, and seat back angle. *Keyboard height.* Keyboard support surfaces should be adjustable between 585 mm (23 in.) and 710 mm (28 in.) above the floor surface on which the chair sits. *Viewing distance.* Viewing distance from the operator or maintainer's eye to the video display should be between 400 mm (16 in.) and 560 mm (22 in). A viewing distance of between 460 mm (18 in.) and 510 mm (20 in.) is recommended. The top height of the video display should be 1145 mm (45 in.), creating a viewing angle of 10°–20°. Refer to Fig. 6.2, Seated workstation illustration. *Video display height.* Video display height should be adjustable so that the topmost line of display on the screen should not be higher than the operator or maintainer's eyes. If a fixed height is mandated by hardware constraints the top of the screen should be 1145 mm (45 in.) above the surface on which the chair sits. *Knee space.* Knee space under the keyboard support surface should be a minimum of 610 mm (24 in.) for an adjustable surface or 675 mm (26.5 in.) for a non-adjustable support surface. The width of this space should be a minimum of 510 mm (20 in.), and the depth should be a minimum of 380 mm (15 in.). Refer to Fig. 6.2, Seated workstation illustration.

Chapter 7
Manual Valve Operation, Access, Location and Orientation

General

The principles and guidance contained in this chapter apply to the manual operation of valves, including those operated by hand wheels or levers.

Principles

General access. For valves required by function or design to be readily accessible, the valve or its control should be located in a space normally entered without using tools; clear of and protected from obstructions, moving equipment, and hot surfaces that may prevent or delay operation or maintenance; and within arm's reach. Valves should be located so that the operator or maintainer does not have to stand on nearby pipes, cable trays, handrails, equipment, or any object not meant specifically to be used as a standing surface for the operation, maintenance, repair, or replacement of any valve. A minimum of 75 mm (3 in.) clearance should be provided between the outside rim of a valve hand wheel, or the end of a valve lever and any obstacle located throughout the hand wheel or lever's field of travel. Valves should not be mounted in positions where stems on hand wheels or levers will extend into normal walking areas. *Emergency access.* Valves used for emergency operations should not be located below deck gratings or behind covers. If it is absolutely necessary to locate valves behind covers, (e.g., to meet a regulatory requirement) the cover should be capable of being opened without requiring any tools or the removal of any securing fasteners. The cover should be clearly labelled to identify the valve. *Valve operators and indicators.* Valves should close with a right hand (clockwise) motion of the hand wheel or lever when facing the end of the stem. Valves should be provided with a means to determine position. Valve position indicators should be installed so the indicator is directly visible to the operator or maintainer from the normal

A. A. Olsen, *Applying Physical Ergonomics to Modern Ship Design*,
https://doi.org/10.1007/978-3-031-57974-5_7

body position required to open or close the valve. For valves that cannot be operated within the operator or maintainer's reach limits given in the following subsections, they should be operated by mechanical extenders rather than chain operators. For valves fitted for remote control, an independent indicator showing whether the valve is open or closed should be provided with the control. *Labelling.* Labels should be used to identify manual valves. Refer to the section 'Equipment identification labels' in Chap. 8 for guidance on valve labelling. *Maximum force.* The maximum force to initially crack open a manual valve should not exceed 450 N (100 lb).

Categorisation of Valves for Selection of Location

Valve location, accessibility, orientation and ease of use should be determined on the basis of a Valve Criticality Analysis or other analytic means to determine valve location and access based on criticality and method and frequency of use. Identification of individual valve criticality can be of joint participation among vessel designers, operators, and owners. The following three categories define valve criticality. *Category 1 valves.* Category 1 valves include those essential to normal or emergency operations where rapid and unencumbered access and operation or maintenance is essential; thus, permanent access should be provided at a deck level or via a permanent standing surface. If such access is not practical, access by stair or ladder is acceptable. Valves meeting any of the following will qualify as Category 1 valves:

- Valves that are essential for operations.
- Overboard valves.
- Equipment isolation valves.
- Valves that are essential for personnel or process safety, cargo protection, and pollution prevention.
- Valves where there is a high likelihood of failure.
- Valves where the consequence of failure or lack of quick access would result in damage to personnel or property, loss of productivity, or damage to equipment or the environment).
- Valves where an expected operational and/or maintenance frequency is greater than once in a six-month period.

Examples of valves typically included in Category 1 are control valves, their bypasses and isolation valves, relief valves and depressuring valves, trip and anti-surge control valves, and ballast water discharge valves. *Category 2 valves.* Category 2 valves are those that are not critical for normal or emergency operation or maintenance but are used during routine maintenance activities. These should be located with permanent access at deck level, access via stair or other access aid with a purpose-built standing surface/landing. Examples of valves typically in Category 2 are sewage treatment valves, drain valves, service oil valves, battery limit valves, gas freeing connections, manual valves with an expected operating and/or maintenance frequency of less than once per six months, and valves where quick action is not required. *Category 3 valves.* Category 3 valves are those used in particular circumstances on an infrequent or rare basis. While no specific location guidelines are imposed for Category 3 valves, several guidelines exist related to safety of Category 3 valve use and operation. Examples of valves typically included in Category 3 are valves used in dry dock only, valves used in initial commissioning, valves used for decommissioning, valves used to isolate pressure vessels, tanks, etc., for inspection; and valves for pressure test. *Assignment of guidance by valve category.* For Category 1 valves, unless otherwise indicated, all the guidelines stated in this chapter apply to Category 1 valves. Category 2 valve guidelines are indicated in the guidelines statements. Any guideline that applies to Category 2 valves also applies to Category 1 valves. Category 3 valves have no specific access, positioning or orientation guidelines; however, safety of use and operation guidelines are imposed on Category 3 valves. *Valve operators and indicators.* Valves should close with a right hand (clockwise) motion of the hand wheel or lever when facing the end of the stem, or where left-handed valves are provided, direction or turn to close the valve should be indicated on the valve, its label or on the hand wheel itself. (Category 1 and 2). Valve position indicators should be installed on each valve. (Valves where the position of the stem (i.e., rising stem valves) or the position of the handle (i.e., ball valve) provides a direct indicator of the valve position do not require and additional "valve position indicator"). (Category 1 and 2). Valve position indicators should be directly visible to the operator or maintainer from the normal body position required to open or close the valve. (Category 1 and 2). Valves should be capable of being operated without mechanical extenders such as a reach rod or chain operators. (Category 1). For valves fitted for remote control:

- An independent indicator showing the position of the valve (open, closed or midway) should be provided at the control. (Category 1 and 2).
- An independent indicator showing the position of the valve (open, closed or midway) should be provided at the physical location of the valve. (Category 1 and 2).
- Category 1 valves should be provided with redundant controls located near or on the valve body itself (for the cases where remote means of operation is not working).

Labelling, marking and coding. Labels should be provided to identify manual valves. (Category 1 and 2) and labels, markings and coding should be visible to the operator from the normal body position required to open or close the valve. (Category 1 and 2). *Clearances.* Clearance of at least 300 mm (12 in.) for wrenches should be provided adjacent to flanged connections for valves and equipment where bolts can be accessed from one side only. (Category 1 and 2). Clearance of at least 760 mm (30 in.) should be maintained on both sides of pipe where access for work or maintenance is required. Where simply accessing bolts from both sides of the pipe, greater than 300 mm (12 in.) clearance should be provided. (Category 1 and 2); and a 75 mm (3 in.) clearance should be maintained all around valve hand wheels, except that 40 mm (1.5 in.) and smaller valves where operation is by hand or wrist motion may have less clearance, minimum 40 mm (1.5 in.) finger clearance. (Category 1 and 2).

Maximum force. The maximum force to initially crack open a manual valve should not exceed 450 N (100 lb force). (Category 1 and 2). *Turning aids.* Knurling, indentation, high-friction covering or a combination of these should be built into hand wheels to facilitate the operator's grasp for applying maximum force. (Category 1 and 2). *Hand wheel rim dimension.* The hand wheel rim for hand wheels with diameters larger than 150 mm (6 in.) should be cylindrical to facilitate grasping and applying maximum force. (Category 1 and 2); and the rim diameter should be 20 to 40 mm (0.75 to 1.5 in.). (Category 1 and 2). *Human endurance.* The human endurance limit applied should be 100 turns maximum to open or close a valve at a rate between 15 and 60 revolutions per minute. (Category 1 and 2). Where the above guideline is not met, a handheld power (pneumatic, hydraulic, or electric) valve-turning machine should be provided. (Category 1 and 2); and the weight of such a handheld valve-turning machine should not be in excess of 20 kg (44 lbs). (Category 1 and 2).

Preferred Valve Mounting Heights and Orientations

Hand wheel-operated valves. Hand wheels of less than 100 mm (4 in.) in diameter should be provided when intended for one-hand operation. Hand wheels of greater than 150 mm (6 in.) diameter should be provided when intended for two-hand operation. (Category 1 and 2). Hand wheels with diameters between 100 mm (4 in.) and 150 mm (6 in.) should be avoided (since they are too large for one-hand operation and too small for two-hand operation). (Category 1 and 2). Valve hand wheels should be located as shown in the following figures for orientation of valve stem in the vertical, horizontal or angled position. Category 1 valves can be located in either Category 1 or 2 valve locations when:

- Hand wheels are in excess of 600 mm (24 in.) diameter.
- Valves are used for emergency responses, (for example, firefighting or deballasting).
- Stem freeze-up is possible.
- Hand wheel operating torque is in excess of 54 N-m (40 ft-lbs).
- The valve may be expected to be operated by a short male or female.

Category 2 valve locations are reserved for Category 2 valves, and Category 1 valves should not be fitted in those locations. Category 1 or Category 2 valves should not be fitted outside the indicated locations. *Lever-operated valves. Lever*-operated valves oriented with the stem in a vertical position should be provided when the valve lever can be located between 760 mm (30 in.) and 1270 mm (50 in.) above the standing surface. Lever-operated valves oriented with the stem in a horizontal position should be provided when the lever is located between 150 mm (6 in.) and 760 mm (30 in.), or more than 1270 mm (50 in.) above the standing surface. The maximum height above the standing surface to the lever tip should not exceed 1900 mm (75 in.). (Category 1 and 2). Horizontal stem valves should not be located overhead in working areas. (Category 1 and 2).

Alternative Valve Orientations

Valves in overhead position. When it is necessary to locate a manual valve's actuator above a user's head, the operating torque should be no more than 27 N-m (20 ft-lbs). (Category 1 and 2). Valve hand wheels should not be larger than 510 mm (20 in.) in diameter. (Category 1 and 2). *Valves in walkways.* Valve orientation should not place the valve lever into a passageway. (Category 1, 2 and 3). *Valves accessible from one side only.* When access to a lever-operated valve is available from one side only, the valve should be mounted such that the lever moves to and from the accessible side where the operator or maintainer will be positioned, as shown in Fig. 7.1, Direction of travel for valve levers accessible from one side only. (Category 1 and 2). *Valves at or below standing surface.* If a valve is located at or below the standing surface, requiring stooping or squatting to operate the valve, the valve position in relation to an operator's or maintainer's body position should be as shown in Fig. 7.2, Physical reach from a stooping or squatting position. (Category 3 only). When it is necessary to locate hand wheel or lever valves below the operator's or maintainer's standing surface, horizontally- or vertically-oriented valves should be installed as shown in Fig. 7.3, Mounting position for valve levers and hand wheels below standing surface. (Category 3 only). Deck opening sizes to reach and operate levers or hand wheels located below the standing surface and oriented parallel to the standing surface should be as shown in Table 7.1, Access opening and mounting depth dimensions for levers and hand wheels oriented parallel to the standing surface. This table also includes guidance on mounting depths for valves (Category 3 only). *Valves operated from a ladder. Where* valves must be operated from a permanent ladder, they should be limited to those that can be operated with one hand (valves with hand wheels less than 155 mm (6 in.) in diameter and lever valves). (Category 1 and 2). Valve levers should not intrude into the ladder climbing place. (Category 1, 2 and 3). The ladder should be positioned to allow operation of a valve within a 610 mm (24 in.) distance forward of the ladder or 1220 mm (48 in.) distance from the side of the ladder, as shown in Fig. 7.4, Orientation and reach for ladder parallel to valve levers and Fig. 7.5, Orientation and reach for ladder perpendicular to valves (Category 1 and 2). These figures show the required ladder orientation in relation to valves. The valve orientation, direction of valve operation, and distances shown are applicable to both lever and hand wheel-operated valves. When operating lever valves from a ladder, the levers must not protrude into the climb space thus impeding safe access and egress. (Category 1 and 2). *Valve hand wheel accessibility from elevated platforms.* Valve hand wheels operated from elevated platforms 1830 mm (72 in.) or more above grade should be located within the confines of the platform railing (Category 1 and 2).

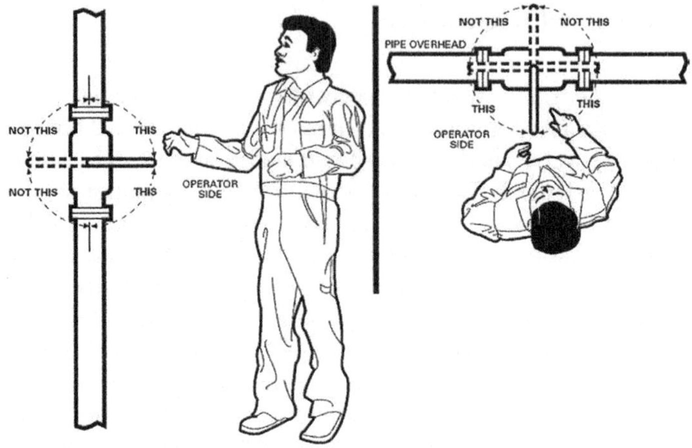

Fig. 7.1 Direction of travel for valve levers accessible from one side only

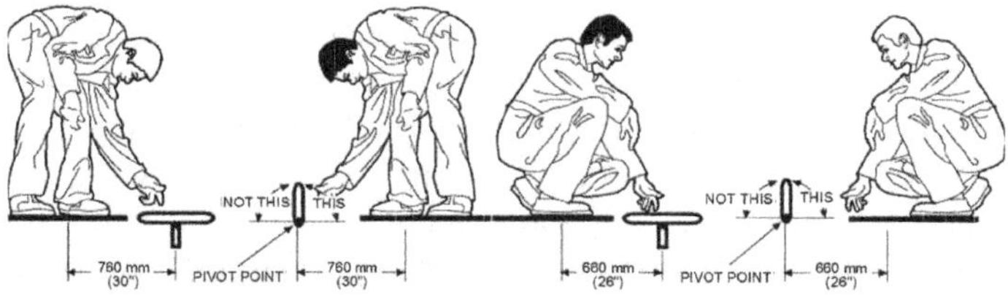

Fig. 7.2 Physical reach from a stooping or squatting position

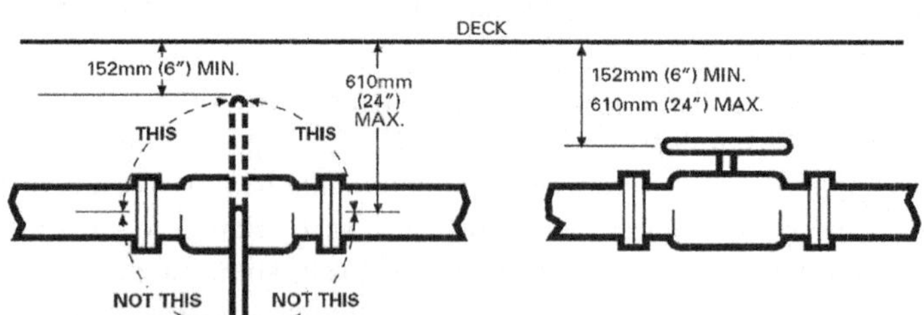

Fig. 7.3 Mounting position for valve levers and hand wheels below standing surface

Valve Manifolds

Where valves are mounted together to create a valve manifold (e.g., fuel oil transfer or cargo oil transfer), the valves handles should be arranged such that as the operator faces the valve manifold, the location of the tank or pump with which the valve is associated replicates, or has direct spatial relationship with, the location of the

Table 7.1 Access opening and mounting depth dimensions for levers and hand wheels oriented parallel to the standing surface

Valve handle (diameter or length)	Depth below deck	Deck opening size (square or diameter)
Hand wheel		
130 mm (5 in.) or less	150–255 mm (6–10 in.)	180 mm (7 in.)
	Greater than 255 mm (10 in.)	215 mm (8.5 in.)
130 mm (5 in.) or more	150 mm (6 in.) to 510 mm (20 in.)	The diameter of the handwheel plus 150 mm (6 in.) with a minimum of 360 mm (14 in.)
Levers		
Any lever length	Any depth	Lever length greater than 50 mm (2 in.)

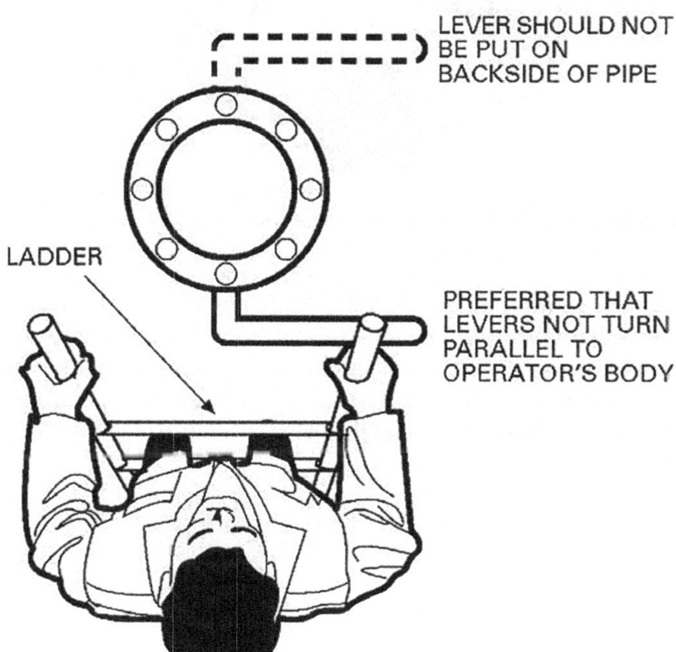

Fig. 7.4 Orientation and reach for ladder parallel to valves

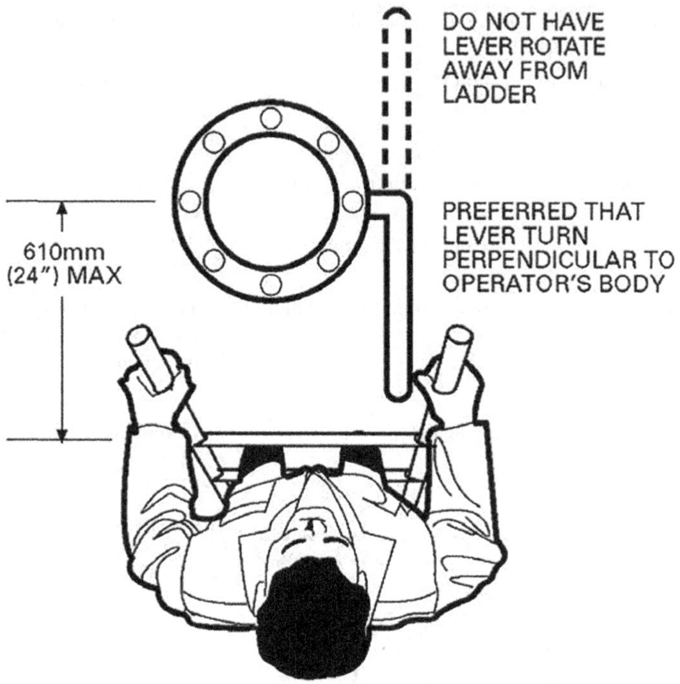

Fig. 7.5 Orientation and reach for ladder perpendicular to valves

associated item as shown in Fig. 7.6, Valve Manifold for tanks located athwartship and Fig. 7.7, Valve manifold for tanks located fore and aft. The overall vertical orientation and horizontal location of the manifold valves should provide a direct spatial relationship between the valves in the manifold and the equipment (e.g., tank, pump, exchanger, etc.) associated with the valves. Refer to Fig. 7.8, Valve manifold for fill, high suction, and low suction.

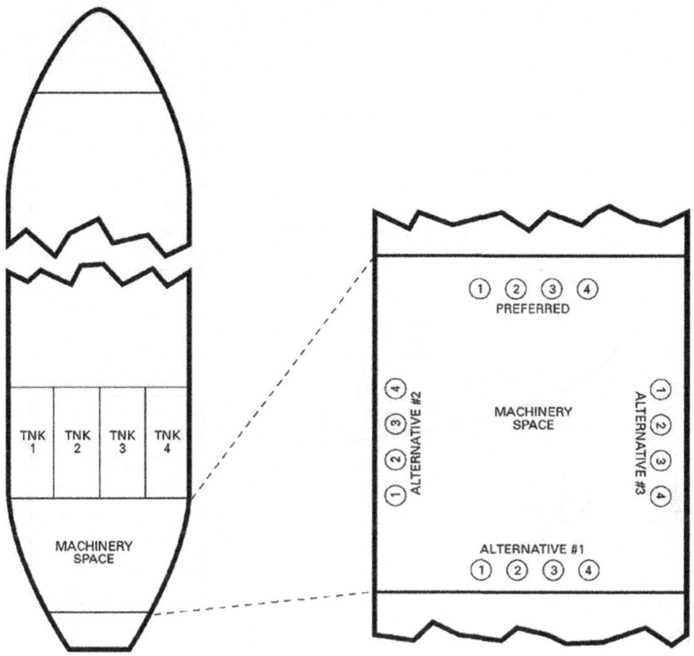

Fig. 7.6 Valve manifold for tanks located athwartship

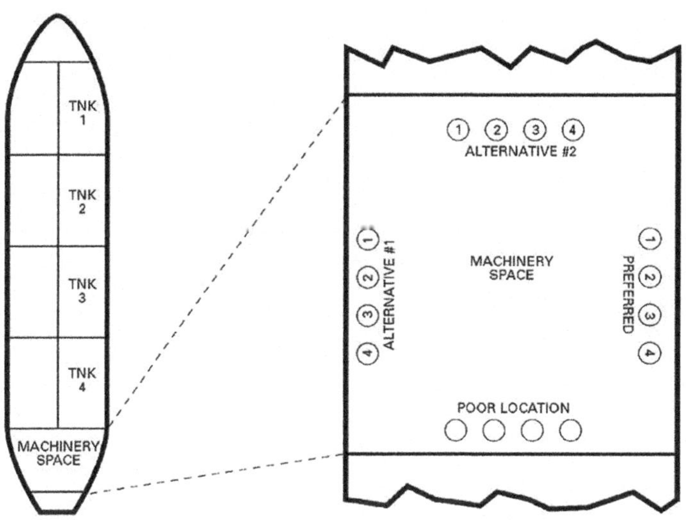

Fig. 7.7 Valve manifold for tanks located fore and aft (In Fig. 7.7, Valve manifold for tanks located fore and aft, the starboard facing orientation of the valve manifold is preferred over the port facing orientation to allow the ordering of the tank numbers to read from left to right. This preference principle is different from that for a mimic control console (such as a ballast control console) where the preferred console orientation is facing forward or to port with the forward most items located on the right-hand side of the console)

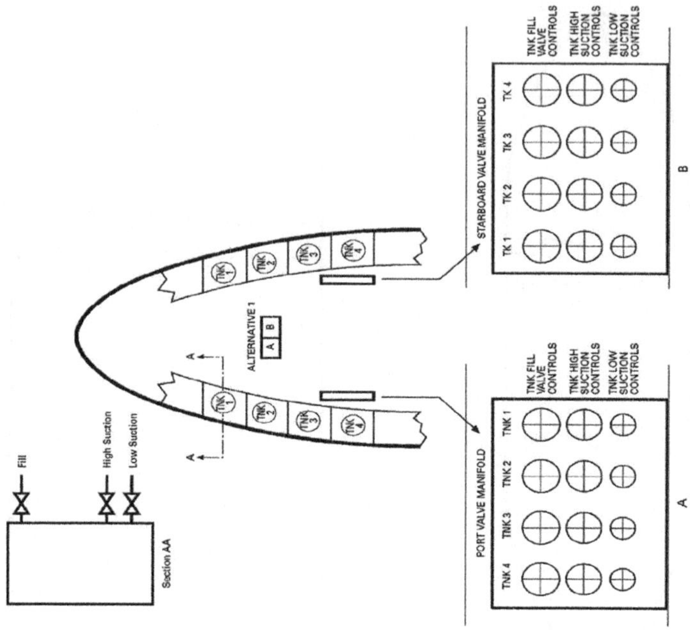

Fig. 7.8 Valve manifold for fill, high suction, and low suction

Chapter 8
Labelling, Signs and Graphics

General

The principles and guidance set out in this chapter apply to any plate, sign, placard, inscription, legend, marking, or combination of these that gives information, warning, or instructions via text, graphics, or symbols. It should be noted that when labels, signs, graphics, or symbols are addressed in regulations from the IMO, Flag State Administrations, classification societies, or other applicable regulatory bodies, those requirements take precedence.

Principles

The following principles apply to the design and placement of labels. *Content.* Labels and signs should use words and abbreviations familiar to personnel. Full words should be used wherever possible. Abbreviations should rarely be used and only if there is insufficient space for the words or if the abbreviation is more commonly known and understood than the actual word (e.g., AFFF rather than Aqueous Film Forming Foam). Trade names, company logos, model numbers, or other information not directly required for operation should be avoided on the face of a display, control, or labels corresponding to these components. No more than three hazard, information, instruction, or graphic labels should be grouped in a row or placed close together. Within the text of a label, references to other documents or sources of information should not be made if the referenced material is required to complete a task, avoid a hazardous situation Nor understand the purpose of the label. *Background colour and characters.* In general, characters should be black on a white background. Letters on labels should be of a simple block-type format using all upper-case letters. Upper- and lower-case letters should be used for extended sentence messages. Numerals should be Arabic (not Roman Numerals). The use of the number “1” in combination

A. A. Olsen, *Applying Physical Ergonomics to Modern Ship Design*,
https://doi.org/10.1007/978-3-031-57974-5_8

with the letters "L" or "I" and the number "0" with the letters "O" or "Q" should be avoided. *Orientation and placement.* Text should be written in the normal orientation for the language used. For example, labels written in Indo European languages would be horizontal and read from left-to-right. Labels, graphical schematics, or diagrams should be positioned to be read from the normal operating or working position. Labels, signs, and other posted information should be placed in locations that will not be covered by movable items such as doors, racks, access openings, or equipment. Care should be taken so that these are not placed in locations where equipment or other items are likely to be temporarily stored. Labels, signs, and other posted information should be mounted on, or immediately adjacent to, the location where the information is used. Labels should not be placed on components that turn where the label could be placed in an unreadable position. Labels, signs, and other posted information should be fully visible to personnel and preferably on flat surfaces. Attached tags should be used, where adequate surfaces are not available for mounting labels. *Multiple labels, signs, graphics, or symbols.* The grouping of more than three hazard, information, instruction, or graphic labels in a row, or close together, should be avoided. *Material.* Labels, signs, or graphics should be made from materials, and manufactured by a process, which will be suitable for all expected conditions in the operating environment. The material should also be compatible with the surface on which it is attached and be resistant to ultraviolet (UV) damage, if exposed to sunlight.

Component Labels on Consoles and Panels

Labels should be used to identify individual controls, displays, or groups of controls and displays on consoles and panels. Each display and control, including their setting positions, should be labelled in accordance with the following guidance. *Label all controls and displays.* Each control, the control setting, and display on every console or panel should be labelled. Also, any individual control or display mounted on a pole, desk, or other structure should be labelled. *Character size.* The character size of labels should follow the guidelines shown in Fig. 8.1, Guidelines for labels on consoles and panels. Character sizes greater than the minimum are acceptable, if applied consistently to all levels of labelling. *Mounting location.* As shown in Fig. 8.1, Guidelines for labels on consoles and panels, the preferable location of labels is above the controls and displays they identify. This facilitates natural eye movement during reading. In addition, placing labels above controls will prevent a person's hand from obscuring the label when using the control. The relative placement of labels to their respective controls and displays should be consistent. In limited circumstances, when placement above a control or display may result in an unreadable height, the label may be placed below the control or display, provided any confusion with other components or labels is minimised. *Content.* The content of labels can be determined by using the following guidelines:

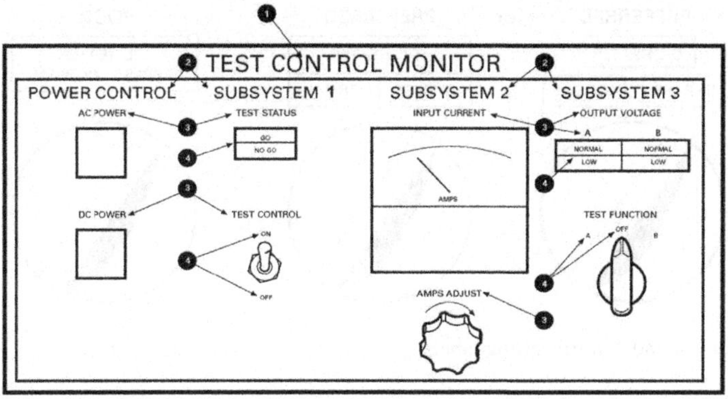

ITEM	LABEL DESIGNATION	CHARACTER SIZE mm/INCHES/POINT	LOCATION
1	PANEL TITLE	12/0.50/36	CENTERED; 6mm (1/4 INCH) FROM TOP EDGE OF PANEL
2	PANEL SUBSECTION	10/0.37/26	CENTERED AT TOP OF SUBSECTION
3	SUBTITLE/SINGLE COMPONENT IDENTIFICATION	5/0.20/14	6mm (1/4 INCH) ABOVE COMPONENTS OR 12MM (1/2 INCH) ABOVE LABELS OF INDIVIDUAL COMPONENTS
4	SWITCH POSITION/ DISPLAY LABEL	3/0.125/9	6mm (1/4 INCH) ABOVE AND/OR BELOW SWITCH

Fig. 8.1 Guidelines for labels on consoles and panels

- *Describe the function.* Control and display labels should indicate the function of the device rather than the technical name for the device. For example:
    1. **VOLTAGE** rather than **VOLTMETER**
    2. **POWER ADJUST** rather than **POWER ADJUSTER SWITCH**

Describe the control movement. Control labels should indicate the result of a control movement by either words or appropriate symbols (e.g., **RAISE**, **START**, +, ↓, →), Include any units of measure. These (e.g., **psig**, **volts**, **kPa**, **mm**) should appear on the face of displays, not on the labels; and.

- *Label components consistently.* Label terminology should be consistent for the same controls and displays on different equipment or systems.

Relationship of control and control setting labels. It is preferred to place control functional and control setting descriptions on separate labels. Alternatively, they may be placed on a single label provided there is no confusion between the control label and the control setting labels. Refer to Fig. 8.2, Control and control setting labels.

Group labels. When controls and displays are grouped together, appropriate labels or panel demarcation lines (or both) should be used to indicate their relationship (e.g., functional, sequential, similar equipment) as shown in Fig. 8.3, Control and display group labels. The label should be located above the group it identifies. When a line is used to demarcate a group and define its boundaries, the label should be centred at the top of the group. The label should be placed either in a break in the demarcation line or just below the demarcation line.

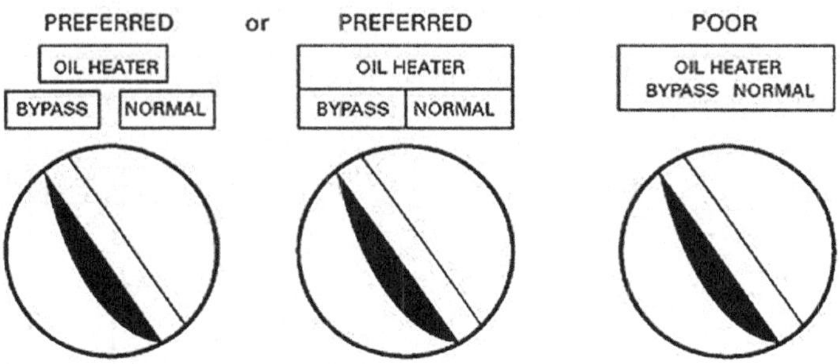

Fig. 8.2 Control and control setting labels

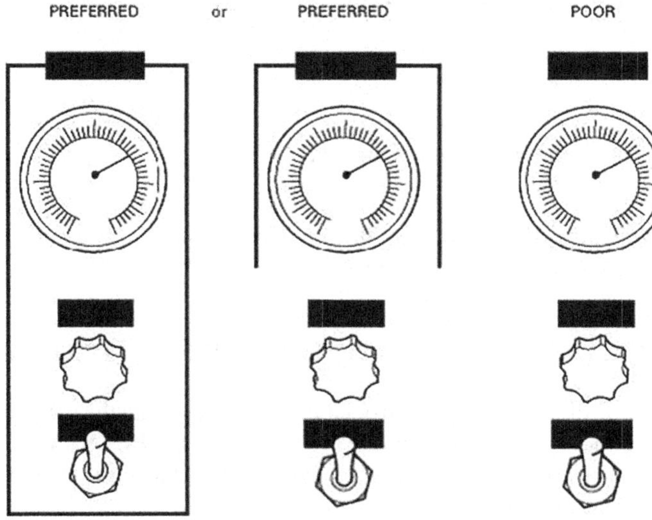

Fig. 8.3 Control and display group labels

Control setting labels for multiple controls. When multiple controls perform the same function for different equipment, and they are grouped together, the settings for each control should be labelled as shown in Fig. 8.4, Control setting labels for multiple controls.

Material. Control and display labels should be made of durable materials such as engraved plastic, reverse Lexan, or flexible vinyl. Where labels may be exposed to sunlight, the label material should be resistant to UV damage.

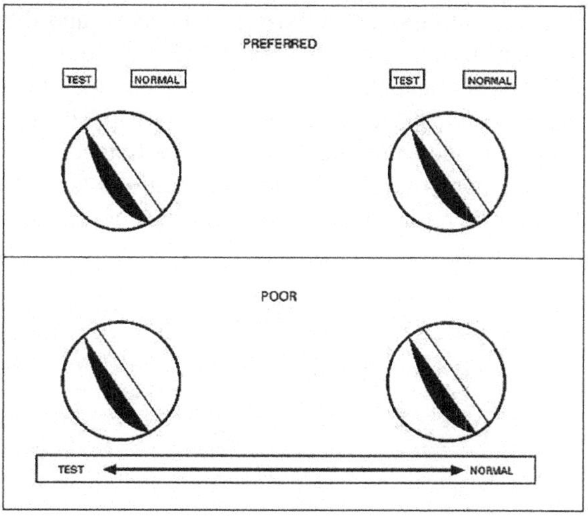

Fig. 8.4 Control setting labels for multiple controls

Equipment Identification Labels

Description and equipment number. Identification labels should be placed on all equipment (e.g., valves, gauges, controls, filters, pumps, pressure vessels, electrical and communication equipment) requiring interface from operations or maintenance personnel. The label should contain a specific functional description of the equipment and a unique identifying number for it. Refer to Table 8.1, Examples of equipment labels.

Table 8.1 Examples of equipment labels

Item	Label
Lube oil pump for generator engine no. 1	LUBE OIL PUMP GENERATOR No.1 PEP-100
Lighting junction box	LIGHTING JUNCTION BOX LJB-313X
Discharge pressure gauge on cargo pump no. 3	CARGO PUMP No. 3 DISCHARGE PRESSURE PI-211
Diesel day tank for generator no. 4	DIESEL DAY TANK GENERATOR No. 4 CAPACITY 445 GAL DDT-4
Port ballast pump	PORT BALLAST PUMP PBE-1000
Flanged ball isolation valve (numbered N350-05) on the steam line leading from a steam separator (NBD-350) to the glycol reboiler (HBG-640)	STEAM ISOLATION TO GLYCOL REBOILER HGB-640 N350-05
Low output pressure sensor on a ballast pump	BALLAST PUMP PRESSURE SAFETY LOW PSL—2001

Format. The number and size of characters and the number of lines per label for different sized labels are shown in Fig. 8.5, Equipment label format. *Mounting location.* For permanently installed equipment, identification labels should be mounted directly on the piece of equipment. Alternatively, they may be mounted immediately adjacent to the equipment provided it is immediately obvious which label and piece of equipment are related. Labels should not be mounted on surfaces or equipment that may be replaced due to maintenance or repair.

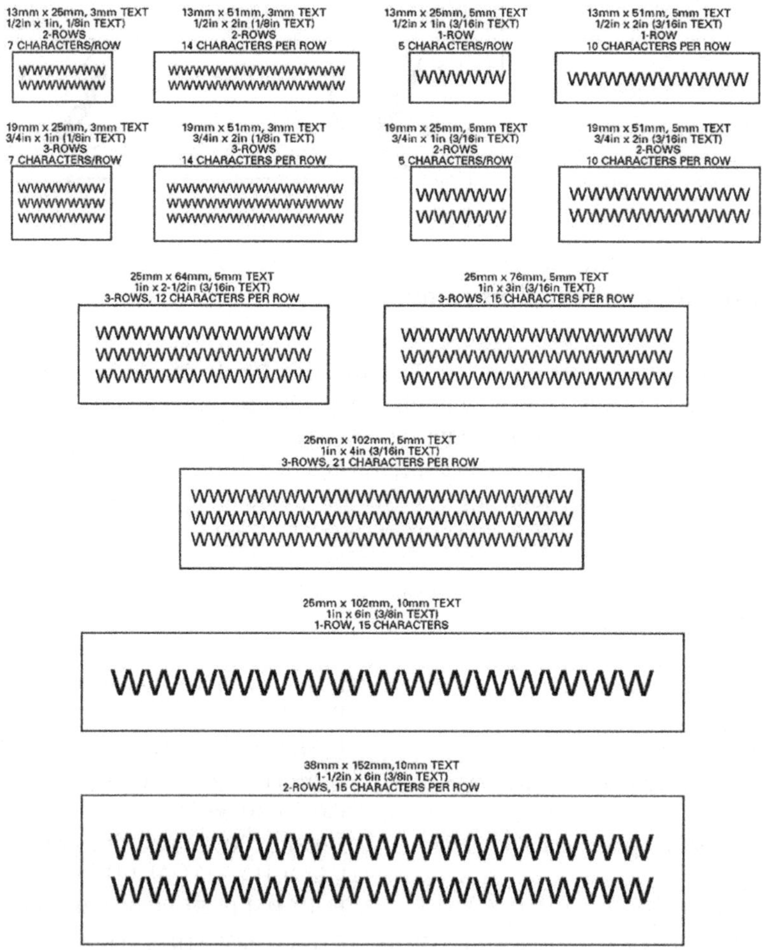

Fig. 8.5 Equipment label format

Fig. 8.6 Sensor label

PORT BALLAST PUMP

PRESSURE SAFETY LOW

PSL-2001

Electrical System Labels

Each electrical cable should be labelled with a unique identifying code as described in the section ‘Component labels’ on consoles and panels. The label should be located at the exit from the main switchboard, the distribution panel at the connection to the load and at the entrance and exit of intermediate panel boards, junction boxes, and controllers. Circuit breakers should be provided with identification labels, as described in the section ‘Equipment identification labels’, including both the written name of the circuit and the circuit code identified on the cable label.

Sensor Labels

When identifying sensors, [e.g., **PRESSURE SAFETY HIGH (PSH), TEMPERATURE SAFETY LOW (TSL), FLOW SAFETY HIGH (FSH)],** the label should identify the unit of equipment to which the sensor is attached. It should also include the sensor’s identification number. The information should be presented according to the following rules. An example is provided at Fig. 8.6, Sensor label.

Room or Space Identification Labels

Identification labels should be mounted on the door leading into a room or space. The style of the labelling, its character size and format should follow Fig. 8.5, Equipment label format. Labels should be placed at or slightly below personnel’s standing eye height. For 50th percentile North American personnel, the label would be located approximately 1650 mm (65 in.) above the standing surface.

Pipe Marker Labels

Usage. Pipe marker labels should contain text to identify pipe contents and use arrows to indicate flow direction. This is recommended in order to identify those pipes carrying firefighting mediums (e.g., water, foam) to aid personnel in damage control or other emergency situations; to identify those pipes which contain flammable liquids; to reduce the chance of injury during maintenance from cutting into a pipe carrying dangerous contents, such as high-pressure gases or corrosive liquids; and to assist personnel in becoming familiar with the piping system. *Format.* Pipe marker labels should either use different coloured backgrounds or bands of colour on a solid neutral coloured background to represent various pipe contents via a colour-coding scheme. Text and flow arrows are placed on the labels to identify pipe contents and its flow direction. Text labels should appear in the centre of the pipe marker label or immediately adjacent the colour band that identifies pipe contents. Flow arrows should be positioned at one, or both, ends of the band. The colours for the text labelling and flow arrows should be of a colour that contrasts with the background colour. Figure 8.7, Pipe marker labels, provides examples of a pipe marker label. Pipe marker label preferred width, length, and associated text character height for various sizes of pipes are shown in Table 8.2, Pipe label format.

Selection of colours. The colours used for the bands or text should be selected to be seen under all levels of lighting likely to be encountered on the vessel or offshore

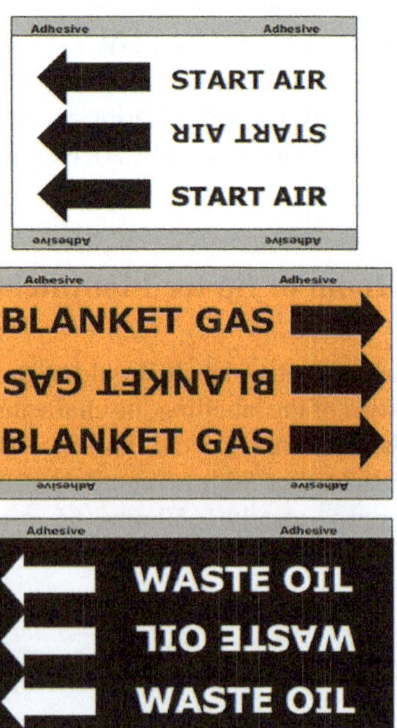

Fig. 8.7 Pipe marker labels

Table 8.2 Pipe label format

Marker type	Pipe nominal size	Length along pipe	Width around pipe	Character height
Wrap around	10–15 mm (0.375–0.50 in.)	125 mm (5.0 in.)	N/A	N/A
Wrap around	20–65 mm (0.75–2.50 in.)	205 mm (8.0 in.)	N/A	20 mm (0.75 in.)
Wrap around	50–200 mm (2–8 in.)	305 mm (12.0 in.)	N/A	30 mm (1.25 in.)
Non-wrap around	More than 200 mm (more than 8 in.)	610 mm (24.0 in.)	100 mm (4.0 in.)	90 mm (3.50 in.)

installation. The colours should be compatible with other colour codes and the scheme used for pipe marker labels should be the same throughout the vessel or offshore installation. Where possible, the colours chosen should be related to the contents carried by the pipe (e.g., fire water pipes are red, drilling mud pipes are brown, sea water pipes are green). An example of a colour coding scheme is shown in Table 8.3, Example colour-coding scheme for vessel piping. For consistency of colour, the chromaticity coordinates of the suggested colours are provided Table 8.4, chromaticity coordinates for colour coding of vessel/offshore installation piping.

Use of two colours or two bands of colour. Additional colours or secondary colour bands may be used to distinguish between several different types of the same contents, such as various grades of fuel or different types of waste. For example, one colour would be chosen to represent "waste". However, the various types of waste could

Table 8.3 Example colour-coding scheme for vessel piping[a]

Main colours	Medium
Black	Waste media (e.g., wastewater, black water, grey water waste oil, exhaust gas)
Blue	Fresh water
Brown	Fuel
Green	Sea water[a]
Gray	Non-flammable gases
Maroon	Masses/bulk materials (dry and wet)
Orange	Oils other than fuels
Silver	Steam
Red	Firefighting and fire protection
Violet	Acids, alkalis
White	Air in ventilation system
Yellow-ochre	Flammable gases

[a] Colours adopted from ISO 14726-1

Table 8.4 Chromaticity coordinates for colour coding vessel/offshore installation piping[a,b,c,d]

Colour		Chromaticity coordinates of corner points determining the permitted tighter colour area as given in CIE-Publication 15.2							
		1		2		3		4	
		x	y	x	y	x	y	x	y
Black	(BK)	0.385	0.355	0.300	0.270	0.260	0.310	0.345	0.395
Blue	(BU)	0.078	0.171	0.196	0.250	0.225	0.184	0.137	0.038
Brown	(BN)	0.510	0.370	0.427	0.353	0.407	0.373	0.475	0.405
Green	(GN)	0.313	0.682	0.313	0.453	0.209	0.383	0.013	0.486
Gray	(GY)	0.350	0.360	0.300	0.310	0.290	0.320	0.340	0.370
Maroon	(MN)	0.302	0.064	0.307	0.203	0.374	0.247	0.457	0.136
Orange	(OG)	0.610	0.390	0.535	0.375	0.506	0.404	0.570	0.429
Silver	(SR)	CIE chromaticity luminance factor ß: ß > 0.50							
Red	(RD)	0.690	0.310	0.595	0.315	0.569	0.341	0.655	0.345
Violet	(VT)	CIE chromaticity x and y, luminance factor ß: y < 0.17x + 0.233; y < 2.6x–0.49; y > 0.559–0 0.394x y > 7x–1.845 0.36 < ß < 0.50							
White	(VH)	0.350	0.360	0.300	0.310	0.290	0.320	0.340	0.370
Yellow-ochre	(YEO)	0.522	0.477	0.470	0.440	0.427	0.483	0.465	0.534

[a] For vessels with mixed navigation (sea-river vessels) all outside waters
[b] Excluding fire extinguishing materials
[c] As given in IEC 60 757
[d] Colours and chromaticity coordinates adopted from ISO 14726-1

be differentiated using a second colour presented as a band or stripe of colour or two bands/stripes of colour. When used, colour bands should be of 25 mm (1 in.) to 50 mm (2 in.) width and of a colour that contrasts with the primary colour band (light band with a dark primary colour band and dark with light colour bands). Colour bands should be placed at each end of label text. Refer to Fig. 8.8, Pipe marker labels with two colours for examples.

Mounting location. Pipe marker labels should be placed at or near the inlet or outlet connection to equipment (e.g., pumps, pressure vessels, filters), at the termination of a pipe run (e.g., loading station, hose reels), and approximately every 5 m (16.5 ft) on straight runs of pipe. On vertical pipes, pipe marker labels should be placed approximately 1830 mm (72 in.) above the standing surface. They should be placed on both sides of a bulkhead, deck, or deck head penetration, unless the label is readily visible from both sides of the penetration. Where pipe marker labels are used on two or more pipes in a group of pipes located side-by-side, such as in a pipe rack, all the pipe marker labels should be installed side-by-side so they can be easily scanned at one time. *Material.* Pipe marker labels may be self-adhesive, or the label can be painted directly on the pipe. Self-adhesive pipe marker labels should be made

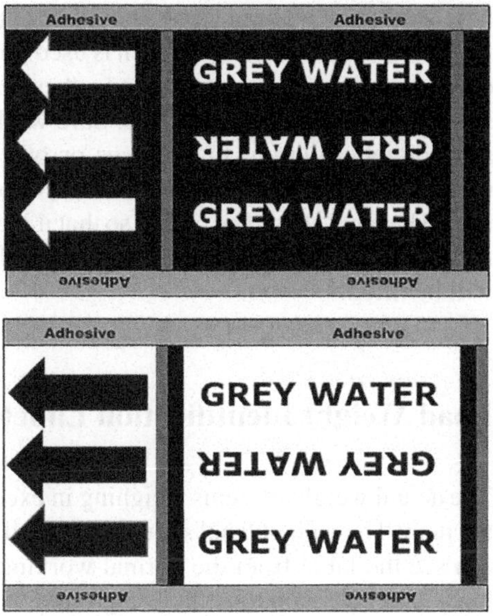

Fig. 8.8 Pipe marker labels with two colours

from an abrasion- and chemical- resistant vinyl or polyester-type material. Where the pipe marker label may be exposed to sunlight, the material should be resistant to UV damage. The material should be durable enough to resist fading, chipping, or cracking. (***Note:*** *most self-adhesive labels are manufactured such that the label can adhere to itself when it is wrapped around a pipe rather than to adhere to piping. This allows the marker to be easily removed without affecting the pipe, its surface, or its insulation*). For fibreglass, copper-nickel materials, or other pipes for which self-adhesive pipe marker labels are not satisfactory, the text labels and flow arrows may be painted directly onto the pipe without the use of a colour band. For copper-nickel pipe material, the colour yellow or white is preferred for the text labels and flow arrows. For black, green, or red unpainted fibreglass pipe material, the colour white is preferred.

Safe Working Load Identification Labels

Applicability. Safe working load identification labels should specify the maximum safe load for a lifting device. They should include the appropriate units (e.g., kilos, pounds, tons, tonnes). *Colour and wording.* The labels should have black letters on a yellow background. An example of the text on a safe working load identification label could be **CAPACITY 20 TONNES**. *Character size.* Lettering and numbers on the labels should be large enough to be read from the standing surface on which personnel will be operating the lifting mechanism and/or reading the load limit label. In no case should the lettering or numbering be smaller than of 25 mm (1 in.) height. *Material.*

Safe working load limit labels should be of a flexible vinyl, adhesive backed, stick-on material. If the lifting mechanism is used outdoors, it should be resistant to ultraviolet (UV) damage. Label material and adhesive should be guaranteed for a minimum of seven years. *Mounting locations.* Safe working load limit labels should be mounted immediately adjacent to pad-eyes or other fixed lifting points, or on the web of monorails, traveling cranes, jib booms, or other such structures supporting a moving load. The label should be placed so that it is directly visible and oriented for horizontal reading by personnel when viewed from the standing surface from which the load will be lifted.

Load Weight Identification Labels

The actual weight of items weighing in excess of 20 kg (45 lbs) should be marked on the item. The weight label should be placed in a location on the item where personnel can see the label from the normal working/lifting position. The labels should have black characters on a white background. Numbers should be a minimum of 13 mm (0.5 in.) height.

Hazard Identification Signs

Regulatory requirements. A hazard identification sign should be used to identify and provide information about conditions that may pose hazards to personnel, equipment, or the environment. Often, Flag State Administrations or other regulatory bodies may specify when such signs are required and the format and content of signs. Where this is true, those regulatory requirements take precedence over any guidance presented in this chapter. *Signs with text and symbols.* To be effective, hazard identification signs should be appropriate for the personnel expected to use the information. It should be remembered that personnel who live and work on vessels or offshore installations are often from different cultures, may speak different language and represent a wide range of educational backgrounds. To accommodate these factors, hazard identification signs should use simple and understandable safety symbols accompanied with explanatory text. The more traditional "word only" signs remain effective for homogeneous and literate crews, when those signs are worded in the crew's native language. Regardless of the approach taken, the following guidance applies to design of hazard identification signs:

- Formats and styles should be consistent, including organisation of information, use of font styles and sizes, and use of colours.
- Technical wording and symbols should be consistent with other instances of written information. For example, if a specific hazard symbol is used on an alarm, or in a written procedure, hazard signs should use the same symbols and meanings.

- The language and wording of hazard identification signs should be compatible with the expected population of readers and written in their native language.
- Where there is a mix of people of different spoken languages on a vessel or offshore installation, text-only signs should be avoided.

The guidance that follows, presents design information for text-only signs and for the text portion of signs that also use warning symbols. For information concerning the design of symbols for hazard identification signs, see the section on 'Symbols'.
Text content. Hazard labels should contain the following:

- The signal word (**DANGER**, **WARNING**, or **CAUTION**) at the top of the label.
- A brief statement of the hazard, in as specific detail as possible (e.g., **600 VAC**, not **HIGH VOLTAGE**).
- A brief statement on how to avoid the hazard, in detail if possible. If more than one step is involved, the hazard avoidance procedure should be provided in step-by-step sequence.
- A brief statement of the possible consequences if the avoidance instructions are not followed (e.g., serious injury could occur, fire or explosion could occur, death could result).

Text signal levels. Three signal levels are recommended by ANSI Z535.2, Environmental and Facility Safety Signs, for use with hazard identification labels. This standard specifies signal levels based on the potential effect of hazards to personnel. Additional guidance has been provided in the following signal level descriptors to include potential effects to property or the environment:

- **DANGER**: indicates an imminently hazardous situation that, if not avoided, will result in death or serious injury. This signal word is to be limited to the most extreme situations.
- **WARNING**: indicates a potentially hazardous situation, which, if not avoided, could result in death or serious injury. A **WARNING** signal word may be used on a sign to signify situations that could result in serious damage to vital equipment, or a major pollution problem.
- **CAUTION**: indicates a potentially hazardous situation which, if not avoided, may result in minor or major injury. A **CAUTION** signal word may be used on a sign without symbols to signify situations where property damage or a minor pollution problem is possible.

Header colour and format. Hazard label headers, including signal words, are shown in Fig. 8.9, Hazard signal word headers. Signal words should appear at the top of the sign in a header. The signal word should be written in upper case letters and be preceded by an "attention" symbol located within a triangle. A simple sans-serif character font, like Arial or SmartSigns Clearview® should be used. Colours associated with the various signal words and symbols are as follows:

Fig. 8.9 Hazard signal word headers

- The signal word **DANGER** should be white letters on a red background. The attention symbol exclamation point should be red. The equilateral triangle surrounding the exclamation point should be white,
- The signal word **WARNING** should be black characters on an orange background. The attention symbol exclamation point should be orange. The equilateral triangle surrounding the exclamation point should be black, and
- The signal word **CAUTION** should be black characters on a yellow background. The attention symbol exclamation point should be yellow. The equilateral triangle surrounding the exclamation point should be black.

Message text format. Message text, other than the signal word header, on a hazard identification sign should use upper- and lower-case letters. Either black characters on a white background or white characters on a black background should be used. The character font should be a simple sans-serif character font, like Arial or SmartSigns Clearview®. Figure 8.10, Examples of text and symbols on signs provides sample signs including text.

Message text. The character height of the message text should be based on the label size as shown in Table 8.5, Message text character heights. The character height of the signal word should be at least 50% higher than that of the message text. There should be at least two-line widths separating the signal word and the first line of the hazard description and one line width separating each of the other two statements.

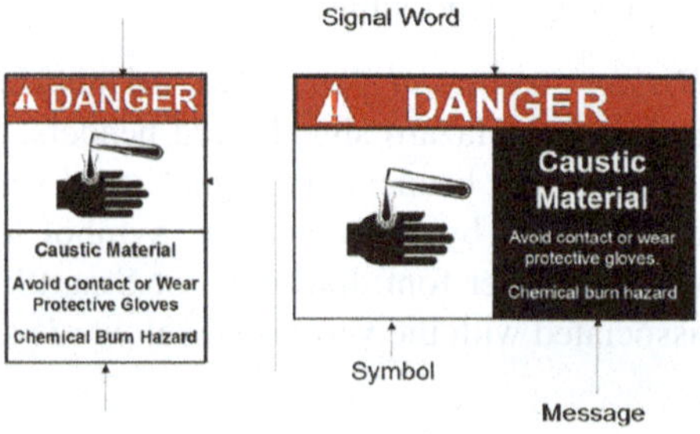

Fig. 8.10 Examples of text and symbol on signs

Table 8.5 Message text character heights

Label size	Character height
90 mm × 130 mm (3.5 in. × 5.0 in.)	3.2 mm (0.125 in.)
130 mm × 180 mm (5.0 in. × 7.0 in.)	3.2 mm (0.125 in.)
180 mm × 250 mm (7.0 in. × 10.0 in.)	8.0 mm (0.312 in.)
250 mm × 360 mm (10.0 in. × 14.0 in.)	14.0 mm (0.56 in.)

Mounting location. Hazard labels should be mounted at or immediately adjacent to the source of the hazard. The labels should be readable before personnel reach the point of the hazard source. Preferably, no more than three hazard labels should be mounted adjacent to each other.

Information Signs or Placards

Information labels should be used to present non-procedural information of a general nature and should only be used to provide information that will significantly contribute to the safety and/or efficiency of personnel. Information may be presented in step-by-step format. Where there are several items to be presented, they should be presented in list form. Subheadings within the list should be used as appropriate to organise and identify related information. *Character height and line spacing.* The height of the message text should be the same as defined in the section 'Hazard identification signs'. The signal word **NOTICE** should be incorporated in the placard and should be at least 50% higher than the height of the sub-headings (if any are used), which in turn should be 25% higher than the height of the text. If no sub-headings are used, then the signal word should be 50% higher than the message text. *Character colour.* The signal word **NOTICE** should be in white italicised characters on a blue rectangular background. Message text should be black characters on a white background. Figure 8.11, Example of an information sign provides a sample sign including text.

Instruction Labels

General. All instruction labels should be titled at the top of the label to identify the instructions as operating or maintenance, or both, and should identify the equipment or system for which the instructions have been prepared. *Format.* The instruction labels should be prepared as follows:

Fig. 8.11 Example of an information sign

- Instructions should be provided in a numbered step-by-step format with each step stated as briefly as possible. For instructions written in English, a command form should be used where the subject "you" is understood and not written in the instruction.
- The instruction should avoid the use of specific equipment model names or numbers, or other aspects of the equipment that may change at some future date and thus mandate the preparation of new instructions. Note that assigned equipment names and numbers that are specific to a vessel, or an offshore installation should be included to aid the personnel in matching the instructions to the labels on equipment.
- Hazard and safety information that pertains to a specific step in the instructions should be provided immediately preceding the step in the instruction and should be headed by the appropriate signal word, (i.e., **DANGER**, **WARNING**, or **CAUTION**). Hazard and safety information that pertains to the procedure in general should appear at the beginning of the instruction.
- Instructions should be provided on a separate label from diagrams, schematics, charts, or other types of graphic labels that may be referenced in the instructions. However, the referenced labels should be immediately adjacent to the instruction label to provide easy transfer between the two by personnel.
- If both operating and maintenance instructions are provided, they should be provided on separate postings or signs. They should be clearly separate from one another and labelled with the appropriate headings.
- Where instructions related to emergencies are provided, the following additional criteria should be applied:
- Provide a title at the top to identify the instructions as emergency instructions and identify the type of emergency for which the instructions have been prepared.
- Instructions should be concise, precise, specific, and explicit and should not require interpretations or assumptions.
- Symbols should be used as appropriate to enhance universal comprehension.

Text. Text on instruction labels should be orientated normally for the language used (e.g., for English, text would be oriented horizontally). The instructions, including any text and associated graphics or schematics, should be visible and legible under all expected lighting conditions including bright sunlight and artificial lighting. Emergency instruction labels should also be visible and legible under reduced emergency lighting conditions. *Character size.* The height of label text should be sized as required for the reading distance, but in no case should it be less than 3.2 mm (0.125 in.). The label title should be at least 50% higher than the height of the sub-headings (if any are used), and the sub-headings should be 25% higher than the height of the smallest text. *Mounting location.* Instruction labels, including emergency instruction labels, should be located so that they are fully visible at all times and free from obstructions including doors or covers in open positions. *Material.* Instruction labels may be made of any of the following materials:

- Reverse screened Lexan.
- Vinyl cut letters on a plastic or metal backing plate.
- Silk screened vinyl with laminate overlay.
- Engraved plastic.
- UV resistant vinyl mounted on a stainless-steel backing.
- Fibreglass.

Graphical Schematics, Charts and Diagrams

Graphics are used to present information through line schematics, diagrams, charts, tables, or pictures. All graphics should be titled at the top to identify the equipment, system, or information being presented. If more than one schematic, diagram, or chart is being presented on a single label, each should be labelled with a heading. The following guidelines apply to graphics. *Schematics and diagrams.* All graphic symbols and special nomenclature used on a graphic label should be defined in a legend box. Descriptive text should be brief as possible and not obscure the diagram or schematic. Information should be provided on the label that indicates location and/or orientation of the equipment or system. *Charts.* All charts of the same type or function, such as lubrication charts, should be formatted with the same headings, units of measurement, organisation, and presentation details. *Character size.* The height of label text should be sized as required for the reading distance, but in no case should it be less than 3.2 mm (0.125 in.). The label title should be at least 50% higher than the height of the sub-headings (if any are used), and the sub-headings should be 25% higher than the height of the smallest text. *Line size.* Line widths on schematics and diagrams should be a minimum of 1.5 mm (0.06 in.). *Mounting location.* Graphics should be mounted as close as possible to the equipment, system, or component they represent. Graphics which show piping schematics or equipment arrangements should be located so the spatial layout of the label directly matches the arrangement of the actual piping and equipment when both are viewed by the operator or maintainer facing the label.

Orientation Plans

Format. The orientation plan labels should be prepared as follows:

- All orientation plans should be titled at the top to identify the specific area and location where the plan or sign is located.
- All orientation plans should provide a realistic pictorial/graphical presentation of the vessel or installation layout.
- Orientation plans should be positioned so that they are spatially orientated in terms of the vessel or offshore installation from the perspective of personnel viewing the plan.
- Orientation plans should orient personnel using the plan by marking where the person is located when viewing a particular sign (i.e., **YOU ARE HERE**).

Mounting location. Orientation plans should be located in strategic areas around the vessel or installation (e.g., at muster areas, main stair landings, close to firefighting areas, and at emergency exits/entrances, as applicable). They should be clearly visible and legible, under all expected lighting conditions, including bright sunlight, artificial lighting, and reduced emergency lighting conditions.

Symbols

Independent symbols and internationally recognised symbols, such as those of the IMO, should be used. Flag State Administrations or regulatory bodies may also require the use of particular symbols to convey information. Sources of internationally accepted graphic symbols for hazard identification signs and placards include:

- American National Standards Institute Z535.3, Criteria for Safety Symbols.
- British Standard 5499: Graphical symbols and signs.
- IMO Resolution A.1021(26): Code on Alerts and Indicators.
- IMO Resolution A.654 (16): Graphical symbols for fire control plans.
- IMO Resolution A.760 (18): Symbols related to life-saving appliances and arrangements.
- International Labour Organisation: Accident prevention aboard ship at sea and in port.
- International Maritime Dangerous Goods Code (IMDG Code).
- Marine Safety Committee Circular 451 (MSC/Circ.451): Guidance concerning the location of fire control plans for the assistance of shoreside firefighting personnel.

Chapter 9
Stairs, Ladders, Ramps, Walkways, Platforms and Hatches

General

The principles and guidance provided in this chapter apply to the design of stairs, vertical ladders, ramps, walkways, work platforms, and hatches. Criteria are also provided for handles.

Principles

Use. Stairs, vertical ladders, and ramps should be provided whenever operators or maintainers must change elevation abruptly by more than 300 mm (12 in.). These structures should be used, when appropriate, for passage over low objects (e.g., pipes, lines, and ridges). *Selection.* The selection of stairs, vertical ladders or ramps should be based on the purpose, frequency of use, and angle of ascent. Refer to Table 9.1, Selection of access type for related guidance on angle of inclination. *Operations requirements.* Stairs are the most appropriate means for changing from one walking or work surface to another when space is available. Examples of where stair access is recommended include:

- When operations activities require regular travel between levels or decks by personnel.
- When access is required to elevated work platforms (e.g., mezzanines) daily, or at least once a shift.
- When quick escape may be needed from elevated work areas or platforms where personnel could be exposed to caustics, chemicals, gases, or other toxic materials.
- When employees could be hand-carrying heavy or bulky tools and/or equipment.
- Ramps may be preferred in the following situations when the change in vertical elevation exceeds 610 mm (24 in.) in height:

A. A. Olsen, *Applying Physical Ergonomics to Modern Ship Design*,
https://doi.org/10.1007/978-3-031-57974-5_9

Table 9.1 Selection of access type

Type	Angle of inclination (in.)	Preferred angles of inclination (in.)
Stairs	30°–50°	38°
Inclined ladders	50°–60°	Not used
Vertical ladders	75°–90°	75°–85°
Ramps for personnel	7°–15°	7°–15°
Ramps for materials handling	4°–7°	4°

1. To move people, vehicles, or materials. In these circumstances, the angle of inclination should be 15° or less for foot traffic and 7° or less for vehicles and materials.
2. When a ramp would allow more efficient egress along an emergency access/egress route, as long as the angle of inclination is 15° or less.

- When space is unavailable for stairs or a ramp, access may be provided using separate steps or a vertical ladder.

Maintenance requirements. Stairs are recommended when access is required to elevated work platforms (e.g., mezzanines) one or more times per day; where the maintainer must carry bulky loads or loads in excess of 13 kg (29 lbs), a ramp or elevator should be provided. Alternatively, access may be provided for personnel via stairs, steps, or ladders, as long an adequate lifting mechanism is provided for the load; ramps, elevators, or equivalent means should be provided when maintainers must transport heavy or bulky equipment; and ladders should not be used when maintainers carry equipment because both hands should be free to grasp and climb ladders. *Emergency access.* Stairs should be the primary mean of egress from spaces. However, in areas where two means of egress (e.g., mezzanine deck) are provided, a vertical ladder can be used as the secondary egress means.

- *Restrictions.* A ladder leading to a deck scuttle is not used as a means of escape except on vessels less than 19.8 m (65 ft) in length, or as not more than one of the means of escape from any crew accommodations space or workspace,
- *Dimensions.* Each ladder used as a means of escape should have the following design features:

1. The ladder is mounted at least 205 mm (8 in.) from the nearest permanent object in the back of the ladder.
2. Rungs are at least 410 mm (16 in.) in width.
3. Rungs are between 275 mm (11 in.) and 300 mm (12 in.) apart.
4. Rungs are uniformly spaced for the length of the ladder.

Stairs

General. In respect to the design and installation of stairs, the following general requirements apply:

- Stairs are appropriate means for changing from one walking surface to another when the change in vertical elevation is greater than 600 mm (23.5 in.).
- Stairs should be provided in lieu of ladders or ramps in accommodations spaces, office spaces or to the navigational bridge.
- The angle of inclination should be sufficient to provide the riser height and tread depth that follows, with a minimum angle of 38° and maximum angle of 45°.
- Stairs exposed to the elements should have additional slip resistance due to potential exposure to water and ice.
- Stairs should be used in living quarters instead of inclined ladders.
- No impediments or tripping hazards should intrude into the climbing spaces of stairs (for example, electrical boxes, valves, actuators or piping).
- No impediments or tripping hazards should impede access to stair landings (for example, piping runs over the landing or combings/retention barriers) (Fig. 9.1).

Stair risers and treads. Stair risers and treads should have the following design:

- A riser height should be no more than 230 mm (9 in.) and a tread depth of 280 mm (11 in.), including a 25 mm (1 in.) tread nosing (step overhang).
- For stairs, the depth of the tread and the height of riser should be consistent.
- Minimum tread width on one-way (where there is expected to be only one person transiting, ascending or descending stairway) stairs shall be at least 700 mm (27.5 in.).

Fig. 9.1 Internal stairs

Fig. 9.2 Stair step riser and tread design

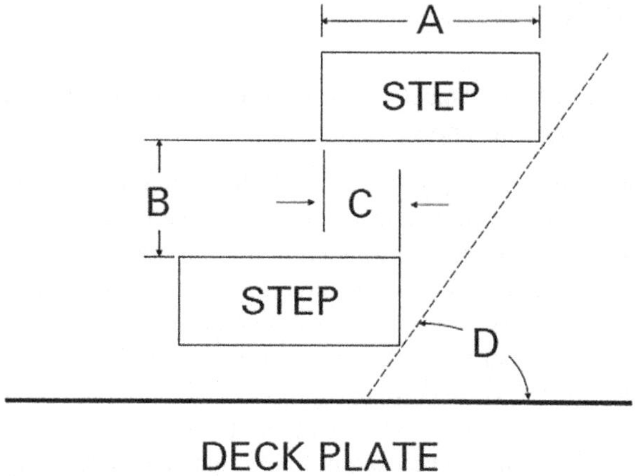

- Minimum tread width on two-way (where there may be two persons, ascending and descending, or passing in opposite directions) stairs shall be at least 900 mm (35.5 in.).
- Once a minimum tread width has been established at any deck in that stair run, it should not decrease in the direction of egress.
- Nosings should have a non-slip/skid surface that should have a coefficient of friction (COF) of 0.6 or greater measured when wet.

Stairs serving accommodations and those in the escape route should be a minimum of 1120 mm (44 in.) in width. Other stairs should be a minimum of 915 mm (36 in.). However, infrequently used stairs may be 710 mm (28 in.) in width (Fig. 9.2).

Stair landings. (1) *Clear landing.* A clear landing at least as wide as the tread width and a minimum of 915 mm (36 in.) long should be provided at the top and bottom of each stairway. Intermediate landing. An intermediate landing should be provided at each deck level serviced by a stair, or a maximum of every 3500 mm (140 in.) of vertical travel for stairs with a vertical rise of 6100 mm (240 in.). (2) *Change in climb direction.* Any change of direction in a stairway should be accomplished by means of an intermediate landing at least as wide as the tread width and a minimum of 915 mm (36 in.) long. (3) *Maximum angle of inclination.* Stairways should have a maximum angle of inclination from the horizontal of 45°. (4) *Accommodating stretchers.* Where stairs change directions, intermediate landings along paths for evacuating personnel on stretchers should be 1525 mm (60 in.) or greater in length to accommodate rotating the stretcher. (5) *Stair handrails.* Stair handrails should be fitted as follows:

- Stairs with three or more steps should be provided with handrails, as detailed in Table 9.2, Stair handrail arrangements.
- A single-tier handrail to maintain balance while going up or down the stairs should be installed on the bulkhead side(s) of stairs.

Table 9.2 Stair handrail arrangements

Arrangement	Handrail requirement
1120 mm (44 in.) or wider stair width with bulkhead on both sides	Single tier handrail on both sides
Less than 1120 mm (44 in.) stair width with bulkhead on both sides	Single tier handrail on one side, preferably on the right-side descending
1120 mm (44 in.) or wider stair width, one side exposed, one with bulkhead	Two tier handrails on exposed side, single tier on bulkhead side
Less than 1120 mm (44 in.) stair width, one side exposed, one with bulkhead	Two tier handrails on exposed side
All widths, both sides of stairs exposed	Two tier handrails on both sides

- A two-tier handrail to maintain balance and prevent falls from stairs should be installed on non-enclosed sides of stairs.

Dimensions. Handrails should be constructed with a circular cross section with a diameter of 40 mm (1.5 in.) to 50 mm (2.0 in.). Square or rectangular handrails should not be fitted to stairs. The height of single tier handrails should be 915 mm (36 in.) to 1000 mm (39 in.) from the top of the top rail to the surface of the tread. Two-tier handrails should be two equally spaced courses of rail with the vertical height of the top of the top rail 915 mm (36 in.) to 1000 mm (39 in.) above the tread at its nosing. A minimum clearance of 75 mm (3 in.) should be provided between the handrail and bulkhead or other obstruction. *Individual steps.* Individual steps comprised of tread surfaces attached directly to a structure may be used to change vertical elevations where a stair or vertical ladder is not practical. All dimensions for such steps should comply with the stair treads described in the section on 'Risers and treads'. Handrails, as described in the section on 'Stair handrails', should be provided for stairs with three or more steps. *Spiral stairs.* Spiral stairs should not be permitted, except on tanks or other round structures whose diameter is greater than 2.4 m (8 ft), and where a normal stair design is inappropriate. The stair should ascend in a clockwise direction to allow the stair rail to be on the right-hand side during descent. *Removable stairs.* Stairs may be pinned at the top and bottom for easy removal in locations where removable stairs would enhance ease of equipment removal for maintenance or replacement. However, stairs that serve as a required means of egress in emergencies should be permanently installed (Fig. 9.3).

Vertical Ladders

This section outlines the requirements related to the design of the different attributes of vertical ladders, inclined ladders and individual rung ladders. The requirements included in the figures and tables below provides the design and dimension requirements. Requirements related to topside stairs are provided in the section on 'Stairs'.

Fig. 9.3 Steps with landing

The considerations listed below apply to the design of vertical and inclined ladders and are not represented in the following figures. Vertical ladders, inclined ladders or ramps should be provided whenever operators or maintainers must change elevation abruptly by more than 300 mm (12.0 in.). (Requirements relating to ramps are found in the section on 'Ramps'). These structures, or a simple step, should also be used for passage over low objects (e.g., pipes, lines, ridges). Vertical ladders and inclined ladders should be provided with skid/slip resistant on the rungs that shall have a COF of 0.6 or greater measured when wet. Refer to Table 9.3, Inclination of ladders for requirements on angle of inclination (Fig. 9.4).

General. In respect to vertical ladders, the following General requirements apply:

- Permanent vertical ladders should be attached to a permanent structure.
- Fitted with a maximum distance from the ladder's centreline to any object that must be reached by personnel from the ladder not exceeding 965 mm (38.0 in.).
- Should be provided with skid/slip resistant on the rungs that should have a COF of 0.6 or greater measured when wet.
- Located so as not to interfere with the opening and closing of hatches, doors, gratings or other types of access.

Table 9.3 Inclination of ladders

Requirement	Angle requirement
Inclined ladders	45°–60°
Vertical ladders	80°–90°

Fig. 9.4 Vertical ladder with 'back scratcher'

- No impediments should intrude into the climbing space (for example, electrical boxes, valves, actuators or piping).

Design loads. The design loads listed in the following sub paragraphs are design load requirements; however, if requirements for design loads specified by other regulatory bodies (e.g., Flag State Administrations and Port State Authorities) are more limiting, those requirements take precedence over the guidance provided in this book. In this book, we define "design load" as the maximum intended load, being the total of all loads including the weight of the personnel, materials, equipment and means of access structure, and should be as follows:

- Ladder handrails should withstand anticipated loads but not less than 90 kg (200 lbs) at any point and in any direction when applied to the top rail.
- For vertical ladders, the design load should be determined by the anticipated usage of the ladder, but should not be less than a single concentrated live load of 90 kg (200 lbs). The weight of the ladder and attached appurtenances together with the design load should be considered in the design of rails and fastenings.
- Inclined ladders should be designed and constructed to carry a load of at least three times the normal load anticipated.

Ladders exposed to the elements should have additional slip resistance due to potential exposure to water and ice. *Dimensions.* (1) *Ladder rungs.* Rungs should be covered with a non-skid surface to prevent slipping. Where multiple runs of ladders are provided to cover the vertical distance, the same rung spacing should be used in each run. (2) *Ladder stringers.* Ladder stringers should be provided on both sides of the ladder over the full length and should be constructed of pipes with preferable nominal diameter of 40 mm (1.5 in.). Flat bar should be avoided for stringers. For multiple runs of vertical ladders, the stringers should extend 1070 mm (42 in.) above landings or intermediate platforms. (3) *Ladder heights.* Continuous ladders should not exceed 9.1 m (30 ft) in height. Where the vertical height exceeds 9.1 m (30 ft), intermediate landings and separate multiple ladder runs of equal length should be provided. (4) *Clearances.* A minimum of 760 mm (30 in.) horizontal clearance should be provided for the total width and length of the ladder on the climbing side as measured perpendicular to the ladder rungs. A minimum horizontal distance of 180 mm (7 in.) (205 mm (8 in.) is preferred) should be provided behind the rungs and 100 mm (4 in.) behind the stringer. A minimum overhead clearance of 2.1 m (7 ft) should be provided above the step-off rung and a horizontal distance of 380 mm (15 in.) between the ladder centreline and the nearest obstruction on either side. (5) *Top rungs on step-through ladders.* For vertical ladders that personnel use to access a deck, work platform, building roof, or top of a tank by stepping through the ladder to a walking or standing surface, the top (or step-off) rung should be flush with that surface. (6) *Intermediate platforms.* Platforms used to access a vertical ladder should provide a minimum clear standing area of 760 mm (30 in.) in front of the ladder with at least 460 mm (18 in.) of width on each side of the ladder centreline. (7) *Intermediate platforms—use of handrails/toeboards.* Intermediate platforms should be provided with handrails and toeboards, as described in the section on 'Handrails for walkways and work platforms', and on all sides not used to access a vertical ladder. (8) *Horizontal separation between ladder and platform.* The distance from the edge of the ladder stringer to the sidestep platform being accessed should be no more than 300 mm (12 in.), with a preferred distance of 150 mm (6 in.). There should be a rung on the vertical ladder at the same height as the standing surface of the intermediate platform. (9) *Horizontal separation between two ladders.* The horizontal separation between two vertical ladders, stringer to stringer, using an intermediate platform should not be greater than 460 mm (18 in.). (10) *Individual rung ladders.* The requirements listed below apply to individual rung ladders and are not represented in the following figure:

- Individual rungs may be attached directly to a bulkhead, tank or steel structure and used as a vertical ladder but should be limited to changes in vertical elevation of 3.6 m (12.0 ft) or less.
- Each rung should be attached to the structure to fully support a climber and any design loads.
- Rungs should be provided with lateral support for the foot.

Inclined Ladders

Inclined ladders should be attached to a permanent structure. Where inclined ladders change directions and are used to carry personnel on stretchers, intermediate landings along paths should be 1525 mm (60.0 in.) in length to accommodate rotating a stretcher. Inclined ladders and handrails should be located so as not to interfere with the opening and closing of hatches, doors, gratings or manholes. No impediments should intrude into the climbing space (for example, electrical boxes, valves, actuators or piping). *Tread/step design.* Steps should be equally spaced along the entire flight of the inclined ladder. Square bar, round bar, or full steps can be used to form the step. Where square bar is used, treads/steps should be fitted with an edge pointing upward. Where full steps are use, the steps should be parallel to the deck surface. These steps should also be carried through the side stringers and attached by double continuous welding. *Handrail design.* Handrail design loads are presented in the section on Design loads. In summary, square handrails should not be used; and a clearance of at least 75 mm (3.0 in.) should be provided between the handrail and bulkhead or other obstruction.

Cargo Hold Stairs and Ladders

As far as possible, stairs and ladders installed in cargo holds should minimize the likelihood of slips and falls. However, when approved by the appropriate authorities, such as flag Administrations, alternative designs can be used for cargo hold stairs and ladders. Access to cargo hold ladders should be through a trunk way. Access by climbing over the main deck cargo hatch coaming is not recommended. Generally, for cargo vessels, vertical ladders may be fitted where the length of the ladder does not exceed 6.1 m (20 ft). When the length exceeds 6.1 m (20 ft), stairways should be fitted in lieu of ladders. For cargo vessels, a vertical entrance section of up to 2.5 m (8 ft) may be fitted with a vertical ladder, provided the remaining distance to the bottom is fitted with a stairway. As an alternative arrangement, a vertical ladder, with a maximum height of 6.1 m (20 ft), may be substituted at the lower end of the stairway closest to the tank bottom, provided the shortened length of the stairway is not less than 2 m (6.6 ft) vertically (Fig. 9.5).

Guardrails and Climber Safety Devices

This section includes general principles as well as the design requirements for the design and arrangement of handrails and climber safety devices for fall protection, whether by active protection (safety rails and harnesses) or passive protection (safety cages). These requirements include:

- Height and diameter of handrails.
- Span and gap between handrail stanchions.
- Distance between handrail and any obstruction.
- Intermediate rails.
- Climber safety equipment.

Fig. 9.5 Cargo hold staircase

General principles. The principles listed below apply to the design of walkways and ramps and are not represented in the following tables:

- Guardrails should be provided at the exposed side of any walking or standing surface that is 600 mm (23.5 in.) or higher above the adjacent surface and where a person could fall from the upper to the lower surface.
- Where stays are provided for supporting stanchions, they should be fitted so as not to obstruct safe passage.
- Stanchion scantlings can be formed of flat or round bar.
- The top handrail should be round bar.
- Brackets joining the guardrail stanchions to the means of access should be oriented in a way to avoid causing a trip hazard (e.g., parallel to direction of the walkway).

Means should be provided for de-icing of guardrails and climber safety devices. *Provision of safety railings.* Rails should be installed parallel to the deck along deck edges and walkways and around open hatches, elevators, working platforms and along other boundaries wherever there is danger of operators or maintainers falling to a lower level of $\geq$ 600 mm (23.5 in.); or wherever there is danger of operators or maintainers becoming enmeshed with operating machinery; and around openings with a coaming height below 760 mm (30 in.). ***Note:*** *temporary rails can be used around unprotected opening into which a person may slip, trip or fall. Storm rails.* Suitable storm rails/handrails should be provided in all interior passageways where persons on board might have normal access. In these cases, storm rails/handrails should be installed on both sides of passageways that are 1830 mm (72 in.) or more in width. The storm rails/handrails should be 865 mm (34 in.) to 965 mm (38 in.) high and the distance between storm rails/handrails and any obstruction must be 75 mm (3 in.) or greater. *Deck edge, elevated walkway railing design.* Deck edge and elevated walkway railings should have the following design:

- The heights of rails or bulwarks should be at least 1070 mm (42 in.) from the deck.
- Rail courses or equivalents should be installed between a top rail and the deck so that the opening below the lowest course does not exceed 230 mm (9 in.) and the distance between the remaining courses is not more than 380 mm (15 in.).
- Toeboards should be at least 100 mm (4.0 in.) in height and have no more than a 6 mm (0.25 in.) clearance between the bottom edge of the toeboard and the walking surface.
- Vertical stanchions for railings should be spaced no more than 1525 mm (60 in.) apart horizontally.
- At least every third vertical stanchion should be supported by a bracket or stay; and
- Chain or wire rope should not be used as a deck edge or elevated walkway rail.

Safety cages. Below are requirements related to climber safety cage (sometimes referred to as a 'back scratcher') design:

- Cages should be used on vertical ladders over 4.5 m (15.0 ft) in height.
- Climber safety rails or cables should be used on vertical ladders in excess of 6.1 m (20.0 ft).
- Cages should extend 1400 mm (54 in.) above the top-landing surface.
- Cages equipped with intermediate landings should extend 1400 mm (54 in.) above the intermediate landing with the cage open on the side facing the landing.
- Safety cages should be provided for ladders $\leq$ 4.5 m (15.0 ft) in height where a fall to a level or deck below the ladder base is possible (e.g., within 1825 mm (72 in.) of the edge of a deck).

Climber safety rails or cables. Contained here are requirements related to the use of climber safety rails or cables, where provided. A climber safety rail or cable should be installed at each permanently installed topside ladder on masts, king posts and other similar topside structures providing access to fall hazard locations. The climber safety rail should be stainless steel flat bar and equipped with two safety slides, which can be attached to the flat bar or cable. If climber safety devices are used, the ladder stringers at a top landing should be designed to allow personnel to access any associate landing without unfastening. *Safety drop bars.* All fixed ladders serving elevations 760 mm (30 in.) or more above ground, platform or floor level should be equipped with drop bars or safety gates. Drop bars should be attached as follows:

- Side access ladders should hinge at the ladder side.
- Front access ladders should hinge at the right when facing the ladder from the platform side.
- Drop bars should not be placed beyond the outer edge of the platform.
- Chains should not be used in lieu of a drop bar.

Safety gates. Where a self-closing safety gate is provided, the self-closing safety gate should be installed at the top of each ladder and should cover the full width of the opening between the ladder stringers. The gate should open away from the person climbing up the ladder. Safety gates should be sufficiently robust to resist the full weight of a 90 kg (200 lbs) person in both the vertical and horizontal direction; and chains should not be used in lieu of a safety gate.

Fall Protection from Secondary Fall Points

General. Vertical ladders should not be located within 1.83 m (6 ft) of other nearby potential fall points (including the deck edge, cargo holds and lower decks) without additional fall protection such as guardrails. Additional fall protection should be provided for the ladder climber where a vertical ladder (of any height) is located within 1.83 m (6 ft) of another and nearby potential (secondary) fall point (for example overboard or to a lower deck or landing); and if the potential fall distance is greater than 4.6 m (15 ft). (The potential fall distance is the height of the ladder plus the height of the secondary fall); and if no active fall protection is fitted to the ladder (a safety cage is not considered to provide active protection); then Additional fall protection to the ladder climber shall be provided, regardless of whether a climber safety cage is fitted to the ladder. ***Note:*** *Passive fall protection is a safety design or device that requires a person to take no specific action prior to a potential loss, for example, a safety cage permanently fitted to a ladder.* Active fall protection is a safety design or device that actively (or directly) requires a person to take specific actions before a potential loss, for example, donning a fall arrestor fitted to both the ladder and the climber. Protection for vertical ladders without safety cages or climber safety rails/cables. The following should apply to vertical ladders less than of 4.5 m (15.0 ft) that are not fitted with a safety cage or a safety rail. Guardrail requirements are provided in Table 9.4, Guardrail requirements for vertical ladders without safety cages or climber safety rails/cables provide the dimensions for requirements to guardrails near the base of the ladder. Other designs and arrangements that serve to protect personnel from falls may also be deemed to be acceptable.

Protection for vertical ladders with safety cages and without climber safety rails/cables. Guardrail requirements are provided in Table 9.5, Guardrail requirements for vertical ladders with safety cages and without climber safety rails/cables provides requirements for guardrails near the base of the ladder. Other designs and arrangements that serve to protect personnel from falls may also be deemed to be acceptable.

Guardrail requirements for movable vertical ladders. If the vertical ladder is movable (e.g., can rotate because it is mounted on the side of a crane cab), the elevated height of the vertical guardrail should continue horizontally to a distance of 600 mm (23.5 in.) from each side of the centreline of the movable object when it is at its travel limits. *Use of vertical ladder safety gates or bars versus safety chains.* Safety gates or bars,

Table 9.4 Guardrail requirements for vertical ladders without safety cages or climber safety rails/cables

Arrangement	Guardrail requirement
Height of vertical guardrail	Height shall extend to within 760 mm (30 in.) of the top of the ladder
Width of vertical guardrail	Protection shall be provided for a minimum of 1220 mm (48 in.) on each side of the centreline of the ladder, space permitting
Distance between guardrail courses or tiers	A maximum of 460 mm (18 in.) shall be provided between guardrail courses or tiers of the guardrail extension. The measurement shall be taken from the course or tiers outside diameter to outside diameter

Table 9.5 Guardrail requirements for vertical ladders with safety cages and without climber safety rails/cables

Arrangement	Guardrail requirement
Height of vertical guardrail	The height shall extend to within 760 mm (30 in.) of the lower edge of the safety cage
Width of vertical guardrail	Protection shall be provided for a minimum of 1220 mm (48 in.) on each side of the centreline of the ladder, space permitting
Distance between guardrail extension courses or tiers	A maximum of 460 mm (18 in.) shall be provided between guardrails courses or tiers of the guardrail extension. The measurement shall be taken from the course or tier outside diameter to outside diameter
Horizontal distance between ladder and rails	1830 mm (72.0 in.)

rather than safety chains, should be used to protect personnel near the opening at the top of a vertical ladder. Safety gates should comply with the following:

- A self-closing safety gate covering the full width of the opening between the ladder stringers should be installed at the top of each ladder to protect personnel from falling while near the top of a ladder.
- Where ladders provide access to small platforms and that do not provide sufficient space for the self-closing gate to swing horizontally, manually operated gates that open and close via a vertical swing should be installed.
- The gate should open away from the person climbing up the ladder.

Ladder and Access Handles

General. This section contains requirements related to the design of handles. Handles should be designed to accommodate personnel wearing either lightweight and medium weight gloves or cold weather gloves and mittens. Handles should be constructed of round bar. Handles are particularly useful where a vertical ladder comes up to a manhole from the deck below where the ladder does not extend through the platform. Handles should be accessible at all stages during climbing or traversing through access hatches (embarking and disembarking) and within reach of the shortest (e.g., 5th percentile female) user. To provide for safe ascending and descending, while stepping onto or from ladders, individual ladder rungs or steps and through hatches or lightening holes, suitably located handles or hand grabs should be provided. Handles exposed to the elements should have additional slip resistance due to potential exposure to water and ice.

Walkways and Work Platforms

This section includes general principles as well as the design requirements for the arrangement of walkways and ramps and the provision of guardrails. *General principles.* The principles which apply to the design of walkways and ramps are handrails should be provided at the exposed side of any walking or standing surface that is 600 mm (23.5 in.) or higher above the adjacent surface and where a person could fall from the upper to the lower surface. Ramps are used with changes in vertical elevations of less than 600 mm (23.5 in.). Ramps are provided with a non-skid surface that should have a COF of 0.6 or greater measured when wet. Headroom in all walkways should be $\geq$ 2130 mm (84 in.). Toeboards should be provided on elevated walkways, platforms and ramps. Walkways and ramps exposed to the elements should have additional slip resistance due to potential exposure to water and ice. No impediments or tripping hazards should intrude into the transit space (for example, electrical boxes, valves, actuators or piping). No impediments or tripping hazards should impede access to a walkway or ramp (for example, piping runs, hatch covers, deck impediments (e.g., through bolts) or combings/retention barriers). *Design loads.* Where requirements for design loads specified by other regulatory bodies (e.g., Flag State Administrations and Port State Authorities) are greater, those requirements take precedence over the guidance provided in this book. In this book, we define "design load" as the maximum intended load, being the total of all loads including the weight of the personnel, materials and equipment, including the means of access structure. *Guardrails.* Guardrails should withstand anticipated loads but not less than 90 kg (200 lbs) at any point and in any direction when applied to the top rail. *Design loads and maximum deflection.* The minimum design loads for the walkways and ramps are:

- 2.0 kN/m^2 (0.29 lbf/in^2) under uniform load for the structure.

- 1.5 kN (337 lbf) concentrated load applied in the most unfavourable position over a concentrated load area of 200 mm × 200 mm (8 in. × 8 in.) for the flooring.
- When loaded with the design load, the deflection of the flooring should not exceed 1/200th of the span and the difference between the loaded and adjacent unloaded flooring should not exceed 4 mm in height.

Walkway design. The dimensions relating to the design of walkways and ramps are presented in Fig. 9.22, Walkway and ramp design. *Openings in gratings.* The maximum opening in a walkway grating under which the presence of persons is expected should be less than 22 mm (0.9 in.). The maximum opening in a walkway grating under which the presence of persons is not expected should be less than 35 mm (1.7 in.). *Toeboards.* Toeboards should have a height of 100 mm (4.0 in.) and have no more than a 6 mm (0.25 in.) clearance between the bottom edge of the toeboard and the walking surface.

Ramps

Ramps are best used with changes in vertical elevations of less than 610 mm (24 in.), but may be used for any height provided that the angle of inclination to the horizontal complies with Table 9.6, Recommended ramp inclination. Ramps should be used where hand-carrying bulky loads or loads in excess of 13 kg (29 lbs) is required.

Dimensions. The following dimensions are applicable to ramps for pedestrian traffic or materials handling.

- Ramps should have level landings, at least as wide as the ramp with a minimum length of 1525 mm (60 in.), at the top and bottom of each ramp. To provide adequate space for people and equipment to turn where ramps change direction at landings, the minimum landing size should be 1525 mm by 1525 mm (60 in. by 60 in.).
- Ramps should be at least 915 mm (36 in.) wide.
- Handrails, as per the requirements outlined in the section on 'Handrails for walkways and work platforms', should be provided on any open side of a ramp provided the vertical distance from the ramp to the nearest adjacent surface is 610 mm (24 in.) or greater.
- Ramps, especially those with an angle of inclination greater than 8°, should be provided with a non-skid surface.

Table 9.6 Recommended ramp inclination

Ramp use	Recommended incline in degrees
Pedestrian traffic without materials handling	7° (recommended)–15° (maximum)
Materials handling	4° (recommended)–7° (maximum)

Work Platforms

Work platforms should be provided at locations where personnel must perform tasks that cannot be easily accomplished by reaching from an existing standing surface. Work platforms exposed to the elements should have additional slip resistance due to potential exposure to water and ice. The principles listed below apply to the design of work platforms:

- Work platforms more than 600 mm (23.5 in.) above the surrounding surface should be provided with guardrails and handrails.
- Platforms should be of sufficient size to accommodate the task and allow for placement of any required tools, spare parts or equipment.

Design loads. Where requirements for design loads specified by other regulatory bodies (e.g., Flag State Administrations and Port State Authorities) are greater, those requirements take precedence over the guidance provided in this book. In this book, we define "design load" as the maximum intended load, being the total of all loads including the weight of the personnel, materials and equipment, including the means of access structure. *Guardrails.* Guardrails should withstand anticipated loads but not less than 90 kg (200 lbs) at any point and in any direction when applied to the top rail. Design loads and maximum deflection. The minimum design loads for the working platforms and landings are:

- 2.0 kN/m^2 (0.29 lbf/in^2) under uniform load for the structure.
- 1.5 kN (337 lbf) concentrated load applied in the most unfavourable position over a concentrated load area of 200 mm × 200 mm (8 in. × 8 in.) for the flooring.
- When loaded with the design load, the deflection of the flooring should not exceed 1/200th of the span and the difference between the loaded and adjacent unloaded flooring should not exceed 4 mm in height.

Handrails for walkways, ramps, and work platforms. Platform handrails should be provided at the exposed side of any platform surface that is 610 mm (24 in.) or higher above the adjacent surface and where a person could fall from the upper to the lower surface. Handrails should be two-tiered and be provided with a toeboard. Height of the handrail should be 1070 mm (42 in.) from the top of the top rail to the walking surface, with the intermediate rail located 460 mm (18 in.) below the top rail. Vertical stanchions should be provided at a maximum spacing of 2.4 m (8 ft). The handrails should be constructed of 40 mm (1.5 in.) outside diameter pipe. The toeboard should be 100 mm (4 in.) in height and should have no more than 6 mm (0.25 in.) clearance between the bottom edge of the toeboard and the top of the walking surface. *International Convention on Loadlines.* For locations requiring handrails or guardrails in accordance with the International Convention on Loadlines, the handrail or guardrail should be of three tiers. The height of the top tier should be at least 1000 mm (39.5 in.) from the deck, except where this height would interfere with the normal operation of the vessel; a lesser height may be considered. The opening below the lowest course of the guardrail should not exceed 230 mm

(9 in.). The intermediate course should not be more than 380 mm (15 in.) from the other two courses. For additional details concerning the protection of crew, refer to Regulations 25, 26, and 27 of the International Convention on Loadlines. Where any conflicts exist, statutory requirements take precedence over the guidance provided in this book. *Access walkway to tanker bows.* Tankers, including oil tankers, chemical tankers and gas carriers, should be provided with a means to enable personnel to gain safe access to the bow in severe weather conditions. The International Convention for the Safety of Life at Sea refers to IMO Resolution MSC.62 (67), Guidelines for Safe Access to Tanker Bows, for details. Access should be by means of either a walkway on the deck or a permanently constructed gangway at or above the level of the superstructure deck or the first tier of a deckhouse and be in accordance with the following:

- The access should not be less than 1000 mm (40 in.) wide situated on or as near as practicable to the centreline of the vessel and located so as not to interfere with access across working areas of the deck.
- The access should be fitted at each side throughout its length with a toeboard and guardrails supported by stanchions. Such rails should consist of no less than three courses, the lowest being not more than 230 mm (9 in.) and the uppermost being at least 1000 mm (40 in.) above the gangway or walkway, and no intermediate opening more than 380 mm (15 in.) in height. Stanchions should be at intervals of not more than 1500 mm (60 in.).
- The access should be constructed of fire-resistant, non-slip material and have a level surface.
- The access should be provided with openings in the handrails and with ladders, where appropriate, to and from the deck. Openings should not be more than 40 m (131 ft) apart.
- If the exposed length of the access exceeds 70 m (230 ft), it should have shelters of substantial construction at intervals not to exceed 45 m (148 ft). Every shelter should be capable of accommodating at least one person and be constructed to provide weather protection on the forward, port and starboard sides.
- If the access is obstructed by permanently installed pipes or other fittings, means for passage over such obstruction should be provided.

Hatches

This section contains requirements related to the design of hatches. The principles listed below apply to the design of hatches and lightening holes and are not represented in the following figures or tables. For access through horizontal hatches, the dimensions should be sufficient to allow a person wearing a self-contained air-breathing apparatus (SCBA) and personal protective equipment (PPE) to ascend or descend any ladder without obstruction and also provide a clear opening to facilitate the movement of an injured person through the hatch. Where hatch covers are heavy [e.g., above 11 kg (24.3 lbs)] or unwieldy, aids should be provided to assist in lifting or lowering the hatch cover. *Hatch design.* Refer to Fig. 9.6, Hatch design and dimensions. Figure 9.7, Hatch design and dimensions (alternative arrangement) illustrates an IACS approved alternative design for access. This design is subject to the verification of easy evacuation of injured or stretcher-borne personnel. *Horizontal hatch access near a coaming.* When access to a cargo hold is arranged through the cargo hatch, the top of the ladder should be placed as close as possible to the hatch coaming (refer to Fig. 9.8, Ladder distance from hatch coaming). Access hatch coamings having a height greater than 900 mm (35.5 in.) should also have steps on the outside in conjunction with the ladder (refer to Fig. 9.9, Access hatch heights of ≥ 900 mm (35.5 in.).

Horizontal hatch access through a deck. For access to the deck from a ladder below, the top of the ladder should be placed within 50 mm (2 in.) of the leading edge of the hatch opening (refer to Fig. 9.9, Horizontal hatch access through a deck). Minimum dimensions of the opening (round or square) are 810 mm (32 in.).

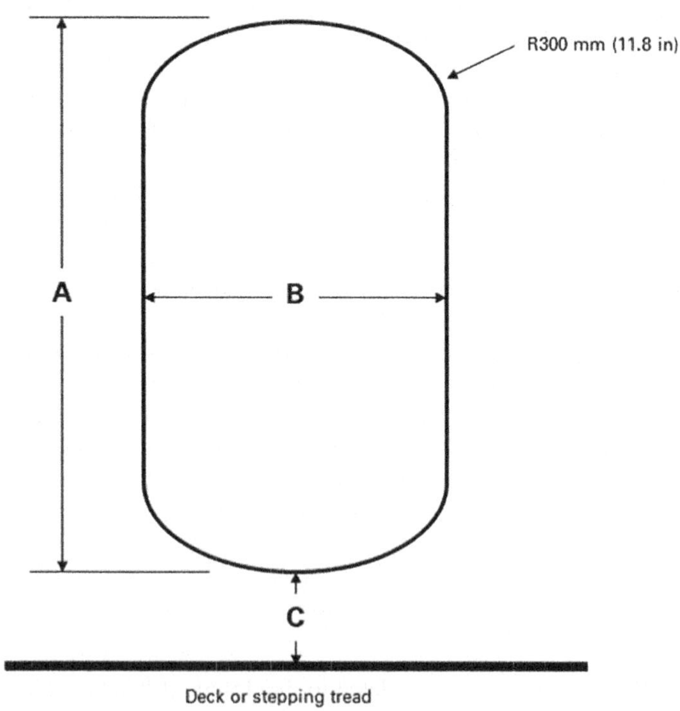

Fig. 9.6 Hatch design and dimensions

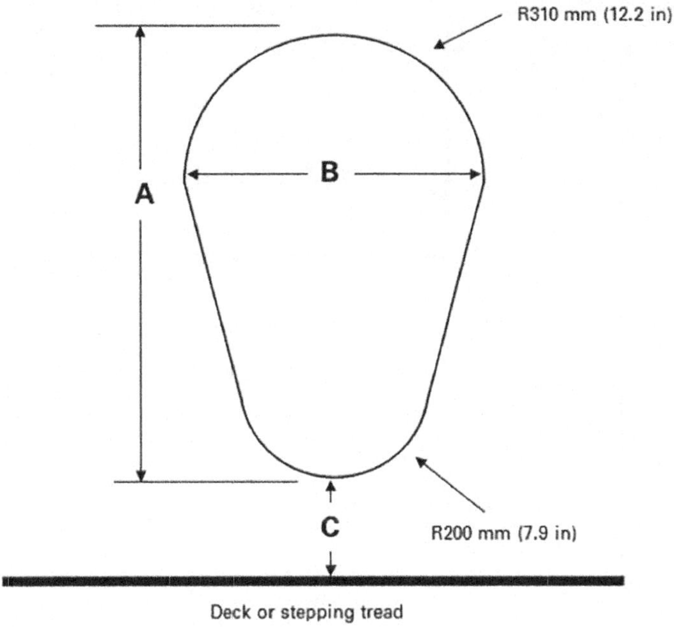

Fig. 9.7 Hatch design and dimensions (alternative arrangement)

Doors and Deck Scuttles

General. This section contains guidance related to the design of doors and scuttles. Means of escape or egress. (1) *Door operation.* Doors, hatches, or scuttles used as a means of escape should be capable of being operated by one person, from either side, in both light and dark conditions. Doors should be designed to prevent opening and closing due to vessel motion and shall be operable with one hand. (2) *Door dimensions.* Doors (other than emergency egress) used solely by crew members must have a clear opening width of at least 710 mm (28 in.). The distance from the deck to the top of the door must be at least 1980 mm (78 in.). (3) *Clear means of escape.* The method of opening a means of escape must not require the use of keys or tools. Doors in accommodation spaces (with the exception of staterooms), stairways, stair towers, passageways, or control spaces, should open in the direction of escape or egress. (4) *Markings.* The means of escape should be marked from both the inside and outside. *Deck scuttles.* Deck scuttles that serve as a means of escape should be fitted with a release mechanism that does not require use of a key or a tool and should have a hold back device to hold the scuttle in an open position. Deck scuttles that serve as a means of escape should have the following dimensions:

- Round—670 mm (26.5 in.) or greater in diameter; and/or
- Rectangular—670 mm (26.5 in.) by 330 mm (13 in.) or greater.

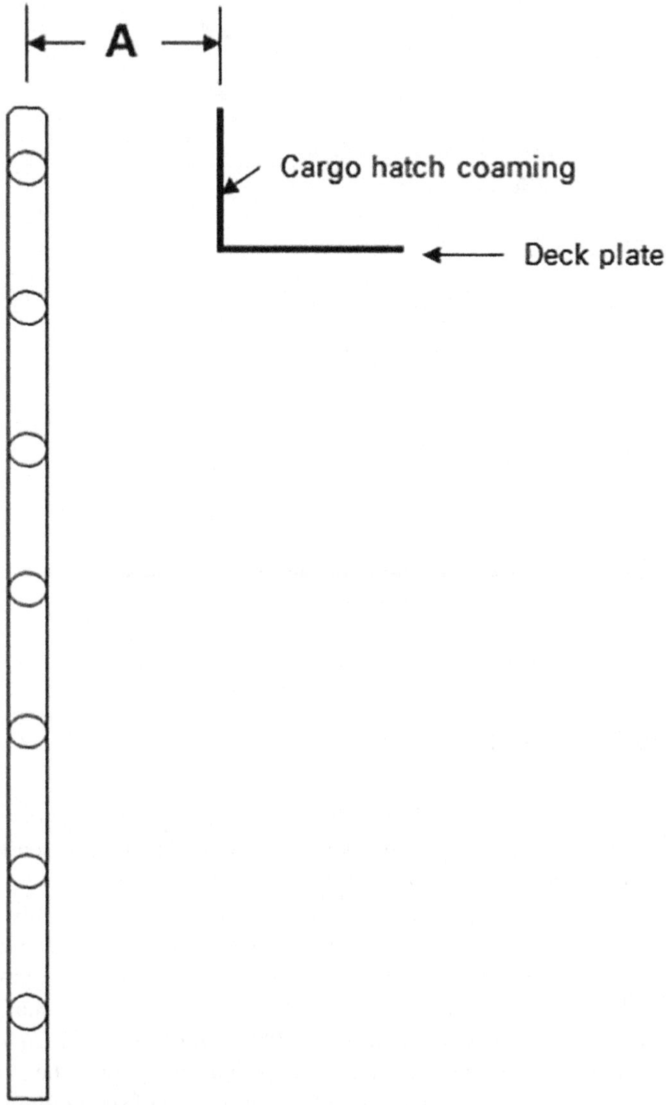

Fig. 9.8 Ladder distance from hatch coaming

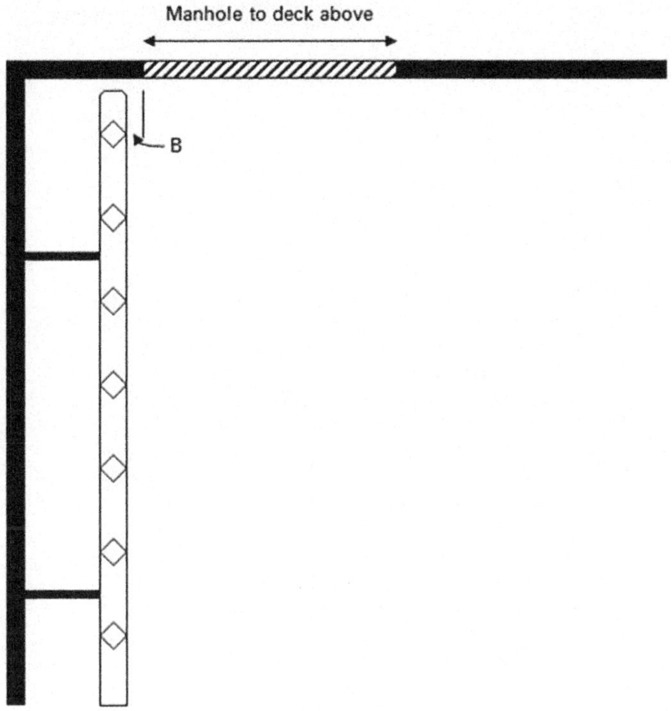

Fig. 9.9 Horizontal hatch access through a deck

Chapter 10
Maintenance and Considerations

General

The principles and guidance contained in this chapter apply to bridge related equipment maintenance, including equipment access and task requirements.

Principles

These maintainability and accessibility principles are directed toward the accomplishment of necessary maintenance tasks quickly, safely, accurately and effectively with minimum requirements for personnel, skills and special tools. Maintenance or operational access criticality analysis. Equipment location and accessibility should be determined on the basis of a maintenance criticality analysis based on the criticality, frequency, and inherent safety of inspection, planned maintenance, or repair. The assessments of safety or system criticality can be achieved by conduct of risk analyses, failure modes analyses, or other commonly accepted and applied techniques. Identification of criticality is used to determine the specific applicability of the requirements in this section to specific access locations. Identification of individual access criticality can be of joint participation among vessel designers, operators, and owners. The following three categories define maintenance criticality. Category 1 maintenance or operational access. Category 1 maintenance or operational criticality should include those maintenance actions that are system and safety critical, meaning a system necessary for the safe operation of the vessel cannot function without that aspect being functional. All the requirements in this section apply to Category 1 maintenance items. Category 2 maintenance or operational access. Category 2 maintenance or operational criticality should include those maintenance actions which are important to maintain operations, or any maintenance actions that are performed frequently. The requirements in this section that apply to Category

A. A. Olsen, *Applying Physical Ergonomics to Modern Ship Design*,
https://doi.org/10.1007/978-3-031-57974-5_10

2 maintenance or operational items are identified as part of each requirement statement. Category 3 maintenance or operational access. Category 3 maintenance or operational criticality include those maintenance actions which are considered to be non-critical to system status and safety, and that do not require frequent access.

Maintenance task access analysis. Analysis. An analysis of maintenance tasks should be performed for the purpose of sizing and outfitting maintenance workspaces. The analysis addresses concerns such as: design of maintenance platforms; work surfaces; work envelopes; tools, spare parts and spent part stowage; and consumables storage at a maintenance site during planned or corrective maintenance. The analysis is also intended to identify requirements for lifting and carrying devices for heavy or awkward loads; and to facilitate maintenance action workflow. This analysis should be applied to all Category 1 and 2 maintenance actions. *Typical analysis steps.* A maintenance area task and requirements analysis typically include the following steps:

- *Step 1.* (1) general and frequent maintenance tasks (such as housekeeping and routine or and recurring maintenance), and (2) list major planned maintenance or corrective maintenance actions:
  1. Number of participants/maintainers involved for each task.
  2. List of major equipment/components to be removed, and replaced, or repaired in place or nearby for each task.
- *Step 2.* For each maintenance task, identify and list the following items requiring space allocation:
  1. Consumables (lubricants, adhesives, connectors).
  2. Test equipment, stowed in place and/or portable.
  3. Tools or special tools.
  4. Replacement parts.
  5. Need for aided or unaided voice communications.
  6. Any required PPE.
  7. Lifting aids required.
  8. Number of people involved in the immediate maintenance area.
  9. Any required procedures, technical manuals, or check lists.
  10. Whether the maintenance action occurs in a workshop, or directly on installed equipment.
- *Step 3.* For each maintenance task, estimate and then sum the space requirements. Allocate maintenance work area space based on the analysis, including (where possible) a space allocation margin of 30% or more.
- *Step 4.* Identify and allocate or fit any requirements for storage surfaces for procedural materials, or ready access to consumables.
- *Step 5.* Identify requirements for parts and materials carts (for deck level access to maintenance areas), or lifting supports (including overhead crane, chain falls, pad eyes, I-beams, etc.) for very heavy components or when in the presence of ramps or stairs to gain maintenance access or transfer. This also requires lifting

from one engine room/deck level to another. (For example, from the one engine room deck to another shop level above or below, or from an engine room deck lifting to a maintenance pedestal or platform,

- *Step 6.* Special space accommodation should be made for access to, and use of, PPE.
- *Step 7.* Identify locations of pre-positioned items (such as tools, special tools, consumables) that can be located with the equipment to be maintained.
- *Step 8.* Develop maintenance area workspace concepts and drawings, considering the above sizing requirements and each maintenance actions activity sequence. This includes provision of suitable space in areas around a piece of equipment (e.g., behind the equipment near a bulkhead) that is needed to remove internal parts, motors, shafts, or attached equipment and the personnel/equipment required to accomplish these tasks.
- *Step 9.* Refine the concepts and drawings and prepare them as part of build-to packages.

General access. Structural components (e.g., stiffeners, stairs, pedestals, etc.) should not prevent access to, or removal of, maintained equipment. *(Category 1 and 2).* Maintenance items that need to remain accessible should not be blocked by removal of panel cases and covers. *(Category 1 and 2).* Maintenance items that are exposed to the external environment should be provided with means of de-icing. *(Category 1 and 2).* Equipment design. Design should include identification (such as labels), orientation, and alignment of components and sub-assemblies (including cables and connectors). *(Category 1 and 2).* Skids (generators, air compressors, etc.) should provide direct means of access to doors and removable panels to provide access to internal components for maintenance. *(Category 1 and 2).*

Access Design

Equipment should allow for access without the removal of other parts or components, and clearance shall be provided to effectively use tools through their full range of motion. *(Category 1 and 2).* Permanent access aids including ladders, stairs, platforms and ramps should be provided for all Category 1 and 2 maintenance items. (Category 1 and 2). Every ladder should have a pad eye for chain fall attachment so repair parts and tools can be hauled up if the overhead crane is not available or is not accessible. *(Category 1 and 2).* Equipment design should permit maintenance from above, in front, from behind and outside with visibility, rather than from underneath or inside components or without complete visibility. *(Category 1 and 2).* A minimum of 2.0 m^2 (18 ft^2) per person minimum should be provided for personnel, their clothing (including required PPE), tools and equipment as well as free space for the movements and activities required to perform maintenance tasks. *(Category 1 and 2).* When maintenance requires the removal of large internal or external parts (e.g., tube bundles from a heat exchanger, cylinder heads), a lay down area should be provided for the component, parts, tools and equipment. *(Category 1 only).* Lateral clearance for hand access between a bulkhead and nearby structures (for example, a skid) should not be less than 425 mm (17 in.). *(Category 1 and 2).* Physical access. When maintenance tasks require the maintainer to squat, kneel, sit down, or stand sideways, access design should accommodate these postures according to the following tables and associated figures (Category 1 and 2 applies to Tables 10.1, 10.2, 10.3, 10.4, 10.5, 10.6, and 10.7):

Table 10.1 Seated, forward reach—both arms

	Dimensions
Depth of reach	500 mm (19 in.) minimum
Breadth of aperture	425 mm (17 in.) minimum
Floor to top of aperture	1125 mm (44 in.) minimum
Floor to bottom of aperture	950 mm (37 in.) minimum
Vertical dimension of aperture	300 mm (12 in.) minimum

Table 10.2 Cross-legged seated, forward reach—both arms

	Dimensions
Depth of reach	450 mm (18 in.) minimum
Breadth of aperture	425 mm (17 in.) minimum
Floor to top of aperture	675 mm (26 in.) minimum
Floor to bottom of aperture	500 mm (20 in.) minimum
Vertical dimension of aperture	275 mm (11 in.) minimum

Table 10.3 Standing, forward reach—both arms

	Dimensions
Depth of reach	575 mm (23 in.) minimum
Breadth of aperture	475 mm (18 in.) minimum
Floor to top of aperture	1700 mm (66 in.) minimum
Floor to bottom of aperture	1450 mm (57 in.) minimum
Vertical dimension of aperture	425 mm (17 in.) minimum

Table 10.4 Standing, single arm forward reach

	Dimensions
Depth of reach	650 mm (25 in.) minimum
Breadth of aperture	300 mm (12 in.) minimum
Floor to top of aperture	1675 mm (66 in.) minimum
Floor to bottom of aperture	150 mm (57 in.) minimum
Vertical dimension of aperture	425 mm (17 in.) minimum

Table 10.5 Standing, single arm lateral reach

	Dimensions
Depth of reach	650 mm (25 in.) minimum
Breadth of aperture	25 mm (10 in.) minimum
Floor to top of aperture	1675 mm (66 in.) minimum
Floor to bottom of aperture	1450 mm (57 in.) minimum
Vertical dimension of aperture	400 mm (16 in.) minimum

Table 10.6 Squatting space dimensions

	Dimensions
Height	1300 mm (51 in.)
Depth	925 mm (6 in.)
Depth	1025 mm (40 in.)

Table 10.7 Kneeling space dimensions

	Dimensions
Depth	1225 mm (48 in.)
Height	1550 mm (60 in.)
Optimum work point	775 mm (27 in.)

- Table 10.1, Seated, forward reach—both arms.
- Table 10.2, Cross-legged seated, forward reach—both arms.
- Table 10.3, Standing, forward reach—both arms.
- Table 10.4, Standing, single arm forward reach.
- Table 10.5, Standing, single arm lateral reach.
- Table 10.6, Squatting space dimensions.
- Table 10.7, kneeling space dimensions.

Access openings. Access openings should be provided, as required, to all equipment or items that require testing, servicing, calibrating, adjusting, removing, replacing or repairing during normal maintenance efforts. *(Category 1 and 2).* Access openings should be large enough to accommodate hands, arms and tools and also provide full visual access to the task area. *(Category 1 and 2). Dimensions.* The dimensions of access openings for arms and hands should be no less than those shown in the following tables. *(Category 1 and 2). Visual access.* Access openings should provide full visual access to the task area, or an auxiliary viewing port should be provided through the use of viewing ports or quick release access. Visual inspections should be able to be done without use of special tools. *(Category 1 and 2)* (Tables 10.8, 10.9, 10.10, 10.11, and 10.12).

Hatches. Round, square, or rectangular hatches should be sized as shown in Table 10.13, Hatch shapes and dimensions. Depending on their shape and orientation (i.e., side, top or bottom entry), these dimensions may require adjustment if personnel are likely to be wearing bulky clothing or personnel protective equipment (e.g., breathing air canisters) when using a hatch for access to areas such as tanks, bilges, ship double bottoms areas or other confined spaces. Where hatch covers are

Table 10.8 Opening dimensions for single hand access with tools

Description of opening	Minimum dimensions mm (in.)			Task description
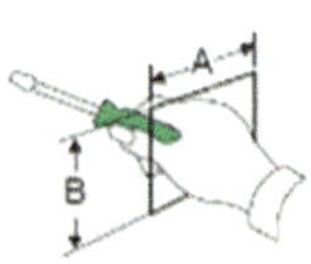	Bare hand	110 (4.25)	120 (4.75)	Using common screwdriver, test probe etc., with freedom to turn hand through 180°
	Gloved	165 (6.5)	180 (7.0)	
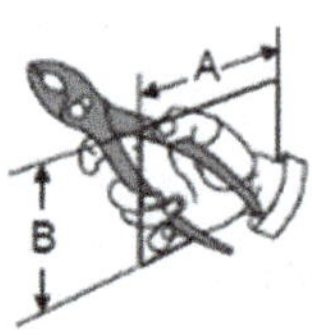	Bare hand	125 (5.0)	115 (4.5)	Using pliers and similar tools
	Gloved	175 (7.25)	200 (7.75)	
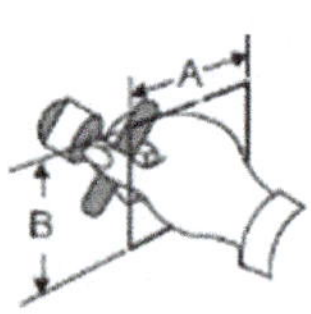	Bare hand	150 (6.0)	150 (6.0)	Using "T" handle wrench, with freedom to turn hand through 180°
	Gloved	190 (7.5)	210 (8.25)	
	Bare hand	275 (11.0)	200 (8.0)	Using open-end wrench, with freedom to turn wrench through 60°
	Gloved	325 (12.75)	260 (10.25)	
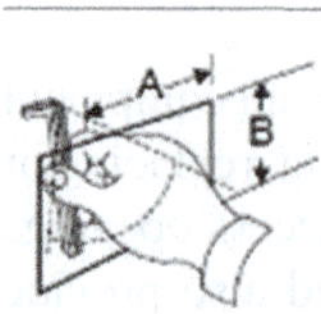	Bare hand	120 (4.75)	150 (6.0)	Using Allen-type wrench, with freedom to turn wrench through 60°
	Gloved	175 (7.0)	200 (8.0)	

Table 10.9 Opening dimensions for single hand access without tools

Description of opening	Minimum dimensions mm (in.)			Task description
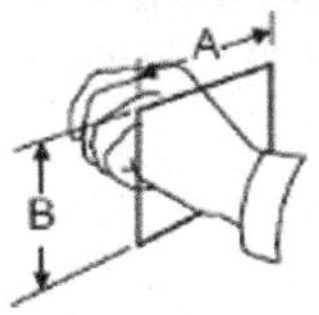	Bare hand	125 (5.0)	100 (3.75)	Empty hand Clenched fist extended to wrist
	Gloved	150 (6.0)	115 (4.5)	
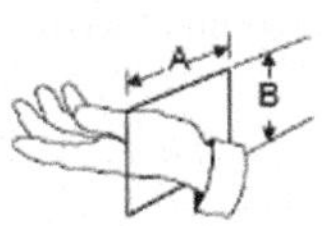	Bare hand	100 (4.0)	60 (2.25)	Empty hand Hand flat extended to wrist
	Gloved	150 (6.0)	100 (4.0)	
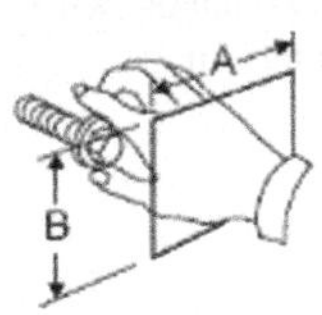	Bare hand	115 (4.5)	120 (4.75)	Grasping small objects (up to 50 mm (2 in.) or wider) with one hand
	Gloved	170 (6.75)	175 (7.0)	
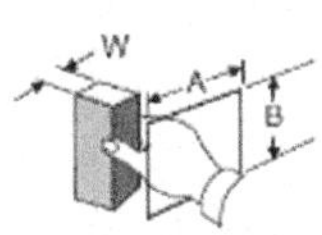	Bare hand	W+45 (W+1.75)	125 (5.0)	Grasping large objects, 50 mm (2 in.) or wider with one hand
	Gloved	W+100 (W+4.0)	175 (7.0)	

Table 10.10 Opening dimensions for arm access without tools

Description of opening	Minimum dimensions mm (in.)		Task description
	Light clothing	100 mm (4.0 in.) × 115 mm (4.5 in.) or 115 mm (4.5 in.) diameter	Arm to elbow
	Cold weather clothing	180 mm (7.0 in.) square or diameter	
	Light clothing	125 mm (6.0 in.) square or diameter	Arm to shoulder
	Cold weather clothing	215 mm (8.5 in.) square or diameter	

heavy [e.g., above 11 kg (25 lbs)] or unwieldy, consideration should be given to providing aids for lifting or handling the hatch cover.

Table 10.11 Opening dimensions for two handed access

Description of opening	Minimum dimensions mm (in.)			Task description
	Bare hand	W+75 (W+3.0)	125 (5.0)	Grasping large objects with two hands, with hand extended through openings up to fingers
	Gloved	150 (6.0)	115 (4.5)	
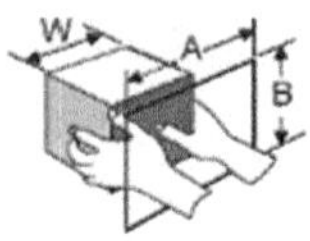	Bare hand	W+130 (W+5.25)	180 (7.0)	Grasping large objects with two hands, with arms extended through openings up to wrists
	Gloved	150 (6.0)	100 (4.0)	
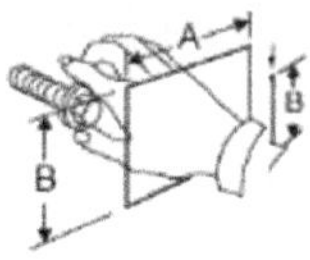	Bare hand	W+150 (W+6.0)	125 (5.0)	Grasping large objects with two hands, with arms extended through openings up to elbows
	Gloved	170 (6.75)	175 (7.0)	

Table 10.12 Two hand access without visual access

Description of opening	Minimum dimensions mm (in.)		
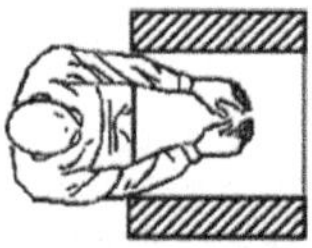	Reaching with both hands to depth of 152 to 489 mm (6.0 to 19.25 in.):		
	Light clothing	Width	200 mm (8 in.) or the depth of reach, whichever is larger
		Height	130 mm (5 in.)
	Cold weather clothing	Width	150 mm (6 in.) plus 3/4 depth of reach
		Height	180 mm (7 in.)
	Reaching full arm's length (to shoulders) with both arms		
	Width	500 mm (19.5 in.)	
	Height	130 mm (5 in.)	
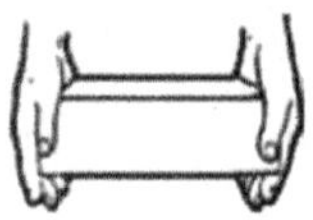	Inserting box grasped by handles on the front		
	13 mm (0.5 in.) clearance around box, assuming adequate clearance around handles		
	Inserting box with Hand on the sides		
	Light clothing	Width	Box plus 115 mm (4.5 in.)
		Height	125 mm (5.0 in.) or 13 mm (0.5 in.) around box
	Cold weather clothing	Width	Box plus 180 mm (7 in.)

Table 10.13 Hatch shapes and dimensions

Access shape and dimensions		Minimum dimensions
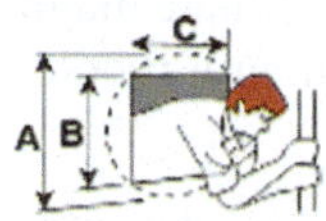		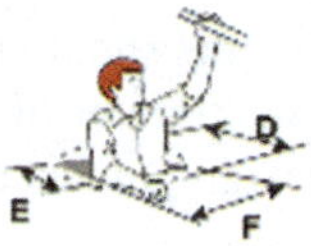
Side access		
Round	–A	685 mm (27 in.)
Square		660 mm (26 in.)
Rectangle	–B	660 mm (26 in.)
Square	–C	760 mm (30 in.)
Top or bottom access		
Round	–D	635 mm (25 in.)
Square		580 mm (23 in.)
Rectangle	–E	330 mm (13 in.)
	–F	610 mm (24 in.)

Labelling

Equipment should be labelled with an accurate description specific to the equipment, including an exclusive identifying number. In addition, signs and graphics, such as schematics, should be placed to provide hazard information and maintenance instructions. Each cable or wire should be labelled or coded throughout its entire length. Labels installed at a bench preparatory site should be placed so they will be visible and in the proper orientation to the reader once the wire or cable is installed. Cables and wires should be labelled with the name of the equipment to which they belong and the connectors to which they mate. Pipes should be colour-coded for identification of content and direction of flow. The colour-coding should also differentiate between grades of fuel. All stored energy devices should be labelled as such and should have a **DANGER** hazard warning sign attached to the device. Procedures for releasing or constraining the energy should be displayed on the unit. Each access cover should be labelled with terminology for items visible or accessible through it, terminology for auxiliary equipment to be used with it, and recommended procedures for accomplishing operations. Accesses should be labelled with hazard signs advising of any hazards existing beyond the access. If instructions applying to a covered item are lettered on a hinged door, the lettering should be properly oriented to be read when the door is open.

Equipment Design

Design. Systems should be designed to facilitate diagnostics and troubleshooting to minimise the amount of time spent locating a problem, thus expediting the repair process. Design guidance includes:

- Equipment design should facilitate rapid and positive fault detection and isolation of defective items to permit their prompt removal and replacement.
- When warranted, equipment should have automatic fault detection and isolation capability.
- Components performing similar functions should be physically close to one another.
- Components should provide easily identifiable physical evidence of their status (e.g., a blown fuse).
- Whenever possible test replacement parts prior to installation.
- Provide self-test fault and error messages:
  - Standardise error messages.
  - Provide accurate, up-to-date, reliable, and understandable error messages.
  - Scale fault messages to meet the range of experience and skill levels of troubleshooters.
  - Failure messages should identify the origin of the failure.

Equipment modularisation. Designing equipment and components into specifically-defined units to aid in troubleshooting and replacement may also enhance the maintenance function. Design guidance suggests that, where possible, equipment be replaceable as modular packages and should be configured for removal and replacement by one person, within structural, functional, and weight limitations. Heavy, large, or complex equipment should be divided into modules. Modularisation of equipment can make it easier to locate and isolate malfunctions, properly allocate maintenance functions and responsibilities; reach, remove, and maintain components; and handle the equipment for installation and repair. Equipment mounting and installation. Equipment should be designed so that it cannot be mounted or installed improperly. Methods to help prevent equipment from being improperly mounted or installed include:

- Measures (e.g., coding) to facilitate identification and interchange of interchangeable items.
- Measures (e.g., alignment pins) to facilitate proper mounting of items.
- Identification, orientation, and alignment provisions for cables and connectors.
- Provision of adequate separation of components to ease access during installation and revaluation.
- Provision of supports and guides for the handling, aligning, and positioning of components:
  - For loads up to 11 kg (25 lbs), provide bottom mounted alignment pins or side mounting brackets to allow for sliding rather than lifting.

- For loads 11 kg (25 lbs) or greater, provide for two-person handling or provide a mechanical means to aid lifting.

Fasteners and connectors. Fasteners are available in a wide variety of types and sizes. However, the diversity of fasteners and connectors used in a specific design should be minimised. Fasteners or connectors requiring the use of special tools should not be used. Extensions connected to equipment. Accessories, utilities, cables, wave guides, hoses or any other protuberance extending from the equipment or item to be lifted or carried should be designed for easy removal or disconnection from the equipment or item before handling. *Requirements-based maintenance design.* Equipment that is part of a system that must operate continuously should be capable of undergoing maintenance without interrupting the operation of the overall system. If continuous operation is required and required maintenance on a unit of equipment would interrupt the operation, redundant equipment should be provided, with the following guidance in mind:

- Equipment should be designed so that components that fail frequently (e.g., lamps and fuses) can be easily replaced.
- Degraded operation should not cause damage to other equipment or components, nor should it aggravate the original fault.
- When warranted by its importance in a system, a unit of equipment that has a partial failure should be designed to operate in a degraded mode while awaiting maintenance.
- Degraded operation and faults should be sensed, and appropriate information identified, displayed, or transmitted to maintainers and, if appropriate, to operators.

Equipment standardisation. Equipment standardisation is the design of systems and equipment to use interchangeable components (e.g., using identical valves in a piping system), measures (e.g., all metric), and processes (e.g., standard operating procedures). The following points regarding standardisation should be followed. Standard parts should be used whenever possible. If an existing design for equipment, modules, components, or parts meets the relevant requirements and the applicable ergonomics criteria in this document, designers should use that existing item rather than design a new one. Units of equipment should be designed to facilitate the interchangeability of modules, components, and parts. Units of equipment or modules that are similar in size and shape to other items but different from them in functional purpose should be easily identifiable and distinguishable. In addition, they should not be physically interchangeable (e.g., different connector types for breathing air versus nitrogen). Equipment should be designed to minimise the numbers and types of auxiliary equipment and tools required to accomplish maintenance tasks.

Equipment Handling

Rests and stands. When required to support operations or maintenance functions, rests or equipment stands on which units can be placed, including space for test equipment, tools, technical orders and manuals, should be provided. Such rests or stands should be part of the basic unit, rack or console chassis. Space around the rests or stands should be provided to both support the equipment or item and allow for the number of personnel required to lift or work on the item.

Tools

Uncommon or specially designed tools should be used only when common hand tools do not satisfy the guidelines or when the special tools provide a significant advantage over common hand tools.

Safety

Elevated work platforms and handrails. Work platforms should be provided at locations where personnel must perform tasks that cannot be easily accomplished by reaching from an existing standing surface. Platforms should be of sufficient size to accommodate the task and allow for placement of any required spare parts or equipment. Work platforms should be no less than 760 mm (30 in.) wide and 915 mm (36 in.) deep or larger to accommodate tools, equipment and additional maintenance personnel, except for platforms used exclusively for standing which may be no less than 380 mm (15 in.) wide and 460 mm (18 in.) in depth. Work platforms more than 610 mm (24 in.) above the surrounding surface should be provided with handrails and toeboards. Work platform handrails should be provided at the exposed side of any platform surface that is 610 mm (24 in.) or higher above the adjacent surface and where a person could fall from the upper to the lower surface. Handrails should be two-tiered and be provided with a toeboard. The height of the handrail should be 1070 mm (42 in.) from the top of the top rail to the walking surface with the intermediate rail located 460 mm (18 in.) below the top rail. Vertical stanchions should be provided at a maximum spacing of 2.4 m (8 ft). The handrails should be constructed of 40 mm (1.5 in.) outside diameter pipe; and the toeboard should be 100 mm (4 in.) in height and should have no more than 6 mm (0.25 in.) clearance between the bottom edge of the toeboard and the top of the walking surface. *Fall protection.* Where feasible, designs should minimise the requirement for work to be performed in elevated locations. Permanent barriers such as handrails should be provided to protect routinely-used workstations and maintenance platforms. Handrails and related barriers will withstand a minimum of 200 lbs of lateral force. Where safety cages are installed, the safety cages should be installed on vertical ladders where the vertical height exceeds 4.5 m (15 ft). Consideration should be given to providing safety cages for ladders less than 4.5 m (15 ft) in height as well where a fall to a level or deck below the ladder base is possible (e.g., within 1830 mm (72 in.) of the edge of a deck). The cage should start 2.1 m (7 ft) above the bottom-landing surface and should extend to 1400 mm (54 in.) above the top-landing surface; and cages used on vertical ladders equipped with intermediate landings should extend 1400 mm (54 in.) above the intermediate landing with the cage open on the side facing the landing.

For ladders over 6.1 m (20 ft), a climber safety rail or cable should be considered. Additional consideration should be given to adopting such devices based on the consequence of a fall (e.g., falls onto equipment, other decks, overboard). Climber safety rails or cables should begin approximately 915 mm (36 in.) above the standing surface and extend to approximately 1070 mm (42 in.) above the standing surface of the upper landing. A climber safety rail should be stainless steel flat bar and equipped with two safety slides, which can be attached to the rail or cable. The ladder stringers at a top-landing where climber safety devices are used should be designed to allow personnel to access the landing without unfastening. From the centreline of the rail or cable to the inside of each ladder stringer should be a minimum distance of 380 mm (15 in.). In addition to the fall protection provided by handrails required for stairs, ladders, walkways and platforms, fall protection tie-off points where maintenance personnel can attach a safety harness should be provided. Tie-off points should be located anywhere in the ship or maritime structure where personnel could be working in an elevated location in excess of 1524 mm (60 in.) above the deck. Anchorage points should be reachable by hand without need for an extension device or unprotected access. Anchorage points should be designed to withstand a minimum of 3600 lbs of force. In areas where anchorage points for fall protection or for lifting or hoisting are required, two sets of anchorage points should be provided, one for equipment and one for personnel. Both will be designed to withstand the test force required by the more stringent criteria of the two (that required for lifting of equipment or that required for fall protection). A sufficient number of anchorage points should be provided so personnel are always tied-off. Double lanyards should be used. For those locations where stairs are required to come up from a deck below and pass through an open horizontal hatch and eventually connect to a vertical stanchion on the deck just above the hatch, the use of chains, ropes or wire to serve as temporary handrails should not be used. This also applies to walkways to a deck where permanent handrails are not allowed to protrude above deck. Instead, regular stair handrails should be designed to allow the handrails to be retracted, removed or stowed below the hatch opening or flight deck level until needed, but then can be easily and quickly installed to provide a rigid and sturdy solid handrail. Removable handrails should not be used unless there is an approved operational or maintenance need, but are not be used for ease of fabrication. Removable solid handrails may be used to facilitate the removal of equipment or material handling, provided the safety of personnel is not jeopardised. Bolted posts or sockets can be used for removable hand railing and should be stored in such a place as to be easily accessible when needed for hatch opening.

Tripping hazards. Elevation changes in work areas or walkways should be avoided. Any elevation change in excess of 6 mm (1/4 in.) should be clearly visible or marked to alert personnel of a potential tripping hazard. The handle on all vertical stem valves should not rotate into a walkways or working areas so as to become a tripping or knee knocker hazard. Deck hatches provided in structures to access and transfer equipment should be flush with the adjacent floor or deck surface so as not to create a tripping hazard. Hatch lifting aids should not become a tripping or safety hazard. *Hot or cold surfaces.* Hot surfaces having an external surface temperature sufficient to burn human tissue with momentary contact and located within 2134 mm (7 ft) vertically from walking or working surfaces or 381 mm (15 in.) horizontally from floor, working level, stairs, ramps or fixed ladders should be insulated or otherwise guarded against accidental contact. Generally, metal surfaces with a surface temperature of 60 °C (140 °F) or more should be insulated or guarded. Insulation is the preferred means of protection from hot surfaces, unless heat retention is undesirable and should be of sufficient thickness to reduce the surface temperature to a maximum of 60 °C (140 °F). Guarding can also be used to provide protection from hot surfaces. Cold equipment should be evaluated to prevent "cold" burns. *Guarding.* If a hazardous condition (such as exposed, high voltage conductors) exists behind the access, the physical barrier over the access should be equipped with an interlock that will de-energise the hazardous equipment when the barrier is open or removed. Both the presence of the hazard and the fact that an interlock exists should be noted on the equipment case or cover such that it remains visible when the access is open. *Hazard signs.* Hazard signs should be used to identify and provide information about conditions that may pose hazards to personnel, equipment or the environment. These hazards include, but are not limited to those listed in Table 10.14, Common shipboard hazards.

Hazard identification signs should use simple language and understandable icons. *Access aids.* Maintenance access guidelines are as follows. Permanent barriers such as handrails should be provided to protect routinely-used workstations and maintenance platforms. *(Category 1 and 2).* Stairs should be provided when access is required to elevated work platforms (e.g., mezzanines) one or more times per day. *(Category 1 and 2).* Ladders should not be used when maintainers carry equipment. In all cases, both hands should be free to grasp and climb ladders. *(Category 1 and 2).* Items to be maintained from a ladder should require only one hand and should not be located more than 965 mm (38 in.) from the ladder's centreline. *(Category 1*

Table 10.14 Common shipboard hazards

Common shipboard hazards	
High noise areas	Confined space hazards
Fire/explosion hazards	Fall hazards
Steam (drains, etc.) hazards	Crane hazards
High temperature hazards	Access hazards
Electrical hazards	Toxic/caustic hazards
Mechanical hazards	Other maintenance related hazards

and 2). Maintenance on masts or antenna, lights or other equipment mounted on the mast, or other elevated structures, should not be performed while suspended by a safety harness, and permanent structural accesses (such as platforms) should be provided. *(Category 1 and 2)*. Vertical access should be provided with adequate clearance. If this is not achievable, masts should be hinged and be able to be lowered for maintenance; otherwise, a rotating base should be provided. *(Category 1 and 2)*. Portable ladders should not be used to access permanent work platforms to perform temporary, infrequent maintenance. *(Category 1)*. Where the maintainer must carry bulky loads or loads in excess of 13 kg (29 lbs), a ramp or elevator should be provided. Alternatively, access may be provided for personnel via stairs, steps or ladders, as long an adequate lifting mechanism is provided and used; and ramps, elevators or equivalent means should be provided when maintainers must transport heavy or bulky equipment. *Crawl spaces*. Crawl spaces should conform to the guidelines expressed in Table 10.15, Minimum dimensions for crawl spaces below. *(Category 1 and 2)*.

Table 10.15 Minimum dimensions for crawl spaces

Supine workspace (lying on back)	
Height—visual inspection	350 mm (14 in.)
Height—hand arm reach	550 mm (22 in.)
Length	1875 mm (74 in.)
Crawling space	
Height	425 mm (17 in.)
Length	1525 mm (60 in.)
Prone work or crawl space	
Height	775 mm (31 in.)
Length	2875 mm (113 in.)

Chapter 11
Materials Handling

General

The principles and guidance provided in this chapter apply to materials handling. Both manual and assisted materials handling are addressed, though particular emphasis is placed upon aspects and factors affecting manual lifting and carrying capacities.

Principles

There are many factors that affect personnel's ability to safely and effectively perform materials handling tasks, including the design of the task, the form of the material handled, the design of the working area, the availability of assisted lifting devices, and the physical/physiological characteristics of the personnel themselves. *Key points related to acceptable materials handling and lifting.* A general approach to designing for materials handling and lifting tasks include avoiding manual materials handling altogether, whenever possible; redesigning the load; redesigning the lifting/carrying task; redesigning the working environment; and introducing assisted lifting devices. *One-versus two-person lifting and carrying.* Limits on lifting and carrying should never be exceeded; accordingly, items weighing up to 20 kg (45 lbs) or less may be lifted or carried by one person; and items weighing more than 20 kg (45 lbs) and less than 40 kg (90 lbs) should be lifted or carried by two persons, provided the lifting load is distributed equally between the two. *Assisted lifting.* Assisted lifting devices (e.g., pad eyes, chain falls, come-alongs, rail cranes, mono-rails, etc.) should be provided for items or devices under the following conditions:

- Load weight in excess of 11 kg (25 lbs), for loads which are lowered from or placed in locations greater than 1525 mm (60 in.) high.

A. A. Olsen, *Applying Physical Ergonomics to Modern Ship Design*,
https://doi.org/10.1007/978-3-031-57974-5_11

- Load weight in excess of 11 kg (25 lbs), which needs to be supported during removal, replacement, or installation.
- Load weight in excess of 20 kg (45 lbs), where the load is not distributed evenly or is unwieldy.
- Load weight in excess of 40 kg (90 lbs).

Materials Handling Factors

There are numerous factors that impact the safety and efficiency of materials handling tasks. For this reason, pre-planning of materials handling tasks should be undertaken before lifting, carrying, or moving any load. Key factors include the actual materials handling task, the size and weight of the load, how the load will be lifted (e.g., manually or with assisted lifting devices), how the load will be moved or transported, the working area, personnel capabilities, and safety considerations. Each factor, either alone, or in combination with other factors, can significantly impact materials handling tasks. Task-related factors. Task-related factors for both manual materials handling and assisted lifting that should be considered as a part of pre-planning include vessel and offshore installation layout in terms of porches, hatches, removable plates, crane placement, pathway size, and location for moving loads; vessel or installation motions or movement, as well as motions or movement associated with support or supply vessels; the material's location with regard to access, pathways, hatches, doors, porches, and any obstructions (overhead or on the deck) in the path of movement; the distance the load needs to be moved, from the origin of the lift to its final destination; and the type of lifting method (e.g., manually or with assisted lifting devices). Moreover, task-related factors related solely to manual materials handling that should also be considered during the pre-planning stage include the need for support equipment (e.g., carts, trolleys, overhead rail or conveyor systems); how personnel will handle the load (e.g., lifting, lowering, carrying, pushing, pulling, and static lifting/loading); the duration and frequency of the handling task (e.g., how long and how often will personnel be performing the handling task?); the required body postures for lifting and placing (e.g., reaching, bending, and twisting of the torso and flexion of joints such as the wrist, elbow, neck, and knees); the distance of the load from the torso of the worker; the load height at lift origin and termination, as well as carry distance; the effects of gloves and/or other personal protective equipment; and any appropriate rest and recovery time.

Load-related factors. Load-related factors, for both manual materials handling and assisted lifting that should be considered as a part of pre-planning include the weight, size, and dimensions of the load; the load's apparent "unwieldiness" and the potential difficulty in securing (e.g., strapping) or manually grasping the load; the apparent stability or instability of the load (e.g., likelihood of the load to shift, including within a container, during movement); and the centre of gravity of the load. *Working-area related factors.* Working-area related factors for both manual materials handling and assisted lifting that should be considered as a part of pre-planning include the vessel and offshore installation layout in terms of space availability for the securing (e.g., strapping) or manually grasping the load; the material's location with regard to access, pathways, hatches, doors, porches, and any obstructions (overhead or on the deck) in the path of movement; the design of stairs, ladders, pathways, hatches, and doors; any extremes of temperature, humidity, and/or air movement; and area lighting where the loads are secured, grasped, lifted, and placed. Working-area factors related solely to manual materials handling that should also be considered during pre-planning include the walking/working surfaces (e.g., slippery, unstable, moving, variable, inclined); and the design and condition of handling aides (e.g., hand trucks, carts, etc.). *Personnel capacity-related factors.* The factors related to the personnel involved with or performing the manual or assisted materials handling tasks that should be considered in relationship to the task, the load, and the work area factors during pre-planning include anthropometrics (e.g., international population differences); individual strength; physical condition (e.g., cardiovascular and muscular endurance); flexibility (range of motion); health habits (e.g., smoking and or alcohol consumption); history of previous injury or predisposition for injury; and manual and assisted materials handling knowledge, training, and experience.

Safety considerations. Safety should be considered for both manual and assisted materials handling. Safety concerns should be examined within the context of the task, load, working area, and the personnel involved with or performing the handling task. Safety considerations include any clearance for equipment through walkways, passages, doorways, etc., and any obstructions (overhead or on the deck) in the path of movement; availability of motorised/mechanised equipment to support inspection and handling; vehicles and equipment securing (e.g., brakes, chocks, etc.) while loading and unloading; surfaces likely to be effected by wind gusts (e.g., sheets of boarding); noise levels (e.g., ambient and impulse); obstruction of vision (e.g., crane operator or other load handling personnel); housekeeping requirements (e.g., is the work area provided with amenities for keeping the area free from clutter or obstructions); load swinging; flammable, corrosive, or reactive substances; surface characteristics for personnel or placement of temporary assisted lifting equipment (e.g., even surfaces, those that are likely to abrade, puncture, cut or burn); exposure to vibration and shock for personnel or the load; personnel slip, trip, and fall hazards; and the consequences associated with dropping the load.

Manual Materials Handling Planning Tools

There are many tools available for the evaluation and design or re-design of manual materials handling tasks. These tools can assist with planning activities by allowing manual materials handling tasks to be prioritised and by determining when assisted lifting may be required. The majority of these tools are based on a North American population, but some do exist for international populations. Table 11.1, Material

Table 11.1 Material handling planning and analysis tools

Analysis tool	Uses	Source
NIOSH lifting equation	Analysis of lifting with several constraints on its application	Waters, T., Putz-Anderson, V., and Garg, A. (1994). Applications Manual for the Revised NIOSH Lifting Equation. US Department of Health and Human Services (NIOSH) Publication No. 94–110
Psychophysical tables	Analysis of lifting, lowering pushing, pulling, carrying	Snook, S. H., and Ciriello, V. M. (1991). The design of manual handling tasks: Revised tables of maximum acceptable weights and forces. Ergonomics 34:1197
Psychophysical lifting capacity for Chinese subjects	Lifting capacity of Chinese subjects, as well as anthropometric considerations	Wu, Swei-Pi. (1999). Psychophysically determined infrequent lifting capacity of Chinese participants, Ergonomics 42(7)
Job stress index	Analysis of lifting	Mital, A., Nicholson, A., and Ayoub, M. (1993). A Guide to Manual materials Handling, Taylor and Francis, Washington, DC
Energy expenditure model	Analysis of lifting, lowering, and carrying	Garg, A., Chaffin, D., and Herrin, G.: Pre- diction of Metabolic Rates for Manual Materials Handling Jobs, American Industrial Hygiene Journal, 38, 661–674 (1978)

handling planning and analysis tools, identifies several tools, their uses and their sources.

Manual Materials Handling Planning

The values in Table 11.2, Design weight limits for lifting and Table 11.3, Design weight limits for carrying provide maximum values based on the "ideal" lift or carry for manual material handling tasks. The "ideal" lift or carry includes ergonomic consideration of all the factors and considerations listed in the section 'Materials handling factors'. The "ideal" lift is a lift of a box with a stable load and good handholds, with the centre of the load placed 150 mm (6 in.) away from the lifter's body, at elbow height, carried to the final destination, less than 10 m (33 ft) away, over a dry deck, and placed on a surface at elbow height. *Personnel lifting limits.* The weight limits in Table 11.2, Design weight limits for Lifting should be used as maximum values in determining the design weight of items requiring one person lifting with two hands. The weight limits can be doubled for two person lifts, provided the load is uniformly distributed between the lifters and both lifters have the same handles or handholds. If the weight of the load is not uniformly distributed, the weight limit applies to the heavier lift point. *Object load size.* The weight limits in Table 11.2, Design weight limits for lifting apply to an object with uniform mass distribution and a compact size not exceeding 460 mm (18 in.) high, 460 mm (18 in.) wide, and 300 mm (12 in.) deep (refer to Fig. 11.1, Ideal object size and carrying mode). This places the centre of the load at half the object's depth, or 150 mm (6 in.) away from the body. If the depth of the object being lifted exceeds 300 mm (12 in.), refer to Table 11.4, Lifting and carrying multipliers for weight limit adjustments. Lifting in the presence of obstacles. The weight limits in Table 11.2, Design weight limits for lifting assume that there are no obstacles between the person lifting and the surface onto which the object is to be placed. If there is an obstacle, such as a lower shelf, refer to Table 11.4, Lifting and carrying multipliers for weight limit adjustments. *Lifting frequency.* The weight limits in Table 11.2, Design weight limits for lifting are not for repetitive lifts as found in the loading or unloading of supply vessels. If the frequency of lifts exceeds one lift in five minutes or 20 lifts per eight hours, refer to Table 11.4, Lifting and carrying multipliers for weight limit adjustments. *Personnel carrying limits.* The weight limits in Table 11.3, Design weight limits for carrying should be used as the maximum value in determining the design weight of items requiring one person carrying of objects. The weight limits can be doubled for two person carries, provided the load is uniformly distributed between the carriers. If the weight of the load is not uniformly distributed, the weight limit applies to the heavier lift point. For carrying activities, it is assumed that the object is first lifted from the floor, carried, and placed on the floor or on another surface not higher than 915 mm (36 in.).

Carrying frequency. The weight limits in Table 11.3, Design weight limits for carrying are not for repetitive carries. If the frequency of carry exceeds one carry in five minutes or 20 carries per eight hours, refer to Table 11.4, Lifting and carrying multipliers for weight limit adjustments. *Object carry size.* The weight limits in Table 11.3, Design weight limits for carrying apply to an object with uniform mass distribution and a compact size not exceeding 460 mm (18 in.) high, 460 mm (18

Table 11.2 Design weight limits for lifting

Handling function	Male and female handlers	All male handlers
Lift an object from the floor and place it on a surface not greater than 1525 mm (60 in.) above the floor	16.8 kg (37 lbs)	20.00 kg (45 lbs)
Lift an object from the floor and place it on a surface not greater than 915 mm (36 in.) above the floor	20.0 kg (45 lbs)	20.0 kg (45 lbs)

Table 11.3 Design weight limits for carrying[a],[b]

Handling function	Male and female handlers	All male handlers
Carry an object 10 m (33 ft) or less	19.0 kg (42 lbs)	20.0 kg (45 lbs)
Carry an object more than 10 m (33 ft):		
Object carried at side with one hand (tool chest, container with handles, etc.)	9.5 kg (21 lbs)	10 kg (22.5 lbs)
Object with irregular sides (electronic equipment chassis, etc.)	11.4 kg (25 lbs)	14.0 kg (35 lbs)
Object or other item with two hands	14.0 kg (35 lbs)	20.0 kg (45 lbs)

[a] Weight limits are based on a North American population
[b] Refer to Table 11.4, Lifting and carrying multipliers for weight limit adjustments

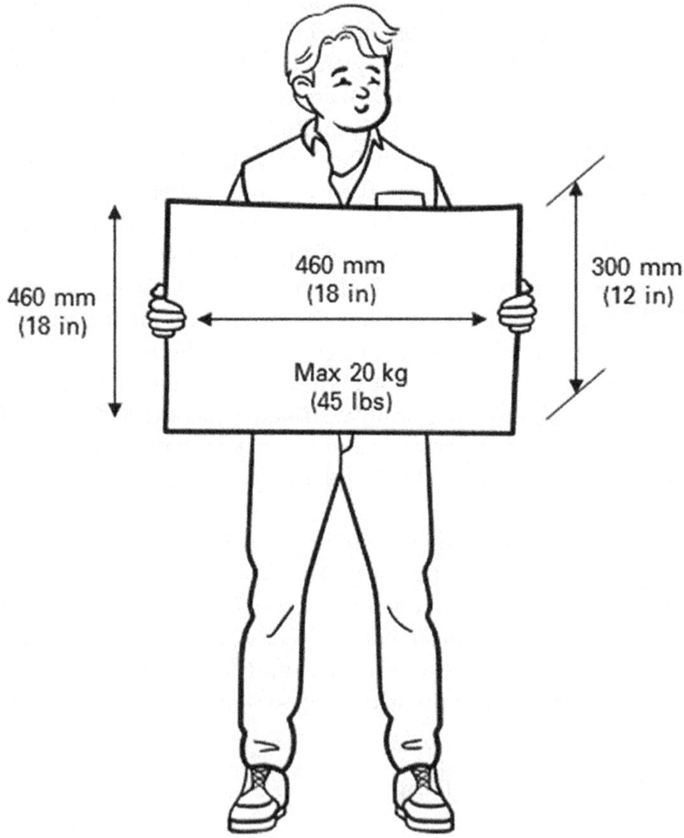

Fig. 11.1 Ideal object size and carrying mode

Table 11.4 Lifting and carrying multipliers

Lifting or carrying situation	Multiplier
Age—for personnel over 50 years old	0.80
Asymmetrical lifting—when a worker must twist the torso 45° or more	0.70
Asian male	0.80
Handholds—in the absence of handles or with poor handholds	0.90
Lifting or carrying frequency—if the frequency of lift or carry exceeds one in five minutes or 20 lifts or carries per eight hours, the weight limits should be reduced by the factor (8.33 × LF)/100, where LF is the lift frequency in lifts per minute. For example, if the lift frequency is 6 lifts per minute, then the maximum permissible weight is reduced by (8.33 × 6)/100, which equates to 0.5	(8.33 × LF)/100
Lifting or carrying load size	
• If the depth of the object exceeds 61 cm (24 in.)	0.66
• If the depth of the object exceeds 91 cm (36 in.)	0.50
• If the depth of the object exceeds 122 cm (48 in.)	0.33
Limited headroom—where personnel must remain "bent" at the waist	0.65
Obstacles—if a lower protruding shelf or other obstacle limits the lifter's approach to the desired surface	0.66
Temperature—for temperatures greater than 32 °C (90°F)	0.88

in.) wide, and 300 mm (12 in.) deep (refer to Fig. 11.1, Ideal object size and carrying mode). This places the handholds at half the depth, or 150 mm (6 in.), away from the body. If the load size exceeds these dimensions, refer to Table 11.4, Lifting and carrying multipliers for weight limit adjustments.

Reducing or eliminating manual materials handling. As mentioned in the section 'Key points related to acceptable materials handling and lifting', where possible, manual materials handling tasks may be reduced or eliminated by avoiding manual materials handling altogether, whenever possible; redesign the load; redesigning the lifting/carrying task; redesigning the working environment; and introducing assisted lifting devices. Assisted lifting aids include devices such as cranes, hoists, counter-balancing mechanisms, trolleys, mono-rails, come-alongs, pad eyes, A-frames, etc. Regardless of the choice, in order to use assisted lifting devices, sufficient space is needed either to permanently install a new lifting device or to temporarily place removable equipment in addition to space needed for the load, its manipulation, and personnel assisting with the materials handling task.

Materials Handling Planning for Assisted Lifts

In cases where large (e.g., 40 kg (90 lbs) or greater) loads are anticipated, it is best if a materials handling study is undertaken during the design stage of a vessel or offshore installation in order to determine which equipment or items are likely to need removal or replacement; the dimensions and weights associated with the equipment or items; the location of the equipment or items within the vessel or installation; the frequency of materials handling for these; the assisted lifting methods and devices, including:

1. Single lift, multi-staged lift, lift to a vessel or dock, etc.
2. Whether devices will be permanently installed or moveable.
3. The best locations for the permanently installed assisted lifting devices.
4. The storage locations for moveable or erectable assisted lifting devices.
5. The availability of assisted lifting devices that must be obtained from outside sources.

It is also important to consider the most appropriate transport methods to be used within the vessel or offshore installation and the transport method off the vessel or offshore installation. In summary, materials handling studies should determine all lifting, carrying, moving, and transporting activities related to operations or maintenance in all modes of operation from regular replenishment activities for food stores or supplies to emergency situations. Where such studies are not performed during the design stage, the same types of factors should be considered within the context of the existing vessel or offshore installation when planning an assisted materials handling task.

Chapter 12
Crew Habitability

General

The principles and guidance of this chapter apply to the living and working conditions aboard vessels and offshore installations. The guidance in this chapter provides target minimum levels. These are consistent with overall values and goals of this book in providing specific assistance while improving human performance. The criteria should not be taken as minimum values for regulatory or classification purposes. Where conflicts exist between guidance provided in this chapter and regulations or rules from the IMO, Flag State Administrations, classification societies, or other regulatory bodies, the regulations or rules take precedence.

Principles

Crew habitability typically comprises five aspects of vessel or offshore installation design and layout. These are the design and layout of accommodations spaces, as well as the ambient environmental characteristics of human whole-body vibration, noise, indoor climate, and lighting. Each of these aspects has the ability to affect performance and quality of life. To help maintain acceptable levels of performance, the following principles apply to the personnel who serve on vessels and offshore installations: (1) maintenance of enhanced levels of mental and physical fitness is important to promote reliable human performance by reducing the potential for fatigue and human errors; (2) extreme levels of whole-body vibration, either low-frequency or high-frequency, can expose the human body to acceleration vectors of varying directions and magnitudes, which can consume energy and induce fatigue; (3) high-noise levels from environmental or mechanical sources, can significantly affect voice communications, audibility of signals and alarms, and the ability to rest; (4) appropriate lighting levels are critical to personnel performance because vision is

A. A. Olsen, *Applying Physical Ergonomics to Modern Ship Design*,
https://doi.org/10.1007/978-3-031-57974-5_12

the dominant sensory channel for receiving information; and (5) indoor climate variables can influence performance when conditions either change or become extreme resulting in fatigue and reduced personnel performance.

Accommodations Spaces

One factor that influences personnel performance is facility design. The design of the accommodations (where personnel sleep, eat, and relax) influences job performance and quality of life. Quality of life is influenced by the location, arrangement, and outfitting of accommodations spaces. Where an owner/operator wishes to gain access to accommodations design guidance or to have crew habitability assessed and/or obtain the Habitability class notation, reference should be made to Part 3 of this book.

Human Whole-Body Vibration

Working and/or living aboard vessels or offshore installations impose a series of low- and high-frequency vibrations, as well as single-impulse shock loads, on personnel. Loads are transmitted through the feet of standing personnel, the buttocks, back, and feet of seated personnel, or the supporting area of recumbent personnel in the frequency range of 0.5–80 Hz. Low-frequency vibrations are created by various sea states, wind actions, vessel speed, manoeuvring, etc. Low-frequency vibrations may result in motion sickness, motion-induced instabilities/ interruptions and/or motion-induced muscle fatigue. Higher-frequency vibrations (up to 80 Hz for those frequencies affecting humans) are generally associated with high-speed rotating machinery. High-frequency vibrations induce corresponding motions and forces within personnel, creating discomfort and resulting in a likely reduction of their performance and health. Other factors that affect the severity of this type of vibration include the size of the vessel or offshore installation, location of personnel, sea conditions, speed, direction of travel, personnel's body posture, duration of exposure and the availability of stabilising, ballasting or vibration-dampening equipment. Prolonged exposure to excessive levels of whole-body vibration can affect work quality, productivity, safety, health, comfort, and can induce motion sickness. Vibration can also modify an operator or maintainer's perception, influence control motions, cause motion-induced instabilities and interruptions, and impair speech. These factors are likely to result in increased reaction and response times, as well as an increase in the probability of human error. Human whole-body vibration levels should be measured in weighted root-mean-square acceleration. For more information on maximum vibration levels and measurement procedures, refer to Part 3 of this book.

Noise

Another factor that can significantly influence personnel performance is noise. Prolonged exposure to noise can interfere with speech, communications, impair concentration, and result in a sensation of annoyance. Owners and operators of vessels and offshore installations control noise levels to conserve their personnel's hearing. However, additional steps should be taken to further reduce noise levels to promote reliable personnel performance, communications, and reduce the potential for human errors. *Hearing conservations.* To prevent the occurrence of potentially hazardous noise levels, vessels and offshore installations should be designed to comply with appropriate maximum recommended exposure limits for noise, such as the IMO Assembly Resolution A.468(XII)—Code on Noise Levels on Board Ships. The maximum recommended noise levels for specific spaces within a vessel or offshore installation should not exceed those indicated in Table 12.1, Maximum recommended noise levels. For those spaces where the maximum noise level is greater than 85 dB(A), personnel should wear hearing protection when in the space. Refer to the section 'Noise exposure limits' for guidance on noise exposure limits and hearing protection.

Noise exposure limits. Permissible daily and occasional noise exposure limits are shown in Fig. 12.1, Permissible noise exposure limits and discussed below:

- *Maximum exposure with protection, Zone A* of Fig. 12.1, Permissible noise exposure limits. No personnel should be exposed to levels exceeding 120 dB(A) or to an $L_{eq}(24)$ (24-h equivalent continuous sound level) exceeding 105 dB(A) even while wearing hearing protectors.
- *Occasional exposure, Zone B* of Fig. 12.1, Permissible noise exposure limits. Only occasional exposure should be allowed in Zone B and both earmuffs and ear plugs should be used unless the exposure duration is restricted to not more than 10 min when only earmuffs or plugs are required.
- *Occasional exposure, Zone C* of Fig. 12.1, Permissible noise exposure limits. In Zone C, only occasional exposures should be allowed and earmuffs or plugs should be required.
- *Daily exposure, Zone D* of Fig. 12.1, Permissible noise exposure limits. If personnel routinely work with daily exposure in spaces with noise levels within Zone D, hearing protectors should be worn.
- *Maximum exposure without protection, Zone E* of Fig. 12.1, Permissible noise exposure limits. For exposures of less than eight hours, personnel without hearing protection should not be exposed to noise levels exceeding 85 dB(A). When personnel remain for more than eight hours in spaces with a high noise level, an $L_{eq}(24)$ of 80 dB(A) should not be exceeded. Consequently, for at least a third of each 24 h the personnel should be subject to an environment with a noise level not exceeding 75 dB(A).
- 24-h *equivalent continuous sound level*, permissible noise exposure limits. As an alternative to Fig. 12.1, Permissible noise exposure limits, the personal exposure to a 24-h equivalent continuous sound level should be no greater than 80 dB(A).

Table 12.1 Maximum permissible noise levels

Space	Noise limit dB(A) maximum
Workspaces	
Machinery space (continuously manned)	90
Machinery space (not continuously manned)	110
Machinery control rooms	75
Workshops	85
Non-specified spaces	90
Navigation spaces	
Navigation bridge and chartroom	65
Listening post, including bridge wings and windows	70
Radio rooms	60
Radar room/areas	65
Accommodation spaces	
Cabins and hospitals	60
Mess rooms and recreation rooms	65
Open recreation areas	75
Offices	65
Service areas	
Galleys	75
Serveries and pantries	75
Normally unoccupied spaces	
Spaces not specified	90

Adapted from IMO Assembly Resolution A.468(XII)

Each individual's daily exposure duration in spaces requiring the use of hearing protectors should not exceed four hours continuously or eight hours in total.

Task performance and comfort noise levels. In addition to conserving personnel's hearing, noise levels can be provided that promote task performance. This is accomplished by ensuring noise levels in a workspace allow it to be used for its intended purpose with minimal interference or annoyance from noise. Additionally, appropriate noise levels for living spaces can be provided that promote relaxation and rest. Task performance and comfort noise criteria for typical vessel and offshore installations spaces are given in Part 3 of this book. These criteria have been selected to improve personnel performance and to facilitate communication, rest and sleep in appropriate vessel or offshore installation spaces. The suggested noise criteria provided herein are given as equivalent continuous A-weighted sound pressure levels (L_{Aeq}) and are expressed in decibels [dB(A)].

Hearing protection and warning information. Hearing protection should be used upon the entrance of all spaces with a noise level of 85 dB(A) and greater. The use

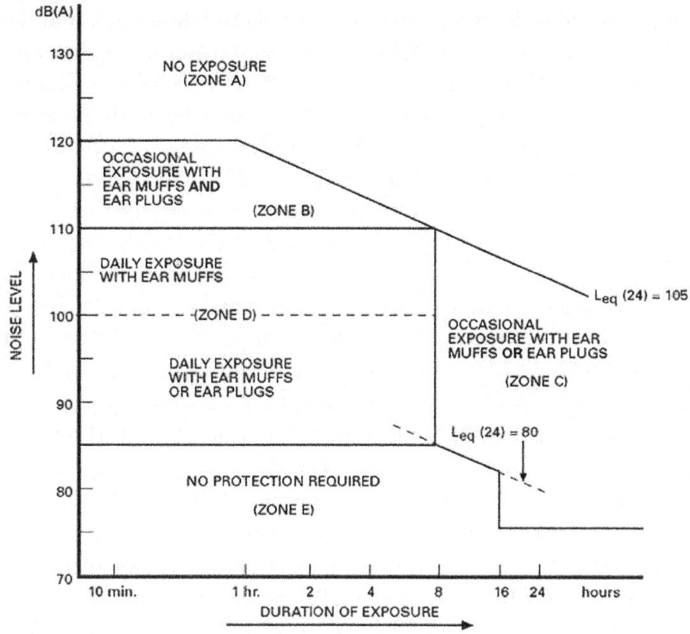

Fig. 12.1 Permissible noise exposure limits

of earplugs, earmuffs, or both is based on the noise level and length of exposure. Hearing protectors should provide at least the following noise attenuation at the ear:

- Ear plugs 20 dB(A).
- Earmuffs 30 dB(A).
- Ear plugs and earmuffs 35 dB(A).

When noise levels exceed 85 dB(A) in any space, a symbol and supplementary warning sign should be posted at each entrance. If hand tools or other portable equipment produce noise levels above 85 dB(A) under normal working conditions, a hazard identification sign should be posted at the work site to identify the hazard and require the use of hearing protection. Further guidance with respect to hazard identification signs is provided in Chap. 8, Labelling, signs, graphics, and symbols.

Indoor Climate

The effect of indoor climate is not perceived by personnel until conditions either change or become extreme. Thermal comfort is subjective, and thus it is difficult to specify an ambient indoor climate that will satisfy everyone. Thermal comfort is largely determined by the interaction of ambient environmental factors such as air temperature, mean radiant temperature, air velocity, humidity, and the personal factors of activity and clothing. The heating, ventilation and air conditioning (HVAC) system on a vessel or offshore installation should be designed to effectively control the indoor climatic variables to within limits. This should be accomplished within the range of external environmental conditions to which the vessel or offshore installation may be subjected. HVAC systems that are adjustable provide a further enhancement for comfort by allowing for a range of personal preferences to be met. Indoor climate criteria and assessment methodologies are included in Part 3 of this book. *Air temperature, humidity, thermal gradients and air velocity criteria.* Guidance for desired levels of air temperature, humidity, thermal gradient and air velocity are given in Part 3 of this book. This is provided for personnel wearing typical indoor clothing occupied in light, primarily sedentary activities and resulting in a thermal environment acceptable to at least 80 percent of the occupants. *Ventilation.* Adequate ventilation should be assured by introducing fresh air into all normally manned spaces. For small spaces (e.g., enclosed volume of 4.25 m^3 (150 ft^3) or less per person), a minimum of 0.85 m3 (30 ft3) of ventilation air-per-minute-per-person should be provided with approximately two thirds of the volume being outside air. For larger enclosures, the air supply-per-person should be in accordance with the curves shown in Fig. 12.2, Large enclosure ventilation requirements. The rate of air change for enclosed spaces should be at least six complete changes-per-hour. Care should be taken to provide outside air from a location away from process equipment, hazardous areas, exhausts, contamination sources or sources of odour. Ventilation should not produce air velocities exceeding 30 m/minute (100 ft/minute), measured on average. An exception would be in work locations where spot cooling of personnel is provided. In these cases, air should move past personnel at a velocity of less than 60 m/minute (200 ft/minute). In areas where manuals or other loose papers are used, the air velocity preferably should be 20 m/minute (65 ft/minute) but no more than 30 m/minute (100 ft/minute). Engine room ventilation should comply with ISO 8861, Engine room ventilation in diesel engine ships—Design requirements and basis of calculations, and ISO 8862, Air conditioning and ventilation of machinery control—rooms on board ships—Design conditions and basis of calculations. To reduce potential health hazards caused by exposure to oil and inhalation of oil mist, consideration is to be given to the installation, location, and ventilation of equipment, which may produce oil vapour and mist. Intakes for ventilation systems should be located to minimise the introduction of contaminated air from sources such as exhaust pipes and incinerators.

Lighting. Sufficient lighting is required for effective and efficient personnel performance. For most tasks, vision is the main sense for receiving information and lighting is therefore a critical element in design. The lighting of work and accommodations spaces should facilitate visual task performance and the movement of people in the space and enhance the visual environment. Lighting design integrates adequate lighting for safe working, well-being of personnel and for the tasks encountered on the vessel or offshore installation. Visual discomfort can impair visual performance. Visual tasks encountered on vessels and offshore installations vary widely. In addition to the lighting level, other factors such as contrast with the background, reflectance,

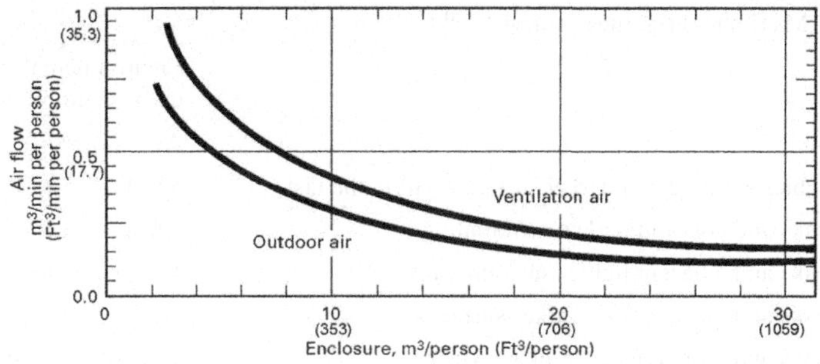

Fig. 12.2 Large enclosure ventilation requirements

object size, brightness and the time available for viewing or recognition help to determine the visibility of an object within the visual field. Consideration should also be given to task duration, visual fatigue, task criticality, glare, reflections, shadows and the age and visual acuity of the likely observer. Aesthetics, colour and the psychological effects of lighting should also be considered. These factors are all interrelated, so no single aspect of lighting should be considered in isolation. *Lighting design.* Lighting should be designed and located to avoid glare from working and display surfaces, as viewed from the normal working position. The maximum luminance ratio between any two different sources of luminance light within an operator or maintainer's field of view should not exceed 5:1. To reduce glare, non-reflective or matte finished surfaces should be provided on consoles, panels, and other work surfaces. Placement of smooth, highly polished surfaces within 60° of a person's normal visual field should be avoided. When night vision is required for extended periods of time, low-level white lighting is preferred to low-level red lighting. Red lighting should not be used in areas where colour recognition is required. Brightness. The brightness ratio between the lightest and darkest areas or between a task area (e.g., computer screen) and its surroundings (e.g., bulkhead behind the screen) should be no greater than specified in Table 12.2, Maximum brightness ratios. *Reflectance.* Surface areas in working and living areas should be covered with non-saturated light colours such as tints, pastels, whites, and light shades of grey. Preferred surface reflectance values for spaces such as offices, control rooms, and maintenance areas are shown in Fig. 12.3, Surface reflectance values. *Lighting criteria.* For further criteria on lighting, refer to Part 3 of this book. *Emergency lighting criteria.* Emergency lighting requirements exist in Statutory, Class and other regulatory documents, the IMO Convention for the Safety of Life at Sea (SOLAS), and the *Mobile Offshore Drilling Unit (MODU) Code.* As a result, we will not discuss the criteria related to emergency lighting in this book. Readers are strongly advised to consult the SOLAS, MODU and appropriate Class rules specific to their vessel or installation.

Table 12.2 Maximum brightness ratios[a,b,c,d,e]

	Environmental classification		
Comparison	A	B	C
Between lighter surfaces and darker surfaces within the task	5 to 1	5 to 1	5 to 1
Between tasks and adjacent darker surroundings	3 to 1	3 to 1	5 to 1
Between tasks and adjacent lighter surroundings	1 to 3	1 to 3	1 to 5
Between tasks and more remote darker surfaces	10 to 1	20 to 1	b
Between tasks and more remote lighter surfaces	1 to 10	1 to 20	b
Between luminaries and adjacent surfaces	20 to 1	b	b
Between the immediate work area and the rest of the environment	40 to 1	b	b

[a] Adopted from DOT/FAA/CT-96/1—Human Factors Design Guide
[b] Interior areas where reflectance of entire space can be controlled for optimum visual conditions
[c] Areas where reflectance of nearby work can be controlled, but there is only limited control over remote surroundings
[d] Areas (indoor and outdoor) where it is completely impractical to control reflectance and difficult to alter environmental conditions
[e] Brightness ratio control is not practical

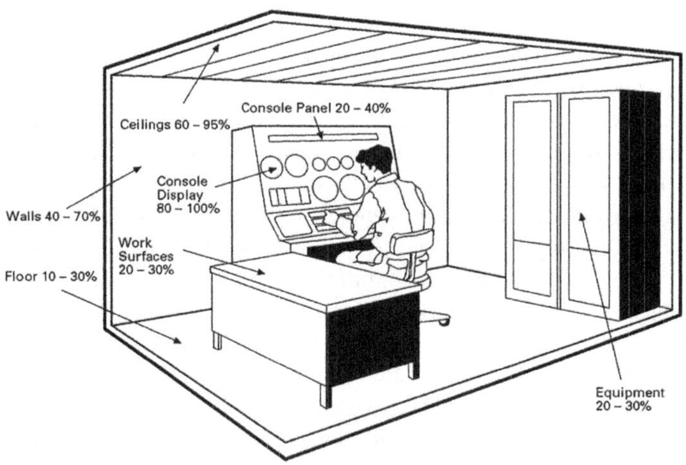

Fig. 12.3 Surface reflectance values

Chapter 13
Anthropometric Design Principles and Dimensions

General

The principles and data in this chapter apply to the sizing of equipment, accesses, and clearances to provide physical compatibility of the human body (in terms of physical dimensions) and design of interfaces such as: hardware, chairs, hatches, doorways, consoles, panels, and clothing. Sizing of these interfaces is usually done within the context of performing general tasks, such as reaching a control, viewing over a console, sitting in a chair, or exiting through a hatch. Throughout this Part 1 we have been presented with extensive guidance that relates to anthropometric dimensions, and specifically, in respect to Chap. 5, Integration of controls, displays, and alarms; Chap. 6, Computer workstation design; Chap. 7, Manual valve operation, access, location, and orientation; Chap. 9, Stairs, ladders, ramps, walkways, platforms and hatches; and Chap. 10, Maintenance considerations. Building on what we have already discussed in these previous chapters, much of the anthropometric guidance presented in the main body of this chapter is applicable to the size and strength characteristics of North American males. Since this book is intended to have international appeal, this chapter is provided so that the main body anthropometric guidance can be modified for use in designs to be operated and maintained by peoples of other regions with different body dimensions. The science of anthropometry focuses on the measurement of human variability of body dimensions and strength as a function of gender, race, and regional origin. The application of anthropometry to design establishes limits (or boundary conditions) for sizing equipment for human use. In essence, it defines size limits in design based on the dimensions of the anticipated population of operating and maintenance personnel. By imposing size limits in design (e.g., designing so the shortest expected operator or maintainer can reach all controls), it follows that personnel who are less demanding in their requirements will also be accommodated (e.g., have greater reach than the limiting personnel). For any body dimension, the 5th percentile value indicates that 5% of the population will be equal to or smaller than that value, and 95% will be larger. On

A. A. Olsen, *Applying Physical Ergonomics to Modern Ship Design*,
https://doi.org/10.1007/978-3-031-57974-5_13

the other hand, the 95th percentile value indicates that 95% of the population will be equal to or smaller than that value, and 5% will be larger. Therefore, use of a design range from the 5th to the 95th percentile (for either male or female populations, but not both) values will theoretically provide coverage for 90% of that (male or female) population using those limiting dimensions, and only those smaller than the 5%, and larger than the 95% will be excluded by design. Using limiting dimensions for males and females (5th percentile female and 95th percentile male) will accommodate approximately 94% of the entire design population (since over 99% of males are larger than the 5th percentile female, and over 99% of females are smaller than the 95th percentile male, so few small males, or large females, are excluded). Note that the notion of the "average person" is misleading in that an individual will vary among different anthropometric dimensions. For example, individuals who are of average (50%) stature, can be comparatively smaller or larger on other dimensions, such as arm length.

Principles

In general, there are four principles of applied anthropometrics in design. These are:

1. Design for the smallest.
2. Design for the largest.
3. Design for the average.
4. Design for the range.

Design for the smallest. This principle applies primarily to application of physical force and vertical and lateral reach distances. For example, the forces required to pull, push, or turn a handle. Usually, the maximum force that can be readily applied by the 5th percentile person for that movement is used as the criterion. Similarly, the reach of the 5th percentile person is often used as the criterion. Design for the largest. This principle applies primarily to clearances, such as escape hatches, maintenance accesses, lifeboats, walkways, and overhead clearances. Clearances generally are such that at least 95%of the expected population is accommodated. In some cases, persons whose body size exceeds the designed clearances are precluded from selection for the system. Design for the average. This principle applies to workstations that are not adjustable (e.g., fixed height tables, desks, or other work surfaces). In these situations, designing for the average person better accommodates the entire population. Design for the range. This principle is applied to determining the amount of adjustability that should be built into such things as variable height work surfaces and workstation seating (e.g., horizontal and vertical adjustability). In general, the dimension criteria used for designing adjustability readily accommodates the middle 90% of the population.

Anthropometric Data

The guidance and data provided herein can be used in designing human-system interfaces which are required to be "easily accessible", "readily operable", "based on sound ergonomic principles", and similar phrases used extensively in maritime and industry related documentation. The following summarises the figures and tables containing anthropometric data.

- Table 13.1, Differences in 50th percentile male stature for 12 regions, in mm and in, from the average stature of 1717 mm (67.6 in).
- Table 13.2, Differences in 50th percentile female stature for 11 regions, in mm and in, from the average stature of 1593 mm (62.7 in).
- Table 13.2, Essential international anthropometric data for females.

The remaining tables present three critical anthropometric dimensions for each of these twenty regions. Also note that Fig. 13.3, Anthropometric illustrations provides a graphical representation of the various measurement points associated with the anthropometric dimensions provided in the following tables:

- Table 13.3, Anthropometric geographical regions define 20 geographic regions.
- Table 13.4, Standing height dimensions.
- Table 13.5, Sitting Eye height dimensions.

Clothing and Postural Effects

The effect of clothing, special equipment, and PPE should be accounted for in the design for the operation and maintenance of a vessel or offshore installation (Table 13.6). Most anthropometric data have been gathered from minimally clothed or nude subjects. Clothing, special equipment, and PPE can affect clearances, ranges of motion, reach envelopes, dexterity, mobility, strength, tactile sensitivity, and grasping capability. Given this, the effect of shoes, boots, hard hats, coats, gloves, other "standard" clothing, respirators, firefighting equipment, and barrier suits, as applicable, should be accommodated by design. Postural effects should also be accommodated by design, in particular postural slump. Non-specific clothing and posture slumping data is provided in Table 13.7, Clothing and postural effects. Modifications to clothing, special equipment, and PPE allowances may be needed depending on the particular design and should be measured and accounted for before design.

Table 13.1 Essential international anthropometric data for males[a,b,c,d]

Males (mm and in)	North America (US)			Japan			Southeast Asia			Europe		
	5%	50%	95%	5%	50%	95%	5%	50%	95%	5%	50%	95%
Stature (floor to top of head)	1644	1760	1876	1593	1687	1781	1530	1630	1720	1604	1728	1847
	64.7	69.3	73.9	62.7	66.4	70.1	60.2	64.2	67.7	63.1	68.0	72.7
Seated stature (buttocks to top of head)	862	923	984	855	913	970	790	840	900	836	894	959
	33.9	36.3	38.7	33.7	35.9	38.2	31.1	33.1	35.4	32.9	35.2	37.8
Shoulder breadth (deltoid to deltoid)	449	509	569	413	448	484	380	410	430	420	456	496
	17.7	20.0	22.4	16.3	17.6	19.1	15.0	16.1	16.9	16.5	18.0	19.5
Standing eye height	1529	1643	1758	1474	1568	1661				1509	1613c	1720c
	60.2	64.7	69.2	58.0	61.7	65.4				59.5c	63.4c	67.7c
Seated eye height (buttocks to eye)	733	793	853	726	775	832	680	730	780	740	796d	851d
	28.9	31.2	33.6	28.6	30.5	32.8	26.8	28.7	30.7	29.1d	31.3d	33.5d
Back of knee height (approximate maximum seat height)	404	450	496	369	408	447	380	415	445	425	473	521
	15.9	17.7	19.5	14.5	16.1	17.6	15.0	16.3	17.5	16.7	18.6	20.5
Forward functional reach (back of shoulder to fingertips)	842	909	975	758	820	882	730	780	820	711d	771d	827d
	33.1	35.8	38.4	29.8	32.3	34.7	28.7	30.7	32.3	28.0d	30.4d	32.6d
Forward functional reach (back of shoulder to fingertips)	842	909	975	758	820	882	730	780	820	711d	771d	827d
	33.1	35.8	38.4	29.8	32.3	34.7	28.7	30.7	32.3	28.0d	30.4d	32.6d
Standing overhead reach (floor to tip of extended middle finger)	2067	2237	2407	1959	2121	2282						
	81.4	88.1	94.8	77.1	83.5	89.8						
Arm length (fingertip to shoulder)	737	796	856	682	735	787				711	771	827
	29.0	31.3	33.7	26.9	28.9	31.0				28.0	30.4	32.6

(continued)

Table 13.1 (continued)

Males (mm and in)	North America (US)			Japan			Southeast Asia			Europe		
	5%	50%	95%	5%	50%	95%	5%	50%	95%	5%	50%	95%
Hip breadth	312	372	432	299	328	357				310c	344c	368c
	12.3	14.6	17.0	11.8	12.9	14.1				12.2c	13.5c	14.5c

[a] Represented by Brunei, Indonesia, Laos, Malaysia, Philippines, Singapore, Thailand, and Vietnam. Source: International Data on Anthropometry. International Labour Organisation, 1990. The data in this source dates to the middle 1960s and may be out of date. For males, regional growth in the area since that time is approximately 29 mm in stature, or an increase of about 2%. Given this, users of the Southeast Asia data for males may consider scaling stature, leg length, and reach dimensions by a factor of 1.02. The data contained in the above table are true to the source document

[b] Represented by data from Sweden, Germany, Italy, and France. Source: ADULTDATA: The Handbook of Adult Anthropometric and Strength Measurements—Data for Design Safety. Institute for Occupational Ergonomics, University of Nottingham, UK. 1998

[c] German population data

[d] French population data

Adapting Data for Various Populations

To illustrate how the data in this chapter can be used, the following examples are provided.

Example 1—Specifying overhead clearances in a hard hat area. Design of the decks on an offshore installation is underway and the walking overhead clearance needs to be specified. There will be overhead pipes, pipe hangers, cable trays, and ductwork in this particular area. The area will require the wearing of hard hats and steel-toed shoes. Personnel for this offshore installation are expected to predominantly be Northern European men. Chap. 9, Stairs, ladders, ramps, walkways, platforms, and hatches, states that … "headroom in all walkways should be ≥ 2130 mm (84 in)". To specify (or verify) a walking overhead clearance for 95th percentile Northern European men:

- Identify a regional population of expected users. In this case it will be Northern Europeans.
- Select a percentage of the population to be accommodated, thereby establishing a preliminary limiting dimension. In this case, the 95th percentile male is chosen, since overhead clearance is a safety concern, and selecting this will protect 95% of be Northern European men, and virtually all females. This limiting dimension is 1910 mm (75.2 in) according to Table 13.4, Standing height dimensions.
- Identify clothing allowances, in this case, steel-toed shoes and hard hat. For footwear, there is an allowance of 25 mm (1 in), and for the hard hat, there is an allowance of an additional 75 mm (3.0 in). These allowances are added to the limiting dimension, and thus stature becomes 2010 mm (79.2 in).
- Allow for the dynamic characteristics of walking and starting and a clearance safety factor.

Table 13.2 Essential international anthropometric data for females[a,b,c,d]

Females (mm and in)	North America (US)			Japan			Southeast Asia			Europe		
	5%	50%	95%	5%	50%	95%	5%	50%	95%	5%	50%	95%
Stature (floor to top of head)	1512	1627	1742	1474	1558	1641	1440	1530	1620	1500	1610	1710
	59.5	64.1	68.6	58.0	61.3	64.6	56.7	60.2	63.8	59.1	63.4	67.3
Seated stature (buttocks to top of head)	802	861	920	793	848	902	750	800	850	794	852	904
	31.6	33.9	36.2	31.2	33.4	35.5	29.5	31.5	33.5	31.3	33.5	35.6
Shoulder breadth (deltoid to deltoid)	381	474	567	376	409	441	340	380	410	366	401	447
	15.0	18.7	22.3	14.8	16.1	17.4	13.4	15.0	16.1	14.4	15.8	17.6
Standing eye height	1406	1517	1628	1370	1448	1527				1402	1502c	1596c
	55.4	59.7	64.1	53.9	57.0	60.1				55.1c	59.1c	62.8c
Seated eye height (buttocks to eye)	677	735	794	679	721	764	660	700	740	703	751d	801d
	26.7	28.9	31.3	26.7	28.4	30.1	26.0	27.6	29.1	27.7d	29.6d	31.5d
Back of knee height (approximate maximum seat height)	354	400	446	331	362	393	365	385	405	380	431	481
	13.9	15.7	17.6	13.0	14.3	15.5	14.4	15.2	15.9	15.0	17.0	18.9
Forward functional reach (back of shoulder to fingertips)	746	811	876	696	753	810	690	730	780	652d	704d	761d
	29.4	31.9	34.5	27.4	29.6	31.9	27.2	28.7	30.7	25.7d	27.7d	30.0d
Forward functional reach (back of shoulder to fingertips)	1890	2057	2225	1796	1935	2075						
	74.4	81.0	87.6	70.7	76.2	81.7						
Standing overhead reach (floor to tip of extended middle finger)	649	718	788	613	660	706				652	704	761
	25.6	28.3	31.0	24.1	26.0	27.8				25.7	27.7	30.0

(continued)

Table 13.2 (continued)

Females (mm and in)	North America (US)			Japan			Southeast Asia			Europe		
	5%	50%	95%	5%	50%	95%	5%	50%	95%	5%	50%	95%
Arm length (fingertip to shoulder)	307	404	500	291	316	342				314c	358c	405c
	12.1	15.9	19.7	11.5	12.4	13.5				12.4c	14.1c	15.9c
Hip breadth	1512	1627	1742	1474	1558	1641	1440	1530	1620	1500	1610	1710
	59.5	64.1	68.6	58.0	61.3	64.6	56.7	60.2	63.8	59.1	63.4	67.3

[a] Represented by Brunei, Indonesia, Laos, Malaysia, Philippines, Singapore, Thailand, and Vietnam. Source: International Data on Anthropometry. International Labour Organisation, 1990. The data in this source dates to the middle 1960's and may be out of date. For males, regional growth in the area since that time is approximately 29 mm in stature, or an increase of about 2%. Given this, users of the Southeast Asia data for males may consider scaling stature, leg length, and reach dimensions by a factor of 1.02. The data contained in the above table are true to the source document

[b] Represented by data from Sweden, Germany, Italy, and France. Source: ADULTDATA: The Handbook of Adult Anthropometric and Strength Measurements—Data for Design Safety. Institute for Occupational Ergonomics, University of Nottingham, UK. 1998

[c] German population data

[d] French population data

- Walking, starting, and some gaits are typically assumed to add a variability of approximately 50 mm (2 in). Adding a safety clearance factor is a matter of assumption and judgment. In this case, a factor of 75 mm (3 in) is reasonable. Adding these, the minimum overhead clearance should be approximately 2135 mm (84 in), providing a dynamic walking clearance of approximately 75 mm (3 in) for 95% Northern European males and over 99% of be Northern European females wearing steel-toed footwear and hard hat.
- Specify a minimum overhead clearance of 2135 mm (84 in) for all areas where personnel are expected to walk (without crouching or slumping). That minimum includes potential overhead hazards such as the bottom of a pipe, cable tray, or their hangers, lever-operated valves in any position, hanging wires, etc.

As a matter of interest, 2.1 m (7 ft) is a typical overhead clearance specified internationally. Some points to make in the above sample:

- Many of the dimensions are easily specified based on the tables in this chapter. However, it is important to select an appropriate group as even adjacent regions can offer very different anthropometric values. Looking at 95th percentile European males in Table 13.1, Essential international anthropometric data for males, reveals a stature of 1847 mm (72.7 in). Looking at 95th percentile Northern European males in Table 13.4, Standing height dimensions, reveals a stature of 1910 mm (75.2 in): a difference of 63 mm (2.5 in),
- Frequently, assumptions need to be made, as in the example above, for a clearance safety factor. Safety factors need to be identified and considered in light of safety risks. For example, risk associated with a taller person having to bend at the waist

Table 13.3 Anthropometric geographical regions

Geographic region	Countries included in the geographic region
Northern Europe	Denmark, Finland, Germany, Iceland, Netherlands, Norway, Sweden
North America	United States, Canada
Australia	Australia, New Zealand (European population)
France	France
Central Europe	Austria, Belgium, Eastern Germany, Czech Republic, Slovakia, Switzerland, United Kingdom, Luxembourg
Eastern Europe	Poland, Russian Federation
Latin America	Argentina, Belize, Brazil, Chile, Costa Rica, Guyana, Surinam, Uruguay, Caribbean Island States (European and Negroid populations)
Iberian Peninsula	Portugal, Spain
Southeast Europe	Bulgaria, Greece, Hungary, Italy, Israel, Malta, Romania, Slovenia, Croatia, Bosnia and Herzegovina, Serbia and Montenegro
Japan	North Korea, South Korea, Japan
North Asia	China (Northern), Mongolia, Russian Federation
North Africa	Algeria, Chad, Egypt, Ethiopia, Libya, Mali, Morocco, Niger, Sudan, Tunisia
Near East	Afghanistan, Bahrain, Iraq, Iran, Jordan, Kuwait, Lebanon, Oman, Yemen, Saudi Arabia, Syria, Turkey, United Arab Emirates
West Africa	Benin, Cameroon, Congo, Guinea, Gabon, Ghana, Liberia, Nigeria, Sierra Leone, Zaire
Southeast Africa	Kenya, Malawi, Madagascar, Mozambique, Rwanda, Somalia, South Africa, Uganda, Tanzania, Zambia, Zimbabwe
North India	Bangladesh, India (Northern), Nepal, Pakistan
South India	India (Southern), Maldives, Sri Lanka
Southeast Asia	Brunei, Indonesia, Laos, Malaysia, Philippines, Singapore, Thailand, Vietnam
Latin America (Indian population)	Bolivia, Columbia, Ecuador, El Salvador, Guatemala, Honduras, Mexico, Nicaragua, Panama, Peru, Paraguay, Venezuela
South China	China (Southern), Hong Kong, Macao, Taiwan

to access a control situated to accommodate shorter personnel pales to the risk of taller personnel moving rapidly in areas with marginal overhead clearances.

- Dynamic aspects of motion need to be taken into account. Anthropometric data are often based on static values, with measurements taken under strictly adhered to postures and procedures. Dynamic characteristics are much harder to quantify, and therefore assumptions need to be made.

Table 13.4 Standing height dimensions[a,b]

Geographic region	Male		Female		Min	Max
	5%	95%	5%	95%		
Northern Europe	1710 mm (67.3 in)	1910 mm (75.2 in)	1580 mm (62.2 in)	1791 mm (70.5 in)	1580 mm (62.2 in)	1910 mm (75.2 in)
North America	1664 mm (64.7 in)	1876 mm (73.9 in)	1510 mm (59.5 in)	1742 mm (68.6 in)	1510 mm (59.5 in)	1876 mm (73.9 in)
Australia	1661 mm (65.4 in)	1890 mm (74.4 in)	1560 mm (61.4 in)	1770 mm (69.7 in)	1560 mm (61.4 in)	1890 mm (74.4 in)
France	1661 mm (65.4 in)	1890 mm (74.4 in)	1530 mm (60.2 in)	1740 mm (68.5 in)	1530 mm (60.2 in)	1860 mm (74.4 in)
Central Europe	1669 mm (65.7 in)	1860 mm (3.2 in)	1550 mm (61.0 in)	1750 mm (68.9 in)	1550 mm (61.0 in)	1860 mm (73.2 in)
Eastern Europe	1661 mm (65.4 in)	1849 mm (72.8 in)	1539 mm (60.6 in)	1720 mm (67.7 in)	1539 mm (60.6 in)	1849 mm (72.8 in)
Latin America	1651 mm (65.0 in)	1849 mm (72.8 in)	1519 mm (9.8 in)	1720 mm (67.7 in)	1519 mm (59.8 in)	1849 mm (72.8 in)
Iberian Peninsula	1,580 mm (62.2 in)	1,829 mm (72.0 in)	1,509 mm (59.4 in)	1,699 mm (66.9 in)	1,509 mm (59.4 in)	1,829 mm (72.0 in)
Southeast Europe	1687 mm (66.4 in)	1829 mm (72.0 in)	1529 mm (70.2 in)	1720 mm (67.7 in)	1529 mm (60.2 in)	1829 mm (72.0 in)
Japan	1593 mm (62.7 in)	1781 mm (70.1 in)	1474 mm (58.0 in)	1641 mm (64.6 in)	1474 mm (58.0 in)	1781 mm (70.1 in)
North Asia	1560 mm (61.4 in)	1821 mm (71.7 in)	1501 mm (59.1 in)	1669 mm (65.7 in)	1501 mm (59.1 in)	1821 mm (71.7 in)
North Africa	1580 mm (62.2 in)	1811 mm (71.3 in)	1501 mm (59.1 in)	1720 mm (67.7 in)	1501 mm (59.1 in)	1811 mm (1.3 in)
Near East	1621 mm (63.8 in)	1801 mm (70.9 in)	1539 mm (60.6 in)	1699 mm (66.9 in)	1539 mm (60.6 in)	1801 mm (70.9 in)
West Africa	1560 mm (61.4 in)	1791 mm (70.5 in)	1440 mm (56.7 in)	1621 mm (63.8 in)	1440 mm (56.7 in)	1791 mm (70.5 in)
Southeast Africa	1590 mm (62.6 in)	1781 mm (70.1 in)	1481 mm (58.3 in)	1662 mm (65.4 in)	1481 mm (58.3 in)	1781 mm (70.1 in)
North India	1580 mm (62.2 in)	1770 mm (69.7 in)	1450 mm (57.1 in)	1631 mm (64.2 in)	1450 mm (57.1 in)	1770 mm (69.7 in)
South India	1529 mm (60.2 in)	1720 mm (67.7 in)	1389 mm (54.7 in)	1600 mm (63.0 in)	1389 mm (54.7 in)	1720 mm (67.7 in)
Southeast Asia	1529 mm (60.2 in)	1720 mm (67.7 in)	1440 mm (56.7 in)	1620 mm (63.8 in)	1440 mm (56.7 in)	1720 mm (67.7 in)

(continued)

Table 13.4 (continued)

Geographic region	Male		Female		Min	Max
	5%	95%	5%	95%		
Latin America (Indian population)	1519 mm (59.8 in)	1709 mm (67.3 in)	1389 mm (54.7 in)	1560 mm (61.4 in)	1389 mm (54.7 in)	1709 mm (67.3 in)
South China	1610 mm (63.4 in)	1709 mm (67.3 in)	1430 mm (56.3 in)	1590 mm 62.6 in)	1430 mm (56.3 in)	1709 mm (67.3 in)

[a] International Data on Anthropometry. International Labour Organisation, 1990
[b] ADULTDATA: The Handbook of Adult Anthropometric and Strength Measurements—Data for Design Safety. Institute for Occupational Ergonomics, University of Nottingham, UK. 1998

Example 2—Design a door label mounting height. There is a general need to identify spaces and equipment on vessels and offshore installations, to accommodate new personnel, visitors, technical representatives, contractors, and the like. In Chap. 8, the section Room or space identification labels states … "Identification labels should be mounted on the door leading into a room or space. Labels should be placed at or slightly below personnel's standing eye height. For 50th percentile North American male personnel, the label would be located approximately 1650 mm (65 in) above the standing surface. To specify an effective compartment/space entrance labelling height suitable for a different population of personnel, the following provides an example of how to determine where to situate a door label:

- Identify a regional population of expected uses. In this case it will be Japanese, since the vessel under design is a Japanese fishing vessel.
- Select a percentage of the population to be accommodated, thereby establishing a preliminary labelling height above the deck. In this case, the 50th percentile male is chosen, since there are no safety concerns associated with extreme dimensions (5th or 95th percentiles) in this task and choosing the 50th percentile will accommodate the greatest number of personnel, who are expected to be, for the most part, Japanese males.
- According to Table 13.1, Essential international anthropometric data for males, the standing eye height of the target population is 1568 mm (61.7 in).
- Identify clothing allowances, in this case, shoes: an allowance of 25 mm (1 in).
- There are no dynamic characteristics for which to account, and there are no safety margins that need to be considered, therefore, door label heights should be approximately 1593 mm (63 in).

Example 3—Reassessment of water survival craft seat width. The *International Life-Saving Appliance Code* [IMO Resolution MSC.48(66)] specifies anthropometric requirements for seating on lifeboats. The objective of the IMO in generating these requirements was to optimise seating on lifeboats to anthropometrically accommodate as many passengers as possible, while maximising the human payload capacity. The IMO Resolution specifies a seat accommodation of 215 mm (17 in). This seat size coincides with the 95th percentile North American male hip width as reported by AdultData (1998) (presumably, the IMO figure is based on the 95th percentile

Table 13.5 Sitting height dimensions[a,b]

Geographic region	Male		Female		Min	Max
	5%	95%	5%	95%		
Northern Europe	770 mm (30.3 in)	871 mm (34.3 in)	711 mm (28.0 in)	818mm (32.3 in)	711mm (28.0 in)	871mm (34.3 in)
North America	733 mm (28.9 in)	853 mm (33.6 in)	677 mm (26.7 in)	794mm (31.3 in)	677mm (26.7 in)	853mm (33.6 in)
Australia	760 mm (29.9 in)	861 mm (33.9 in)	711 mm (28.0 in)	810mm (31.9 in)	711 mm (28.0 in)	861 mm (33.9 in)
France	749mm (29.5 in)	851mm (33.5 in)	691mm (27.2 in)	780mm (30.7 in)	691mm (27.2 in)	851mm (33.5 in)
Central Europe	749 mm (29.1 in)	871 mm (33.5 in)	710 mm (27.6 in)	820 mm (31.9 in)	701 mm (27.6 in)	851 mm (33.5 in)
Eastern Europe	729 mm (28.7 in)	851 mm (33.5 in)	671 mm (26.4 in)	790 mm (31.1 in)	671 mm (26.4 in)	861 mm (33.5 in)
Latin America	749 mm (29.5 in)	871 mm (34.3 in)	710 mm (27.6 in)	820 mm (32.3 in)	701 mm (27.6 in)	871 mm (34.3 in)
Iberian Peninsula	719 mm (28.3 in)	851 mm (33.5 in)	681 mm (26.8 in)	810 mm (31.9 in)	681 mm (26.8 in)	851 mm (33.5 in)
Southeast Europe	739 mm (29.1 in)	841 mm (33.1 in)	681 mm (26.8 in)	780 mm (30.7 in)	681 mm (26.8 in)	841 mm (33.1 in)
Japan	726 mm (28.6 in)	832 mm (32.8 in)	679 mm (26.7 in)	764 mm (30.1 in)	679 mm (26.7 in)	832 mm (32.8 in)
North Asia	739 mm (29.1 in)	851 mm (33.5 in)	701 mm (27.6 in)	790 mm (31.1 in)	701 mm (27.6 in)	851 mm (33.5 in)
North Africa	701 mm (27.6 in)	820 mm (32.3 in)	681 mm (26.8 in)	810 mm (31.9 in)	681 mm (26.8 in)	820 mm (32.3 in)
Near East	729 mm (28.7 in)	820 mm (32.3 in)	711 mm (28.0 in)	790 mm (31.1 in)	711 mm (28.0 in)	820 mm (32.3 in)
West Africa	650 mm (25.6 in)	780 mm (30.7 in)	620 mm (24.4 in)	620 mm (24.4 in)	620 mm (24.4 in)	780 mm (30.7 in)
Southeast Africa	706 mm (27.8 in)	790 mm (31.1 in)	681 mm (26.8 in)	780 mm (30.7 in)	681 mm (26.8 in)	790 mm (31.1 in)
North India	711 mm (28.0 in)	800 mm (31.5 in)	650 mm (25.6 in)	729 mm (28.7 in)	650 mm (25.6 in)	800 mm (31.5 in)
South India	660 mm (26.0 in)	739 mm (29.1 in)	620 mm (24.4 in)	719 mm (28.3 in)	620 mm (24.4 in)	739 mm (29.1 in)
Southeast Asia	681 mm (26.8 in)	780 mm (30.7 in)	660 mm (26.0 in)	739 mm (29.1 in)	660 mm (26.0 in)	780 mm (30.7 in)

(continued)

Table 13.5 (continued)

Geographic region	Male		Female		Min	Max
	5%	95%	5%	95%		
Latin America (Indian population)	709 mm (28.3 in)	800 mm (31.5 in)	660 mm (26.0 in)	739 mm (29.1 in)	660 mm (26.0 in)	800 mm (31.5 in)
South China	691 mm (27.2 in)	790 mm (31.1 in)	650 mm (25.6 in)	739 mm (29.1 in)	650 mm (25.6 in)	790 mm (31.1 in)

[a] International Data on Anthropometry. International Labour Organisation, 1990
[b] ADULTDATA: The Handbook of Adult Anthropometric and Strength Measurements—Data for Design Safety. Institute for Occupational Ergonomics, University of Nottingham, UK. 1998

hip width of an international population and was not based on North American male dimensions). According to *Safety Alert No. 192, Water Survival Craft* (January 2001) of the United States Minerals Management Service, it was later noted that in the Gulf of Mexico offshore workers were too large to fit in lifeboats. Field anthropometric measurements of Gulf of Mexico offshore workers revealed an appropriately clothed hip width of 533 mm (21 in): 100 mm (4 in) greater than the design specification. This deviation of specified seat width compared to the hip width of Gulf of Mexico offshore worker suggests that lifeboat manufacturers would overestimate maximum lifeboat occupancy in the Gulf of Mexico by about 20%. Similarly, the *International Life Saving Appliance Code* specifies an average weight of 75 kg (165 lbs) for sizing lifeboats. In the Gulf of Mexico, the observed average weight was 95 kg (210 lbs), a difference of 20 kg (44 lbs), potentially affecting buoyancy and stability. This example makes some strong points about the use of anthropometric data. First, sub populations may vary dramatically from the parent population, as was the case for Gulf of Mexico offshore workers when compared to North American males. Second, these variations can present significant safety concerns, (as evidenced by the possible 20% overestimation of lifeboat occupancy in the Gulf of Mexico). Finally, where the application of anthropometric data to design can have serious safety implications, particularly in special or unique environments, data verification studies should be considered.

Table 13.6 Forward functional reach[a,b]

Geographic region	Male		Female		Min	Max
	5%	95%	5%	95%		
Northern Europe	820 mm (32.3 in)	930 mm (36.6 in)	740 mm (29.1 in)	870 mm (34.3 in)	740 mm (29.1 in)	930 mm (36.6 in)
North America	842 mm (33.1 in)	975 mm (38.4 in)	746 mm (29.4 in)	876 mm (34.5 in)	746 mm (29.4 in)	975 mm (38.4 in)
Australia	800 mm (31.5 in)	920 mm (36.2 in)	740 mm (29.1 in)	860 mm (33.9 in)	740 mm (29.1 in)	920 mm (36.2 in)
France	800 mm (31.5 in)	910 mm (35.8 in)	730 mm (28.7 in)	830 mm (32.7 in)	730 mm (28.7 in)	910 mm (35.8 in)
Central Europe	800 mm (31.5 in)	890 mm (35.0 in)	740 mm (29.1 in)	840 mm (33.0 in)	740 mm (29.1 in)	890 mm (35.0 in)
Eastern Europe	800 mm (31.5 in)	890 mm (35.0 in)	740 mm (29.1 in)	820 mm (32.3 in)	740 mm (29.1 in)	890 mm (35.0 in)
Latin America	790 mm (31.1 in)	890 mm (35.0 in)	710 mm (28.0 in)	830 mm (32.7 in)	710 mm (28.0 in)	890 mm (35.0 in)
Iberian Peninsula	760 mm (29.9 in)	880 mm (34.6 in)	720 mm (28.3 in)	820 mm (32.3 in)	720 mm (28.3 in)	880 mm (34.6 in)
Southeast Europe	790 mm (31.1 in)	880 mm (34.6 in)	740 mm (29.1 in)	830 mm (32.7 in)	740 mm (29.1 in)	880 mm (34.6 in)
Japan	758 mm (29.8 in)	882 mm (34.7 in)	696 mm (27.4 in)	810 mm (31.9 in)	696 mm (27.4 in)	882 mm (34.7 in)
North Asia	780 mm (30.7 in)	900 mm (35.4 in)	720 mm (28.3 in)	850 mm (33.5 in)	720 mm (28.3 in)	900 mm (35.4 in)
North Africa	800 mm (31.5 in)	920 mm (36.2 in)	750 mm (29.5 in)	870 mm (34.3 in)	750 mm (29.5 in)	920 mm (36.2 in)
Near East	780 mm (30.7 in)	860 mm (33.9 in)	740 mm (29.1 in)	815 mm (32.1 in)	740 mm (29.1 in)	860 mm (33.9 in)
West Africa	790 mm (31.1 in)	900 mm (35.4 in)	720 mm (28.3 in)	820 mm (32.3 in)	720 mm (28.3 in)	900 mm (35.4 in)
Southeast Africa	810 mm (31.9 in)	950 mm (37.4 in)	740 mm (29.1 in)	860 mm (33.9 in)	740 mm (29.1 in)	950 mm (37.4 in)
North India	760 mm (29.9 in)	850 mm (33.5 in)	700 mm (27.6 in)	780 mm (30.7 in)	700 mm (27.6 in)	850 mm (33.5 in)
South India	730 mm (28.7 in)	840 mm (33.1 in)	670 mm (26.4 in)	770 mm (30.3 in)	670 mm (26.4 in)	840 mm (33.1 in)
Southeast Asia	730 mm (28.7 in)	820 mm (32.3 in)	690 mm (27.2 in)	780 mm (30.7 in)	690 mm (27.2 in)	820 mm (32.3 in)

(continued)

Table 13.6 (continued)

Geographic region	Male		Female		Min	Max
	5%	95%	5%	95%		
Latin America (Indian population)	730 mm (28.7 in)	820 mm (32.3 in)	670 mm (26.4 in)	750 mm (29.5 in)	670 mm (26.4 in)	820 mm (32.3 in)
South China	760 mm (29.9 in)	840 mm (33.1 in)	690 mm (27.2 in)	760 mm (29.9 in)	690 mm (27.2 in)	840 mm (33.1 in)

[a] International Data on Anthropometry. International Labour Organisation, 1990
[b] ADULTDATA: The Handbook of Adult Anthropometric and Strength Measurements—Data for Design Safety. Institute for Occupational Ergonomics, University of Nottingham, UK. 1998

Table 13.7 Clothing and postural effects

Description	Effect
Effect of footwear in standing height measures	+ 25 mm (1 in)
Effect of clothing in sitting height measures	+ 5 mm (0.2 in)
Effect of clothing in breadth	+ 15 mm (0.6 in)
Effect of clothing in foot length	+ 30 mm (1.2 in)
Effect of hard hat in stature	+ 75 mm (3.0 in)
Effect of gloves on hand length/breadth	+ 7 mm (0.3 in)
Effect of postural slump in standing height	– 20 mm (0.8 in)

Chapter 14
Application of Human Elements to Maritime Design

General

In the previous chapters of Part 1 of this book, we have examined and discussed some of the main human factor and ergonomic principles, guidelines, and criteria for both vessels and offshore installations with the intention of assisting naval architects, designers, and engineers with integrating ergonomics into marine system design. To apply that information systematically, a human-system interface design process should be used. This chapter offers a practical way to implement such a process. It outlines a simplified and structured approach for addressing ergonomics within the context of engineering design through three sets of activities: (1) analysis, (2) design, and (3) verification and validation.

Objective of Integrating Human Elements and Ergonomics into Marine Engineering Design

Human factors and ergonomics focusses on and represents operator and maintainer needs and requirements throughout the life cycle of systems. The goal is to minimise human error, thereby maximising human and total system safety and effectiveness. This can be accomplished through:

- The application of ergonomic principles, guidelines and criteria.
- Conduct of appropriate analyses and solicitation of operator or maintainer input to derive task requirements and needs.
- Application of a logical, practical, human-system interface design process.

This chapter presents a basic human-interface design process that emphasises the infusion of ergonomics principles and criteria into engineering design. This process can be performed during the design or modification of maritime systems. It starts

A. A. Olsen, *Applying Physical Ergonomics to Modern Ship Design*,
https://doi.org/10.1007/978-3-031-57974-5_14

with analysis, moves on to design activities and ends with verification and validation processes. The process described is written to acquaint those unfamiliar with human-interface design processes and is not meant to provide a detailed approach. Such details can be found in the following:

- ASTM F 1337, Standard Practice for Human Engineering Program Requirements for Ships and Marine Systems, Equipment, and Facilities.
- ISO 13407. Human-Centred Design Processes for Interactive Systems.
- Carey, M. Proposed Framework for Addressing Human Factors in IEC 61508. Sudbury, Suffolk, England: HSE Books, 2001.
- Hendrikse, E.J, et al., Effectively including Human Factors in the Design of New Facilities.
- Proceedings of the 2nd International Workshop on Human Factors in Offshore Operations: Demystifying Human Factors – Practical Solutions to Reduce Incidents and Improve Safety, Quality and Reliability. Houston, Texas: April 8–10, 2002.
- Thomas, J., et al., Application of Human Factors Engineering in Reducing Human Error in Existing Offshore Systems. Proceedings of the 2nd International Workshop on Human Factors in Offshore Operations: Demystifying Human Factors – Practical Solutions to Reduce Incidents and Improve Safety, Quality and Reliability. Houston, Texas: April 8–10, 2002.

Ergonomic Design Activities

The effective integration of ergonomic considerations into system development is structured, systematic, interdisciplinary, and requirements-driven. It is a top-down design approach that starts with high-level goals that are divided into the functions necessary to achieve those goals. Functions are allocated to human and system resources. Human functions are then divided into discrete human tasks. Tasks are grouped into meaningful jobs and the human-system interface is designed to support task requirements. A basic human-interface design approach is comprised of three phases:

1. Analysis of ergonomic design requirements.
2. Development of interface designs.
3. Performance of design verification and validation.

Each of these phases are discussed below. The design and evaluation process described is generic and highly simplified. These general activities are appropriate to any specific design project, be it large-scale (e.g., design of a new vessel, offshore installation, or navigation bridge) or small (e.g., the design of a single piece of equipment, such as an electronic chart display).

Analysis of Ergonomic Design Requirements

There are five related issues to consider in the analysis of human-interface design requirements, each of these are discussed below:

1. *Analysis of previous designs.* One of the first human-interface design activities to be performed in any system design is the analysis of previous designs. The objective of this activity is to identify the successes and failures of previous designs to provide a basis for improvements in the new design or modification. The scope of the analysis generally includes:
   - Experiences with ergonomics from similar previous designs.
   - Interviews with subject-matter experts, such as designers, engineers, operators, and maintainers.

     The basic result of this effort is a list of lessons learned, in terms of both positive aspects of previous designs (including such things as coding and established expectancies) that should be carried forth, and awkward or clumsy design aspects that should be changed or improved.
2. *Analysis of human functions and tasks.* One of the earliest and most critical human-interface design activities is the analysis of human functions and tasks. Task analysis focuses on understanding the functions and tasks that personnel will perform with the system in terms of:
   - The inputs and outputs for those tasks with regard to required information and control actions,
   - Feedback provided to an operator after the manipulation of a control.
   - Dependencies between tasks.
   - Sequence of task performance.
   - Time and frequency of performance.
   - Accuracy requirements.
   - Criticality or importance of the tasks.
   - Error tolerance.

     By performing a task analysis, the designer develops an understanding of the how a workstation or work area can be designed and arranged to support reliable task performance. Task analysis also focuses on understanding the flow and sequence of human activity as work is performed. It can be used to understand how personnel move between controls, displays, consoles, panels, workstations, equipment, and work areas in performing a series of tasks. This information is used to arrange consoles, panels, and workstations based on task sequences, frequencies, and criticality.
3. *Characteristics of personnel.* Given that ergonomics concentrates its efforts on the human-system interface, it is necessary to identify the characteristics of the system users, namely, the operators and maintainers. This requires that personnel capabilities and limitations be identified, including:

- Training, certifications, and licenses
- Experience levels.
- Physical characteristics, such as eyesight, strength, and physical size.
- Cultural characteristics related to behavioural expectancies.
- Management styles.
- The characteristics of personnel need to be identified, as these can influence design (e.g., determining colour-coded meanings that can vary as a function of nationality) and influence training, certification, and licensing requirements.

4. *Environmental analysis.* Since living and workplace environmental conditions influence human performance, the environmental conditions to which personnel will be exposed need to be identified. Environmental analyses are performed to characterise external and internal factors, such as expected platform motions and sea states, temperature extremes, exposure to high noise and vibration, and lighting. All these factors have implications for design. The analyses are typically tabletop exercises where environmental conditions are identified for the expected operating areas and associated seasonal variations.
5. *Identify worst-case situations.* In most design efforts, it is important to predict what the likely worst-case operational and maintenance scenarios might be. For example, tasks may be required to be performed under extreme heat or cold conditions with poor lighting, or relief personnel may have lesser/inadequate training for task performance. In other situations, personnel may need to perform while stressed and under emergency conditions where some system functions are unavailable or severely degraded. As a result, the design may be made as simple as possible, or instructions may be mounted with equipment. In other cases, functions may be automated, or special tools or protective equipment provided to allow personnel to accomplish tasks under extreme conditions.

Development of Human System Interface Designs

The objective of design activities that integrate ergonomics principles, guidelines, and criteria is to provide human-system interfaces to support specific functions or tasks that will need to be performed by operators or maintainers. The ergonomics design activities are directed towards integrating function, task and staffing requirements identified with analysis into human-system interface designs consistent with the detailed ergonomic design guidance contained in standards, specifications, and guidelines. Basically, design proceeds within the context of human-system interface requirements identified during analysis, the guidance contained in publications such as this book and a simple framework of human control, such as that presented in Fig. 14.1, System control process. Depending on the function or to be performed, personnel will be involved in monitoring, deciding, planning, controlling activities, or some combination of these. Once this is known, determining the basic category of interface to provide becomes simpler. For example, where a person may be monitoring a system to verify that all essential parameters are within specified limits, the

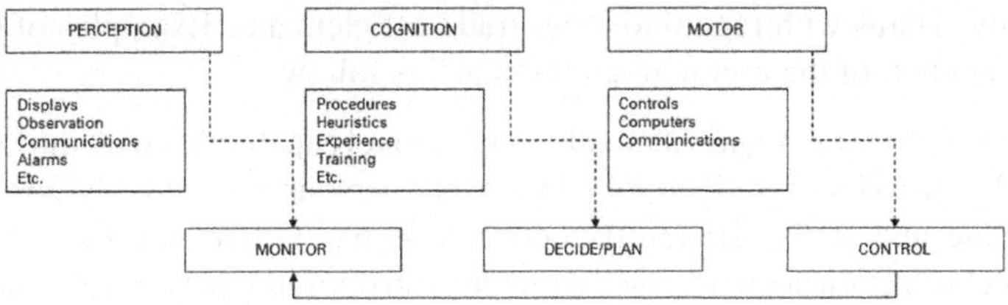

Fig. 14.1 System control process

individual will be using their perceptual abilities. To support the monitoring task, displays and alarms can be provided according to the guidance in Chap. 3, Displays and Chap. 4, Alarms. Furthermore, Chap. 5, Integration of Controls, Displays, and Alarms would serve to guide how the displays and alarms might be presented together to personnel.

Whether the design effort generally entails assembling commercial-off-the-shelf (COTS) equipment (e.g., radar sets or diesel generator sets) or represents a completely new design (e.g., a new turret disconnect panel), the human-system interface should result from a process that considers the human-system interface requirements identified during analysis activities. Perform trade-off analyses. In the practice of applying ergonomic principles in design, an ongoing process of considering trade-offs will occur within and among the elements of the ergonomic model:

- Design and layout (e.g., controls, displays, workspace arrangements).
- Ambient environment (e.g., noise, vibrations, temperature, humidity, design of accommodations).
- Personnel (e.g., training and certifications, health, anthropometry, susceptibility to fatigue).
- Management and organisation (e.g., policies, written procedures, safety plans).
- Trade-offs allow designers to compensate across these domains to provide an appropriate level of human performance and safety. Ergonomic design of human-system interfaces will influence job complexity, which in turn affects personnel knowledge, skill requirements, training requirements, etc. There are of course, other significant factors contributing to the outcome of trade-off decisions, including:
- Cost (e.g., engineering, construction, commissioning, operating, decommissioning).
- Risk tolerance (e.g., business, human safety, environmental, technical, financial, societal reputation).
- Statutory (e.g., laws and agreements that govern functionality and design).
- Competitiveness.

There are certainly others. However, the point here is that human element and ergonomic concerns should represent a significant component of the trade-off and design process. In other words, human element and ergonomic concerns should have a competitive status with regard to other trade-off elements. Examples of trade-offs within the context of the ergonomic model are as follows:

- *Personnel selection.* A way of matching personnel capabilities with the knowledge and skill demands of a given job is to select persons who already possess those knowledge and skills. Trade-offs between human-system interface design and personnel selection are performed to achieve an optimum person-job match. (Note that inadequate ergonomic design of human-system interfaces can induce human errors, and these likely cannot be compensated for by selection).
- *Training.* Ergonomic design principles can be used to introduce consistent expectations in the design of the human-system interface. Where consistently applied, this reinforces the learning experience and reduces the duration and complexity of training. Note as above, training is ineffective in overcoming the error-inducing characteristics of inadequate ergonomic design.
- *Job aids.* Design and use of job aids can reduce workload and knowledge and skill demands. Job aids have also proven highly effective in reducing human error, enhancing task performances, reducing training and lowering minimum selection requirements.

Develop and explore the human-system interface concept. As in most situations, there are often multiple human-system interface alternatives that could be used to meet a particular need. For this reason, it is necessary to develop and explore various design concepts based on functions, requirements, human allocated tasks and selected equipment. The basic steps for concept development are illustrated in Fig. 14.2, Concept development and exploration, and are briefly described here. *Develop concepts for human-system interface designs.* Concepts are typically developed among a group comprised of designers, engineers, ergonomists, and experienced operators and maintainers. Various approaches to human-system interface design will be developed based on the output from the analyses discussed above. In some cases, a "mind mapping" process can be employed where requirements are integrated, and design concepts are developed in the context of a trade-off process. An aspect of the trade-off process is that of selection of components (e.g., video displays, joysticks, radar sets, etc.) that will be arranged into a human-system interface on a console, panel, etc. Device selection generally occurs as a result of a review of human-system interface requirements, and comparison of those requirements (trade-off criteria) to the availability of competing commercial devices (e.g., hard wired displays versus video displays from different manufacturers). *Develop prototypes.* Prototypes can be prepared for the various human-system interface design concepts. Initially these may consist of no more than drawings, and they can progress to more complex mock-ups using cardboard, plywood, or computer simulations. One advantage of a prototype is that designers, engineers, ergonomists, operators, and maintainers can look at the physical representation and perform evaluations based on the model. Another advantage is that the various personnel involved with developing design concepts can interface with the potential user (i.e., operators and maintainers) to consider alternatives to design aspects that can be improved. In addition, informal "tests" of design concepts can be conducted by simulating task performance based on

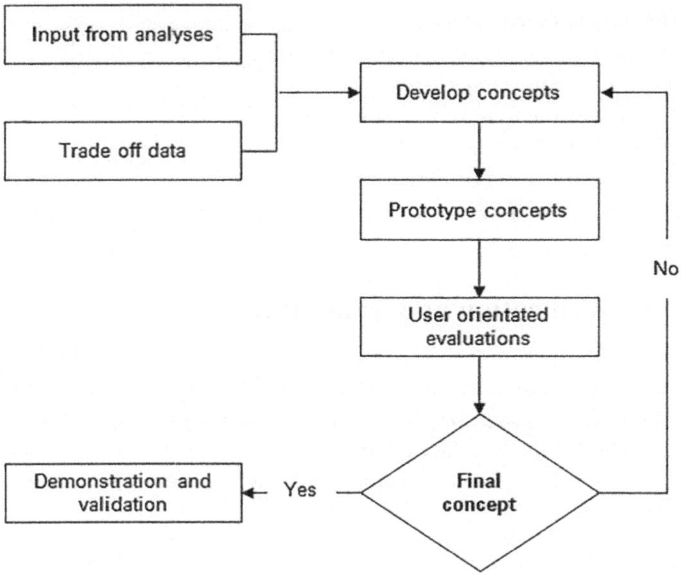

Fig. 14.2 Concept development and exploration

representative scenarios. *Evaluate the concepts.* Whether prototypes are developed or not, the various human-system interface concepts should be evaluated from an ergonomics perspective so that a satisfactory human-system interface is developed. Types of evaluations that can be conducted include subject-matter expert assessments, operator evaluations, walk through assessments and formal usability testing; and the application of check lists related to ergonomic design guidance. The results of these evaluations can then be used to improve design concepts.

Develop maintainability concepts. Design activities often focus on operations and operator needs. Maintenance issues are also critical to overall system safety and performance. To promote system reliability, availability, and safety, maintenance activities should also be analysed and provisions made to incorporate ergonomics data and maintenance requirements into design for maintainability. Designing for maintainability should address any time constraints to repair/ reinstate equipment. Repair times may need to be assumed and any assumptions should be explicitly expressed in design documents where maintainability concepts are discussed. In addition, the resources provided for maintenance, in terms of both staffing and equipment, should be sufficient to meet this goal. Throughout this book, we have discussed and examined various maintainability design principles, guidance and criteria. However, before applying that information, the following should be identified and considered:

- Maintenance concepts and goals.
- Task-based maintenance interface design.
- Location and design of maintenance accesses.

- Provisions for maintenance safety.
- Tool requirements, including any special tools.
- Provisions for logistics and spares.
- Schedules for preventative maintenance.
- Testing and calibration requirements.
- Training and certification of maintainers.
- Necessary materials handling devices and processes.

Ergonomics Verification and Validation

The objective of ergonomics verification and validation (or test and evaluation) is to determine that human-system interface designs conform to human-interface design principles and whether the interface enables personnel to perform tasks while meeting safety and operational goals. Below are typical steps of an ergonomics verification and validation process:

- *Task-support verification.* A check to confirm that the human interface addresses all identified personnel tasks.
- *Ergonomics design verification.* A check to determine whether the design of each individual interface reflects ergonomics principles, standards, and guidelines.
- *Integrated system validation.* Performance-based evaluations of the integrated design to confirm that human interfaces support safe and efficient operation of the vessel or installation.
- *Final design verification.* The final product, as built, conforms to the verified and validated design that resulted from the ergonomics design process.

Subjective evaluations. Designs can be evaluated subjectively by knowledgeable subject-matter experts (generally a team of engineers, operators, maintainers, and ergonomists). This can begin with drawings early in the design cycle and follow through to the construction phase of a project. As the design proceeds, mock-ups, functional prototypes, and simulations can be used to support these evaluations. The main objective of the subjective evaluation is to obtain operator and maintainer input while modifications to the design are possible. Checklist. Drawings, mock-ups, and prototype equipment can be evaluated using check lists that contain ergonomics criteria, or other ergonomics standards, specifications, or guidelines. Where design characteristics are found to deviate from criteria/guidance in a checklist, a review of the tasks and their importance to safe and efficient operations, can be performed to assess whether a design modification is needed. Performance-based tests. Performance-based tests involve collecting data on human-system performance upon which to base objective evaluations of design concepts. For example, when considering which of two chart displays to acquire, having a sample of mariners try the two displays during simulated task performance can provide usability data for each device. This data on usability and performance (as well as subjective opinion) can be used to Compare relative performance. The results of those comparisons can be used to select the better designs. Documentation of the interface design includes the technical basis for human-system interface design requirements, the detailed description of the human-system interface design, including its form, function, and performance characteristics; and the basis for the human-system interface design characteristics with respect to operating experience, trade-off studies, engineering evaluations and experiments, and benchmark evaluations.

Integrating Ergonomic Principles, Criteria, and Processes

This section discusses the systematic and simplified human-systems interface design process. The process described earlier in this Section can be broken down at its most basic level to three sets of activities: (1) analysis, (2) design, and (3) verification and validation. In this section, a review of these is provided in the context of integrating the process with ergonomic principles and guidance contained throughout this book. Analysis. For any human-system interface, three factors should be analysed: the task to be undertaken, who will perform the task and where the task will be undertaken. These issues can be analysed at a high level by addressing the questions presented in Table 14.1, Candidate analysis questions. Human-system interface design. Once the various analyses have been completed to specify which functions and tasks will be performed by humans and under what conditions, work can begin to design the human-system interfaces using ergonomics principles and criteria. Early decisions concern the level of automation anticipated for the vessel or offshore installation. These decisions influence what functions and tasks will be conducted by personnel, and what will be assigned to equipment. Based on automation decisions, design can proceed to determine the controls needed. This guidance can then be used to design control interfaces (refer to Chap. 2, Controls). Displays and alarms can be selected to provide feedback to personnel for control actions or system interventions (refer to Chap. 3, Displays and Chap. 4, Alarms). Selected devices can then be integrated using Chap. 5, Integration of controls, displays, and alarms on or near actual equipment, on local panels or in centralised locations, such as navigation bridges or in control rooms. Where appropriate, control or feedback (displays and alarms) can be provided via video-display units placed on local panels or in centralised areas. Chap. 6, Computer workstation design will be helpful for these types of designs. For equipment located in engineering spaces, on open decks, in the hull or topsides, different design considerations will be needed. Issues such as the placement of equipment, including manual valves, are also part of human-system interface design efforts. For such equipment, the physical location, identification, and access to the location are important. To assist with decision-making, designers and engineers may refer to the principles and guidance located in:

- Chapter 7, Manual valve operation, access, location, and orientation.
- Chapter 8, Labelling, signs, graphics, and symbols.
- Chapter 9, Stairs, ladders, ramps, walkways, platforms and hatches.

Table 14.1 Candidate analysis questions

Type of analysis	Candidate questions
Task	What are the specific tasks required of personnel to operate and maintain the equipment or system? What are the worst-case conditions under which the task will be performed? • Under time pressure? • During an emergency? • Without vital systems? What is the likelihood of a human error while performing this task? Are errors easily reversible? What are the potential consequences associated with a human error while performing this task? • Personnel safety? • Environmental impacts? • Property damage? • Efficiency?
Personnel	Who will be operating or maintaining the equipment? What abilities will personnel need to apply to the task? • Perceptual skills for monitoring • Cognitive skills for making decisions or planning • Motor skills for controlling or maintaining What skills and knowledge, including special training, certifications, or licenses, will these personnel need? Who will these personnel be in terms of nationality, cultural background, behavioural tendencies, past work experience? Will personnel be company employees, contractors, or a combination?
Environment	What will the working environment be under which the tasks are performed? • On an open deck subject to various weather conditions? • On a vibrating surface? • In high noise conditions? • In hot or cold or humid conditions? • At night? • Inside a tank or void space?

Up to now, this section has focused on operability issues, but to promote system reliability and availability, human-system interfaces for maintenance should also be addressed. Many of the human-system interfaces design issues for maintenance are similar to those for operability. For example, controls will need to be provided for some maintenance tasks along with maintenance-related displays and alarms. To avoid confusion during normal operations, such interfaces are often covered or otherwise protected, but easily accessible to maintainers, as needed. Various sections of this book provide maintenance-specific guidance along with guidance directed at operations support. Two sections, however, focus on particular maintenance considerations. Chapter 10, Maintenance considerations gives designers and engineers guidance related to maintenance access and task support, while Chap. 11, Materials handling addresses manual and assisted lifting and materials handling concerns. In addition to workplace concerns, designers and engineers of vessels and offshore installations are likely to be involved with designing accommodation spaces. Chap. 12, Crew habitability has been provided to give basic information related to accommodation design and ambient environmental concerns. It also refers to sources of specific guidance that should prove useful for designing both living and working spaces. This guidance is then expanded on in Part 3.

Trade-offs. Throughout the activities described in this section, trade-offs will need to be made. Designers, engineers, ergonomists, and even construction yard personnel will find themselves in the position of constantly trying to find a balance between the structural/equipment/system design requirements and the following human element concerns:

- Which personnel should be selected for tasks?
- How much training personnel can be assumed to have or to need?
- What jobs aids, in terms of procedures, labelling, signs, and instructions, will suffice?

Furthermore, these human element concerns will need to be weighed against other business considerations including capital and operating costs; risk tolerance; statutory, regulatory and company requirements, and competitiveness. *Verification and validation.* Once a design has been established, ergonomics criteria can be used during design reviews to check the suitability of the design for operability, maintainability, and compliance with ergonomics criteria. The criteria provided throughout this book can used as a part of verification and validation exercises. The criteria can be arranged into a checklist in order to make assessments easier to conduct. Beyond the application of ergonomics criteria, design reviews should be completed by operators and maintainers. These reviews will help to determine where modifications are needed before the start of construction. Final verification and validation activities can be completed in the construction yard. Here the constructed interfaces and physical layout can be checked for compliance with drawings and design intentions. A checklist can be used as well. While in the construction yard, all field-run equipment (e.g., cable trays, piping, ducting) would be checked to verify that equipment, access ways, aisle ways, egress paths, overhead clearances, or assisted lifting devices are accessible and do not intrude into walkways or areas occupied by personnel.

Example 1—Application to equipment design. This example involves the design and arrangement of a control station for three mooring winches located near the bow of a bulk carrier vessel. One winch is to the port side of centreline, the second is on the vessel centreline, and the third is to the starboard of centreline. All three winches may be used for mooring the vessel on either the port or starboard side.

- *Step 1.* Define the tasks that will be required of the operator or maintainer. In this instance the operator's tasks involve controlling the direction of winch rotation to pay out and take in the mooring lines while observing both the shore connection (e.g., T-Bitt) and the person-in-charge of the mooring operation on the deck of the vessel. The operator must monitor the hydraulic pressure at each winch to confirm that sufficient pressure is present to rotate the windlass drum. Also, the operator must monitor the tension of the mooring wire to prevent over-tensioning that can cause the wire to break and possibly backlash across the vessel's deck. In this case, the designer must recognise that the provided controls need to satisfy the operator's expectation between control movement and winch rotation. A movement of the control away from the operator's body should pay out the winch cable and a movement toward the operator's body should take in the cable. Further, this operational expectation supports the use of lever type controls for these winches. A control station for all three winches should be provided on both the port and starboard sides of the vessel to satisfy the visual tasks of observing both the shore connection and the person-in-charge. Further, it is possible that an elevated platform may assist with the visual tasks on shore as well as on the vessel's deck.

- *Step 2.* Identify the personnel who will perform the task, including nationality and regional origin.
- *Step 3.* Identify the operating environments that the controls may be used in (e.g., hot and humid tropical climates to very cold winter conditions) or during day and night hours. This dictates whether or not the controls need to be designed for use by operators wearing personnel protective equipment (e.g., mittens or gloves) and if the controls and displays should be illuminated.
- *Step 4.* Identify the potential "worst case" under which the system could be operated, such as a combination of extreme weather, (e.g., cold temperatures and high winds) and the vessel's condition (e.g., fully loaded). With this background, the designer can now define the design requirements, which are:

  – Use of lever-type controls for each winch with control movement to match the operator's expectations.
  – Use of an analogue gauge to display the hydraulic pressure at each winch.
  – Design of the controls for use by gloved hands.
  – Locate controls that provide the operator with visual contact with the shore connections and person-in-charge on deck.
  – Mount the controls and displays at a height to be operated by personnel of various regional origins.
  – Illuminate the displays and control labelling to be read at night.
  – Provide an automatic torque limiting control on the winches to prevent over-stressing the mooring line.

With the design requirements defined, the designer can now go to the relevant chapter and section of this book, or other sources of ergonomic guidance, for specific principles to consider in the development of the design. Chapter 2, Controls provides pertinent paragraphs on control actuators, control movement expectancy, and control actuator integration for control panels. To establish the proper mounting height for the controls, Chap. 2, Fig. 2.2, Control actuator mounting height for standing operator, indicates that the controls should be 860 mm (34 in) to 1350 mm (53 in) above the operator's standing surface. Control and display labelling requirements are provided in Chap. 8, Labelling, signs, graphics, and symbols. Following the above process will result in ergonomically designed control panels for the three mooring windlasses.

Example 2—Application to Class machinery rules. In this example, we will use, Piping Systems of the ABS Rules for Building and Classing Steel Vessels (SVR), as it provides the requirements for gravity drain systems through the ship's side. Specifically, SVR 4-6-4/3.3 requires that "…the discharge pipe may have two non-return valves without positive means of closing, provided the inboard non-return valve is always accessible for examination under service conditions". An accepted interpretation of this requirement is that the valve should be located above the deepest load waterline and that it should not be in a space that is inaccessible, such as a ballast tank.

- *Step 1.* Define the tasks to meet the stated SVR rule. In this instance, personnel must be able to make a visual inspection of the valve and its flanged connections. Sufficient lighting and placement of the valve must be provided to be capable of both visual inspection and physical contact for operation and maintenance. However, if the valve is leaking, or if there is evidence that seawater is backing up the line, the valve may need to be repaired or replaced. This will require additional personnel tasks such as removing bolts at the flange, lifting the old valve out and replacing it with a functioning valve. These tasks require physical exertion. This implies that a suitable working surface should be provided, since both hands are necessary to do the job.
- *Step 2.* Identify the personnel who will perform the task. If it is known who will be crewing the vessel (e.g., North Americans, Asians, Northern Europeans), the task of identifying the physical dimensions of personnel (i.e., the anthropometric dimensions) becomes easier. However, as this is unknown at this stage, and it is likely that different nationalities will be involved, it is preferable to design for an anthropometrically smaller population (e.g., South-East Asian),
- *Step 3.* Identify the operating environment. In this example, the tasks are largely physical (as compared to mental) and of a short duration. Thus, the effect of the physical work environment might not be of critical importance (e.g., temperature, humidity, and noise). However, the one environmental issue that could be of importance is the lighting level of the work area.
- *Step 4.* Identify the potential "worst case" in which tasks may be performed. The worst case might be a combination of the following:

  - The crew member normally trained to complete a task is not available for whatever reason, and the task should be completed by someone with less training or experience.
  - That the task is required to be performed under stressful (i.e., emergency) conditions.
  - That it be completed under severe operating conditions (e.g., rough weather or extreme environmental limits).

These "worst case" conditions should be anticipated at the time of vessel design in order to reduce the chance of human error if these conditions exist. The following design objectives may then be established:

- Adequate lighting levels at the valve
- Means to allow personnel to access the valve.
- If desired, a work platform to allow valve repair operation.

Lighting can be provided by standard engine room fixtures or dedicated lighting. In this particular case, lighting may also be provided by a portable lighting source (e.g., a flashlight). As an accepted minimum, an illumination level of at least 300 lx for machinery control rooms and 150 lx for other machinery spaces (e.g., generator rooms, fan rooms) should be provided. Considering the relatively simple nature of the task to be completed (i.e., visual or physical inspection of the valve flanges), the lower lighting level should be adequate. To enable personnel to see or reach the valve, the standing surface should be within the extended overhead arm reach of the 5th percentile South-East Asian male. Chapter 7, Fig. 7.3, Mounting heights for hand wheel valves with angled stems shows that valves may be mounted overhead up to a height of 1980 mm (78 in). This dimension is based on North American males and any adjustments to this height should be based on a review of the anthropometric data contained in Chap. 13, Anthropometric design principles and dimensions. However, if that standing surface is also used to access the valve for repair or replacement, it should be closer to the valve, since applying torque to loosen bolts and/or lifting a valve above the head can be difficult and hazardous operations. If the valve height above a permanent deck or other standing surface exceeds that discussed above, a ladder or other means should be provided to allow personnel to safely reach the valve. Suggested design criteria for ladders are contained in Chap. 9, Stairs, ladders, ramps, walkways, platforms and hatches. If vertical access were required just for valve examination, then a vertical ladder would probably be sufficient, since the task requires only one hand. If, however, the vertical access was to be used for valve repair or replacement, some form of work platform would be needed below the valve, since this task would require the use of both hands. Suggested criteria are contained for both in Chap. 9, Stairs, ladders, ramps, walkways, platforms and hatches.

Part II
Applying Human Elements and Ergonomics to the Maritime Environment

Chapter 15
Navigational Bridge Layout and Arrangements

General

This chapter provides general guidance relating to the bridge arrangements which apply to vessels possessing valid SOLAS certificates, and having a bridge designed and equipped so to enhance the safety and efficiency of navigation. The design and layout of navigational equipment should always be based on sound ergonomic principles.

Operational Assumptions

The guidance contained in this book are therefore based on a number of important assumptions, some of which are discussed here. Plans for emergencies and the conditions under which the vessel is intended to operate must be clearly defined in a Bridge Operations Manual (BOM), which is deemed acceptable in terms of scope and detail by the vessel's Flag State Administration. The BOM is to clearly state the bridge crew composition required under any set of circumstances. The composition and qualifications of the crew remains the responsibility of the vessel's Flag State Administration. The requirements of STCW and other applicable statutory regulations are to be complied with at all times. Upon request, optional notations may be assigned by Class for vessels where the navigational bridge design, the design and arrangement of navigational equipment and bridge operational procedures are developed based on accepted IMO Resolutions to aid in improving their level of safety and to reduce the risk of grounding, collision and weather damage, according to the level of compliance.

A. A. Olsen, *Applying Physical Ergonomics to Modern Ship Design*,
https://doi.org/10.1007/978-3-031-57974-5_15

Optional Class Notations

*Navigational bridge layout (***NBL***).* Although each individual Class will have their own distinct set of notations, they generally follow an equivalent process. For example, ABS may assign the notation **NBL** for a vessel having its navigational equipment designed and constructed based on sound ergonomic principles and its navigational bridge equipped in compliance with the requirements in Section "General", as applicable, and Section "Operational Assumptions" of the ABS Guide for Bridge Design and Navigational Equipment and Systems (2022). Where the vessel has been constructed and installed under the supervision of ABS, the vessel will be assigned the notation **NBL** or equivalent. *Navigational bridge layout and equipment/systems* (**NBLES** and **NBLES+**). Alternatively, a vessel having its navigational bridge equipped in accordance with the provisions set out in Sections "General" and "Optional Class Notations" of the ABS Guide for Bridge Design and Navigational Equipment and Systems (2022) may be assigned the notation **NBLES** where Class (in this instance, ABS) determines the navigational bridge layout, equipment, and systems sufficiently meet the requirements of the **NBLES** notation as per in this chapter, Navigational equipment for **NBLES** notation. Where the vessel has met the requirements to achieve an **NBLES** notation but is also equipped with additional equipment on the bridge wings as specified in this chapter, Additional navigational equipment for **NBLES+** notation or equivalent, the vessel may be assigned an **NBLES+** notation or equivalent. *Navigational bridge layout and equipment/systems for coastal and offshore service* (**NBLES (COS)**). For vessels engaged in coastal and offshore operations, additional Class requirements apply in respect to navigational bridge layout and equipment and systems. This is predominantly due to the different types of hazards and dangers that coastal and offshore vessels must endure (such as the constant risk of close-range navigational hazards and smaller operational envelopes) which are not as frequently experienced by vessels engaged in long distance trades. For this reason, a vessel having its navigational bridge equipped according to Chaps. 13–17 of this book and which has been constructed and installed under the supervision of Class survey may be assigned the notation **NBLES (COS)** or equivalent. *Navigational integrated bridge system* (**NIBS**). Alternatively, vessels equipped with an integrated bridge system (IBS) for navigational purposes in compliance with IMO document SN.1/Circ.288, Guidelines for bridge equipment and systems, their arrangement and integration (BES), and which comply with the Class requirements outlined in Chaps. 13–17 of this book may be assigned the notation **NIBS** or similar.

Regulations

For the purposes of this book, the *International Regulations for Preventing Collisions at Sea, 1972,* (COLREG), and all other relevant regulations relating to radio and safety of navigation required by Chapters IV and V of 1974 SOLAS, as amended, must be fully complied with. Valid statutory certificates issued by the pertinent Flag State Administration must be maintained on board the vessel and made available to Class approved surveyors or other duly appointed officials upon request.

Flag State Administration and National Authorities

Vessel owners and/or other interested parties (such as vessel operators and charterers) are strongly encouraged to consult and engage with the vessel's Flag State Administration and relevant National Authorities concerning the required manning levels on the bridge and any additional requirements which may be imposed by them for the legal, safe and efficient operation of the vessel.

Plans and Data to Be Submitted to Class

When applying for a Class notation, the vessel designer must submit the relevant plans and data for Class review and approval. These plans should be submitted electronically to Class, however hard copies will also generally be accepted. For most standard notations relating to bridge design and navigational equipment and systems, the following provisions should be clearly and unambiguously included in the plans submitted to Class:

- Arrangements of windows, including dimensions and angles of inclination, dimensions of frames, height above deck surface of upper and lower edges, type of glass, and details of clear view arrangements (wipers, freshwater wash systems, de-icing/de-misting systems, and sun blinds).
- Fields of vision from the bridge workstations, including any blind sectors caused by obstruction outside of the wheelhouse.
- The location and arrangement of workstations, including dimensions of consoles, layout of instrumentation and controls, handrails, and seating.
- Clearances between floor and ceiling, or between floor and the underside of ceiling mounted instruments, throughout the wheelhouse.
- Arrangements for the general illumination of the bridge and the individual illumination of workstation instruments and controls.
- Details of wheelhouse ventilation and heating systems.
- Details of internal communication systems operable from the bridge.
- Arrangements/details of exterior catwalk in front of bridge windows.

- Details of non-slip flooring.
- Details of wheelhouse doors, including hold-back arrangements.
- Location of washroom facilities.
- Arrangements for the drainage of the bridge decks.
- Arrangements/details of the measures to minimise hazards to personnel.

Class submissions related to **NBLES**, **NBLES +** , **NBLES (COS)** *and* **NIBS** *notations and equivalent.* (1) *List of equipment.* Furthermore, a list of navigational equipment to be installed on the bridge must be submitted for Class review. This is to include the manufacturer's name and model number for each equipment item, together with copies of any relevant type approval certificates. (2) *Alarms and alarm display.* A complete operational description of the relevant monitoring systems including a list of alarms and displays must be included. This may be achieved by means of simplified block diagrams of navigation equipment, internal communications systems and watch monitoring and alarm transfer systems, and central alarm panel (where provided) including a comprehensive list of alarms. (3) *One-line diagrams.* Where simplified one-line diagrams of the relevant systems are provided, these should include the power supplies to the bridge equipment, circuit protection ratings and settings, cable sizes, the ratings of connected loads, detailed descriptions and interactions. (4) *Operations manuals.* Operational and technical manuals for the installed navigational equipment and systems must be written and submitted for Class approval. Usually only one copy of each document is required. For sea trial-test plans, one copy must be sent to Class for review. Sea trial plans and test programme. A sea trial plan, test programme and schedule must be submitted for Class review. Usually only one copy need be provided. Refer to Chap. 17, Requirements for Class notation **NBLES** and **NBLES+** or equivalent for more detail.

Applicability to **NBLES (COS)** *notations and equivalent.* (1) *Documentation denoting vertical and horizontal field of visions from bridge and bridge workstations.* (2) *Documentation from the navigational equipment suppliers* related to the compliance of the Human Machine Interface (HMI) requirements contained in Chap. 18, Bridge arrangements and the working environment for coastal and offshore vessels. (3) *Failure mode and effects analysis (FMEA) or an equivalent* and acceptable national/international standard for bridge operation during manoeuvring the vessel is to be carried out and the report to be submitted for consideration. An FMEA study showing how a single failure of an equipment or system will impact the bridge operation during manoeuvring or transit. IEC standard 60812 may be used as reference document for conductance of the FMEA. (4) *Manoeuvring data booklet.* The manoeuvring data booklet including the results of all manoeuvring trials, identifying the mode of equipment and applied corrections. Applicability to **NIBS** notation and equivalent. (1) *Workstation details.* Details and arrangements of the workstations and systems as described in Chap. 19, Requirements for Class notation **NIBS** or equivalent. (2) *Programme for sea trials.* The sea trial programme is expected to include test details of the electronic chart display and information systems (ECDIS) and IBS.

Chapter 16
Requirements for Class Notations Relating to Bridge Layout

General

The design of navigational equipment should be based on sound ergonomic principles. They are to be constructed of robust, durable and flame retardant materials incorporating the required degree of enclosure protection (i.e., IP 20 for bridge installation and IP 56 for open deck installation) required by the vessel's Flag State Administration and Class requirements.

Equipment Design and Construction

Fault isolation. Circuits are to be designed to permit the isolation of a fault while maintaining the functionality of the remaining circuits or sub-components (i.e., using printed circuit cards, or modules, etc.) should allow the easy and safe replacement of the faulted portion of the circuit. Replacement of components. Replaceable components should be designed and arranged so that it will not be possible to connect them erroneously or to use incorrect replacements. Self-support. Workstations, panels, cabinets, and so forth, are required to be secured to a solid foundation with the sides and back suitably protected. They are to be self-supported or braced to the bulkhead or ceiling. Where the last method is used, the means of bracing is to be sufficiently flexible so to allow deflection of the deck without buckling the assembly structure. Configuration of devices. Alarms, displays and control devices must be arranged in a functional and logical manner to allow the operator an easy and clear means of identification of each of the components or systems included therein. Grouping of like system alarms, displays, and devices, and the use of labels and colour schemes are some appropriate methods which may be used to realise this intent. Precautions must be taken to prevent the inadvertent operation of controls that may lead to critical

A. A. Olsen, *Applying Physical Ergonomics to Modern Ship Design*,
https://doi.org/10.1007/978-3-031-57974-5_16

situations, i.e., exercising care in the identification and location of switches, activation controls, and handles, the use of recessed or covered switches and controls, and the arrangement for sequential operation. Instruments and controls. Instruments and controls should be designed to permit the easy and correct reading of displays by day and night and so fitted as to minimise glare or reflection or being obscured by strong light. In this sense, the following should be applied:

- *Digital readouts.* Digital readouts should not be used where the reading changes rapidly so as to preclude the operator from reading its changing value (i.e., numbers change is performed by snap action rather than by continuous movement).
- *Circular scale.* For an index moving relative to a circular scale, the index should move clockwise (or the scale is to move counter-clockwise) for increasing readings.
- *Linear scale.* For an index moving relative to a linear scale, the index should be horizontal or vertical with the pointer moving to the right or upwards for increasing readings. Deviation from this norm may be considered for special applications such as for water depth measurements.
- *Distinction controls.* Distinction Controls or combined controls/indicators should be visually and tactually distinguishable from elements that only provide some form of indication (i.e., rectangular buttons may be used for control elements and round lights for indicator elements).
- *Mechanical control.* The shape of mechanical controls should clearly indicate the method of operation of the control. Rotary finite-position controls (e.g., stepped switches) should have toggles or levers, whereas rotary continuous-position controls (rheostats) should have knobs or wheels except the steering control.
- *Light contrast.* Instruments providing information are to be presented on backgrounds with a high contrast, emitting as little light as possible by night. They are to be designed to show a light text on a dark non-reflecting background at night.
- *Illumination and lighting.* All instruments and controls should be provided with some means of illumination. Such illumination must be adjustable to zero, except for the illumination of alarms and dimmer controls. Such items must remain visible at all times. For the illumination of displays and alarms, red light (i.e., with a wavelength of 620 mm (24.4 in.) or higher) should be used.

With respect to earthing, non-current carrying metal parts must be effectively earthed to avoid risk of electrocution and/or current surging.

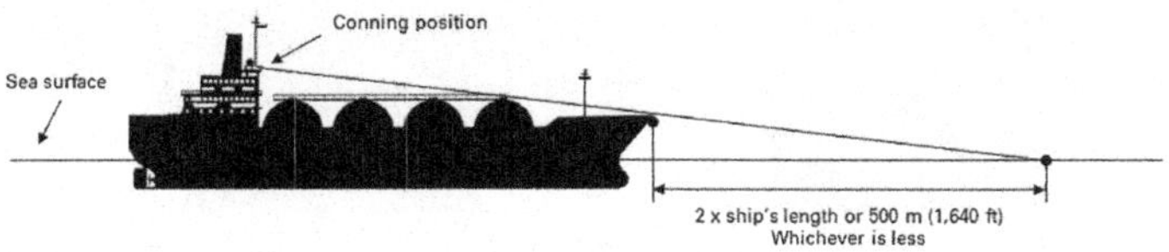

Fig. 16.1 Diagram showing the conning position and expected fields of visibility,[1,2]

Bridge Arrangement and Working Environment

Navigation bridge visibility. The first and foremost important factor in navigational bridge design is that of visibility. Vessels with the keel laid or in similar stage of construction on or after 1 July 1998, are required to meet the following requirements with respect to the visibility from the navigation bridge, unless they are solely engaged in navigating the Great Lakes of North America and their connecting and tributary waters as far east as the lower exit of the St. Lambert Lock at Montreal in the Province of Quebec, Canada. Fields of vision. (1) Conning position. The view of the sea surface from the conning position on the navigation bridge must not be obscured by more than 2 × length Overall (LOA) or 500 m (1640 ft), whichever is less, forward of the bow to 10° on either side for all conditions of draught, trim and deck cargo under which the vessel is expected to operate. Refer to Fig. 16.1, Diagram showing the conning position and expected fields of visibility. No blind sector caused by cargo, the cargo gear or any other obstructions outside of the wheelhouse forward of the beam which obstructs the view of the sea surface as seen from the conning position may exceed 10°. The total arc of blind sectors must not exceed 20°. The clear sectors between each blind sector are expected to be a minimum of at least 5°. However, in the view described above, each individual blind sector must not exceed 5°. The horizontal field of vision from the conning position should extend over an arc of not less than 225°, that is, from right ahead to not less than 22.5° abaft the beam on either side of the vessel. Refer to Fig. 16.1, Diagram showing the conning position and expected fields of visibility.

Bridge wings. From each bridge wing, the horizontal field of vision should extend over an arc of at least 225°, that is, from at least 45° on the opposite bow to right ahead and then from right ahead to right astern through 180° on the same side of the vessel. Refer to Fig. 16.2, *Demonstration of minimum bridge wing dimensions.* The vessel's side is to be visible from the bridge wing at all times. This requirement may be accomplished when:

[1] A conning position is a place on the bridge with a commanding view and which is used by navigators when commanding, manoeuvring and controlling a vessel.

[2] Attention is drawn to Flag State Administrations requiring lengths of less than 2 × LOA.

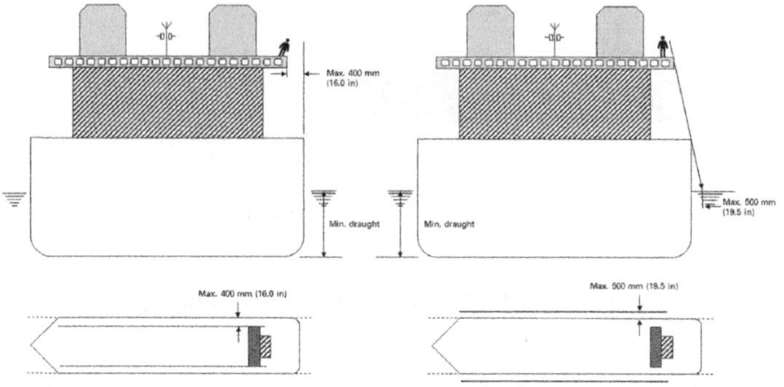

Fig. 16.2 Demonstration of minimum bridge wing dimensions

- A view from the bridge wing plus a distance corresponding to a reasonable and safe distance of a seafarer leaning over the side of the bridge wing, which needs not to be more than 400 mm (16 in.), to the location vertically right under the maximum beam of the vessel at the lowest seagoing draught is not obscured; or
- The sea surface at the lowest seagoing draught and with a transverse distance of 500 mm (19.5 in.) and more from the maximum beam throughout the vessel's length is visible from the side of the bridge wing. Refer to Fig. 16.3.

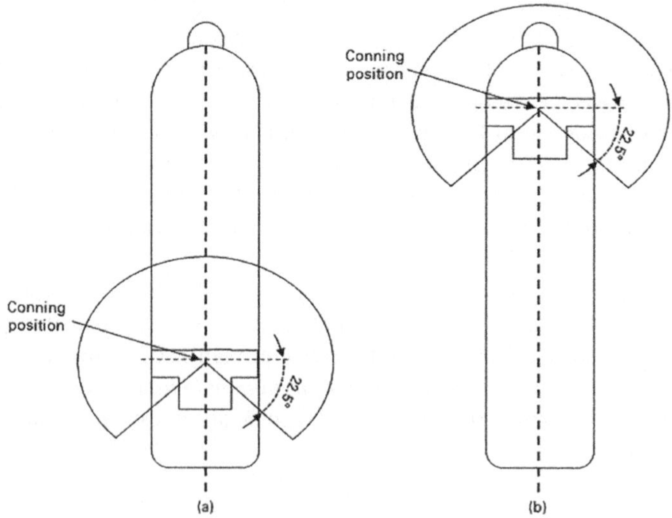

Fig. 16.3 Horizontal field of vision

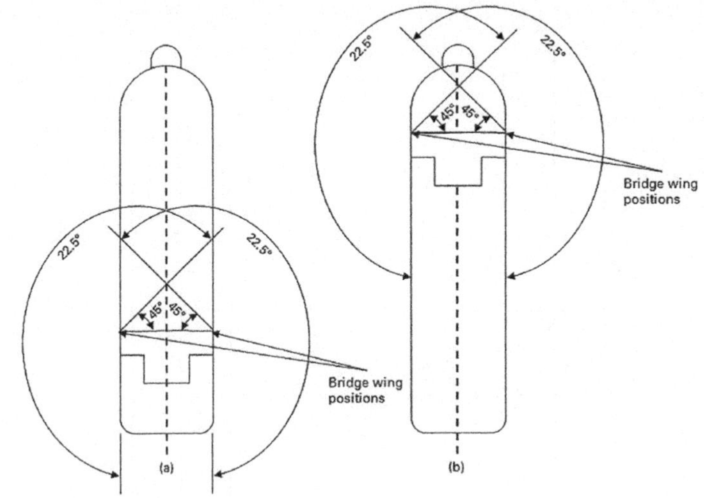

Fig. 16.4 Horizontal field of vision

For particular vessel types, such as tug/tow boat, offshore support and supply vessels (OSV), rescue ships, work ships (e.g., floating crane ships), etc., that are designed such that, in normal operations, they come along side, or operate in close proximity to, other vessels or offshore structures at sea, the bridge wing visibility requirement may be met provided the bridge wings extend at least to a location from which the sea surface, at the lowest seagoing draught and at a transverse distance of 1500 mm (59 in.) from the maximum beam throughout the vessel's length, is visible. If this ship type is changed to a type other than those addressed in this paragraph, then this would no longer apply.

Main steering position. From the main steering position, the horizontal field of vision should extend over an arc from right ahead to at least 60° on each side of the vessel. Refer to Figs. 16.4 and 16.5.

Remote camera systems. The use of a remote camera systems may be accepted for vessels of unconventional design, other than those mentioned above, as means for achieving the view of the vessel's side from the bridge wing, provided that:

1. The installed remote camera system is to be redundant from the circuit breaker to the camera and screen, including communication cables, i.e., the system is to provide on each side of the vessel redundancy of:
   - The power cables and circuit breakers from the main switchboard to the camera and the screen.
   - The camera.
   - The screen.

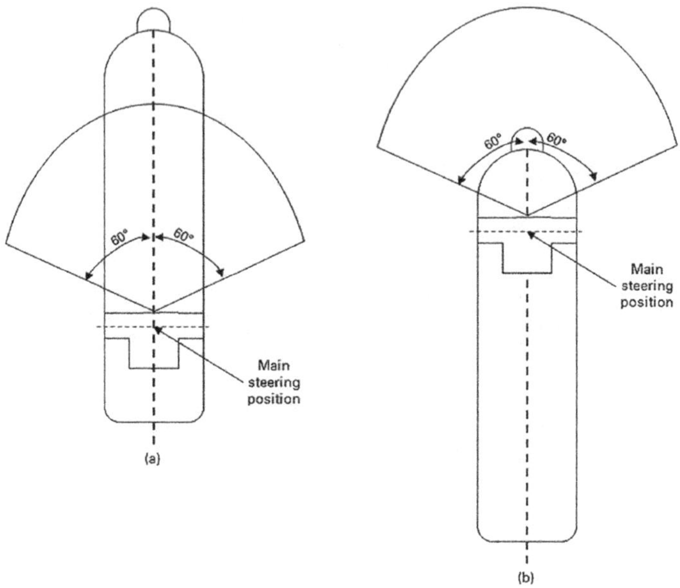

Fig. 16.5 Horizontal field of vision

- The transmission lines from the camera to the display screen.
- The components associated with these lines and cables.

2. The remote camera system is powered from the vessel's main source of electrical power and is not required to be powered by the emergency source of electrical power.
3. The remote camera system is capable of continuous operation under environmental conditions as per Table 16.1 Type tests for control, monitoring and safety equipment and Table 16.2 Tests for unit certification of control, monitoring and safety equipment.
4. The view provided by the remote camera system is analogous to that from the bridge wing so the vessel's side is to be visible and is also displayed at locations where the manoeuvring of the vessel may take place.
5. The upper edge of the vessel's side abeam is directly visible from locations where the manoeuvring of the vessel may take place.

Windows and their arrangements. Windows and their arrangements must meet the following minimum requirements:

- *Framing*. The framing between navigation bridge windows is best kept to a minimum to meet the structural strength and stiffness requirements and must not be installed immediately in front of any workstations.

Table 16.1 Type tests for control, monitoring and safety equipment

No.	Test	Procedure	Test parameters			Other information
1	Power supply variations (a) electric	-	AC supply			
			Voltage variation permanent (%)		Frequency variation permanent (%)	
			1	+ 6	+ 5	
			2	+ 6	– 5	
			3	– 10	– 5	
			4	– 10	+ 5	
			Voltage transient 1.5 s (%)		Frequency transient 5 s (%)	
			5	+ 20	+ 10	
			6	– 20	– 10	
			DC supply			
			Voltage tolerance continuous		+ 10%	
			Voltage cyclic variation		5%	
			Voltage ripple		10%	
			Electric battery supply:			
			+ 30 to –25% for equipment connected to charging battery or as determined by the charging/discharging characteristics, including ripple voltage from the charging device			
			+ 20 to –25% for equipment not connected to the battery during charging			
2	Power supply variations (continued) (b) pneumatic and hydraulic	–	Pressure: ± 20% Duration: 15 min			

(continued)

Table 16.1 (continued)

No.	Test	Procedure	Test parameters	Other information
3	Dry heat	IEC 60068-2-2	Temperature: 55 °C (131 °F) ± 2 °C (3.6 °F) Duration: 16 h Or Temperature: 70 °C (158 °F) ± 2 °C (3.6 °F) Duration: 16 h [refer to Note 1]	Equipment operating during conditioning and testing Functional test during the last hour at the test temperature For equipment specified for increased temperature the dry heat test is to be conducted at the agreed test temperature and duration
3A	Dry heat—Higher temperature (refer to Note 8) (optional test)	IEC 60068-2-2	Temperature: 70 °C (158 °F) ± 2 °C (3.6 °F) Duration: 16 h [See Note 8]	Equipment operating during conditioning and testing. Functional test during the last hour at the test temperature

(continued)

Table 16.1 (continued)

No.	Test	Procedure	Test parameters	Other information
4	Damp heat	IEC 60068-2-30—Test Db	2.0 (+3/-0) Hz to 13.2 Hz—amplitude ± 1 mm (0.039 in.) 13.2 to 100 Hz—acceleration ± 0.7 g For severe vibration conditions, e.g., on diesel engines, air compressors, etc.: 2.0 to 25 Hz—amplitude ± 1.6 mm (0.063 in.) 25.0 to 100 Hz—acceleration ± 4.0 g ***Note:*** *more severe conditions may exist for example on exhaust manifolds or fuel oil injection systems of diesel engines. For equipment specified for increased vibration levels, the vibration test is to be conducted at the agreed vibration level, frequency range and duration* *Values may be required to be in these cases 40–2000 Hz—acceleration ± 10.0 g at 600 0 °C duration 90 min*	Duration: 90 min at 30 Hz in case of no resonance conditions Duration: 90 min for each resonance frequency at which $Q \geq 2$ is recorded During the vibration test, functional tests are to be carried out Tests to be carried out in three mutually perpendicular planes It is recommended as guidance that Q does not exceed 5 Where sweep test is to be carried out in- stead of the discrete frequency test and a number of resonant frequencies are detected close to each other duration of the test is to be 120 min Sweep over a restricted frequency range between 0.8 and 1.2 times the critical frequencies can be used where appropriate ***Note:*** *critical frequency is a frequency at which the equipment being tested may exhibit*: • Malfunction and/or performance deterioration • Mechanical resonances and/ or other response effects occur, for example, chatter

(continued)

Table 16.1 (continued)

No.	Test	Procedure	Test parameters	Other information
5	Vibration	IEC 60068-2-6, Test Fc	Static 22.5°	1. Inclined at an angle of at least 2.5° to the vertical 2. Inclined to at an angle of at least 22.5° on the other side of the vertical and in the same plane as in (1) 3. Inclined to at an angle of at least 22.5° to the vertical and in plane at right angles to that used in (1) 4. Inclined to at an angle of at least 22.5° on the other side of the vertical and in the same plane as in (3) ***Note:*** *The duration of testing in each position should be sufficient to fully evaluate the behaviour of the equipment*
6	Inclination	IEC 60092-504	Dynamic 22.5°	Using the directions defined in (1) to (4) above, the equipment is to be rolled to an angle of 22.5° each side of the vertical with a period of 10 s The test in each direction is to be carried out for not less than 15 min

(continued)

Table 16.1 (continued)

No.	Test	Procedure	Test parameters				Other information
7	Insulation resistance	–	Rated supply voltage (V)	Test voltage (DC voltage) (V)		Minimum insulation Resistance	Insulation resistance test is to be carried out before and after: damp heat test, cold test, salt mist test, and high voltage test: • Between all phases and earth • And where appropriate, between the phases 'Un' is the rated (nominal) voltage
					Before test (MΩ)	After test (MΩ)	
			Un ≤ 65	2 × Un (min. 24 V)	10	1.0	
			Un > 65	500	100	10	
8	High voltage	–	Rated voltage Un (V)		Test voltage [AC voltage 50 or 60 Hz] (V)		Separate circuits are to be tested against each other and all circuits connected with each other tested against earth Printed circuits with electronic components may be removed during the test Period of application of the test voltage: 1 min
			Up to 65		2 × Un + 500		
			66–250		1500		
			251–500		2000		
			501–690		2500		
9	Cold	IEC 60068-2-1	Temperature: + 5 °C (41 °F) ± 3 °C (5.4 °F) Duration: 2 h Or Temperature: – 25 °C (– 13 °F) ± 3 °C (5.4 °F) Duration: 2 h [refer to Note 2]				Initial measurement of insulation resistance Equipment not operating during conditioning and testing, except for functional test Functional test during the last hour at the test temperature Insulation resistance measurement and the functional test after recovery

(continued)

Table 16.1 (continued)

No.	Test	Procedure	Test parameters	Other information
10	Salt mist	IEC 60068-2-52 Test Kb	Four spraying periods with a storage of 7 days after each	Initial measurement of insulation resistance and initial functional test Equipment not operating during conditioning of the test specimen Functional test on the seventh day of each storage period Insulation resistance measurement and performance test: 4–6 h after recovery [refer to Note 3] On completion of exposure, the equipment shall be examined to verify that deterioration or corrosion (if any) is superficial in nature

(continued)

Table 16.1 (continued)

No.	Test	Procedure	Test parameters	Other information
11	Electrostatic discharge	IEC 61000-4-2	Contact discharge: 6 kV Air discharge: 2, 4, 8 kV Interval between single discharges: 1 s Number of pulses: 10 per polarity According to test level 3	To simulate electrostatic discharge as may occur when persons touch the appliance The test is to be confined to the points and surfaces that can normally be reached by the operator Performance Criterion B [refer to Note 4]
12	Electromagnetic field	IEC 61000-4-3	Frequency range: 80 MHz to 2 GHz Modulation*: 80% AM at 1000 Hz Field strength: 10 V/m Frequency sweep rate: $\leq 1.5 \times 10–3$ decades/s (or 1% / 3 s) According to test level 3	To simulate electromagnetic fields radiated by different transmitters The test is to be confined to the appliances exposed to direct radiation by transmitters at their place of installation Performance criterion A [refer to Note 5] * If for tests of equipment, an input signal with a modulation frequency of 1000 Hz is necessary, a modulation frequency of 400 Hz may be chosen

(continued)

Table 16.1 (continued)

No.	Test	Procedure	Test parameters	Other information
13	Conducted low frequency		AC: Frequency range: rated frequency to 200th harmonic Test voltage (rms): 10% of supply to 15th harmonic reducing to 1% at 100th harmonic and maintain this level to the 200th harmonic, minimum 3 V (rms), maximum 2 W DC: Frequency range: 50–10 kHz Test voltage (rms): 10% of supply, maxi- mum 2 W	To simulate distortions in the power supply system generated for instance, by electronic consumers and coupled in as harmonics Performance criterion A [refer to Note 5] For keeping max. 2 W, the voltage of the test signal may be lower
14	Conducted radio frequency	IEC 61000-4-6	AC, DC, I/O ports and signal/control lines: Frequency range: 150–80 MHz Amplitude: 3 V rms [refer to Note 6] Modulation **: 80% AM at 1000 Hz Frequency sweep range: $\leq 1.5 \times 10{-}3$ decades/sec. (or 1% / 3 s) According to test level 2	Equipment design and the choice of materials are to simulate electromagnetic fields coupled as high frequency into the test specimen via the connecting lines Performance criterion A [refer to Note 5] ** If for tests of equipment, an input signal with a modulation frequency of 1000 Hz is necessary, a modulation frequency of 400 Hz should be chosen
15	Electrical fast transients/burst	IEC 61000-4-4	Single pulse rise time: 5 ns (between 10 and 90% value) Single pulse width: 50 ns (50% value) Amplitude (peak): 2 kV line on power supply port/earth 1 kV on I/O data control and communication ports (coupling clamp) Pulse period: 300 ms Burst duration: 15 ms Duration/polarity: 5 min According to test level 3	Arcs generated when actuating electrical contacts Interface effect occurring on the power supply, as well as at the external wiring of the test specimen Performance criterion B [refer to Note 4]

(continued)

Table 16.1 (continued)

No.	Test	Procedure	Test parameters		Other information
16	Surge	IEC 61000-4-5	Test applicable to AC and DC power ports Open-circuit voltage: Pulse rise time: 1.2 μs (front time) Pulse width: 50 μs (time to half value) Amplitude (peak): 1 kV line/earth 0.5 kV line/line Short-circuit current: Pulse rise time: 8 μs (front time) Pulse width: 20 μs (time to half value) Repetition rate: ≥ 1 pulse/min Number of pulses: 5 per polarity Application: continuous According to test level 2		Interference generated for instance, by switching "ON" or "OFF" high power inductive consumers Test procedure in accordance with Fig. 10 of the standard for equipment where power and signal lines are identical Performance criterion B [refer to Note 4]
17	Radiated emission	CISPR 16–2-3	For equipment installed in the bridge and deck zone:		Procedure in accordance with the standard but distance 3 m (10 ft) between equipment and antenna Alternatively, the radiation limit at a distance of 3 m from the enclosure port over the frequency 156 MHz to 165 MHz shall be 30 dB micro-V/m peak
			Frequency range: 0.15–0.3 MHz 0.3–30 MHz 30–2000 MHz Except for: 156–165 MHz	Quasi peak Limits: 80–50 dBμV/m 60–54 dBμV/m 54 dBμV/m Except for: 24 dBμV/m	

(continued)

Table 16.1 (continued)

<table>
<tr><th>No.</th><th>Test</th><th>Procedure</th><th colspan="2">Test parameters</th><th>Other information</th></tr>
<tr><td rowspan="4">18</td><td rowspan="4">Conducted emission</td><td rowspan="4">CISPR 16–2-1</td><td colspan="2">Test applicable to AC and DC power ports
For equipment installed in the bridge and deck zone:</td><td rowspan="4"></td></tr>
<tr><td>Frequency range:Limits:
10–150 kHz
150–350 kHz
350–30 MHz</td><td>Limits:
96–50 dBμV
60–50 dBμV
50 dBμV</td></tr>
<tr><td colspan="2">For equipment installed in the general power distribution zone:</td></tr>
<tr><td>Frequency range:
10–150 kHz
150–350 kHz
0.5–30 MHz</td><td>Limits:
120–69 dBμV
79 dBμV
73 dBμV</td></tr>
</table>

(continued)

Table 16.1 (continued)

No.	Test	Procedure	Test parameters	Other information
19	Flame retardant	IEC 60092–101 or IEC 60695–11-5	Flame application: 5 times 15 s each Interval between each application: 15 s or 1 time 30 s Test criteria based upon application The test is performed with the Equipment Under Test (EUT) or housing of the EUT applying needle-flame test method	The burnt out or damaged part of the specimen by not more than 60 mm long No flame, no incandescence or in the event of a flame or incandescence being present, it shall extinguish itself within 30 s. of the removal of the needle flame without full combustion of the test specimen Any dripping material shall extinguish itself in such a way as not to ignite a wrapping tissue. The drip height is 200 mm ± 5 mm

Notes:

1. Equipment to be mounted in consoles, housing, etc. together with other equipment are to be tested with 70 °C (158 °F)
2. For equipment installed in non-weather protected locations or cold locations, test is to be carried out at – 25 °C (– 13 °F)
3. Salt mist test is to be carried out for equipment installed in weather exposed areas
4. Performance criterion B (for transient phenomena): the equipment under test (EUT) is to continue to operate as intended after the tests. No degradation of performance or loss of function is allowed as defined in the technical specification published by the manufacturer. During the test, degradation or loss of function or performance which is self-recoverable is, however, allowed but no change of actual operating state or stored data is allowed
5. Performance criterion A (for continuous phenomena): The equipment under test is to continue to operate as intended during and after test. No degradation of performance or loss is allowed as defined in relevant equipment standard and the technical specification published by the manufacturer
6. For equipment installed on the bridge and deck zone, the test levels are to be increased to 10 V rms for spot frequencies, in accordance with IEC 60945 at 2, 3, 4, 6.2, 8.2, 12.6, 16.5, 18.8, 22, 25 MHz
7. Alternative equivalent testing procedures may be accepted, provided the requirements in the other columns are complied with
8. When requested, equipment which has undergone the higher temperature and duration test will be recognised accordingly in the PDA certificate. The purpose of introducing the optional 3HT test is for the convenience of equipment manufacturers should their clients request evidence that the equipment has been tested to the higher temperature requirements noted in Item 3A of the Table
9. As used in this document, and in contrast to a complete performance test, a functional test is a simplified test sufficient to verify that the EUT has not suffered any deterioration caused by the individual environmental tests

Table 16.2 Tests for unit certification of control, monitoring and safety equipment

No.	Test	Procedure	Test parameters	Other information
1	Visual inspection	–	–	Conformance to drawings, design data Quality of workmanship and construction
2	Performance testing	Manufacturer's performance test programme based upon specification and relevant Rule requirements When the EUT is required to comply with an international performance standard (e.g., protection relays), verification of requirements in the standard are to be part of the performance testing required in this initial test and subsequent performance tests after environmental testing where required by Table 16.1	Standard atmosphere conditions Temperature: 25 °C (77 °F) ± 10 °C (18 °F) Relative humidity: 60 ± 30% Air pressure: 96 kPa (0.98 kgf/cm2, 13.92 psi) ± 10 kPa (0.10 kgf/cm^2, 1.45 psi)	Confirmation that operation is in accordance with the requirements specified for particular system or equipment Checking of self-monitoring features Checking of specified protection against an access to the memory Checking against effect of erroneous use of control elements in the case of computer systems
3	External Power supply failure	–	3 × interruptions during 5 min; switching-off time 30 s each case	The time of 5 min may be exceed- ed if the equipment under test (EUT) needs a longer time for startup, for example, booting sequence For equipment which requires booting, one additional power supply interruption during booting is to be performed Verification of: • The specified action of equipment upon loss and restoration of supply • Possible corruption of program or data held in programmable electronic systems, where applicable

Notes

1. Alternative equivalent testing procedures may be accepted, provided the requirements in the other columns are complied with

- *Inclination angle.* The bridge front windows are required to be inclined from a vertical plane top out, at an angle of not less than 10° and not more than 25°. Refer to Fig. 16.6, Arrangement and inclination angle of navigation bridge windows.
- *Glass.* Polarised and/or tinted windows are not permitted.
- *Clear view.* At all times, regardless of the weather conditions, at least two of the navigation bridge front windows must provide a clear view, and in addition, depending on the bridge configuration, an additional number of windows are to provide a clear view. To this end, the following, or equivalent, should be provided:

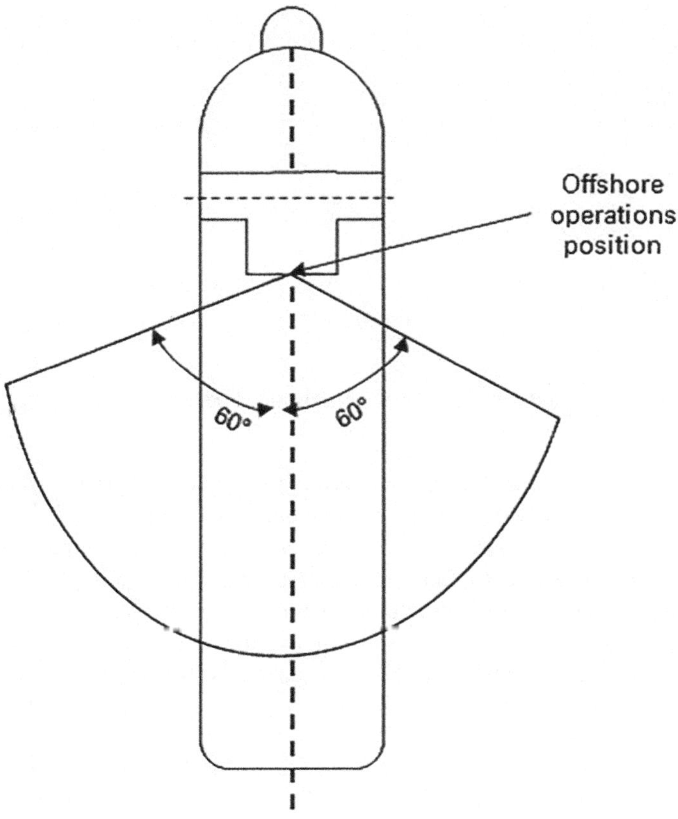

Fig. 16.6 Horizontal field of vision (offshore vessels)

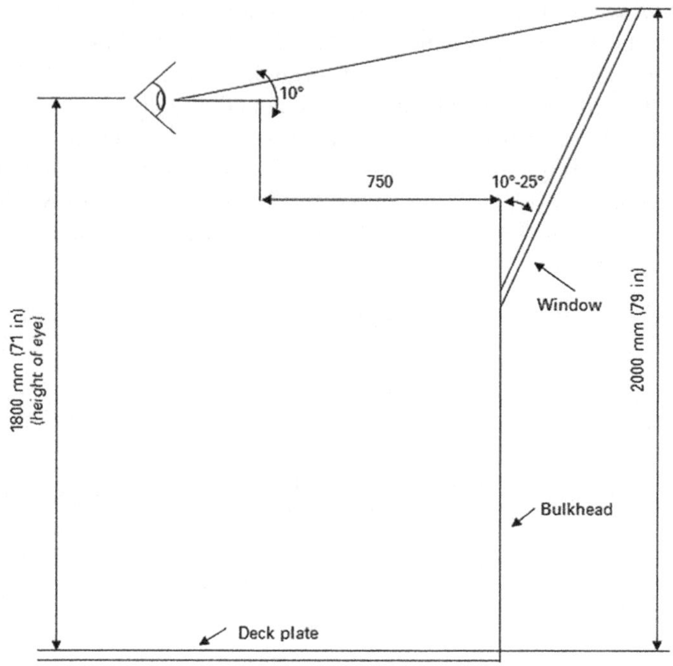

Fig. 16.6 Arrangement and inclination angle of navigation bridge windows

- *Sun blinds.* Sun blinds with minimum colour distortion. These sun blinds are to be readily removable and not permanently installed.
- *Wipers and fresh water washing systems.* Heavy-duty wipers, preferably provided with an interval function, and fresh-water wash systems. These wipers are to be capable of operating independently of each other.
- *De-icing and de-misting systems.* De-icing and de-misting systems.
- *Fixed catwalk.* A fixed catwalk with guardrails, fitted forward of the bridge windows, to enable manual cleaning of windows in the event of failure of the above systems.
- *Lower edge.* The height of the lower edge of the navigation bridge front windows above the bridge deck is to be kept as low as possible. In no case is the lower edge to present an obstruction to the forward view.
- *Upper edge.* The upper edge of the navigation bridge front windows should allow a forward view of the horizon, for a person with a height of eye of

1800 mm (5 ft 11 in.) above the bridge deck at the conning position, when the vessel is pitching in heavy seas. Class, where satisfied that an 1800 mm (5 ft 11 in.) height of eye is unreasonable and impractical, may permit a reduction of the height of eye but usually not to less than 1600 mm (5 ft 3 in.).

Vessels with an unconventional bridge design. For vessels with an unconventional design which cannot comply with the above requirements, arrangements must be submitted to Class for consideration. If the submitted arrangement achieve the satisfaction of Class in respect of the level of visibility that is as near as practical to those prescribed here, an exception notice may be issued. Articulated tug-barge units. Tugboats designed to push barges as part of an Articulated Tug-Barge unit (ATB), are required to meet the requirements set by Class. The length overall is based on the length from the barge's stem to the stern of the tug when operating as a combined unit. The visibility is to be determined based on the largest barge that the tugboat is designed to operate with as an articulated unit. Control of the vessel. The relevant workstations should be designed and positioned so that navigational and traffic surveillance/manoeuvring, docking and other tasks may be performed by the officer of the watch (OOW) in cooperation with other persons manning individual workstations. Workstations used for navigating and traffic surveillance/manoeuvring, manual steering, voyage planning and communication must not cover a working area with an axis longer than 15 m (49 ft 3 in.). Furthermore, where workstations are widely separated, talk back facilities should be provided so that unhindered communications between workstations can be achieved under all operating conditions. Conning position. An adequate conning position should be provided in proximity to the forward centre window and arranged such so as to enable the navigator(s) to view the area immediately in front of the bridge superstructure and observe all relevant information required to maintain a safe course and speed of the vessel in narrow waters, harbour areas and during final passages without interfering with the tasks of the bridge personnel. The rudder, propeller, thrust, pitch and operational mode indicators, or other means to determine and display rudder angle, propeller revolutions, direction of thrust and, if applicable, the force and direction of lateral thrust and the pitch and operational mode, are all to be readable from the conning position(s). However, if the view in the centre-line is obstructed by large masts, cranes, or other hindrances, two additional conning positions giving a clear view ahead are to be provided, one on the port side and one on the starboard side of the centre-line. These must be no more than 5 m (16 ft 5 in.) apart from each other.

Navigation and traffic surveillance/manoeuvring workstation. The navigation and traffic surveillance/manoeuvring workstation should be arranged so as to enable the OOW to carry out the required tasks and to provide them with all necessary information so that they can carry out their functions from a seated or standing working position but without being restricted to a specific location. In addition, this workstation is to be designed, arranged and located within an area having sufficient space for not less than two operators, but which would allow the workstation to be operated efficiently by one operator. *Route planning workstation.* The route planning workstation is to enable the navigator to plan the intended voyage without interfering with the actual navigation or manoeuvring of the vessel. *Monitoring workstation.* From the monitoring workstation, it should be possible to see and hear the persons at the navigation and traffic surveillance/manoeuvring workstation and steering workstations. Manual steering workstation. The workstation for manual steering is preferably to be located on the vessel's centreline. If the workstation for manual steering is located off the centreline, special steering references are to be provided (e.g., sighting marks forward). If the view ahead is obstructed by large masts, cranes, etc., the steering workstation is to be located a distance to starboard of the centreline, sufficient to obtain a clear view ahead. *Docking workstations.* The workstations for vessel docking are to enable the navigator, together with a pilot, to observe all relevant external and internal operations which will affect the safe docking of the vessel. Talk back facilities must be provided between the docking workstations and the wheelhouse when the distance from the wing extremity to the wheelhouse centreline is greater than 10 m (32 ft 10 in.).

Offshore Operations Bridge Visibility

All offshore support vessels are to be provided with an unobstructed as possible and practical view abaft from the Offshore Operations Control Area towards the cargo or work decks and machinery or specialist equipment. On supply vessels it is necessary to control both, the station keeping as well as cargo loading/unloading operations. On anchor handling/towing vessels it is necessary to control the heading, position, speed as well as deck machinery engaged for towing and/or anchor handling operations. On other types of offshore support vessels it may be necessary to control the station keeping as well as the specific service equipment installed on work deck. Where direct visual observation of machinery and equipment is unavoidably obstructed, a CCTV system suitable for use in the marine environment would be considered acceptable. Field of vision. The view of the cargo or work deck area, including loading/unloading ports on supply vessels, as well as view of the sea surface in close vicinity of the vessel shall not be obstructed. From the Offshore Operations Control Area, the horizontal field of vision should extend over an arc from right astern to at least 60° on each side of the vessel, as per Fig. 16.6. Aft window arrangements. Aft windows and their arrangements are to meet the following requirements: (1) *Framing.* Framing between navigation bridge aft windows is to be kept to a minimum to meet the structural strength and stiffness requirements and is not to be installed immediately in front of any main commanding/manoeuvring and controlling positions; and (2) *inclination angle.* The bridge aft windows are to be inclined from a vertical plane top out, at an angle of not less than 5° and not more than 15°; refer to Fig. 16.7, Arrangement and inclination angle of aft facing windows. *Glass.* Polarised and tinted windows are not permitted. *Clear view.* At all times, regardless of the weather conditions, at least two of the above aft windows are to provide a clear view, as described above. The following, or equivalent, are to be provided as well: (1) *sunscreens.* Sunscreens are to be with minimum colour distortion; are to be readily removable and are not to be permanently installed; (2) *wipers and freshwater wash systems.* Heavy-duty wipers provided with an interval function and freshwater wash systems, being capable of operating independently of each other; (3) *de-icing and de-misting systems.* De-icing and de-misting systems are to be provided; (4) *fixed catwalk.* A fixed catwalk with guardrails, fitted aft of the bridge windows is to be provided, enabling manual cleaning of windows in the event of failure of the above systems; and (5) *impact protection for windows.* Suitable protection of the aft windows against impact from any cargo being lifted on board the vessel is recommended. *Lower and upper edges.* The lower edge of the navigation bridge aft windows is to be kept as close to the bridge deck as practical. The upper edge of the navigation bridge aft windows is to be kept as close to the overhead deck as practical.

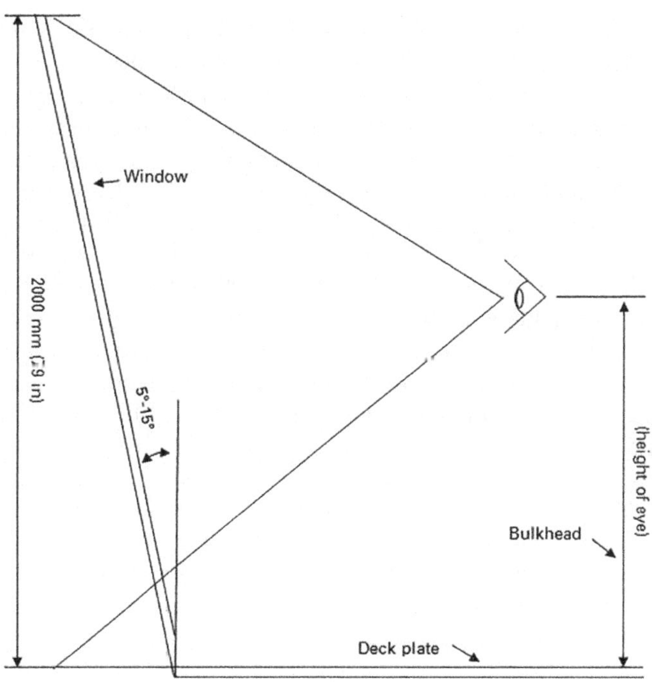

Fig. 16.7 Arrangement and inclination angle of aft facing windows

Routes and Working Clearances

Across the wheelhouse. A clear route across the wheelhouse from bridge wing to bridge wing is to be provided and its width is to be at least 1.2 m (3 ft 11 in.). However, the width may be reduced to not less than 700 mm (27.5 in.) at any single point of obstruction subject to the following:

1. There is to be clear visibility for individuals on either side of the obstruction to see each other as they approach the area of reduced width.
2. The length of the obstruction along the passageway is not to exceed 1 m (39.4 in.).
3. There is sufficient room on either side of the obstruction to allow one individual to step aside and allow another individual to pass through [i.e., at least 1.2 m (3 ft 11 in.)].
4. The above actions will not interfere with any crew member at a station required to be continuously Manned.
5. The attending Class approved surveyor is satisfied that two individuals travelling in opposite directions can pass through the area of reduced width with only a very brief pause by one of the two.

From lower decks. There are to be no obstructions between the points of entry to the bridge wings and wheelhouse from lower decks and the clear route required above. *Between workstations.* The distance between adjacent workstations is to be sufficient to allow unobstructed passage. To this end, the free passageway should be at least 0.7 m (2 ft 4 in.) wide. The workstation operating area must form part of the workstation and not of the passageway. *Front passage.* The distance from the front bulkhead, or from any workstation and installations placed against the front bulkhead, to any workstation or installations placed away from the bridge front is to be sufficient for two persons to pass each other. This distance is preferably to be 1 m (3 ft 3 in.) but in no case less than 0.8 m (2 ft 7 in.). *Clear height.* The clear ceiling height in the wheelhouse is to be designed with respect to the installation of overhead panels and instruments. To this end, the clear height between the bridge deck surface covering and the underside of the deck head beams is to be at least 2.25 m (7 ft 5 in.). The lower edge of deck head mounted equipment is to be at least 2.1 m (6 ft 11 in.) above the deck in open areas, passageways and at standing workstations. *Workstations.* Instruments providing visual information to more than one person on duty are to be located for easy viewing by all users concurrently, or if this is not possible, the instruments are to be duplicated. Instruments displaying information to more than one workstation may be located above the front windows if dimensions allow; such instruments are those denoting the vessel's heading, wind, water depth, speed, rate of turn, rudder angle, propeller revolutions (r/min), propeller pitch and time. Configuration and dimensions as outlined below do not apply to radar consoles. *Configuration.* In general, workstations are to be divided into two parts where possible:

- *Vertical part.* Instruments dealing with information/presentation of data are to be placed in the vertical part.
- *Horizontal part.* Controls of the relevant equipment are to be placed in the horizontal part.

Dimensions

- *Height.* The height of workstations is not to interfere with the navigating bridge window's view requirements as discussed above.
- *Width.* Based on sound ergonomic principles, the width of workstations designed for single person operation should not exceed 1.6 m (5 ft 3 in.).
- *Chart table.* The chart table is to be large enough to accommodate all chart sizes normally used internationally for marine traffic. The dimensions of the chart table are to be as follows: width, not less than 1.2 m (3 ft 11 in.); depth, not less than 0.85 m (2 ft 9 in.); height, not less than 0.9 m (2 ft 11 and not more than 1 m (3 ft 3 in.). Additionally, the chart table is to be provided with 10 mm (0.3 in.) openings in the front and back of the table to accommodate charts which are larger than the chart table.

Instruments and controls

- *General.* Instruments and controls are to be grouped according to their main functions such as navigating and traffic surveillance/manoeuvring, and communication.
- *Line of sight.* Each instrument and control are to be placed with its face normal to the navigator's line of sight, or to the mean value if the navigator's line of sight varies through an angle.
- *Glare.* To avoid glare, all instruments and controls are to be positioned relative to the operator considering the surrounding light sources.
- *Transparent covers.* Transparent covers fitted over instrument(s) are to minimise reflections.
- *Symbols/labels.* The purpose of each control is to be clearly illustrated by symbols where standard symbols have been internationally adopted or indicated by a label in English.

Lighting and illumination

- *General.* A satisfactory level of lighting is to be provided to enable personnel to complete required bridge tasks at sea and port in both daytime and night. To this end, individual task areas are to have a greater luminance than the general lighting level.
- *Lighting in dark hours.* Relevant equipment fitted on the bridge is to be able to be discerned during hours of darkness. This should be achieved via internally or externally located lighting.

Red light is to be used to maintain dark adaptation whenever possible in areas or on items of equipment, other than the chart table, requiring illumination in the operational mode. This is to include instruments and controls on the bridge wings. Additionally, to prevent red lights in the wheelhouse from being mistaken for navigation lights by another vessel, indirect low level red lighting should be fitted at deck level, especially for internal doors and staircases. *HVAC system.* An adequate HVAC system is to be provided in order to maintain the temperature of the wheelhouse within the comfort range of 14 °C (57 °F) to 30 °C (86 °F). *Sound signals.* Fixed sound signals are to be placed as high as practicable and if possible, forward of the bridge. External sound signals from vessels and fog signals that are audible on open deck, are to be also audible inside the wheelhouse; to this end, a sound reception system (of a recommended frequency range of 70 Hz to 700 Hz) is to be provided to reproduce such signals inside the wheelhouse (the opening of doors or windows is generally not to be accepted as an equivalent solution). *Noise levels.* The noise level on the bridge is not to interfere with verbal communication, mask audible alarms or be uncomfortable to the bridge personnel. In this respect, the ambient noise level on the bridge in calm weather is not to exceed 65 dB(A). The noise of ventilation fans, engine intake fans and other noise sources are to be dampened from the bridge operational area by suitable locations for fans and their associated housing.

Surfaces

- Glare-free. All prepared surfaces are to be glare-free.
- Non-slip. The flooring throughout the bridge is to be provided with non-slip surfaces, effective in both wet and dry conditions.

Drainage. Bridge decks outside, including the wings, are to be provided with means for drainage. *Toilet facilities.* Toilet facilities are to be provided on or adjacent to the bridge, on the same level. *Doors.* Doors to the bridge wings are to be capable of being operated with one hand. Means are to be provided to hold the doors open. *Refreshment facilities.* Refreshment facilities and other amenities provided for the bridge personnel are to include means for preventing damage to bridge equipment and injury to personnel resulting from the use of such facilities and amenities.

Safety of personnel

- *Sharp edges and protuberances.* There are to be no sharp edges or protuberances which could cause injury to personnel.
- *Handrails or grab rails.* Sufficient handrails or grab rails are to be fitted to enable personnel to move or stand safely in bad weather.
- *Seat securing.* Where provisions for seating is made in the wheelhouse, means for securing same are to be provided, having regard to storm conditions.

Tests and Sea Trials

During sea trials, the navigational equipment and systems must be tested to the satisfaction of the attending Class approved surveyor and carried out in accordance with the approved test programme.

Chapter 17
Requirements for Class Notations Relating to Bridge Layout and Equipment

General

Vessels complying with the provisions set out in Chaps 13–15 of this book may be assigned the Class notation **NBLES** (Navigational bridge layout and equipment/systems) or the Class notation **NBLES+** (or equivalent). The equipment required for **NBLES** is listed in Table 17.1 Navigation equipment for **NBLES** notations (found at end of the chapter) and for **NBLES+** are listed in Tables 17.1 Navigation equipment for **NBLES** notations and 17.2, Additional navigation equipment for **NBLES+** notations, also at the end of this chapter.

Documentation, Type Approval and Performance Standards of Navigational Equipment

The manufacturer or assembler of the relevant navigational equipment as stipulated throughout this and the preceding chapters as they apply to **NBLES** and **NBLES+** Class notations (or equivalent) are required to provide documented evidence indicating that the equipment meets the Class criterion. *Type-approved equipment.* Navigational equipment is to be type approved to the satisfaction of the vessel's Flag State Administration in conformity with appropriate performance standards specified in the respective IMO Resolutions and Circulars. *IMO's performance standards.* In general, relevant navigational equipment is required to comply with IMO Resolution A.694(17), "General requirements for shipborne radio equipment forming part of the global maritime distress and safety system (GMDSS) and for electronic navigational aids", and those discussed throughout this Chap 15 and in this chapter. For ready reference, the reader may wish to review the compilation of performance standards in IMO Publication 978–88.04.E (1988), "Performance Standards for Navigational Equipment".

A. A. Olsen, *Applying Physical Ergonomics to Modern Ship Design*,
https://doi.org/10.1007/978-3-031-57974-5_17

Table 17.1 Specific equipment or system requirements

Course information system	Bridge navigational watch alarm system
Automatic steering system	Route planning system
Speed measuring system	Vessel's automatic identification system
Depth measuring system	Telephone system
Radar system	Sound reception in bridge (if fitted)
Automatic traffic surveillance system	Radio communication system
Position-fixing system	

Manual Mode of Operation

Navigational systems intended for automatic operation are required to be fitted with manual mode provisions which enable the officer of the watch to take the appropriate action in the event of failure of the automatic system. For automatic systems for which this is not possible, an alternative system is to be provided.

Alarm Systems

Characteristics. Alarm systems are to be of the self-monitoring type and designed so that a fault in the alarm system is to cause it to fail to the alarmed condition. Additionally, they are not to react to normal transient conditions or spurious signals. Alarms are to be both audible and visual and are to flash when first activated. *Audible alarm circuits.* A fault in the visual alarm circuits must not affect the operation of the audible alarm circuits. *Acknowledgment.* Alarms are required to be acknowledged by manually changing the flashing display of the incoming alarm to a steady display and by silencing the audible signal; the steady state light display is to remain activated until the fault condition is rectified. Alarming of other faults that may occur during the acknowledgment process is not to be suppressed by such action and is to be alarmed and displayed accordingly. *Disconnection and resumption of functions.* Alarm circuits may be temporarily disabled for maintenance purposes or during initial start-up of machinery provided that such action is clearly indicated to the officer of the watch. However, such alarms are to be automatically reactivated after a pre-set time has elapsed. *Built-in testing.* Alarm systems are to be provided with effective means for testing all audible and visual alarms and indicating lamps without disrupting the normal equipment or system operation.

Table 17.2 Navigational equipment for NBLES (or equivalent)

Workstation for	Main functions to be performed	Item	Equipment	Remarks
Navigation and Traffic Surveillance/ manoeuvring [refer to Note 1]	Observation of all vessels and objects	A1	Gyro compass heading indicator	For **NIBS** or similar notations, two independent gyro compasses are to be provided on the bridge Refer to Note 2
	Recognising dangerous situations	A2		
	Deciding on collision avoidance actions	A3		
	Checking vessel's own signal selector switch	A4	Rudder pump	
	Checking own course and speed	A5	Steering mode selector switch	
	Keeping and/or changing own course and speed (track keeping)	A6	Steering position indicator	
	Checking own position	A7	Rudder angle indicator	
	Handling own internal communication on board	A8	Pitch indicator	For controllable-pitch propeller
	Handling communication vessel/ vessel, and vessel/shore (VHF)	A9	Rate-of-turn indica- tor and controller	For vessels 50,000 GT or greater
	Releasing alarms	A10	Speed and distance indicator	For **NIBS** or similar notations, the speed measuring system is to be independent of the position fixing systems
	Perception of group alarms with aids for decision-making	A11	Depth water indicators with adjustment controls	
	Observation of weather and seaway	A12	9 GHz radar	For vessels 3,000 GT and above, an additional independent radar together with a change-over switch is to be provided

(continued)

Table 17.2 (continued)

Workstation for	Main functions to be performed	Item	Equipment	Remarks
	Acknowledging watch check alertness alarm	A13	Automatic traffic surveillance system including ARPA	
	Keeping deck log (a dictaphone may be used)	A14	Position fixing equipment/system including automatic. Visual position indicator	Two types of receivers are to be provided. One of the systems is to be GPS or equivalent, and the other: Decca, Loran-C, GLONASS, or other means
	Sounding signals	A15	Officer of the watch check alertness acknowledgment device	
		A16	Back-up navigator call alarm device	Two-way communication wireless portable device to be provided
		A17	Facilities for use of navigational charts	This may be separated from the navigation and traffic surveillance/ manoeuvring workstation
		A18	Vessel's AIS	
		A19	Propulsion engines/thrusters controls including emergency stops	
		A20	Propulsion engine revolution	If reduction geared engine
		A21	Propeller revolutions indicator	
		A22	Wind direction and velocity indicator	
		A23	Air and water temperature indicator	
		A24	Telephone system	
		A25	Radio communication equipment	

(continued)

Table 17.2 (continued)

Workstation for	Main functions to be performed	Item	Equipment	Remarks
		A26	NAVTEX automatic receiver and recorder	For navigational and meteorological warning purpose. To comply with IMO Resolution A.617(15) - "Implementation of the NAVTEX system as a Component of the Worldwide Navigational Warning Service"
		A27	Signal transmitter for: • Ship's whistle • Automatic device for fog signal • General alarm • Morse signalling light	
		A28	Search light controls	E.g., searchlight on/off switch
		A29	Controls for windscreen wiper, washer, and heater	
		A30	Night vision equipment	
		A31	Sound reception system	
		A32	Workstation lighting control device	
		A33	HVAC controls	
		A34	Clock	
		A35	Group alarms and reset controls	See also item B17 of this table
Monitoring [refer to Note 1]	Observation of all vessels and objects	B1	Gyro compass heading indicator	
	Recognising dangerous situations	B2	Rudder angle indicator	
	Handling own internal communication onboard	B3	Pitch indicator	For controllable-pitch propeller

(continued)

Table 17.2 (continued)

Workstation for	Main functions to be performed	Item	Equipment	Remarks
	Handling communication vessel/ vessel, and vessel/shore	B4	Rate-of-turn indica- tor and controller	For vessels 50,000 GT or greater
	Perception of group alarms with aids for decision-making		Speed and distance indicator	For **NIBS** or similar notations, the speed measuring system is to be independent of the position fixing systems
	Releasing alarms		Depth water indicator	
	Observation of weather and seaway		Radar	
	Acknowledging watch check alertness alarm		Officer of the watch check alertness acknowledgment device	
	Keeping deck log		Propulsion engine revolution	
	When workstation is occupied by an additional navigator, provides assistance to naviga- tor at the navigation and traffic surveillance/ manoeuvring workstation		Propeller revolutions indicator	
	When workstation is occupied by a pilot, advises to vessel's command			
		B11	Telephone system	
		B12	Radio communication equipment	
		B13	Signal transmitter for whistle	
		B14	Controls for windscreen wiper, washer, and heater	

(continued)

Table 17.2 (continued)

Workstation for	Main functions to be performed	Item	Equipment	Remarks
		B15	Workstation lighting control device	
		B16	Clock	
		B17	Required alarms and reset controls	In addition to the alarms/ indicators which may be required by the various IMO Resolutions referenced in this book and pertinent Flag State Administration, the following conditions are to be alarmed at the monitoring workstation: (a) Off-heading (b) Off-track (c) Planned route deviation (d) Pre-warning of approach waypoint, and closest point of approach (e) Off-preset water depth (f) Gyro compass failure (g) Failure of alarms (h) Failure of power supply to distribution panel serving relevant equipment (Alarming of the above conditions at the monitoring workstation is not a substitute for alarming at the required relevant workstations)
Manual steering (Helmsman's) [refer to Note 1]	Steering vessel according to rudder angle orders	C1	Gyro compass heading indicator (repeater)	For **NIBS** or similar notations, two independent gyro compasses are to be provided on the bridge
	Steering vessel according to course instruction	C2	Magnetic compass heading indicator	
	Steering vessel following land-mark/sea marks	C3	Course reminder (set course) indicator	

(continued)

Table 17.2 (continued)

Workstation for	Main functions to be performed	Item	Equipment	Remarks
	Acknowledging watch check alertness alarm	C4	Manual steering with override and selector control switches including steering wheel/ steering lever	
		C5	Rudder angle indicator	
		C6	Rate-of-turn indicator	For vessels 50,000 GT or greater
		C7	Watch check alertness acknowledgment device	
		C8	Telephone system	
		C9	Controls for windscreen wiper, washer, and heater	

Notes:
1. As the navigation and traffic surveillance/manoeuvring, monitoring and manual steering workstations are functionally interrelated and usually installed in proximity from each other, considerations will be given to the omission of duplicate required equipment at each of the aforementioned workstations
2. Master gyrocompass may be located in the electrical/instrumentation room and the gyrocompass repeaters on the bridge to meet this requirement

Computerised Equipment

Where computerised equipment is interconnected through a computer network, failure of the network must not prevent individual equipment from performing their individual functions.

Power Supply

Power sources. Electrically operated systems and equipment covered in Chap 15 and in this chapter of this book are to be connected to distribution panels placed in an accessible position on, or adjacent to, but at the same level of the bridge; each item of equipment is to be individually connected to its distribution panel. These panels are to be supplied by two exclusive circuits: one fed from the main source of electrical power and one fed from an emergency source of power. The power supplies to the distribution panels are to be arranged with automatic changeover facilities between the two sources. *Emergency service.* The vessel's emergency source of power is to be of sufficient capacity to supply the navigational related loads discussed in this book, in addition to other electrical loads as required by the vessel in her normal operating conditions. *Loss of power supply.* Following a loss of power which has lasted for 45 s or less, navigational equipment/systems essential for the performance of primary bridge navigational functions (those are: gyro compass, radar, position-fixing system and electronic chart system) are to be automatically reinstated to their pre-power loss configuration upon recovery from blackout. All others are to be readily reinstated within five minutes, with minimal operator intervention, by virtue of the emergency source of power and, where necessary, by an uninterruptible power system (UPS). Loss of power to the distribution panels is to activate an alarm.

Navigational Systems

In general, navigational equipment/systems installed on board vessels are to be arranged such that the failure of one piece of navigational equipment cannot reduce the vessel's ability to perform the functions specified in this chapter. *Heading information system.* The vessel is to be provided with continuous heading information at the appropriate workstations and at the main steering position. To this end, a magnetic compass and a gyro compass having the capability to determine the vessel's heading in relation to the geographic (true) North are to be provided. The magnetic compass and the gyro compass are to comply with the IMO's Resolution A.382(X), "Performance Standard for Magnetic Compass" and the IMO's Resolution A.424(XI), "Performance Standard for Gyro Compass", respectively. In addition, a pelorus or compass bearing repeater, or some other means to take bearings over an arc of the horizon of 360°, and a gyro compass heading repeater is to be provided. The following is to be complied with. To ensure the availability of heading information the vessel is to be fitted with an independent gyro compass. Some means for taking optical bearings should be made available on board the vessel. Means are to be provided for the correction of errors induced by speed and latitude. When the position of the vessel cannot be received, the heading of the vessel is to be maintained and such condition is to be alarmed.

Steering system. Means for manual and automatic steering of the vessel are to be provided. The steering system is to comply with the provisions discussed in this section. The automatic pilot with an on/off indicator is to comply with IMO Resolution A.342(IX), as amended by Annex 3 to MSC.64(67) "Performance Standard for Automatic Pilots". The automatic pilot equipment is to be monitored by an off-heading alarm, which is to activate, when the actual heading deviates from a pre-set heading beyond a pre-set limit, in the wheelhouse. This alarm is to be derived from a system independent from the automatic steering system. The off-heading alarm is not to be released when setting a new course reference. The off-heading alarm may receive input from the gyrocompass, provided the compass is independent of the automatic steering gear. The magnetic compass may be used as a signal input, provided that same is used as a back-up to the gyro compass. A heading control system is to work together with a track control system, adjusting its heading for drift. The heading control system is to comply with IMO Resolution MSC.64(67), Annex 3 "Recommendation on Performance Standards for Heading Control Systems" and the track control system is to comply with IMO Resolution MSC.74(69), Annex 2 "Recommendation on Performance Standards for Track Control Systems". An overriding control device is to be provided at the navigation and traffic surveillance/ manoeuvring workstation. The override control is to enable instant take-over from the autopilot as well as from the manual steering station. For vessels of 50,000 GT DWT and above, a rate of turn indicator is to be provided. The rate-of-turn indicator is to comply with IMO Resolution A.526(13), "Performance Standards for Rate-of-Turn Indicators". For vessels of 50,000 GT DWT and above, a track controller, or other means to automatically perform turns with a preset radius or rate of turn is to be provided.

Speed measuring system. The vessel is to be fitted with the means for measuring speed and distance through the water. Vessels above 50,000 GT DWT are also to be capable of measuring speed in the forward, and athwartship directions. The speed log is to comply with IMO Resolution MSC.96 (72), Annex 14 "Performance Standards for Devices to Indicate Speed and Distance" as amended by IMO Resolution 334(90), "Adoption of Amendments to Performance Standards for Devices to Indicate Speed and Distance". *Depth measuring system.* The vessel is to be fitted with an echo sounder or equivalent for measuring the water depth under the keel. An alarm is to be initiated when the water depth beneath the vessel is less than the predetermined value. The echo sounder is to comply with IMO Resolution MSC.74(69), Annex 4 "Performance Standards for Echo-Sounding Equipment". *Radar system.* A 9 GHz radar or other means is to be provided to determine and display the range and bearing of search and rescue transponders (SART) and of other surface craft, obstructions, buoys, shorelines and navigational marks to assist in navigation and in collision avoidance. The radar installation is to comply with IMO Resolution A.477(XII) as amended by Annex 4 to MSC.64(67), "Performance Standards for Navigational Radar Equipment" and the following:

- If the vessel's minimum reflective cross section area is less than 100 m2 (328 ft2), a radar reflector or other means is to be provided so that the vessel can be detected by another vessel navigating by radar. The radar reflector must comply with IMO Resolution 164(78) "Revised Performance Standard for Radar Reflectors".
- For vessels of 3,000 GT DWT and above, a second radar, independent of the 9 GHz radar, is to be provided.

Additionally, a change-over switch between the radars together with the means to bypass the switch is to be provided. *Automatic traffic surveillance system.* The vessel is to be fitted with an automatic traffic surveillance system to automatically plot the range and bearing of other vessel's ability to determine collision risk. This traffic surveillance system is to comply with IMO Resolution A.823(19), "Performance Standards for Automatic Radar Plotting Aids (ARPA)". The ARPA function may be independent or built into the radar equipment. The system is to be based on the assumption that all floating objects may come into a collision course with own vessel if the object's course is changed by 45° with its speed maintained. An alarm is to be given to the navigator at a time which is to be adjustable in the range of 6–30 min, having regard to the danger, time to closest point of approach (T)CPA). The system is to feature the following capabilities:

- True motion and relative motion modes.
- Daylight-visible display.
- Guard zone system, featuring adjustable parameters, notably alarm set for the CPA and TCPA.
- Simulator function showing the likely effects of a course or speed change in relation to tracked targets.
- Incorporated self-checking properties.

For vessels of 3000 GT DWT and above, automatic acquisition and tracking of 20 radar targets and the means to simulate a trial manoeuvre are to be provided. Additionally, a heading or track controller or other means to automatically control and keep to a heading and/or track is to be provided and for vessels of 10,000 GT DWT and above, means to automatically control and keep the vessel in a straight track is to be provided. *Position fixing system.* The vessel is to be fitted with at least two types of automatic position fixing systems for the waters she is to navigate. These systems are to be capable of automatically and continually determining and displaying the vessel's position. One of the systems is to be GPS enabled or equivalent, and the other either Decca, LORAN-C, GLONASS, or an equivalent means. The positioning equipment/system is to comply with IMO Resolution A.819(19), "Performance Standards for Shipborne Global positioning System (GPS) Receiver Equipment" and the following:

- Means for manually inputting required data in case of sensor failure, and the means to indicate the system's mode of operation i.e., manual or automatic, are to be provided.
- A means to self-test the major function(s) of the system is to be provided.
- Decca receivers are to comply with IMO Resolution A.816(19), "Performance Standard for Shipborne Decca Navigator Receivers"; Loran-C receivers are to comply with IMO Resolution A.818(19), "Performance Standard for Shipborne Loran-C and Chayka Receivers"; GLONASS receivers are to comply with IMO Resolution MSC. 113(73), Annex 26 "Performance Standard for Shipborne GLONASS Receiver Equipment".

Bridge navigational watch alarm system (BNWAS). Means to monitor the alertness of the officer of the watch and alarm other bridge personnel if disability occurs is to be provided. The BNWASs are to comply with IMO Resolution 128(75), "Performance Standards for a Bridge Navigational Alarm System". Additionally, conditions of danger to navigation caused by traffic or improper course-keeping in relation to planned route are to be monitored and such adverse conditions are to be alarmed at the bridge and at the locations specified herein. *Officer of the watch alertness check system.* (1) *General.* A system is to be provided to monitor the alertness of the OOW present on the bridge. This system is not to cause undue interference with the performance of bridge function(s) and it is to be designed and arranged that it cannot be operated in a unauthorised manner (i.e. bypassed). The system is to be connected to the alarm transfer system. (2) *Periodic verification.* The system used for periodic verification of the watch alertness system is to be adjustable up to 12-min intervals, and it is to be arranged so that only the vessel's master has access for enabling and disabling it (i.e., removing the fuses or keeping the acknowledgment button permanently depressed) and for setting the appropriate intervals for a periodic verification. (3) *Acknowledgment of alertness check alarm.* The system is to provide for the acknowledgment by the officer of the watch at the navigation and traffic surveillance/manoeuvring workstation and at the monitoring workstation. (4) *System failure alarm.* An alarm is to operate on the bridge and at the spaces described later in this chapter in the event of a failure of the alertness check alarm system.

Alarm transfer system. (1) *General.* A fixed alarm transfer system is to be provided and connected to all vessel navigating officers' cabins and public rooms. (2) *Transfer of alarms.* Alarms per the "Remark" column of item B17, a through h, in Table 17.1 are to be automatically transferred to the master's cabin if not acknowledged at the bridge within 30 s. Additionally, a selector switch is to be provided in the event the master deems it necessary to also transfer the aforementioned alarms to the selected back-up navigator's cabins and public rooms. (3) *Back-up navigator call-alarm.* Provisions are to be made at the bridge to activate the back-up navigator call-alarm. This alarm is to be audible in all the spaces described later in this chapter. A fixed installation may serve this purpose. (4) *Portable communication device.* A wireless portable device allowing two-way communication with the officer of the watch is to be provided for use by the back-up navigator when attending locations not connected to the fixed installation.

Passage planning. The vessel is to carry official charts sufficient to enable passage planning and monitoring for the intended voyage. An ECDIS is also accepted as meeting the chart carriage requirement. The ECDIS is to comply with IMO Resolution MSC.232(82), "Adoption of the Revised Performance Standards for Electronic Chart Display and Information Systems (ECDIS)". Ships engaged on international voyages are to comply with carriage requirements for ECDIS as specified in SOLAS V/19 Para 2.10. However, where the charting function is partially or fully effected via electronic charts, a back-up means is to be provided. In addition, an alarm is to be given in case of deviation from the planned route, which is to be adjustable having regard to the time to danger of grounding. *Vessel's automatic identification system.* A vessel's automatic identification system (AIS) is to be fitted on board the vessel to provide automatically to appropriately fitted shore stations, other vessels and aircraft, needed navigational related information such as vessel's identity, type, position, course, speed, navigational status, etc., and other safety related information, and to automatically receive such information from similarly fitted vessels and to monitor and track vessels and to exchange data with shore based facilities.

The AIS is to comply with IMO Resolution MSC.74(69), Annex 3 "Recommendation on Performance Standards for an Universal Shipborne Automatic Identification System". *Propulsion engine and thruster controls.* Some form of means for controlling the propulsion engines/thrusters are to be provided in the navigational bridge.

Telephone system. The vessel is to be fitted with an automatic telephone system, which is to comply with the following. In addition, a backup telephone system (e.g., sound powered telephone, etc.) that can operate independently of the power supply from the vessel's main or emergency system is to be installed. The telephone system here may be combined with voice communication system. Both automatic and backup telephone systems are to enable two-way communication between the bridge and the:

- Master's and chief engineer's accommodations.
- Radio room (when located outside the bridge area).
- Steering gear compartment.
- Local control position of propulsion machinery/thruster.
- Engine control room (ECR).
- Fire control station (FCS).
- Cargo control room, if provided.
- Emergency generator room (EGR).

Only an automatic telephone system is required for two-way communication between the bridge and navigating officer's cabins and public rooms. The automatic telephone network is to be designed to carry at least two simultaneous calls. The automatic telephones on the bridge and propulsion machinery control room are to have priority function over any other extension. A list of all relevant telephone extensions is to be permanently posted and clearly displayed adjacent to each telephone. The back-up telephone system is to be independent of the automatic telephone system and both telephone systems are not to be affected by a failure of either system.

Nautical Radio Communications Systems

Vessels are required to be fitted with means for nautical radio communication with other vessels as well as means for communication with tugboats and mooring stations aboard and ashore.

Workstations (Required Equipment)

As a minimum, in addition to the alarms/indicators invoked in the various IMO resolutions referenced throughout this book, the equipment listed in Table 17.1 are to be fitted at the various workstations to enable the officer of the watch and other operators to carry out the required tasks.

Operating/Technical Manuals

An operating/technical manual which is consistent with the information and criteria upon which the notations **NBLES** and **NBLES+** are based is to be placed on board the vessel for the guidance of the operating personnel. The operating/technical manual is to give clear guidance to the vessel's personnel in respect of the vessel's capability, limitations and procedures to follow when navigating the vessel with the required manning on the bridge. The operating/technical manual is to include the following, as a minimum:

- The vessel's registered name and vessel IMO registration number.
- Simplified diagrams of the systems described throughout this chapter.
- The vessel's navigating and manoeuvring capabilities (i.e., particulars of propulsion machinery and steering system, vessel's speed, vessel's stopping ability, vessel's turning ability, etc.).
- Navigational procedures including transfer of alarms to the back-up navigator, and details of the routines, duties and responsibilities of each of the relevant personnel associated with the bridge operation of the vessel.
- The periodical testing procedures for relevant navigational equipment/systems.

The operating/technical manual must be submitted to Class for review solely to ensure the presence of the above information which is to be consistent with the vessel's design information and navigational capabilities. The operation of the vessel is not a condition of the assigned Class notations. Any modifications made to the approved bridge layout, field of views and navigational equipment/systems are to be approved by Class. The operating/technical manual is to be updated accordingly and submitted to Class for review.

Tests and Sea Trials

During sea trials, navigational equipment and systems are to be tested to the satisfaction of the attending Class approved surveyor in accordance with the approved test programme. The test programme is required to include the following test details.

Applicable to all relevant navigational equipment. Prior to testing, all relevant navigational equipment/systems are to be satisfactorily checked, calibrated and operated by the representative of the manufacturer or the equipment supplier who is to issue an affidavit to such effect for the review of the attending Class approved surveyor. Automatic resumption of primary bridge navigational equipment/systems functions are to be demonstrated following a blackout simulation of a period of 45 s. Similarly, resumption of all other relevant non-primary bridge navigational equipment/systems functions are to be satisfactorily returned to normal operating functionality following a blackout simulation period of five minutes. *Specific equipment or system.* The test plan and programme should provide testing details for the following equipment or systems (Table 17.3).

Table 17.3 Navigational equipment for NBLES+ (or equivalent)

Workstation for	Main functions to be performed	Item	Equipment	Remarks
Docking (bridge wings)	Giving instructions, performing and controlling change of course	D1	Gyro compass heading indicator	For **NIBS** or similar notations, two independent gyro compasses are to be provided on the bridge Refer to Note 1
	Giving instructions, performing and controlling change of speed	D2	Steering position indicator	
	Giving instructions, performing and controlling change of thruster	D3	Rudder controls	
	Handling communication with manoeuvring stations	D4	Rudder angle indicator	
	Handling communication with tugs, pilot boat	D5	Pitch indicator	
	Watching water surface along vessel's side	D6	Rate-of-turn indicator	
	Releasing signals	D7	Propulsion engines/ thrusters controls	
	Acknowledging watch check alertness alarm	D8	Propulsion engine revolution	

(continued)

Table 17.3 (continued)

Workstation for	Main functions to be performed	Item	Equipment	Remarks
		D9	Propeller revolutions indicator	
		D10	Lateral thrust and lateral movement of vessel indicator	
		D11	Longitudinal movement of vessel, indicator	
		D12	Wind direction and velocity indicator	
		D13	Depth water indicators	
		D14	Officer of the watch check alertness acknowledgment device	
		D15	Whistle controls	
		D16	Search light and Morse lamp controls	
		D17	Telephone system	
		D18	Radio communication Equipment	
		D19	Workstation lighting control device	

Notes:

1. Master gyrocompass may be located in the electrical/instrumentation room and the gyrocompass repeaters on the bridge to meet this requirement

Chapter 18
Bridge Arrangements and Working Environment for Coastal/Offshore Vessels

General

This chapter outlines the various requirements for the navigation bridge and wheelhouse design, including field of vision, blind sectors, bridge windows, bridge configurations, wheelhouse arrangement, workstation configuration and location of equipment within workstations on vessels engaged in coastal and offshore operations. Vessels which comply with the provisions set out in Chaps. 15–18 of this Part 2 may be assigned the Class notation **NBLES (COS)** (Navigational Bridge Layout and Equipment/Systems for Coastal and Offshore Services) or equivalent. This notation improves and optimises the work environment within the bridge area to assist the bridge officer of the watch/navigator/master/pilot to more easily navigate the vessel in coastal and restricted waters, therein reducing the opportunities for potential risks of collision, grounding or weather damage.

Plans and Data to Be Submitted to Class

The plans and data as per Chap. 1 are to be submitted for Class approval and/or information as appropriate when requesting the assigning of Class notation **NBLES (COS)** or equivalent. Moreover, the manufacturer or assembler of the relevant navigational equipment installed on the ship's bridge is required to provide documented evidence indicating that the equipment meets the criteria specified in Chap. 15.

A. A. Olsen, *Applying Physical Ergonomics to Modern Ship Design*,
https://doi.org/10.1007/978-3-031-57974-5_18

Navigation Bridge Design and Layout

This chapter outlines the various requirements for the navigation bridge and wheelhouse design, including fields of vision, blind sectors, bridge windows, bridge configurations, wheelhouse arrangement, workstation configuration and location of equipment within workstations. These requirements apply to ships where visibility procedures are to be established as prescribed in the COLREGs, as amended (1972, Rules 5, 6, 7 and 8) with respect to maintaining safe speed, posting of lookouts and watches and use of sound signals and collision avoidance using radar and plotting devices. *Field of vision around the vessel.* It must be possible for the officer of the watch to observe all objects necessary for navigation, such as ships and lighthouses, in any direction from inside the bridge. There is to be a 360° field of view for an observer moving within the bridge. If the ship is unable to satisfy the requirements of a 360° field of view around the vessel for an observer moving within the wheelhouse an alternative equivalent means may be acceptable such as cameras covering any obstructed views or blind sectors. Alternatively, other technologies such as sensors may also be considered, provided they are shown to be equally effective. *Navigating and manoeuvring workstation field of view—vertical view.* Above the horizontal plane, a vertical angle of view of not less than 5° above a horizontal line, extending from height of eye of 1800 mm (71 in.) in forward direction, is to be provided irrespective of any special equipment, helicopter decks or other obstructions outside of the wheelhouse. Below the horizontal plane, any elevated structure/equipment or cargo obstructing the sea surface close to the ship in excess of 1000 m (3280 ft) within the 180° sector forward of the athwart ship is to be considered as a blind sector and necessary calculations are to be conducted (height of eye of 1800 mm (71 in.). ***Note:*** *if a height of eye of 1800 mm (71 in.) is unreasonable and impractical, it may be reduced, but not to less than 1600 mm (63 in.). Navigating and manoeuvring workstation field of view—obstruction of sea surface.* While standing at the workstation in the conning position during navigation and manoeuvring of the vessel, a person is to be able to observe all objects, the shoreline and water surface to a distance at least 500 m (1640 ft) from the hull within the sectors from 10° on either side of the bow to 90° on both sides, under all conditions of draught, trim and with deck cargo. Refer to Fig. 18.1, Field of vision—obstruction of sea surface. If the ship is unable to satisfy the above requirements, an alternative means may be acceptable such as camera covering any obstructed views or blind sectors.

Navigating and manoeuvring workstation field of view—horizontal view astern. Leading lights and markers are to be provided for visual reference at the stern of the ship while navigating and manoeuvring from the workstation to help to avoid grounding and provides a view the sea surface in close vicinity of the vessel stern. A horizontal field of vision of the vessel stern is to extend over an arc from right astern to at least 5° on each side. No blind sectors are to occur within this 10° field of vision sector. *Navigating and manoeuvring workstation field of view—vertical view astern.* The lower edge of the navigation bridge aft windows is to be kept as close to the bridge deck as practical so the vertical view through the window from the workstation during navigating and manoeuvring enables a view of the sea surface in all conditions at a distance not more than 2000 m (6561 ft) abaft the stern. The upper edge of the navigation bridge aft windows is to be kept as close to the overhead deck as practical, and the upper edge of the window(s) is not to be less than 2000 mm (78 in.) above bridge deck surface. The maximum height from deck to the lower edge of the window(s) is to be 1000 mm (39.3 in.) or at the least the lower part of the window is to allow a clear view of the sea surfaces aft of the vessel while viewing from the navigating and manoeuvring workstation. Only the superstructure or deck may be concealed when viewed from normal operating position at this workstation. *Navigating and manoeuvring workstation field of view—vertical view from the docking workstation.* During vessel manoeuvring, while standing at a docking workstation (e.g., console from the bridge wing), the navigator is to be capable to observe the parallel ship's side, both forward and aft of the vessel (height of eye 1600 mm (62.9 in.). The total length of the visibility of the ship's side is to be not less than 2× LOA of the vessel. *Navigating and manoeuvring workstation field of view—main steering position.* From the main steering position (i.e., workstation for manual steering) the horizontal field of view is to extend over an arc from direct forward to at least 60° on each side of the vessel. The helmsman is to be able to observe the vertical steering references in the fore ship while manoeuvring the vessel from the main working position. *Navigating and manoeuvring workstation field of view—GMDSS and additional functions.* Workstations used for carrying out various functions such as safety monitoring, GMDSS and other bridge functions e.g., chartable/voyage planning, communication etc. are to have horizontal field of vision extending a 90° arc on port bow covering through forward to 22.5° abaft the starboard beam. *Navigating and manoeuvring workstation field of view—blind sectors.* If needed, the blind sectors from the front bulkhead and bridge wings bulwark are to be reduced to maintain the required field of vision while monitoring, navigating and manoeuvring the vessel from different workstations. The obstruction of the view of the sea surface from the GMDSS workstations and additional bridge functions caused by each individual blind sector are not to exceed 10°.

Bridge Windows

This section relates to the detailed design requirements for the bridge windows. *Height of lower edge of front windows.* The height of the lower edge of the front windows is to allow a forward view over the bow, from which a person seated at the workstations to monitor, navigate, and manoeuvre by maintaining the line of sight. In no case, the lower window edge is to present an obstruction from the forward to 90° on each side view. The height of the lower edge of front windows above the deck are to be kept as low as possible and it is to be kept not more than 1000 mm (39 in.) above the deck. For arrangements where the navigator is normally positioned further back from the bridge-front bulkhead, the same eye height is to be used to determine the height of the lower edge of the front windows. *Breadth of the windows.* The width of the bridge windows is to be kept as wide as practicable to maintain an unimpeded view at all times. A minimum width of 1200 mm (47 in.) within the field of vision is to be maintained from the workstations. In exception, the width of a window may be less than 1200 mm (47 in.) only where necessary to avoid obstructions from the field of vision of workstations caused by the divisions/stiffeners of windows. *Windows of enclosed bridge wings.* A ship with enclosed bridge wings may be provided either vertical side windows where one side window can be opened to view the ship's side at the water surface level provided there is no other window providing a downward view in the deck or inclined side windows where the bridge wing deck does not extend to the full width of the ship. Alternative solution may be accepted upon special consideration by the vessel's Class, including installation of camera systems (with a backup system) covering an adequate area. *Window framing.* Divisions/frames between windows are to be kept to a minimum. No frames, including the centreline, are to be installed immediately forward of any workstation. The frames between front windows must not exceed 150 mm (6 in.) in width. If stiffeners between windows are to be covered, coverings are not to cause further obstructions of the field of view from any position inside the wheelhouse. If stiffeners are used, frames are not to exceed 100 mm (4 in.) in width and 120 mm (4.7 in.) in depth. Horizontally sliding windows are not permitted to be used. *Windows with clear view arrangements.* At all times, regardless of the weather conditions, all front windows in the navigation bridge are to be provided with a clear view to maintain a clear field of vision for monitoring, navigating, and manoeuvrings the ship. The following systems or components or equivalent are to be provided:

- Sun blinds with minimum colour distortion; sun blinds are to be readily removable and not permanently installed.
- Heavy-duty wipers, preferably provided with an interval function, and freshwater wash systems. Wipers are to be capable of operating independently of each other.
- De-icing and de-misting systems on all applicable windows.
- A fixed catwalk with guardrails fitted in front of bridge windows for manual cleaning of windows in the event of failure of the above systems.
- Where heated glass panels are installed in bridge windows, they are to be in accordance with ISO 3434.

Enclosed bridge wing windows with clear view arrangements. Heavy-duty wipers with freshwater wash systems are to be provided in the front and rear windows to maintain a clear field of vision from the workstations during navigation, manoeuvrings and docking. Similar wipers may be installed on the side windows (fixed type) located at the bridge wings. All wipers are to be capable of operating independently of each other.

Bridge Design and Configuration

Bridge wing. The height of the bridge wing bulwark is not to exceed 1000 mm (39.3 in.) to reduce the blind sectors for a clear field of vision from the workstations. A handrail of not less than 1200 mm (47.2 in.) height is to be fitted on top of the bulwark where the opening between bulwark and handrail is not less than 120 mm (4.7 in.). Where wind deflectors are fitted at the front side of bridge wing, the length of the deflector is not to obstruct an arc of more than 10° as seen from the operating position of workstations. *Clear heights of ceiling and entrances.* The bridge ceiling clearance height between the bridge deck surface covering and the underside of the deck head beams is to be at least 2.25 m (7 ft 5 in). The lower edge of deck head mounted equipment is to be at least 2.1 m (6 ft 11 in.) above the deck in open areas, passageways and at standing workstations. The height of entrances and doors to the wheelhouse from adjacent passageways is not to be less than 2.0 m (6 ft 7 in.). *Doors and other accesses.* All wheelhouse doors are to be operable with one hand. Bridge wing doors are not to be self-closing. Means are to be provided to hold the bridge wing doors open. Vessels having enclosed bridge wings are required to have at least one direct access route from the adjacent bridge deck area. Convenient access from the bridge deck to the compass deck is to be provided from the vicinity of the wheelhouse. *Visibility of area in front of bridge.* The navigator is to be able to view the area immediately in front of the bridge superstructure from the wheelhouse. There is to be a close approach access to at least one front window. If this requirement is met by combining "an adequate conning position", a second close approach access, besides the access to the position described above, is to be provided, or the width of the total access is to be sufficient to accommodate two persons. This defined position is not to affect the view of the helmsman to right ahead, as the helm is normally positioned on the centreline.

Bridge Arrangements and Workstation Configurations

The design of navigational equipment is to be based on sound ergonomic principles. Construction is to be of robust, durable, and flame-retardant material incorporating the required degree of enclosure protection (IP 20 for installation inside the wheelhouse and IP 56 for open deck installation). Replaceable components for navigational equipment are to be designed and arranged so that it will not be possible to connect them incorrectly or install incorrect replacements. Alarms, displays and control devices are to be arranged in a functional and logical manner to allow the operator an easy and clear means of identification of each of the components or systems included therein. The colour schemes of system alarm(s), displays, and devices are to be in accordance with international standards. Precautions are to be taken to prevent the inadvertent operation of controls that may lead to critical situations. Care is to be taken in the identification and location of switches, activation controls, and handles, the use of recessed or covered switches and controls, and arrangement for sequential operation. The layout of the bridge, including location and layout of the individual workstations is to be such that the field of view required for each function is available from the workstation. Figure 18.1, Typical bridge arrangement, presents a typical bridge layout.

General requirements for workstations. Workstations for voyage planning and/ or safety may be eliminated on a case-by-case basis upon special consideration by the vessel's Class. Alternative systems may be accepted on special purpose Vessel in lieu of the manual steering workstation where exceptional means of steering (e.g., multiple azipod or azimuth thrusters) require a specific expertise for operation. The following requirements are mandated:

- A manual steering device is to be provided at the navigation and manoeuvring workstation and at least one other workstation.
- The above steering devices are to be independent at two workstations and be able to be operated by personnel other than the bridge officer without interfering with other navigational equipment.

The design and details of the manual steering system are to be submitted to the vessel's Class for approval. *Workstations with additional functions.* The workstations for additional functions may be installed on the bridge provided the functionality of these additional workstations does not interfere with the primary bridge functions, e.g., navigating, manoeuvring, monitoring, manual steering, voyage planning, and communication. The workstations for additional functions may include:

- Dynamic positioning (DP).
- Monitoring and control of machinery.
- Extended communication functions.
- Monitoring and control of ballasting and cargo operations.
- Monitoring and control of hull openings.
- Monitoring and control of domestic systems.

The dimensions of the workstations for additional functionality are not to obstruct the required field of visions of the sea surface from the workstations for monitoring, navigating, and manoeuvring for safe operation of the vessel. The primary function of the navigation bridge is to navigate, monitor, manoeuvre, communicate and provide other functions essential for the safe operation of the ship. Other workstations such as those for machinery, cargo operation, or hull opening are to be considered as effective and supportive integrated bridge resource management. The requirements mentioned above may be exempted from any specific type of vessel, e.g., a dynamic positioned vessel which has a dedicated workstation in the wheelhouse for an industrial mission. Such vessels may have a congested bridge and cannot meet the requirements of distance, visibility etc. Alternative designs may be considered for special types of vessels on a case-by-case basis upon submission to the vessel's Class. *Passageways in the bridge wings.* The distance between bulwarks, bulkheads and consoles in the bridge wings are to be as small as possible but it is to be wide enough for one person to pass the console comfortably. The passageway width is preferably to be at least

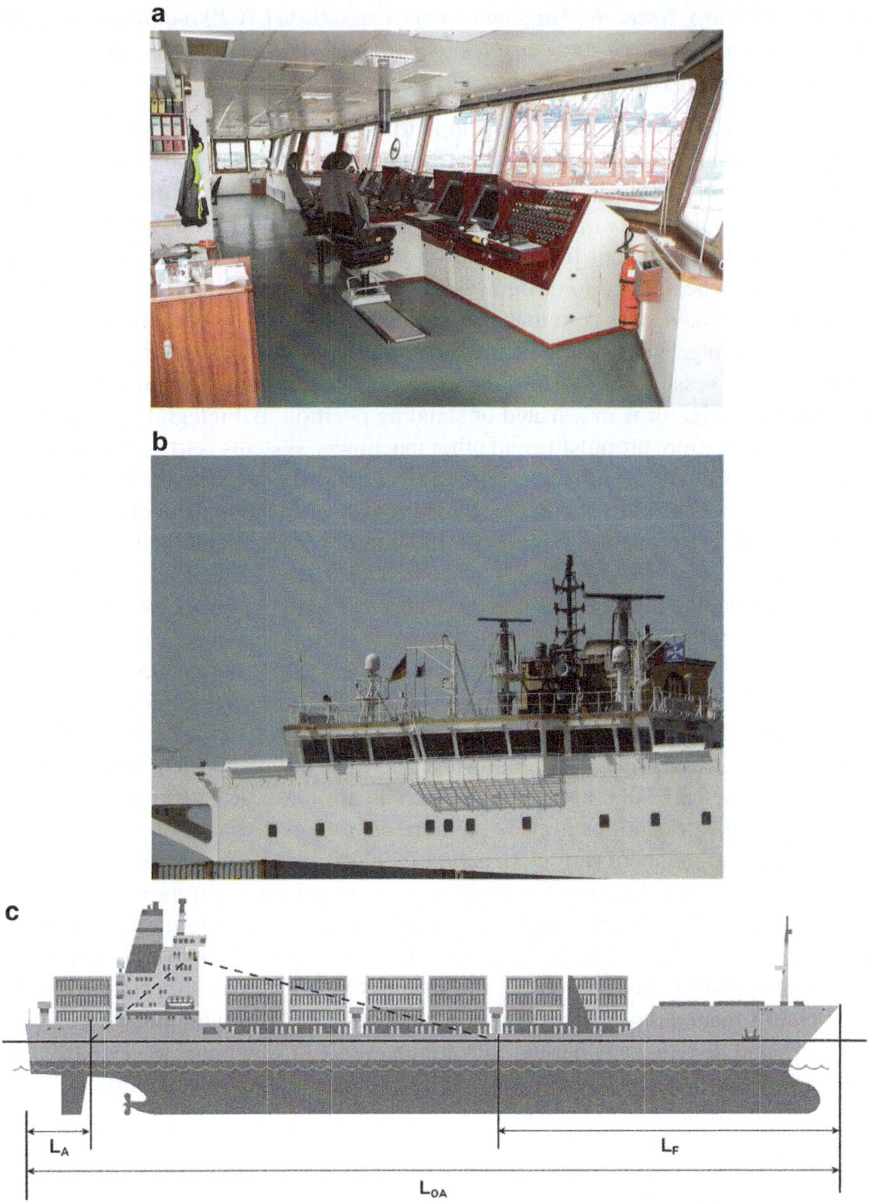

Fig. 18.1 Typical bridge arrangement (**a**) interior. Typical bridge arrangement (**b**) exterior. (**c**) Field of vision—obstruction of sea surface

600 mm (23.6 in.). ***Note:*** *the Panama Canal Commission (ACP) requires a minimum of 1 m clearance between any consoles and adjacent bulkheads/bulwarks. ACP has the authority to provide any relaxation upon request.*

Workstation configuration. The console is to be designed so that from the normal working position, the total required left-to-right viewing angle is not to exceed 190°. This angle is to be reduced whenever possible through appropriate control-display layout. The requirements of IMO Resolution A.694(17), General requirements for shipborne radio equipment forming part of the GMDSS and for electronic navigational aids is to be applied to design such equipment. The workstation in the bridge from where the navigator is to manoeuvre the vessel through water traffic by monitoring, navigating, and communicating based on the displayed course, speed, and position of the vessel, is required to be located with optimum surrounding visibility whether the navigator is in a seated or standing position. All integrated information related to navigation, propulsion and other machinery systems is to be displayed and readily available at or near this location. In all situations, the navigator is to be able to manoeuvre and control the vessel easily from this workstation. The upper leg room of the console is to have a minimum of 450 mm (18 in.) in depth and the lower leg room a minimum of 600 mm (25 in.) in depth. Alternatively, the different dimensions of minimum depth requirements may be accepted in case-by-case basis based on the acceptable justifications and supporting documentation. The top of the consoles of workstation for navigating, monitoring, and manoeuvring is not to exceed a height of 1300 mm (51 in.) and in no case the height of the workstations is not to interfere with the navigating bridge window's view requirements. The top of the consoles of workstations other than navigating, monitoring, and manoeuvring which are intended to be operated by the bridge navigating officers is not to exceed 1300 mm (51.1 in.). If any of these workstations impede the horizontal field of vision of the navigating, monitoring, and manoeuvring workstation then the console height is not to exceed 1200 mm (47.2 in.). Chairs at workstations designed for a sitting position are to be adjustable in height and capable of being arrested on the floor, and capable of rotating with the footrest being arrested. Chairs are to be capable of being moved out of the operating area, if needed. *Provision of manoeuvring information display.* The vessel's manoeuvrings characteristics and relevant information are to be tested, recorded, and displayed as per the Resolution A.601 (15) adopted on 19 November 1987, Provision and Display of Manoeuvring Information on board ships.

Bridge Equipment

General. The basic complement of bridge equipment should include:

- Computerised equipment.
- Power supply.
- Navigational system:

  - Propulsion fixing system.
  - BNWAS.
  - Route planning.
  - AIS.
  - Propulsion engine/thruster controls.
  - Telephone system.
  - Nautical radio communication.

- Weather Information System:
  - Sound reception system (SRS).
  - Public address system (PAS).

Requirements. The design and installation of the bridge equipment is to meet the following additional requirements:

(1) *Computerised equipment.* For installed automation systems and related computerised equipment, the vessel is to comply with the relevant Class and Flag State rules for the vessel's country of registration. The principal consideration in every case is that the crew has the systems, facilities, or functions necessary to satisfy ship navigation or piloting safety requirements, with fall back capabilities for the possible cases in which primary automation functions fail. In addition, the following requirements are applicable:

- Automated systems' False Alarm Rate (FAR) is to be kept as low as reasonably possible,
- Checks are to be performed for validity of the interface and input data, and for user clarity for safety, even in the presence of false alarms, and documented with the system,
- Network connections and remote access to primary essential systems are to comply with the vessel's Cyber security Risk Management System, if implemented, or in the ship's Safety Management System (SMS). Refer to IMO Resolution MSC.428(98) and IMO MSCFAL.1/Circ.3,
- Software development lifecycle must comply with the Class rules for internally developed, safety–critical software. An FMEA recommended for these systems, though not strictly necessary, to inform cyber security related safety procedures in the SMS,
- Software maintenance and updates are required to comply with the Management of Change (MOC) procedures in a Software development lifecycle (SDLC); and

(2) *Power supply.* The vessel is to be fitted with means for electrical power supply in accordance with the following:

- Essential equipment such as radar, ECDIS installed at workstation for navigating and manoeuvring, GPS, conning display, bridge alert management system, navigational watch monitoring system, speed log etc. are to be provided with a transitional source of power such as an UPS or battery to provide a minimum of 10 min of power. Any UPS supplying power to bridge equipment are to be capable to bypass automatically.

- A minimum of one telephone system is to be supplied by a transitional emergency power source capable of operating for a minimum of 30 min. Alternatively, a sound power telephone system or portable self-contained two-way voice communication device is acceptable.

(3) *Navigational system.* In general, navigational equipment/systems installed on board the vessel are to be so arranged that failure of one piece of navigational equipment will not reduce the vessel's ability to perform the functions specified in Chap. 15, Tables 15.1 and 15.2. Moreover, the vessel is to be provided with two ECDIS and both are to be separate and independent from each other. Both ECDIS are to have separate power source and network, i.e., fully separate interfaces to be provided. An interconnection between both ECDIS is to be provided to exchange data and also to be interfaced to the radar systems. Radar installations are to have a bi-directional interface so that voyage plan can be transferred to and displayed on the radar screens. Also, the voyage planning terminal in to be interfaced with the ECDIS. The vessel is to be equipped with a centralised bridge workstation. The navigation system is to be equipped with a communication system in compliance with MSC Res.302 (87) adopted on 17 May 2010, Performance Standards for Bridge Alert Management. The vessel is to be equipped with a conning information display. Job aids and training aides memoire may be provided for the navigational systems equipment which includes grounding avoidance system courses.

(4) *Heading control and information system.* For a heading control and information system, the vessel is to be in compliance with the following:

- The vessel is to be equipped with one additional gyro compass ($\times 2$ total).
- A gyro repeater is to be provided at all steering control positions.
- Continuous heading information is to be provided to repeaters, radar systems, heading control and track control systems.
- ECDIS failure of the selected compass system is to automatically trigger the backup system without disruption to rudder control.
- Heading is to be maintained if one compass fails. The two compass systems are to be arranged for continuous performance with a minimum of one autonomous compass.

(5) *Steering system.* The vessel is to have means for manual and automatic steering along with the following additional requirements:

- Rate of turn indicators and rudder angle indicators are required and is to be in accordance with the turning ability of the vessel while proceeding at normal seagoing speeds.
- The vessel is required to be equipped with a minimum of two independent rudder angle indicating systems.

(6) *Speed measuring system.* The vessel is to have a speed measuring system in compliance with Chap. 15, Table 15.1 together with the following:

- Speed indicators are to be provided in different locations, i.e., positions required for safe conning, monitoring, navigating, manoeuvring, and docking operations of the ship.
- Vessel is to be fitted with a speed measuring system capable of providing speed over ground in longitudinal and transversal directions.
- The speed measuring system is to provide continuous radar and denote speed through water and speed over ground out simultaneously.
- The speed measuring system components are to be easily accessible for maintenance while the vessel is afloat.

(7) *Radar system.* The vessel is to be fitted with a radar system in compliance with MSC. Resolution 192(79).

(8) *BNWAS.* The vessel is to be fitted with means to monitor the alertness of the officer of the watch and alarm other bridge personnel in association with the following: (a) approved motion sensors or initiating operation of various navigation equipment may be considered in addition to reset devices if the vessel's flag administration does not prohibit its use. External devices such as radar, which can act as reset devices, are to be in position where proper lookout can be performed by the officer of watch; and (b) sensors approved for intrusion and hold up alarm systems in accordance with BS EN 50131 may be used and are to comply with the following:

- Timer resets if a forearm moves 0.5 to 1 m/s at the working positions.
- Tampering or failure of the motion sensor deactivates the timer reset function.
- Timer does not reset due to moving objects, warm surfaces or shifting sunlight.
- Proper masking of sensor coverage is to be prepared inside the motion sensor enclosure when masking is needed; and
  1. Wireless portable devices may be used to sound the alarm to reach the assigned individual to attend the bridge.
  2. The second and third stage remote audible alarms are not to be acknowledged by the motion detection system and activation of the motion detection function is to only reset the first state audible alarm on the bridge.
  3. Emergency calls are to be activated by a single operator action and appropriate protective safety measures are to be taken to prevent activation of any emergency call by mistake.
  4. Unacknowledged alarms are to be transferred and means to immediately trigger an emergency call is required.

AIS. The vessel is to be fitted with an AIS. The AIS is to be capable of supporting interconnection with the radar systems. *Propulsion engine/thruster controls.* The vessel is to be fitted with means for propulsion engine/thruster controls. In addition, control of propulsion and steering including thrusters (if fitted) are to be provided in the workstations from where docking operations are conducted. *Telephone system.* The vessel is to be fitted with a means for telephone system. The automatic telephone network is to be designed to carry at least four simultaneous calls. Incoming calls are to be distinguishable by lights and/or different ring tones. All applicable locations are to be provided with a transceiver readily available for portable two-way voice communication equipment to be used for compliance with the back-up telephone requirements. Areas with an ambient noise level above 75 dB(A) are to possess or utilise noisy environment mitigations such as noise-cancelling headphones. The bridge is to have a minimum of four portable ultra-high frequency (UHF) transceivers that operate in the 457–467 MHz band, and which have the capacity to operate continuously for a minimum of 5 h. The battery charger is to be located within the wheelhouse and capable of re-charging all UHF transceivers. A communication system for mooring operations is to be provided between the wheelhouse, bridge wings, and mooring stations on board that support hands free two-way voice transmission. If portable UHF transceivers are used for mooring operations, then the total of the number of transceivers provided are to be not less than twice times the number of mooring stations on board. However, the total numbers of transceivers are to include the above mentioned 4 portable transceivers required for the bridge. Battery chargers that are able to recharge all transceivers simultaneously are to be installed in the wheelhouse. *Nautical radio communication.* The vessel is to be fitted with means for nautical radio communication. This means the wheelhouse is to be equipped with two fixed and independent very high frequency (VHF) transceivers; means of radio communication on the bridge wings to communicate with other vessels is to be provided; and antennas for the nautical radio communication system are to be located to minimise damage or malfunction of the equipment. These are to be located with warning labels of safe distance with information pertaining to human risk warnings.

Internal communication. The vessel is to be equipped with two independent and separate internal communication systems. The OOW is to be capable of calling the crew members from accommodation, open deck or any noisy area as needed. *Weather surveillance system.* The vessel is to have a weather surveillance system installed that is equipped with an anemometer that displays the relative wind speed and direction in a presentation mode in accordance with international standards. The system is to provide information about wind speed, direction, air humidity, barometric pressure, and air temperature that can be displayed on a weather information system. Wind speed sensors are to operate at a minimum range of 0–100 knots (0–115 mph/0–185.2 kph) with an accuracy and resolution of 2.5 knots (2.87 mph/4.63 kph) or higher. The wind direction sensor is to cover an azimuth of 360° with an accuracy and resolution no less than 5°. The weather sensors are to be located to minimise air flow distortion from the ship's structure. The sensor is not to be located at the same height adjacent to a large radar mast or similar structure. In the event the sensor is located in this manner, a secondary sensor is to be installed on each side of the ship with the processing ability to weigh the dual data outputs, relative to wing direction. Relative humidity is to be measured within 0–100% with an accuracy rate of 5% or higher. Air temperature is to be measured within − 15 °C (5 °F) to 55 °C (131 °F) with an accuracy rate of ± 0.5 °C (32.9 °F) (with > 10 knots (> 11.5 mph/ > 18.5 kph) wind speed and sunlight) or higher. Pertinent humidity and temperature measurements are to be displayed with a resolution not inferior to the accuracy. Barometric pressure is to be measured with accuracy better than 5 mbar (500 Pa) with a minimum resolution of 2 mbar. The system is to be capable of receiving reliable weather information relevant to the vessel's area of operation and displayed in a user-friendly manner. The forecasts received by the system are to last a minimum of five days in time steps not exceeding six hours with a minimum geographical resolution of 37.2 mi × 37.2 mi (60 km × 60 km).

Sound reception system. The vessel is to have a sound reception system that complies with the following:

- The sound reception system is to have the capability of sound detection along with wind and mechanical noise suppression with the audio band range of 70–700 Hz. The system performance is to have microphones in a position so that ambient noise does not exceed 65 dB(A).
- The system is to be capable of muting noise up to 75 dB(A) to manage severe weather ambient noise in increments no greater than 3 dB(A).
- The system is to include filters to minimise background noise other than ship whistles and be able to determine the direction of the sound (port, starboard, forward or abaft) within ± 5°.
- The system is to display sound direction results clearly for the duration of the signal plus 2 s with day and night distinction on a minimum of 2 m (6.5 ft).
- The system microphones are to be arranged to minimise noise and be capable of being muted during the ship's whistle and while the outdoor PAS is in use.
- The volume of the system's loudspeaker is to be adjustable within the wheelhouse with a labelled volume control marked of the position where the sound level in the wheelhouse is at the same level of the outdoor listening post; and PAS.

A public address system on the bridge is to be suitable for flush mounting in the workstation consoles. Every station is to have means of visualising its state of readiness. A reference list of all public access areas is to be permanently posted. *FMEA.* The FMEA required to be submitted to Class for review and approval, and should include the vessel's details and Class notations, a list of all equipment (main and standby), a description of each equipment and associated failure mode with its failure causes relative to operational modes of the item; and a description of the failure of each mode on other items or equipment. The analysis is to show how a single failure in a component or subsystem including their integration will grow and how essential systems will operate during failure to avoid any collision or grounding. Accordingly, a step-by-step sequential operating process is to be provided for crew use in the BOM. This will outline the action to be taken by the crew when a failure occurs. After FMEA is carried out, any corrective action, if required, is to be taken based on the results of the findings. An updated FMEA is to be conducted and a summary report including the FMEA test programme is to be submitted for information and kept on board.

Human Element

The ambient environment in which tasks are performed has a significant influence on human performance and vessel equipment. This section addresses task performance design considerations associated with the navigation bridge work environment and various vessel equipment including vibration; noise; lighting; device and instrument illumination; and HVAC. *Ambient environment criteria.* Table 18.1, Ambient environment criteria for navigational bridges, lists the ambient environmental criteria and testing requirements. With the exception of the vibration criteria, all ambient environment criteria are adapted from the ABS Guide for Crew Habitability on Ships.

Personnel safety. To achieve personal safety for bridge personnel, the bridge is to be designed to offer protection from injury during normal weather conditions as well as rough weather conditions. The following guidelines address improving personnel safety on the bridge. *Bridge Physical Hazards.* The bridge area is to be free of physical hazards to bridge personnel. There are to be no trip hazards on the bridge deck such as curled up carpet edges, loose gratings, duck boards, or equipment. *Securing equipment.* Means are to be provided for securing portable equipment such as chairs and tables. Where integrated bridge systems are provided that present a centralised position from which piloting, navigation and communications can be performed, a permanent seat is to be provided for the persons on watch duty. This seat may be moveable on tracks. *Safety equipment accessibility.* All safety equipment on the bridge is to be clearly marked, easily accessible and have its stowage position clearly indicated. *Radiation hazards.* All equipment posing a radiation hazard is to be protected and located in areas that do not present hazards to personnel. *Emergency lighting.* Emergency lighting is to be provided for the bridge control centre, stairways and exits. *Redundant lighting circuits.* Bridge control centre lighting is to be provided from two separate circuits so that in the event of loss of one circuit, lighting on the bridge is not interrupted. *Handrails and grab rails.* Sufficient handrails or grab rails are to be fitted to enable personnel to move or stand safely during inclement weather. Protection of stairway openings are to be given special consideration. *Bridge finishes.* The surface finishes on the bridge are to be considered an integral part of the structure, layout and environmental design. *Non-slip surfaces.* Wheelhouse,

Table 18.1 Ambient environment criteria for navigational bridges

	Criteria	Test planning	Test requirements	Test reporting
Vibration	6 mm/s (214/mm/s^2)	15	15	15
Noise	Max noise level: 65 dB (A)	16	16	16
Indoor climate	5/5	17	17	17
Lighting	6/5 (Tables 3 and 6)	18	18	18

Note

1. Vibration criteria are based on ISO 21984:2018

bridge wings and upper bridge decks are to have nonslip surfaces. *Robust surfaces.* All surfaces of deck heads, bulkheads, doors and floors are to be robust enough to withstand the daily wear of the oceanic environment. All surfaces are to be capable of withstanding a temperature range of – 20 °C to + 70 °C (– 4 °F to + 158 °F), seawater, oils and solvents common to vessels and ultraviolet light. *Electrical circuit fault isolation.* Circuits for equipment on the bridge are to be designed to allow isolation of a fault without having to take other circuits out of service and to allow safe and easy replacement. *Grounding of metal on the bridge.* All metal parts on the bridge not intended to carry electricity are to be effectively grounded. *Component replacement safeguards.* Replaceable components on the bridge are to be designed and arranged so that it is not possible to connect or replace them incorrectly.

Human machine interface

(1) *Human expectations.* Previous experience influences how humans interpret a display or operate a control. Equipment design is to be consistent with those expectations. Design of control actions and display response (directions of movement) are to be consistent across workstations. When selecting bridge hardware and software from different vendors, consideration is to be given to consistency. *Design within physical and perceptual limits.* Equipment design is to take into account the ability of humans to exert force, reach for and manipulate objects and sense the physical environment. Controls are to be placed within easy reach and adjacent to related displays. Displays that are viewed from a nominal position are to be:

    - Easy to read (e.g., character size compatible with viewing distance).
    - Within an immediate field of view (directly in front of the viewer) and not obscured.
    - Provided sufficient colour, brightness, and contrast.
    - Large enough to be seen from expected viewing distances and under expected ambient environments (day/night and weather conditions).

The physical environment is to be controlled so that ambient conditions do not interfere with visual or audible signals. Controls are to be designed to prevent inadvertent operation. The bridge is to be arranged so that objects can be easily and readily reached and manipulated. Limit memory requirements. Human memory is limited in capacity, is often unreliable and can be affected significantly by factors such as fatigue, stress, and physical health. Equipment and controls are to:

- Provide efficient methods of calling up the display of important or changing information.
- Avoid having to scroll through several displays to access frequently used or critical information—it is to be possible to display time critical information immediately.
- Avoid the need to cross-reference between information displays.
- Clearly identify all controls and displays.
- Use simple and memorable codes that are easily distinguishable.
- Perform rapid information updating in a timely manner to bridge personnel.

- Provide obvious, ongoing display of automated system status (e.g., to indicate when automatic processes have been manually overridden, disabled, or failed).

(2) *Facilitation of human attention.* The following measures are to be considered to mitigate possible human errors due to distraction, complacency, habituation, and high workload:

1. Navigation equipment designers/vendors/shipyards are to submit documentation to ensure the following:
   - Frequently used and important displays (e.g., navigation, helm control, radar) are located in a central viewing area.
   - Alarms are not so frequent that they cease to attract attention.
   - Meaningful information groupings are used to enable bridge personnel to easily cope with a large amount of information.
2. The owners/shipyards are to prepare appropriate document indicating the following:
   - Avoidance of any distractions or unnecessary requirements to perform meaningless tasks (during a watch).
   - Avoidance of unnecessary tasks that compete for attention such as paperwork that is not related to standing watch.
   - Definition of clear task priorities.

(3) *Standardisation of display characteristics.* The vendors or equipment suppliers are to provide documentation denoting the following:

- Colours are to be consistent and have the same meaning across all displays.
- Each icon or symbol represents only one object or function and is easily discriminated from all other icons and symbols.
- Icons are designed to look like the objects, processes, or operations they represent, by use of literal, functional, or operational representations.
- Schematics have similar formats (e.g., process flow is normally from left to right) to maintain, when possible, an exact spatial mapping of one display onto another.
- Computer display and panel display information layouts are spatially compatible with one another.
- The typeface is readable and have true ascenders and descenders. The font is to possess uniform stroke width and uniform aspect ratio.
- When abbreviations or acronyms are used, they are to be meaningful, in common usage and kept to a minimum.
- The units of parameter being measured (voltage, mass, pressure, length, speed, temperature etc.) are clearly displayed.

(4) *Standardisation of control requirements.* Controls requiring frequent adjustments or accurate settings are to be designed and placed as per the Industry Standard. Controls are to be located in positions wherein simultaneous operation of

two controls will not necessitate a crossing or interchanging of hands. The most important and frequently used controls are to have the most favourable position with respect to ease of reaching and grasping (particularly rotary controls and those requiring fine settings). Keys for emergency functions are to have a prominent position and be distinctively marked. The arrangement of functionally similar or identical controls are to be consistent from workstation to workstation and panel to panel throughout the bridge. If more than one person controls the action of a bridge (or related) system, all relevant information is to be simultaneously available to the person responsible for coordinating the task. All the effects of an action or command on the process are to be simultaneously observable on associated displays. If equipment of system response time is slow, feedback is to be provided indicating the action has been initiated and is progressing (e.g., rudder position and rate of turn indicators). When possible, all necessary information is to be provided simultaneously (e.g., in parallel rather than sequentially) when needed to enable a diagnosis or a control decision. Movement of a control forward, clockwise to the right or up, is to be:

- Turn the equipment or component ON.
- Cause the quantity to increase.
- Move the equipment or component forward, clockwise, to the right or up.

The above requirements are to be complied with as per the industry standard. Standardisation of priority alarms and audible indicators. An alarm is a visual and/or audible signal typically indicating an abnormal situation demanding human attention and response. An audible indication is usually a display of status or vessel condition (e.g., tones that indicate a telephone or radio call is incoming or that an equipment function has failed). Alarms and audible indicators are to be clearly coded. All alarms are to be prioritised to allow quick assessment of the importance of simultaneous alarms. Prioritisation schemes include:

- Sequencing alarms in a way that enables the development of abnormal events to be better understood.
- Separation of critical alarm information from information or status indications.
- Use of dedicated computerised alarm displays.
- Provision of distinct audible indicators (separate from alarms) that indicate automatic or semi-automatic actions of the system (e.g., transitioning from autopilot to manual steering).

If a failure by the persons on watch duty to notice the occurrence of an audible indicator can lead to unsafe conditions (e.g., failure to notice that the vessel's gyro has failed or low lube oil pressure in the steering gear) then that condition is to be considered an alarm condition and is not to be considered as a simple audible indication; and The above requirements are to be complied with as per the industry standard.

(5) *Acknowledging alarms.* Critical alarms are not to be dismissed until the initiating condition is resolved. The device is to be conveniently located and readable for

acknowledgement. It may be possible to silence an audible alarm only when that action does not cause a loss of alarm information (e.g., visual display of the alarmed parameters). Alarm requirements. Alarms are to have a unique code (audible signal) if a unique type of response is required. Colour is not to be the only means of distinguishing between alarm and non-alarm conditions. Labelling and other information is to be associated with the alarm indicator for the equipment or condition about which the alarm was triggered.

Vessel Survey Requirements

New construction. For a new construction vessel, navigational equipment and systems are to be tested to the satisfaction of the attending Class approved surveyor in accordance with 3/27, MSC 137 (76). Tests and sea trials. During sea trials, navigational equipment and systems are to be tested to the satisfaction of the attending Class approved surveyor in accordance with a test programme, and the following:

- Automatic resumption of primary bridge navigational equipment/systems functions are to be demonstrated following a blackout simulation for a period of 10 min.
- Review of approved sea trials.
- The operating/technical manual is to be used for verification of the testing programme for relevant navigational equipment/systems and is to be retained on board.

Note: *Suitable means such as navigation system power cut-off to verify appropriate operation without total ship's blackout is acceptable.*

Chapter 19
Requirements for Class Notations Relating to Integrated Bridge Systems

General

The requirements discussed in this chapter relate to those vessels which are fitted with the navigational equipment/systems, as required by Class, and are arranged such as to form an IBS. Vessels complying with the provisions set out in this chapter may be assigned the Class notation NIBS (Navigational Integrated Bridge System) or equivalent. The equipment required for this Notation are listed in Table 19.1, Navigational Integrated Bridge System.

Integrated Bridge Systems

An IBS is to be provided and is to comply with IMO SN.1/Circ.28—"Guidelines for Bridge Equipment and Systems, their Arrangement and Integration (BES)". The integrated navigation system is to be arranged in such a manner that the failure of one subsystem does not affect any other subsystem. In the event of the failure of the integrated navigation system it must be possible to operate the primary bridge navigational equipment/systems functions separately. In particular, an individual dimmer control is acceptable in lieu of a single central dimming functionality called for by IMO SN.1/Circular 288 provided the number of the individual dimmer control switches is minimised as far as practicable. *Centralised bridge workstation.* A centralised bridge workstation is to be provided to enable the navigator to perform the necessary navigational, monitoring/alarming and communication functions as discussed in this book. The equipment required in the navigation and traffic surveillance/manoeuvring workstation and monitoring workstation, specified in Chap. 18, is to be integrated within the centralised bridge workstation. Refer to Table 19.1 for the required equipment to be included on this workstation. *Central alarm panel.* The

A. A. Olsen, *Applying Physical Ergonomics to Modern Ship Design*,
https://doi.org/10.1007/978-3-031-57974-5_19

Table 19.1 Navigational equipment for NIBS (or equivalent)

Workstation for	Main function to be performed	Item	Equipment	Remarks
Centralised bridge	Refer to Table 17.1	A1	Equipment required in Chap. 1 for the navigation and traffic surveillance/ manoeuvring and monitoring workstations	Refer to Table 1.1
		A2	Central alarm panel	
		A3	ECDIS	
Conning information	To allow the navigator the easy reading of the manoeuvring state of the vessel from the conning position	B1	Display panel	
Manual steering (helmsman)	Refer to Table 17.1	C1	Equipment required in Chap. 1 for the manual steering workstation	Refer to Table 1.1
Docking (bridge wings)	Refer to Table 17.2	D1	Equipment required in Chap. 2 for the docking workstation	Refer to Table 1.2
Route planning	Determination of favourable course and optimum speed, taking into account weather conditions, current, etc., and route planning	E1	ECDIS including navigation planning station	
	Giving instructions as to the course and speed	E2	Route planning devices	
	Calculation of tidal data	E3	Chart table	
	Handling nautical records, documents, publications	E4	Position-fixing receiver	
	Handling weather reports	E5	Retaining device for drawing triangles, dividers, magnifying lens, pencils, etc	
	Determination of documentation of position in case of conventional operation	E6	Weather chart plotter	

(continued)

Table 19.1 (continued)

Workstation for	Main function to be performed	Item	Equipment	Remarks
	Control of rate and error of chronometer, deviation, radio deviation, documentation of same	E7	Main clock	
	Keeping deck log	E8	Chronometer with receiving facility for time signals	Note 2
	External communication for planning operation using the chart	E9	Radio direction finder	Note 3
		E10	Log, including distance indicator, and course plotter	
		E11	Officer of the watch check alertness acknowledgment device	
		E12	Barograph	
		E13	Command printer	
		E14	Telephone system	

Notes
[a] Attention is drawn to items A1, B1, B5, and C1 of Table 1 and D1 of Table 2 under "Remarks" column
[b] Chronometer is not required, if official universal time is obtained by other means
[c] Radio Direction Finder is not required, if the vessel is provided with other radionavigation equipment suitable for use throughout its intended voyages

centralised bridge workstation is to be fitted with a central alarm panel for instruments and systems related to the functions specified in the section on integrated bridge systems for easy identification and acknowledgment of the individual alarms. Acknowledgment of an alarm at either the equipment fitted on the required workstation or the central alarm panel is to cancel the audible warning at both sources. Cancellation of the visual warning on the central alarm panel is to only be possible at the pertinent workstation. In addition to required navigational alarms and those alarm conditions listed in the "Remark" column to item B17, a through h, of Table 17.1, the following alarm conditions are to be alarmed at the central alarm panel:

- Position fixing inaccurate/lost.
- Loss of heading input.
- Loss of log input.
- Gyro compass mismatch.
- IBS failure.

Passage planning workstation. The route planning workstation is to enable the navigator to plan the intended voyage without interfering with the actual navigation or manoeuvring of the vessel. It is to be large enough to facilitate the use of two charts concurrently, and adequately fitted for efficient route planning. Refer to Table 19.1 for the required equipment to be included on this workstation. *Conning information display.* A conning information display is to be provided on the bridge and which is to be observable from the conning position(s) and designed for easy reading of the manoeuvring state of the vessel. The information on continuous display is to be restricted to information relevant to the actual phase of the voyage. This display may be included on the centralised bridge workstation. *Navigational system requirements.* The requirements contained herein are in addition to, or subsequently modified by those, in stated Chap. 16:

- *Course information system.* The heading information system is to include a magnetic compass and two independent gyro compasses.
- *Speed measuring system.* The speed measuring system is to be independent of the position-fixing systems.
- *Automatic track-keeping system.* The automatic track-keeping system is to automatically enable the vessel to keep along a pre-planned track and the vessel's position is to be monitored continuously. When the vessel's position cannot be received, the current heading or rate of turn is to be maintained until manually altered by the officer of the watch, and such condition is to be alarmed. The vessel's position is to be cross checked by dead-reckoning based upon speed over ground provided by the vessel's log.
- *ECDIS.* Relevant equipment associated with the ECDIS is to be installed on the centralised bridge workstation and at the route planning workstation. The ECDIS is to comply with IMO Resolution MSC.232(82) or IMO Resolution A.817(XIX), as amended by Annex 5 to MSC.64(67) "Performance Standard for Electronic Chart Display and Information Systems (ECDIS)", and Annex 4 to MSC.86 (70) "Performance Standard for Electronic Chart Display and Information Systems (ECDIS)".

Operating/technical manual. The operating/technical manual is to include the following provisions: simplified diagrams of the ECDIS and IBS; and periodical testing procedures for ECDIS and IBS. *Workstations (required equipment.* The equipment listed in Table 18.1 is to be fitted on the bridge.

Tests and Sea Trials

The vessel's sea trial programme must include test details of the ECDIS and IBS.

Chapter 20
Ergonomic Bridge Design

General

This chapter presents an overview of the principles of human interface design appropriate to the design and use of ship navigation bridges. These principles are applicable to the design of displays, controls and the bridge workspace for persons on watch duty who must conduct and monitor operations and respond to ambient and operational conditions. Applying these principles to bridge design will lead to interfaces and work environments that simplify bridge operations, reduce human error, are maintainable and limit the physical demands imposed on persons on watch duty and other bridge personnel. The following eight principles are discussed:

- Principle 1—Defining the roles and responsibilities of the bridge personnel.
- Principle 2—Designing for human limitations, capabilities and expectations.
- Principle 3—Arrangement of bridge devices, controls and displays to maximise access.
- Principle 4—Design of displays consistent with task requirements.
- Principle 5—Designing simple, direct and easy to use inputs and controls.
- Principle 6—Designing for productive performance and reduction in human error.
- Principle 7—Provision of job aids and training.
- Principle 8—Performance of testing.

Terminology

Achromatic: Designating colour perceived to have zero saturation and therefore no hue, such as neutral greys, white or black.

Alarm: A visual and/or audible signal indicating an abnormal situation.

A. A. Olsen, *Applying Physical Ergonomics to Modern Ship Design*,
https://doi.org/10.1007/978-3-031-57974-5_20

Annunciation: An audible signal indicating a condition or system state (usually of an emergency or off-normal nature) via audible signals.

Expectation: A belief about or mental picture of the future. In ergonomic contexts, an expectation is a belief concerning the effect of an action, especially a control action.

Function: The appropriate action of any special components or part of a system to meet some defined purpose or objective; specific occupation or role; an assigned duty or activity.

Habituation: The reduction of a task performance likelihood following repeated false exposures to the stimulus (e.g., an oft repeated false alarm will reduce its ability to command a timely response).

Human task: A piece of work performed as part of one's duties; a function performed; an objective.

Perception: The process of becoming aware of the immediate environment through any of the senses.

Situation awareness: The degree of accuracy by which one's perception of the current environment mirrors reality.

Warning: A cautionary or deterrent indication of an impending abnormal situation.

Principles of Ergonomic Design

Principle 1—Defining the roles and responsibilities of the bridge personnel

For any given bridge design activity, the relative roles and responsibilities of humans and hardware/software need to be defined. Roles and responsibilities will vary depending on vessel trade, owner/operator objectives and processes, level of vessel automation and other factors. Whatever the specific contexts, the roles and responsibilities of the bridge personnel should be clearly identified. The design process presented in Chap. 13, "Ergonomic Design and Evaluation Process" provides a method to identify bridge design requirements based on human–machine functional allocation. The following provides general guidance on defining roles and responsibilities for bridge personnel. *Identifying bridge functions and conduct requirements analysis.* Identify specific functions that bridge systems (including personnel) need to perform. These should include not only those functions performed by equipment (e.g., depth sounding), but also those that are essentially human (e.g., visual lookout, route planning). The objective of this is to completely characterise what has to be accomplished from the bridge, in terms of:

- Functions (e.g., sea surveillance).

- Means to perform or support the function (e.g., radar sets, binoculars).
- Requirements of functions (e.g., detect target presence, determine range and bearing, determine course and speed, compute closest point of approach, etc.).

For each function, requirements are further analysed to identify (or define by allocation) the specific roles of bridge personnel. Each human role/function will have associated human tasks that need to be identified and listed. *Generation of task-based requirements.* Identify bridge personnel tasks and the objectives of those tasks. Examine each task to determine information requirements, control requirements, as well as the skill and knowledge required to the perform tasks. *Identification of task overlap.* Identify, based on contextual and environmental factors, which tasks may overlap in time or bridge location. Examine the relationship between tasks and the subsequent prioritisation of tasks. *Specification of system failure procedures.* Consider what information and control provisions must be available from the bridge in case of a loss of automated functions. As far as possible, bridge systems should "fail-safe" (meaning that they do not fail leaving the systems in a hazardous or risky condition). *Soliciting of bridge personnel inputs to requirements.* Solicit experienced bridge personnel to identify and verify human tasks, as well as associated information, displays and controls required to perform those tasks. *Designing for efficiency and the control of personnel workloads.* Select and arrange bridge equipment and design jobs and tasks so that bridge personnel:

- Can reliably perform their functions.
- Are productive.
- Are not overloaded with work.

Principle 2—Designing for human limitations, capabilities and expectations

Humans and machines have, of course, very different capabilities. Humans are creative, can make decisions in the face of uncertainty, are highly mobile, learn rapidly, can generalise knowledge to novel situations, are communicative and have massive sensing and sensory processing capability. Machines can exert great force for sustained periods, can do repetitive work flawlessly for long periods, can withstand a wide range of physical environments and do not get bored, complacent, forgetful, tired or angry. The objective in bridge design is to take full advantage of the relative capabilities and limitations of humans and machines. In the broadest sense, humans should be involved in vessel management and planning activities and should communicate those plans and activities to the machines that perform the work (without a sense of tedium). Humans also monitor machinery and intervene when plans are not followed and perform replanning as needed when the environment or material condition of the equipment changes. In other words, humans plan and decide and machines implement the outcomes of those decisions at the bidding of the human. Guidance related to design of navigation bridges consistent with human capability is presented below.

Limiting human data processing. The design of interfaces should be compatible with human capabilities and limitations as information processors. Where possible, machines should perform data manipulation and present the result to the human for decision-making. A common example is the ARPA radar that computes the CPA to other vessels. The human uses that result in the context of knowledge-based decision making (e.g., applying the Rules-of-the-Road, negotiating passing plans and replanning ship's course and speed). Machines should be used for such tasks as storing masses of information or performing complex calculations. Humans should be used as sensors, planners and deciders and monitors of machine performance. *Organising human machine interfaces in harmony with physical layouts.* Where appropriate, control and display layouts representing equipment arrangement (e.g., thruster orientations, ballast tank arrangements, navigation light control panels) should mimic the spatial and functional relationships of the equipment monitored and controlled. Where possible, mimic lines should be used to show the physical relationships (e.g., lines connecting pumps and valves) of the components.

Accommodation of human expectations. Experience influences how humans interpret a display or operate a control. Equipment design should be consistent with those expectations. Design of control actions and display response (directions of movement) should be consistent across workstations. Be concerned where different vendors are used to select/assemble bridge hardware and software (vendors may have designed to different expectations) or where different teams generate designs for different interfaces. Standardisation of display characteristics. Display characteristics should be consistent among locations (or computer display pages):

- Colours should have the same meaning across all displays.
- Schematics should have similar formats (e.g., process flow is normally from left to right) to maintain, when possible, an exact spatial mapping of one display onto another.
- Computer display and panel display information layouts should be spatially compatible with one another.

Facilitation of human attention. Matching the task to the attention strengths and capabilities of the human can mitigate possible errors due to distraction, complacency, habituation and high workload:

- Avoid distractions or requirements to perform meaningless tasks (at least during a watch).
- Place frequently used and important displays (e.g., navigation, helm control, radar) in a central viewing area.
- Ensure alarms are not so frequent that they cease to have an attention-getting effect.
- Use meaningful information groupings so that bridge personnel will be able to easily deal with a large amount of information.
- Avoid tasks that compete for attention (e.g., paperwork that is not related to standing watch and keeping a proper lookout).
- Define clear task priorities.

Designing within physical and perceptual limits. These guidelines address the compatibility of equipment design within the limits of humans to exert force, reach and manipulate objects and sense the physical environment:

- Controls should be placed within easy reach and adjacent to related displays.
- Displays that are viewed from a nominal position should be:
- Easy to read (e.g., character size compatible with viewing distance)
- Within an immediate field of view (directly in front of the viewer) and not obscured
- Provided sufficient colour, brightness and contrast
- Large enough to be seen from expected viewing distances and under expected ambient environments (day/night and weather conditions).
- Control the physical environment so that ambient conditions do not interfere with visual or audible signals.
- Guard against inadvertent control operation.
- Arrange the bridge so that objects can be easily and readily reached and manipulated.

Limiting memory requirements. Human memory is limited in capacity, is often unreliable (i.e., error prone) and can be affected significantly by factors such as fatigue, stress and physical health.

- Provide efficient methods of calling up display of important or changing information.
- Avoid having to page through several displays to access frequently used or critical information – it should be possible to display time critical information immediately.
- Avoid the need to cross-reference between information displays.
- Clearly identify all controls and displays.
- Use simple and memorable codes that are easily distinguishable.
- Include rapid information updating so that bridge personnel do not have to wait.
- Provide obvious, ongoing display of automated system status (e.g., to indicate when automatic processes have been manually overridden, disabled or failed).

Use of colour coding to simplify information search and interpretation. Where carefully applied, colour can be very effective in reducing visual search, simplifying interpretation of displayed information, reducing ambiguity in decision making and reducing requirements to recall information from memory. Inappropriate use of colour can lead to confusion and error. When designing colour codes, consider the following:

- Add colour only after the effectiveness of an interface has been maximised in an achromatic format.
- Do not use colour as the sole coding mechanism to convey information. Colour coding should be redundant to some other coding mechanism (such as symbols or text).
- The use of colour should not reduce readability of displays, labels, maps, etc.

- Colour should be applied to avoid or minimise difficulties for personnel having impaired colour vision.
- Where colour coding is employed to denote discrete conditions (e.g., auto-pilot on/off, pump is running, in standby or off, etc.), no more than seven different colours and associated meaning should be used. If similar hues are used, they should be used only with logically related information.

Principle 3—Arrangement of bridge devices, controls and displays to maximise access

Component grouping to minimise bridge traffic. Group components, consoles and devices according to frequency, importance and sequence of use. Arrange bridge components to minimise the need for the persons on watch duty to move to alternative positions on the bridge. For example, locate vessel-to-vessel communications devices adjacent to radar displays and manoeuvring stations. Arrange components on the bridge to facilitate unobstructed external visibility, especially forward-looking. Ensure that any task (or several tasks that need to be performed simultaneously) can be performed from a single standing or seated position. Arrangement of displays by task association. When possible, group all the information relating to a particular task together in one place.

- Locate related information together.
- Locate related items such that they are easy to associate.
- Locate displays close to (or on) other displays and controls with which they are associated.

Centralise important information. Centralise important information to allow actions to be prioritised. Optimisation of arrangements. Optimise arrangements to support time and safety critical tasks. Layouts should emphasise ease and reliability of performance of safety critical tasks, such as collision avoidance, tug interface, piloting and berthing. *Group information to support operations.* Bridge personnel should not have to page through computer displays to collect information required for a particular operation.

Principle 4—Design of displays consistent with task requirements

Provision of external and internal consistency. Coding should be consistent among software displays, hardware displays, written documentation and job aids. For example, when designing symbols used to represent equipment (e.g., pumps, thrusters, sea chests, impellers) use the same symbols that bridge personnel are familiar with from drawings, labels, procedures, national and industry standards or training materials. *Provision of situation awareness displays.* Provide situation awareness displays to keep a summary check on the whole of the system and environment for which bridge personnel have responsibility. Ensure sufficient time to refresh situation awareness at the following times:

- Watch turnovers.
- When additional staffing must be used (e.g., in emergencies, port approaches and departures).
- Following changes in the use of automation (e.g., changing from auto to manual steering or transitioning to bridge monitoring of engineering functions during times when engine rooms are unattended).
- As changes in requirements for surveillance, navigation and piloting occur (e.g., due to increased sea-lane occupancy, weather changes or system failures).

Base display design on information requirements. Design displays based on the requirements of the task. Avoid prior (hardware or software) display formats as a sole basis for the design of computer displays or back-up hard wired displays:

- Involve bridge personnel in the identification of interface requirements (be advised that their understanding of requirements may be based on prior designs and practices that may have been clumsy).
- Involve bridge personnel in the usability evaluation of interfaces under development.

Provision of sufficiently accurate and precise information. Provide information to the level necessary for accuracy and precision required by the task. For example, if a task requires information display only in whole units, do not present decimal values in displays (e.g., **DEPTH UNDER KEEL 12 FEET**, versus **DEPTH UNDER KEEL 12.214 FEET**). Information accuracy and precision includes how current (up to date) information is (as it changes with the passage of time). *Limit display complexity.* The contents of a display should provide all of the information required to perform tasks in a simple, directly usable form (e.g., without having to transform or extrapolate the information). *Display actual equipment status.* Display equipment status (e.g., valve status or rudder position) directly from the actual equipment and not from the demanded control setting (e.g., actual rudder angle as opposed to ordered angle). *Group information to support task performance.* Group information to make it easy to make data comparisons, discern cause and effect relationships, identify time lags and assess rates of change in processes such as changing vessel course against other targets' movements. *Prioritise alarms and audible indicators.* An alarm is a visual and/or audible signal typically indicating an abnormal situation demanding human attention and response (e.g., a depth under keel alarm). An audible indication is usually a display of status or vessel condition (e.g., tones that indicate a telephone or radio call is incoming or that an equipment function has failed). Priorities of alarms and audible indicators should be clearly coded. Prioritise alarms to allow quick assessment of the importance of simultaneous alarms. Prioritisation schemes include:

- Sequencing alarms in a way that enables the development of abnormal events to be better understood.
- Separating critical alarm information from information or status indications.
- Using dedicated computerised alarm displays.

- Provision of distinct audible indicators (being separate from alarms) that indicate automatic or semi-automatic actions of the system (e.g., transitioning from autopilot to manual steering). If a failure by the persons on watch duty to notice the occurrence of an audible indicator can lead to unsafe conditions (e.g., failure to notice that the vessel's gyro has failed or low lube oil pressure in the steering gear) then that condition should be considered an alarm condition and should not be designed as a simple audible indication.

Avoid nuisance alarms. Alarms are used to signify urgent events and conditions. When alarms are presented, only to be discovered as false or of no real importance, subsequent alarms can tend to be regarded as unimportant, to the point of being dismissed (not investigated) entirely. It is important that audible and visual alarms consistently signify events or conditions that require immediate action. This can be accomplished by the following:

- Use audible alarms for abnormal or emergency conditions only.
- Do not use alarms to indicate equipment status or normal conditions (e.g., watertight doors closed, running lights on, control valve position, etc.).
- It should be possible for every bridge alarm to gain immediate attention (even if personnel are not able to deal with the condition immediately).

Acknowledging alarms. It should not be possible to dismiss critical alarms until the initiating condition is resolved. Locate acknowledgement devices such that audible components of alarms can be acknowledged while allowing alarm information to be read. It should be possible to silence an audible alarm only when that does not cause a loss of alarm information (e.g., visual display of the alarmed parameters). *Avoid alarm ambiguity.* Alarms should have a unique code (audible signal) if a unique type of response is required. Colour should not be the only means of distinguishing between alarm and non-alarm conditions. Labelling and other information should obviously associate the alarm indicator with the equipment or condition about which the alarm was triggered.

Principle 5—Designing simple, direct and easy to use inputs and controls

Provide direct human control. Design the system interaction such that bridge personnel pace the tempo of operations. Ensure that the human controls the pace of control entries by explicit actions and that such entries occur through explicit human actions. This includes the actions of automated systems by allocating to the human the function of determining and inputting plans and activities for automated system performance (such as voyage plans) and by providing the human with the ability to intervene in automatic system performance and functioning. *Clearly identify control modes.* Automated, aided or manual control modes (e.g., auto-pilot steering vs. manual control of the steering wheel) should be clearly indicated by a display associated in close proximity to the controlled component. Operating mode changes are indicated and annunciated. When transitioning between control modes (e.g., from manual to auto-pilot steering or transitioning control of the propulsion system from engineering spaces to the bridge), ensure that:

- Positive action on the part of bridge personnel is required (e.g., by positioning a control or issuing a computer command).
- At the time of transfer, associated displays need to indicate the current, actual control mode and then announce the transfer, with an audible signal that sounds at each affected workstation.
- In the case of the steering wheel, if transfer of helm control to the steering wheel (or joystick or thruster control) is accomplished solely by motion of the device (e.g., a helmsman assumes manual control by directly manipulating the wheel) a positive indication of the mode of control should be provided by conspicuous visual and audible displays.

Provide guidance for human intervention in automated systems. Guidance (in the form of written procedures or placards) for when human invention is needed should be explicit for the following conditions:

- When to take over from automatic functions or control.
- When and how to hand over control from a local control space to another location.
- When to shut down equipment or systems.

Provide direct and immediate feedback for control actions. When possible, locate controls and displays (related by action and effect/feedback) together. All the effects of an action or command on the process should be simultaneously observable on associated displays. If equipment of system response time is slow, feedback should be provided indicating the action has been initiated and is progressing (e.g., rudder position and rate of turn indicators). If more than one person controls the action of a bridge (or related) system, all relevant information should be simultaneously available to the person responsible for coordinating the task. When possible, supply all the necessary information simultaneously (e.g., in parallel rather than sequentially) that is needed for a diagnosis or a control decision. Provide simple computerised display navigation. Navigation through computer displays should not require that bridge personnel remember paths to information. Rather, they should be led to the information through recognition techniques (e.g., by use of pop-up menus, navigational maps or navigation palettes). Help bridge personnel not to get lost when navigating among windows. Limit the depth of window and menu hierarchies to no more than three levels. Provide maps or other cues to provide awareness of "where they are" in the interface. *Response latency and visibility of system status.* Where system responses to control actions are slow or delayed, provide intermediate indications of system response. For example, an indication of a ship's speed of course change can be provided by a Rate of Turn indicator for a ship with a wide turn radius and slow vessel response. This allows the helmsman and deck officer to more immediately verify system response to the control order.

Principle 6—Designing for productive performance and reduction in human error

Provide error prevention and tolerance. To the extent possible, equipment should safeguard against human error. In other words, actions that could directly lead to damage to the vessel, people or the environment should be provided guarded controls or in the case of software, potentially dangerous actions should require a confirmatory action (e.g., click **CONTINUE** to confirm that all watertight doors are to be opened or click **CANCEL** to exit.). Further, and where possible, software should be able to monitor and advise on the safety of human actions. An example of this is a radar system that computes a CPA to a navigational hazard or other vessels navigating in the near vicinity after a change in helm order and alerts the persons on watch duty if there is a danger of collision. Consider task communication needs. Communication systems should be designed to:

- Avoid requiring personnel to frequently move to different locations on the bridge to access communications equipment.
- Consider using portable or wearable, as well as fixed, communication equipment, where the first point above, is considered difficult to comply with.

Avoid control conflict. Avoid simultaneous control of systems or equipment from different workstations. Avoid two or more people being able to simultaneously influence the same part of a process from different control and display locations (e.g., controlling engine speed from an engineering station and from the bridge). In other words, do not require ship personnel to compete for control of a system. *Compatibility of bridge staffing with operational requirements.* The following principles should be applied to ensure compatibility among the bridge staffing complement; the tasks to be performed; the environment (time of day, weather conditions, congestion of waterways, etc.); and the bridge design and its automation. Determine staffing for all modes of operation, based on high workload operations. Consider whether staffing levels, skills and experience will be sufficient for all known operational and failure conditions. Consider staffing implications for worst-case scenarios. Consider staffing for normal conditions. Allocate tasks according to skills and experience.

Principle 7—Provision of job aids and training

Identify required knowledge, skills and abilities. Know what bridge personnel have to do and provide the training to perform the required tasks. Performing a task analysis (refer to Chap. 29, Ergonomic bridge design) is sufficient to identify bridge personnel task requirements. *Identify training needs and requirements.* Define training needs specific to a given bridge design and specific automated functions:

- Identify automated functions and what interventions may be necessary by bridge personnel.
- Identify the conditions dictating automated or human task performance.

- Provide training involving vessel control when automated control functions or information is unavailable.
- Provide training for conditions where important tasks compete for attention.
- Provide hands-on training in transitioning between automatic and manual control functions.

Provide procedures. Provide available written or on-line procedures to guide and document infrequent, complex or safety critical operations. Through procedures, indicate appropriate human actions in case of particular alarm conditions (scenarios), rather than cluttering displays with instructions. Provide procedures to address infrequent or postulated conditions. *Provide on-line help.* Provide adequate labels and warnings. Provide standardised, durable, readable, usable labels for all equipment and components. Labels and marking should be consistent in terms of:

- Coding and colours used.
- Mounting location.
- Language and nomenclature.
- Stylistic design.

Principle 8—Performance of testing

Verify functionality. Verify that the identified functional requirements of the bridge are correctly implemented. Perform usability testing. Observe bridge personnel using hardware and software. Simulate infrequent, uncommon, hazardous and unpredictable tasks. During testing take note of:

- Any observed confusing or error-inducing design aspects.
- Bridge personnel observations (Masters, Mates, lookouts, etc.).
- Unnecessary work or task activities.
- Observed or reported errors in task performance.
- Bridge persons on watch duty comments on the usability of components, tasks or arrangement of components.

Usability testing should occur throughout the design cycle of the bridge and at every instance of bridge equipment modification or upgrade.

Chapter 21
Bridge Functionality

General

Work required of bridge personnel extends well beyond that imposed by primary operational and system requirements. Often, analysis of functions and identification of requirements examine only those tasks directly associated with piloting and navigation, interfaces with engineering and means to address a limited number of off-normal events. Additional duties, however, are required of bridge personnel to meet additional technical, administrative and statutory requirements and so on. These requirements should be examined as well during the design of navigation bridges. Design of navigation bridges should consider the entirety of the watchstander job—whether or not specific responsibilities involve some specific hardware interface. Decisions as to how to apply automation to influence bridge personnel workload, for example, must consider task demands that have no attendant hardware and software or that require general purpose hardware and software (e.g., electronic mail).

Terminology

Alarm: A visual and/or audible signal indicating an abnormal situation.

Annunciation: An audible signal indicating a condition or system state (usually of an emergency or off-normal nature).

Function: The appropriate action of any special components or part of a system to meet some defined purpose or objective; specific occupation or role; an assigned duty or activity.

Human task: A piece of work assigned or done as part of one's duties; a function to be performed; an objective.

A. A. Olsen, *Applying Physical Ergonomics to Modern Ship Design*,
https://doi.org/10.1007/978-3-031-57974-5_21

Situation awareness: The degree of accuracy by which one's perception of the current environment mirrors reality.

Warning: A cautionary or deterrent indication of an impending abnormal situation.

Bridge Functions

Bridge personnel task requirements should consider the types of functions listed in Table 21.1, Sample bridge functions and responsibilities, when performing bridge design tasks such as designing overall layouts and component arrangements, selecting devices to meet requirements, estimating and allocating workload, designing procedures and job aids, determining communications requirements; and determining internal and external visibility requirements. Note in Table 21.1, Sample bridge functions and responsibilities, that not all functions are primarily bridge related. These functions are addressed in the table since bridge persons on watch duty must at least maintain cognisance of those functions. *Growth of bridge functionality.* Allow for update of bridge functions, in terms of both expanded functional requirements (e.g., Voyage Data Recorders (VDR) and system upgrades since bridge requirements and components will change over the life cycle of use and new technology may be employed, replacing older equipment. Provisions for such growth of the functionality of a bridge and its technology updates, should be anticipated and accommodated during initial design. For example, it is likely that new or additional alarms will be required and initial design of alarm systems should anticipate that growth. *Ancillary bridge functions.* Task and functional responsibilities of bridge personnel will change over time. Changing bridge requirements (due to international agreements, port/coastal state requirements, customer demands, changes in owner operating philosophies and practices and so on) will alter the job and duties of bridge personnel. One example of such a change would be when a vessel moves from a manned to an unattended engine space. In this case, transitioning to operations associated with unattended engineering spaces introduces bridge requirements such as extensive monitoring of propulsion systems (operation, start-up, emergency shut down, fuel system, lubrication systems, cooling system, alarm conditions, ballast and trim, etc.); monitoring electrical generators; control of other fire suppression systems; and bilge monitoring and management.

In addition, numerous other control or alarm conditions related to engineering and auxiliary systems may be required to be monitored/controlled from the bridge, including steering gear alarms (e.g., power failure, or a failure to override the autopilot), fire detection and firefighting (fire main displays and alarms, fire detection panels), release of towing hook or lines (for towing vessels), gas monitoring and detecting (hydrocarbon and explosion concentrations limits), loss of ventilation/refrigeration to cargo tanks/containers, liquid levels in oil cargo tanks (high and high-high alarms), steering gear status (failures, location of control, off-normal conditions, etc.); and bilge alarms. The point of this discussion is that task requirements of bridge

Table 21.1 Sample bridge functions and responsibilities

Interfacing with deck operations	Miscellaneous/special operations
Anchoring Bunkering, storing Cargo (load plan, lightering, etc.) Cargo management Direct shore gangs Docking/undocking Line handling Line/wire maintenance Pilot embarkation/debarkation Small boat deployment and recovery Tug and towing operations	Bridge—housekeeping Collision/grounding/stranding Extreme weather and seas Fire—large and small Flooding/ballast control Fuel spills/environment hazards Internal security Loss of propulsion/steering Medical Search and rescue/man overboard Structural maintenance Vessel security
Navigation, piloting, vessel control	**Performing functions related to unattended machinery spaces (UMS)**
Area surveillance—auditory Area surveillance—visual Berthing/unberthing Bridge maintenance Collision avoidance/traffic surveillance Communications—external Communications—internal Helm control/conning Hull performance/station-keeping Lookout Maintain trim and ballast Manoeuvring in congested waters Position fixing Signals and lights Surveillance—sea (state, tides, currents, bottom conditions, hazards to navigation, temperature, etc.) Tug interface Record/chart keeping Voyage track keeping Voyage planning Weather monitoring	Bilge monitoring and management Fire main control Fire suppression activation Monitoring electrical system Monitoring/control of propulsion systems (operation, start-up, emergency shut down, fuel system, lube systems, cooling system, alarm conditions) Pollution monitoring and control Steering gear

(continued)

Table 21.1 (continued)

Awareness of maintenance activities	Administration
Calibrations (Loran, GPS, etc.) Hull Propulsion Schedules/equipment availability Vital systems	Crew assessment—fitness Finance/payroll Health care In-port inspections (P&I, auditors, customs, immigration, maintenance personnel, suppliers, vetting inspectors, port state control, port logs/records, etc.) Log keeping Weather service/reporting Oceanographic reporting Personnel issues (STCW, etc.) Planning (watch schedules, hours of rest) Port entry (pilots, dock/harbour masters, etc.) Ship's meetings Stores logistics and loading

personnel extend beyond those directly related to piloting and seamanship and task requirements will likely change over the lifecycle of a vessel. To the extent possible, design of the navigation bridge should anticipate and accommodate planned or unexpected change and growth in bridge functionality in terms of alarming and annunciating systems, general space allocated to functional areas, potential changes in bridge manpower, design of communications systems, design of computing systems; and persons on watch duty workload and bridge staffing.

Chapter 22
Bridge Arrangement and Layout

General

This chapter specifies the basic function and design requirements for bridge arrangement and layout. It has been developed to help ensure that the designs of ships' bridges adequately provide for the requirements for safe navigation by preventing confusion arising from bridge layout, designing for ease of device access and use and reducing workload and human fatigue.

Terminology

Accessibility: The ability for personnel to easily access equipment that requires maintenance, inspection, removal or replacement while wearing the appropriate clothing, including personal protective equipment and using all necessary tools and test equipment.

Anthropometrics: Data relating to physical body dimensions. It includes body characteristics such as size, breadth and the distances between anatomical landmarks, such as elbow to fingertip.

Anthropometric percentile: Given the range of variability of [human] bodily dimensions, anthropometric data are typically expressed as percentile statistics, such as 5th or 95th percentile. A percentile statistic defines the anthropometric point at which a percentage of a population falls above or below that value. For example, the seated eye height of a 95th percentile North American male is 850 mm (33.5 in.), so by definition, 5 percent of North American males will have a seated eye height of greater that this figure and 95 percent will have a lesser seated eye height.

Bridge: Area from which the navigation and control of the vessel is exercised, including wheelhouse, chartroom area and bridge wings.

A. A. Olsen, *Applying Physical Ergonomics to Modern Ship Design*,
https://doi.org/10.1007/978-3-031-57974-5_22

Bridge wing: Parts of the bridge on both sides of the vessel's wheelhouse that extend to the vessel's sides.

Catwalk: Extension to a deck that is wide enough to allow the passage of a person.

Chartroom area: Part of the wheelhouse situated and equipped for adequate performance of voyage planning/plotting activities.

Commanding view: View without obstructions that would interfere with the ability of a navigator to perform all immediate tasks.

Communications workstation: Workstation for operation and control of the equipment for distress/safety communications and public correspondence communications.

Conning position: Place on the bridge with a commanding view that is used by navigators when commanding, manoeuvring and controlling a vessel's movements.

Field of vision: Angular dimension of a scene that can be observed from a position on the vessel's bridge.

Helmsman: Person who steers a vessel while underway.

Monitoring: Act of constantly checking equipment and surroundings in order to detect changes.

Monitoring workstation: Workstation where formal vessel monitoring activities are performed.

Navigation and manoeuvring workstation: Workstation where vessel's speed and course are considered and controlled.

Navigator: Person navigating, operating bridge equipment and manoeuvring the vessel.

Wheelhouse: Enclosed area of the bridge.

Workstation: Position at which one or several tasks constituting a particular activity are carried out. This position includes a combination of job-related items necessary to fulfil job tasks, such as the console, with all devices, equipment and furniture.

Internal Visibility

Height of lower edge of front windows. The height of the lower edge of the front windows should allow a forward view over the bow, from which a person seated at the workstations can monitor, navigate and manoeuvre. The height of the lower edge of front windows above the deck should be kept as low as possible. It should not, as far as practicable, be more than 1000 mm (39 in.) above the deck plate. *Height of upper edge of front windows.* The height of the upper edge of the front windows should allow a forward view of the horizon for a person in a standing position with a standing eye height of 1800 mm (71 in.) at the navigating and manoeuvring workstation.

1. If 1800 mm (71 in) eye height is unreasonable and impractical, the eye height may be reduced, but should not be less than 1600 mm (63 in).
2. The height of the upper edge of the front windows above the deck plate should be as high as practicable and at least allow a forward view of the horizon when the bow is 10° below its even keel position.
3. The minimum height of the upper edge of the front windows above the deck plate should be 2000 mm (79 in) (refer to Fig. 22.1, Example of height of upper edge of front window in relation to eye height). The dimensions in Fig. 20.1 are based on an eye height of 1800 mm (71 in), for a person of 1900 mm (75 in) stature, at a distance of 750 mm (30 in) from the bridge front bulkhead. As a point of reference, the standing eye height of a 95th percentile Northern European or North American male is approximately 1750 mm (69 in.) and stature of that same 95th percentile male is approximately 1860 mm (73 in.).
4. For arrangements where the navigator would normally stand further back from the bridge front bulkhead, the same eye height should be used to determine the height of the upper edge of the front windows.

Window framing. Divisions/frames between windows should be kept to a minimum. No frames, including the centreline, should be installed immediately forward of any workstation. The frames between front windows should not exceed 150 mm (6 in.). If stiffeners between windows are to be covered, this should not cause further obstructions of the field of view from any position inside the wheelhouse. If stiffeners are used, frames should not exceed 100 mm (4 in.) in width and 120 mm (4.7 in.) in depth. In no case should horizontally sliding windows be used. *Window inclination.* To help avoid reflections, the bridge front windows should be inclined from the vertical plane, top forward, at an angle of not less than 10° and not more than 25°. The rear and side windows should be inclined from the vertical plane top outboard, at an angle of not less than 10° and not more than 25°. Exceptions can be made for windows in bridge wing doors. *Clear view.* A clear view through at least two of the bridge windows and depending on the bridge configuration, an additional number of windows should be provided at all times. A clear view should be provided at all times, regardless of the weather conditions. *Bright sunshine.* To ensure a clear view in bright sunshine, sun blinds with minimum colour distortion should be provided at all windows in front of workstations. Such blinds should be readily removable and not permanently installed. Inclement weather. Heavy duty wipers should be provided for the majority of the front windows. These wipers should do the following:

- Include an interval function and freshwater wash,
- Be in accordance with ISO 3904 if clear view screens are installed,
- Be capable of operating independently of each other,
- Have efficient cleaning, de-icing and de-misting systems installed to ensure a clear view in all operating conditions,
- Have heated glass panels, where installed, be in accordance with ISO 3434,

a

b

Fig. 22.1 **a** Typical bridge arrangement. **b** Typical bridge arrangement

- Include a fixed catwalk with guardrails, fitted under the bridge windows, to allow cleaning of the windows in the event of a failure of other cleaning systems; and
- Glass characteristics. Neither polarised nor tinted glass should be used.

Visibility from the bridge window. It should be possible to watch the area in front of the bridge superstructure from the wheelhouse. Position close to forward centre window for adequate conning. A position close to the forward centre window should be provided with a clear view. If the view in the centreline is obstructed by large masts, cranes, etc., two additional positions giving a clear view ahead should be provided, one on the port side and one on the starboard side of the centreline, no more than 5 m (16.5 ft) apart. *Access to front window.* A second close approach access, besides the forward centre window position described above, should be possible, or the width of the position described above should be sufficient to accommodate two people.

External Visibility

The guidance provided in this subsection generally applies to ships. For floating units which do not move frequently and that normally cannot meet blind view requirements, visibility procedures should be set up as prescribed in the COLREGs, as amended (1972, rules 6, 7 and 8) with respect to safe speed, posting of lookouts and watches and use of sound signals and collision avoidance using radar and plotting devices. *View of sea surface.* The view of the sea surface from the conning position should not be obscured by more than two vessel lengths or 500 m (1640 ft), whichever is less, forward of the bow to 10° on either side irrespective of the vessel's draught, trim and deck cargo (e.g., containers). *Field of vision around the vessel.* It should be possible to observe all objects necessary for navigation, such as ships and lighthouses, in any direction from inside the wheelhouse. There should be a field of view around the vessel of 360° for an observer moving within the wheelhouse. *Navigating and manoeuvring workstation field of view.* At the navigating and manoeuvring workstation and at the conning position, the navigator's field of vision should be sufficient to allow compliance with the COLREGs. The forward horizontal field of view from the navigating and manoeuvring workstation and from the conning position, should extend over an arc from 22.5° abaft the beam on one side of the vessel, forward through the bow centreline to 22.5° abaft the beam on the other side of the vessel at a minimum. From a monitoring workstation, the field of view should extend over an arc from 90° on one bow, through forward, to 22.5° abaft the beam on the other bow at a minimum. The field of view from a workstation on the bridge wing should extend over an arc from at least 45° on the opposite bow through dead ahead and then aft to 180° from dead ahead. *Bridge wing field of view.* From each bridge wing, the horizontal field of view should extend over an arc of at least 225°, that is, at least 45° on the opposite bow through right ahead and then from right ahead to right astern through 180° on the same side of the vessel. *Main steering position field of view.* From the main steering position (i.e., workstation for manual steering) the horizontal field of view should extend over an arc from direct forward to at least 60° on each side of the vessel. Main steering position field of vision. *Blind sectors.* The safe look-out from the navigating and manoeuvring workstation should not be influenced by blind sectors. Blind sectors caused by cargo, cargo gear or other objects forward of the beam that obstruct the view of the sea surface from the navigating and manoeuvring workstations should not exceed 10°. The total arc of blind sectors should not exceed 20°. The clear sector between two blind sectors should be at least 5°. Over an arc from right ahead to at least 10° on each side, any individual blind sector should not exceed 5°. *View of the vessel's side.* The vessel's side should be visible from the bridge wing. Bridge wings should be provided out to the maximum beam of the vessel. The view over the vessel's side should not be obstructed.

Overall Arrangement

The layout of the bridge, including location and layout of the individual workstations should ensure the required field of view for each function. Figure 22.1, Typical bridge arrangement, presents a typical bridge layout. *Vessel control.* The control of the vessel should be allocated to a certain area of the bridge where only instruments and controls necessary for navigating and manoeuvring are located. Navigating and manoeuvring workstations. *Location.* The workstation for navigating and manoeuvring should be laid out, if practicable, at the starboard side close to the centreline. *Design for single or dual operations.* The main workstations for navigating and manoeuvring and the arrangement of instruments pertinent to these stations should be located sufficiently close together to enable a single navigator to cover operations. All necessary information should be provided to perform functions from one working position, but without being restricted to a specific location. The main workstations should be planned, designed and placed within an area spacious enough for no less than two crew members, but close enough to allow the stations to be operated by one person. The consoles, including a chart table if provided, should be positioned so that the instruments they contain are mounted facing a person who is looking forward. *Visibility of area in front of bridge.* The navigator should be able to watch the area immediately in front of the bridge superstructure from the wheelhouse. There should be a close approach access to at least one front window. If this requirement is met by combining "an adequate conning position" and the required access specified in this clause, the width of the total access should be sufficient to accommodate two persons. ***Note***: *This defined position should not affect the view of the helmsman to right ahead, since the helm is normally on the centreline. Manual steering workstation.* The preferred location for the manual steering workstation should be on the vessel's centreline. If the workstation for manual steering is located off the centreline, special steering references for use during day and night should be provided (e.g., lighting marks forward). If the view ahead is obstructed by large masts, cranes, etc., the steering station should be located enough of a distance to starboard of the centreline to obtain a clear view ahead. *Monitoring workstation.* The monitoring workstation should be laid out, if practicable, on the port side of the bridge close to the centreline. *Main stations communication.* From a monitoring workstation, it should be possible to see and hear the people at the navigation, manoeuvring and steering workstations. Where workstations are widely spread, a two-way communications system should be provided so that unhampered communications can be achieved under all operating conditions. *Bridge wing communication.* A two-way internal communication system should always be provided between the navigating and manoeuvring workstation and open bridge wings, particularly when the distance from the wing extremity to the wheelhouse centreline is greater than 10 m (33 ft) (Figs. 22.2 and 22.3).

Fig. 22.2 Typical bridge wing arrangement (open)

Fig. 22.3 Typical bridge wing arrangement (enclosed)

Traffic

Clear route across the wheelhouse. A clear route across the wheelhouse should be provided. The width of the passageway should be at least 1200 mm (47 in.). Width of bridge wing doors should be at least 900 mm (36 in.). Obstructions at the point of entry. There should be no obstructions between the points of entry to the bridge wings and wheelhouse from lower decks and the clear route across the bridge. *Adjacent workstations.* The distance between adjacent workstations should be sufficient to allow unobstructed passage for people not working at the stations. The free passage in passageways between different workstation areas should be at least 700 mm (28 in.). The workstation operating area should be part of the workstation and not part of the passageway. *Passageway distance.* The distance from the bridge front bulkhead or from any consoles or installations placed against the front bulkhead, to any consoles or installations placed away from the bridge front should be sufficient for two people to pass each other. The distance of a passageway between the front bulkhead and any consoles should preferably be at least 1000 mm (39 in.) and not less than 800 mm (31.5 in.). *Doors.* All wheelhouse doors should be operable with one hand. Bridge wing doors should not be self-closing. Means should be provided to hold the bridge wing doors open. *Bridge ceiling clearance height.* The bridge ceiling clearance height in the wheelhouse should be designed with regard to the installation of overhead panels and instruments. The clearance height between the bridge deck surface covering and the underside of the deck head beams should be at least 2.25 m (89 in.). The lower edge of deck head mounted equipment should be at least 2.1 m (83 in.) above the deck in open areas, passageways and at standing workstations. *Main workstations.* Main workstations used for navigating, manoeuvring, manual steering, voyage planning and communication should not cover a working area with a transverse axis longer than 15 m (49 ft).

Sound Signals

Audibility of sound signals inside the bridge. Sound signals should be audible from the interior of the wheelhouse. It should be possible to open some windows in the wheelhouse to hear sound signals. *External sound signals.* External sound signals from ships and fog signals that are audible on the open deck should also be audible inside the wheelhouse. A transmitting device should be provided to reproduce such signals inside the wheelhouse (recommended frequency range: 70–700 Hz).

Chapter 23
Console and Workspace Design

General

This chapter discusses the design and use of consoles and workstations on navigational bridges. The goal of this chapter is to assist in the design of usable and efficient bridge elements. The guidance contained in this chapter should be used in conjunction with Chapter 13, Anthropometric design principles and dimensions.

Terminology

Display: Means by which a device presents visual information to the navigator, including conventional instrumentation.

Console: Structural framework for the integration of devices, equipment and storage that together comprise a workstation.

Layout: Physical arrangement of the parts and components that make up a module or a unit of equipment.

Configuration

The following sections provide guidance for the configuration of consoles exclusive of radar consoles. *Workstation area configuration.* The workstations for navigating and manoeuvring, monitoring and for the bridge wings should be planned, designed and placed within an area spacious enough for not less than two crew members, but close enough for the workstations to be operated by one person. *Single watch stander console.* The console should be designed so that from the normal working

A. A. Olsen, *Applying Physical Ergonomics to Modern Ship Design*,
https://doi.org/10.1007/978-3-031-57974-5_23

position, the navigator can use all instruments and controls necessary for navigating and manoeuvring. The width of consoles designed for single-person operation should not exceed 1600 mm (63 in.). Design of consoles for both seated and standing operation. Figure 23.1, Console configuration and dimensions for standing positions, and Fig. 23.2, Console configuration and dimensions for sitting positions, show the configuration and dimensions of consoles to be used by crew members in both standing and sitting positions. The console profile meets the anthropometric value of the 97.5 percentile (male) and the 2.5 percentile (female) of navigational bridge personnel. (The data in the figures are based on Northern European and North American anthropometrics. For different populations or to establish limiting dimensions using different anthropometric percentiles, refer to Chap. 13, Anthropometric design principles and dimensions and Chap. 14, Application of human factors to design.

Console left-to-right viewing angle. The console should be designed so that from the normal working position, the total required left-to-right viewing angle should not exceed 190°. This angle should be reduced whenever possible through appropriate control-display layout. Refer also to Chap. 24, Design of displays, controls and information on the bridge, Fig. 24.1, Immediate and preferred viewing areas. *Console height.* The top of the consoles should not exceed a height of 1200 mm (47 in.). *Console leg room.* The upper leg room of the console should have a minimum of 450 mm (18 in.) in depth and the lower leg room a minimum of 600 mm (25 in.) in depth.

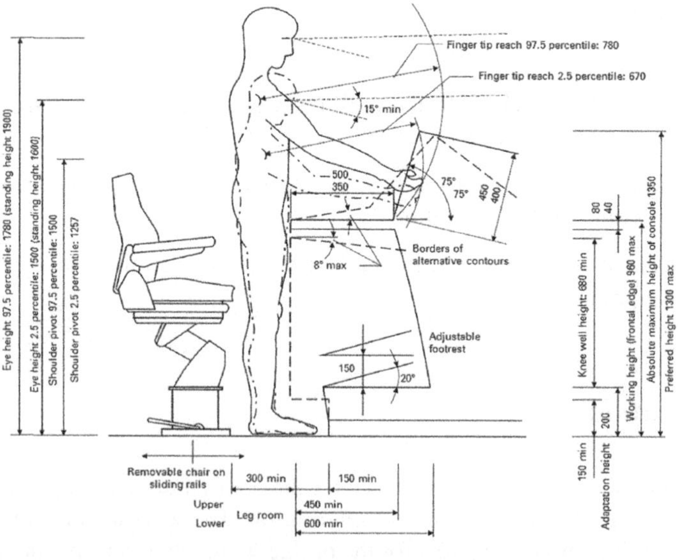

Fig. 23.1 Console configuration and dimensions for standing positions

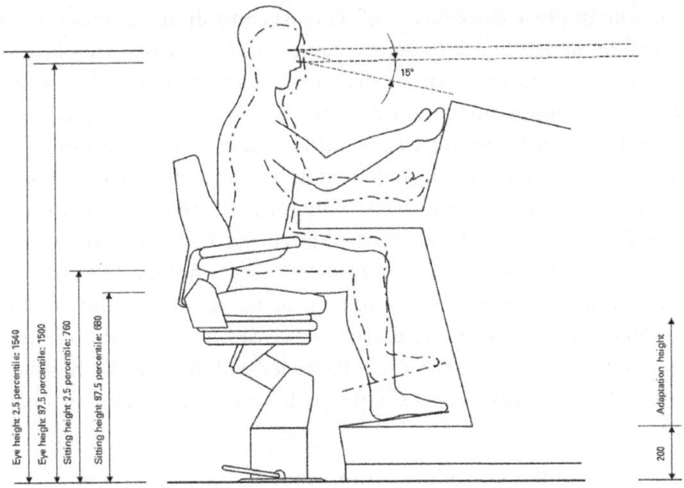

Fig. 23.2 Console configuration and dimensions for sitting positions

Chart table dimensions. The chart table should be large enough to accommodate all chart sizes normally used internationally for maritime traffic. It should also have facilities for lighting the chart. Chart table dimensions should be:

- Width: not less than 1200 mm (47 in.).
- Depth: not less than 850 mm (33.5 in.).
- Height: not less than 900 mm (35.5 in.) and not more than 1000 mm (39 in.).

The chart table should have facilities to accommodate charts larger than the table depth, for example, a 100 mm (0.4 in.) slit along front and back edges of the chart table surface. *Chair design.* Chairs at workstations designed for a sitting position should be capable of rotating with the footrest being arrested, adjustable in height and capable of being arrested on the floor. Chairs should be capable of being moved out of the operating area.

Location of Instruments and Equipment

The guidance in this section is not intended to prevent the use of new control or display techniques, provided the techniques provide equal or better human performance to those included herein. *Location above front window.* Certain instruments displaying information to more than one workstation may be located above the front windows, if such dimensions allow. Such instruments or displays are vessel's heading, wind speed and relative direction, water depth, vessel's speed (e.g., from GPS), rate of turn, rudder angle, propeller revolutions per minute, propeller pitch and time. Instruments and equipment at navigating and manoeuvring workstations. Each workstation should be capable of presenting basic information and should contain the equipment required to enable the navigator to safely carry out the relevant functions and tasks. Ergonomic principles and views of experienced, practicing bridge persons on watch duty should be taken into consideration in the design of workstations. The information system and control possibilities should be made available to the workstations for navigating and manoeuvring in such a way that the tasks at each of these stations can be efficiently carried out. The basic categories of instrument information and equipment for the functions to be performed are described in the following sub paragraphs. *Navigational functions.* Navigation functions include facilities for internal and external communications directly related to navigation of the vessel. External communications include conversing with other vessels in traffic situations, port operations services and vessel movement services (defined in the radio regulations). The navigation functions, controls and displays should be designed to enable bridge personnel to:

- Determine and plot the vessel's position, course, track and speed.
- Alter course.
- Effect internal and external communication related to navigation.
- Monitor time, course, speed and track, propeller revolutions, pitch indication and rudder order and angle.

Manoeuvring functions. Manoeuvring controls and displays should be designed to enable bridge personnel to analyse the traffic situation, decide on collision avoidance, alter course, change speed, effect internal and external communication related to manoeuvring, operate docking aid systems; and monitor time, course, speed and track, propeller revolutions, pitch indicator and rudder order and angle. *Other instrument and equipment locations.* Instruments and equipment should be located to meet bridge personnel needs at each workstation. Typical groupings are as follows. *Navigation workstation.* The navigation workstation should include the instruments and equipment listed in Table 23.1.

Manoeuvring workstation. The manoeuvring workstation, used for collision avoidance/docking, should include the instruments and equipment listed in Table 23.2.

Table 23.2 Manoeuvring workstation instruments and equipment

Navigational radar display(s)	Propeller revolution indicator(s)
Automatic radar plotting aid (ARPA)	Pitch indicator
Engine and thruster controls or telegraphs	Speed and distance indicator
Rudder angle indicator	

Table 23.1 Navigational bridge instruments and equipment

Navigation radar display(s)	Water depth indicator(s)
Position-fixing systems	Navigational chart table with instruments

Instruments and equipment of common interest. Instruments and equipment of common interest to both the stations for navigation and manoeuvring should be located to be visually accessible to both locations without requiring personnel to lean, bend or walk to see the instruments from either position. Such instruments and equipment, if fitted, should include those listed in Table 23.3.

Manual steering workstation. The manual steering workstation should include those instruments and equipment listed in Tables 23.4, 23.5 and 23.6.

Mounting. Instruments, panels and controls should be permanently mounted in consoles or at other appropriate places, taking into account operation, maintenance and environmental conditions. *Portable items*. Portable items, such as safety equipment, tools, lights and pencils should be stored at appropriate places, specially

Table 23.3 Instruments and equipment of common interest

Alarm reset control	Morse light keys
Automatic steering device	Rate of turn indicator
Clock	Rudder angle indicator
Emergency stop controls	Searchlight controls
GMDSS	Speed and distance indicator
Gyro repeater	Steering control
Gyro repeater (bearing)	VHF radiotelephone
Intercommunication systems	Whistle control
Magnetic compass display	Window wiper and washer controls
Manual steering device	

Note Depending on the level of automation, integration of instruments and new methods of display, the workstations for navigation and manoeuvring may be designed as one combined station

Table 23.4 Manual steering workstation

Course indicator	Rate of turn indicator
Gyro repeater	Rudder angle indicator
Magnetic compass display	Talk back to bridge wings
Manual steering device	

Table 23.5 Bridge wing workstations

Communication (external and internal)	Rudder angle indicator
Engine control	Rudder control
Gyro repeater	Sea bottom tracking speed indicator
Morse light keys	Thruster control
Rate of turn indicator	Whistle control

Table 23.6 Bridge wing workstations

Alarms	Radar
Emergency stop controls	Rate of turn indicator
Gyro repeater	Rudder angle indicator
Intercommunication systems	Speed and distance indicator
Monitoring systems	VHF radiotelephone
Propeller revolution indicator(s)	

designed wherever necessary (e.g., in recesses to avoid rolling or falling in heavy seas).

Communication Devices

Internal and external communications. Where communications equipment for internal or external communications is installed on the bridge, it should be installed and used such that it should not interfere with the vessel's navigation. Such equipment, which may be for distress and safety communications or for general communications, is referred to as "additional communications equipment". Additional communications equipment. Additional communications equipment fitted on the vessel's bridge should be fitted in a communications workstation. Those parts of such communications equipment not fitted with operating controls and displays and that, by reason of their size or for other practical considerations, cannot be mounted in this workstation may be mounted elsewhere in the vessel. *Operation.* It should be possible for a navigator to operate the communications equipment situated on the bridge. The communications workstation may be situated adjacent to the navigation workstation. Alternately, the navigation workstation may be fitted with facilities for remote control of the additional communications equipment, so that the navigator may use the equipment in any normal working position. *Navigator unavailable.* Communications equipment on the bridge should be arranged so that whenever the situation does not permit the navigator to operate the additional communications equipment, the navigator can be relieved of this task. The communications workstation or the navigation workstation, if remote control is installed, should include facilities for routing or instantly transferring communications to a position elsewhere in the vessel. Alternately, the navigator should be able to call another person to the communications workstation, whenever the navigator is unable to take care of such communications due to navigational duties. *Radiotelephone public correspondence.* It should be possible to perform radiotelephone public correspondence communications without these communications being audible to the navigator. If communications from the communications workstation might be audible to the navigator, then another place or room fulfilling this requirement should be available for public correspondence. *Reception of written communications.* Equipment on the bridge for reception of written communications (e.g., telex, telefax) should be so arranged that only persons authorised for this task would be able to read such communications.

Chapter 24
Design of Displays, Controls and Information on the Bridge

General

This chapter specifies the design of displays, controls and information on the bridge. The objective of this chapter is to aid in the successful design, organisation and presentation of displays and the devices that control them.

Terminology

Ascenders/descenders: The point size of a given font is measured from the top of the font ascenders (i.e., the top of the letter) to the bottom of its descenders (e.g., the length of the "tail" of the letter "p"). The ascender is the measurement from the top of the lowercase letter to the top of the uppercase letter. The descender is measured from the bottom of the uppercase letter to the bottom of the lowercase letter.

Aspect ratio: The ratio of the font size (i.e., measurement from the baseline to the top of the uppercase letters) to the x-height (i.e., measurement from the baseline to the top of the lowercase letters.)

Control: A device bridge personnel use to input a signal to a display or to change the operating status of the equipment or system.

Icon: A graphical representation of an object or function; a graphic symbol.

Parallax: An apparent change in the direction of an object, caused by a change in observational position that provides a new line of sight.

Stroke width: The width of the line that defines the font.

Visual display: A device that provides readings, status or conditions of machinery, equipment or system operating parameters.

A. A. Olsen, *Applying Physical Ergonomics to Modern Ship Design*,
https://doi.org/10.1007/978-3-031-57974-5_24

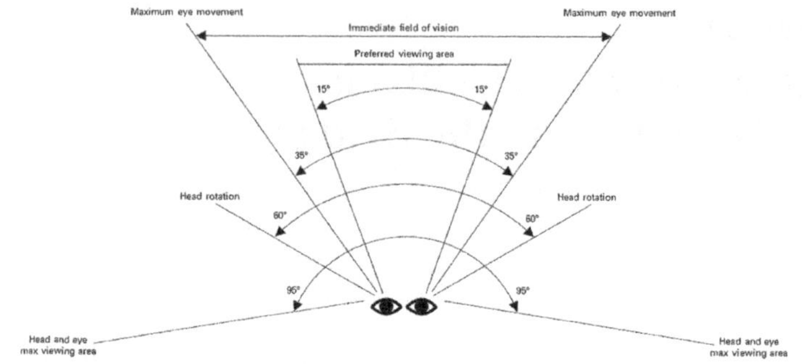

Fig. 24.1 Immediate and preferred viewing areas

Display Arrangement

Immediate field of vision. The most important and/or frequently used displays should be located within the watch stander's immediate field of view, defined as the viewing area with eye rotation only. Refer to Fig. 24.1, Immediate and preferred viewing areas. *Preferred viewing area.* The preferred viewing area should be reserved exclusively for the most important and/or frequently used displays.

Information Display

The guidance in this chapter applies predominantly to the design of computer displays, consoles and display groups. *Lack of ambiguity.* Display indicators should clearly and unambiguously direct and guide the appropriate control response. *Use of digital displays.* Digital displays should be used for the presentation of quantitative data when exact values are required and when continuous trend or rate of change information is not required. *Inappropriate use of a digital readout.* A digital readout should not be used when the information changes with a frequency of more than 0.5 Hz; a higher frequency may be used when the information perception from other displays is not disturbed. *Update of information.* The displayed information should be continuously and automatically updated. *Information duration.* For signals or displays that frequently or consistently change their outputs, the information displayed should have durations of sufficient length to be reliably detected under the expected workload and operational environment. *Display simplicity.* Displays should present the simplest information consistent with their function. Information irrelevant to the task should not be displayed. Extraneous text and graphics should not be present. *Tailoring the display to bridge personnel needs.* Design of data should be tailored to bridge personnel needs, providing only necessary and immediately usable data for any task. *Display fields for the display of important information.* The display fields for the presentation of the most important and/or frequently used information should be assigned exclusively to that information and should not be used to display any other information. *Graphic display enhancement.* When precise reading of a graphic display is required, the display should be annotated with actual data values to supplement their graphic representation. *Indication of scale.* The scale of maps and charts (data) shown on the display should always be indicated. *Aiding distance judgments.* When bridge personnel must judge distances accurately on a map or other graphic display, computer aids should be provided for that judgment. *Multifunction displays.* Multifunction displays, if used, should be colour displays.

Organisation of Visual Information

The guidance in this subsection applies predominantly to the design of computer displays, consoles and display groups. *Consistent screen organisation.* Display screen organisation should be consistent with respect to the location of various system functions (such as a data display zone, control zone or message zone) from one display to another. *Consistent presentation.* The arrangement and presentation of identical visual information should be consistent. *Grouping of information on a display.* Information on a display should be grouped according to obvious principles (e.g., by task, system, function, sequence, etc., based upon requirements in performance of the ongoing task). *Demarcation of information groups.* Information groups should be visually distinct, for example, separated by blanks, lines, colour coding or other means. *High priority display.* Where two bridge personnel must use the same display and the displays have high priority, duplicate sets should be provided, whenever there is adequate space. Otherwise, displays should be centred between bridge personnel. Alternatively, they can be placed so that both can easily monitor them (e.g., above the forward window). *Positioning of shared displays.* Where two bridge personnel must use the same display and secondary displays must be shared, they should be positioned equally between both bridge personnel. If the displays are more important to one watch stander than to any others, they should be placed nearest personnel having the principal requirements for using them. Alternatively, they can be placed so that both can easily monitor them (e.g., above the forward window). Instrumentation grouping. Instruments should be logically grouped according to their functions, such as:

- Navigating.
- Manoeuvring.
- Communication.

Display Design

Font style. A clearly legible font should be used. Fonts should have true ascenders and descenders, uniform stroke width and uniform aspect ratio. *Meaningful abbreviations.* When abbreviations or acronyms are used, they should be meaningful, in common usage and kept to a minimum. *Units of measurement.* The units of measurement (volts, pounds of pressure, inches, etc.) should be labelled. *Icons.* (1) *Appropriate use of icons.* Icons should be designed to look like the objects, processes or operations they represent, by use of literal, functional or operational representations. (2) *Representation and discrimination.* Each icon or symbol should represent only one object or function and should be easily discriminable from all other icons and symbols. (3) *Size.* Icons and symbols should be large enough for bridge personnel to perceive the representation and discriminate it from other icons and symbols. (4) *Highlighting.* An icon or symbol that bridge personnel have selected should be highlighted. *Scaling in standard intervals.* Scales should have graduations at a standard interval of 1, 2, 5 or 10 (or multiples of 10) for labelled divisions, as appropriate. Intervening tick marks to aid visual interpolation should be consistent with the labelled scale interval. *Expansion of graphic displays.* When a graphic display has been expanded from its normal coverage, some scale indicator of the expansion factor should be provided. *Unobtrusive grids.* When grid lines are displayed, they should be unobtrusive and not obscure data elements (e.g., curves and plotted points).

Control Design

Control placement. Controls requiring frequent adjustments or accurate settings should not be placed more than 675 mm (26.5 in.) from the front edge of the console. *Control positioning for simultaneous operation.* Controls should be located so that simultaneous operation of two controls will not necessitate a crossing or interchanging of hands. *Location of primary and frequently used controls. The* most important and frequently used controls should have the most favourable position with respect to ease of reaching and grasping (particularly rotary controls and those requiring fine settings). Keys for emergency functions should have a prominent position and be characteristically marked. *Consistent arrangement.* The arrangement of functionally similar or identical controls should be consistent from workstation to workstation and panel to panel throughout the bridge. *Spacing between controls.* Appropriate spacing between the controls should be provided. *Movement of controls.* Movement of a control forward, clockwise to the right or up, should:

- Turn the equipment or component **ON**.
- Cause the quantity to increase.
- Move the equipment or component forward, clockwise, to the right or up.

Corresponding movements. Controls should be selected so that the direction of movement of the control will be consistent with the related movement of an equipment component or vessel. For example, pushing a propulsion speed control lever forward should cause the vessel speed to increase. The direction of motion of operating elements for manoeuvring equipment should correspond with the direction of the effect on the vessel. *Returning to the navigation monitoring mode.* When a single device is used simultaneously for voyage planning and navigation monitoring it should be possible to revert to the monitoring mode with a single action. *Minimal actions.* Control actions should be simple, particularly for real-time tasks requiring fast response. Control logic should permit completion of a transaction sequence with the minimum number of actions. *Consistency of control actions.* The same functions should be activated on devices by the same control actions, as far as practicable. Feedback. Visual, auditory or mechanical feedback should be provided to indicate that a controller input has been registered. *Ease of operation of controls.* Controls should be easy to identify and operate. *Obscuration of display by the operation of controls.* The operation of a control should not obscure displays where observation of these displays is necessary to allow control adjustments to be made. *Controls for important functions.* (1) *Accessibility of controls for important functions.* A dedicated control should be provided for important and/or frequently used functions and that control should be easily visible and accessible to bridge personnel from the normal working position (e.g., a single variable function key should not be used to control several important or frequently performed functions). (2) *Operation of controls for important functions.* The controls for the most important and/or frequently used functions should require only a single actuation to accomplish their function. *Assignment of controls of important functions.* The controls for the most important and/or frequently used functions should be assigned to only one function. *Prevention of accidental input or actuation.* Control locations and physical arrangements should be designed to prevent the accidental manipulation of those controls that could result in changes to the status of the system, the system functions, components or data, for example, loss of power. One way to accomplish this is through physical protection and/or placement on the equipment. *Control purpose.* The purpose of each control should either be clearly illustrated by symbols where standard symbols have been internationally adopted or labelled. Simple controls and indicators. Controls or combined controls/indicators should be visually and tactually distinguishable from elements that only indicate. Rectangular buttons should be used for control elements and round lights for indicator elements. *Operation of controls.* The shape of a control should provide a visual indication of the method of operation. For example, controls where there are two or three discrete positions should be toggles or levers. Rotary continuous position controls (i.e., rheostats) should have knobs or wheels, except the steering control.

Control and Display Integration

Logical arrangement. Controls and displays should be fitted in a logical arrangement and combined into functional groups. *Location consistency.* Location of recurring functional groups and individual items should be similar from console to console. *Visual information for more than one watch stander.* Displays providing visual information to more than one person on duty should be located for easy viewing by all bridge personnel concurrently. If this is not possible, the displays should be duplicated (e.g., repeaters). *Control and display location.* Controls and their associated displays should be located so that the information on the displays can be easily read during the operation of the controls. *Simultaneous use.* A visual display that must be monitored concurrently with manipulation of a related control should be located so that bridge personnel are not required to observe the display from an extreme visual angle that introduces the possibility of parallax error. *Control/indicator distinctiveness.* Controls or combined controls/indicators should be visually and tactually distinguishable from elements that only indicate. *Readability of instruments.* Instruments should be designed to permit easy and correct reading by day and by night (e.g., that rheostatic or other means are used to brighten or darken displays in accordance with the environmental light conditions). The requirements of IMO Resolution A.574 (XIV) apply to the design of electronic navigational aids. A digital readout should not be used where the reading changes rapidly. For an index moving relative to a circular scale, the index should move clockwise (or the scales move counter-clockwise) for increasing readings. For an index moving relative to a linear scale, the index should be horizontal or vertical and the pointer should move to the right or upward for increasing readings. ***Note***: *there may be special cases where these guidelines do not apply, for example, where an instrument reading may be positive or negative or where depth is indicated.*

Instrument placement. Each instrument should be placed with its face approximately perpendicular to line of sight. *Glare minimisation.* Instruments should be designed and fitted to minimise glare or reflection. Instruments should not be obscured by strong light. All instruments should be placed in positions relative to bridge personnel, considering the surrounding light sources. Where a transparent cover is fitted over an instrument or instruments, it should be designed to minimise reflections. *Principal manoeuvring instruments.* (1) *Readability from main manoeuvring workstation.* The principal manoeuvring instruments should be readable from the main manoeuvring workstation. (2) *Reading distance.* Instruments meant to be operated or positioned in connection with controls should be readable from a distance of at least 1000 mm (39.5 in.). Any other instruments should be readable from a distance of at least 2000 mm (79 in.). (3) *Character height.* Character height in millimetres should be not less than three and a half times the reading distance in metres. With the exception of the number 1 (one) and the letters i, I and l, character width should be 0.7 times the character height, for example:

- Character height for reading distance of 2 m: $2 \times 3.5 = 7$ mm.
- Character width for letter height of 7 mm: $7 \times 0.7 \cong 5$ mm.

- Resulting minimum character size: 7 mm (0.28 in.) height by 5 mm (0.2 in.) width.

Contrast ratio. All information should be presented on a background of high contrast, emitting as little light as possible by night. All bridge instruments should be designed to show a light text on a dark non-reflecting background at night. The contrast ratio should be within 1:3 and 1:10. *Character type*. Instrument character type should be of simple, clear-cut design. Internationally used and recommended character type is Helvetica medium. However, light emitting diode text matrices are acceptable.

Control and Display Failure Prevention

Loss of power. Following a loss of power that has lasted for 30 s or less, all primary functions should be readily reinstated. Following a loss of power that has lasted for more than 30 s, as many as practical primary functions should be readily reinstated. *Computer network failure*. Where computerised equipment is interconnected through a computer network, failure of the network should not prevent individual equipment from performing its individual functions.

Chapter 25
Design of Alarms and Warnings

General

This chapter outlines the design and use of the alarms and warnings present on the navigational bridge. IMO Resolution A.830(19), Code on Alarms and Indicators, (1995) presents "general design guidance [...] to promote uniformity of type, location and priority..." for bridge alarms and indicators. Please refer to that document for tables of required alarms and indicators for ships' bridges and machinery spaces of different functionality, machinery and ship type.

Alarms

(1) *Bridge control system failure.* In the event of a failure of the bridge control system, an alarm should notify the bridge personnel. The speed and direction of the propeller thrust should be maintained until local control is in operation, unless this is considered impracticable. A lack of power (electric, pneumatic or hydraulic) should not lead to major and sudden change in propulsion power or direction of propeller rotation. (2) *Alarm acknowledgement.* (3) *A method of acknowledging all alarms* (silence audible alarms and set visual alarms to steady state), including the indication of the source of the alarm, should be provided at the navigating and manoeuvring workstation. This will help ensure that bridge personnel avoid distraction by alarms that require attention but have no direct influence on the safe navigation of the vessel and that do not require immediate action to restore or maintain safe navigation. (4) Fire and emergency alarms. The alarm indicators and controls of the fire alarm and emergency alarm should be located at the safety workstation. (5) *Failure or reduction of power supply.* Alarms should be provided to indicate failure or reduction in the power supply that could influence the safe operation of the equipment. (6) *Sensor input failure or absence.* Alarms should be provided to indicate sensor input failure

A. A. Olsen, *Applying Physical Ergonomics to Modern Ship Design*,
https://doi.org/10.1007/978-3-031-57974-5_25

or absence. (7) *Alarm status.* Alarm systems should clearly distinguish between alarm, acknowledged alarm and no alarm (normal condition). (8) *Continuation of alarms.* Alarms should continue until they are acknowledged. (9) *Cancellation of alarms.* Alarms and acknowledged alarms should be capable of being cancelled only if the alarm condition is rectified. This cancellation should be possible only at the individual equipment. (10) *Alarm minimisation.* The number of alarms should be minimised. (11) *Alarm testing.* Bridge personnel should be able to functionally test alarms. (12) *Alarm power supply.* Required alarm systems should be continuously powered and should have an automatic changeover to a standby power supply in case of loss of normal power supply. (13) *Indication of alarms.* Alarms should be indicated in order of sequence and provided with aids for decision-making. An explanation or justification of an alarm should be available (on request). (14) *Presentation of alarms.* The presentation of alarms should be clear, distinctive, unambiguous and consistent. (15) *Modes of alarms.* All required alarms should be presented through both visual and auditory means. (16) *Distribution of power supply failure alarm.* Failure of the main power supply to the distribution panels should initiate an audible and visual alarm. (17) *Location of alarms.* Alarms are to be provided at the main control station, engineer accommodation area and on the bridge. Group alarms may be arranged on the bridge to indicate machinery faults. Alarms associated with faults requiring speed reduction or the automatic shutdown of propulsion machinery should be separately identified.

Alarm/Warning Transfer System

(1) *Alarms requiring a response.* Any alarm/warning that requires bridge personnel response that is not acknowledged within 30 s should be automatically transferred to the master and if the master deems it necessary, to the selected back-up navigator and to the public rooms. (2) *Fixed location for transfer.* The alarm and warning transfer should be operated from a fixed installation. (3) *Alarm acknowledgement on bridge.* Acknowledgment of alarms/warnings should be possible only from the bridge. (4) *Two-way speech capability during alarm transfer:*

- *OOW*. At all times, including during blackout, the officer of the watch should have access to facilities enabling two-way speech communication with another qualified officer. The bridge is to have priority over the communication system. Note: the automatic telephone network is acceptable for this purpose, provided that it is automatically supplied with emergency or back-up battery power during blackout situations and that it is available in the locations specified in the paragraph "Alarms requiring a response".
- *Backup navigator*. If, depending on the shipboard work organisation, the back-up navigator may attend locations not connected to the fixed installation(s) described in the paragraph "Alarms requiring a response", the back-up navigator should be

provided with a wireless portable device enabling both the alarm/warning transfer and two-way speech communication with the officer of the watch.

Alarms Concerning Navigation

(1) *Alarm system.* The detection system should initiate audible and visual alarms distinct in both respects from the alarms of any other system not indicating fire, in sufficient places to ensure that the alarms are heard and observed on the navigating bridge and by a responsible engineering officer. When the navigation bridge is unmanned (e.g., when not underway or due to watch stander incapacitation) the alarm should sound in a place where a responsible member of the crew is on duty. Alarms not acknowledged within a specified time (nominally 30 s) on the bridge should sound in a place where a responsible member of the crew is on duty (e.g., cargo control room, master's office, ship's office, and engine control room). (2) *Planned route deviations.* The navigator should be given an alarm when there is deviation from the planned route. This alarm should be adjustable with respect to the time and to the danger of grounding. (3) *Water depth.* An alarm should be initiated when the water depth beneath the vessel is less than a predetermined value. (4) *Hydraulic locking. Where* a single failure can lead to hydraulic lock and loss of steering, an audible and visual alarm, which identifies the failed system, should be provided on the navigation bridge. This alarm should be activated whenever:

- Position of the variable displacement pump control system does not correspond with given orders.
- Incorrect position of 3-way full flow valve or similar device in a constant-delivery pump system is detected.

Visual Alarms

(1) *Discrimination of visual alarms.* Visual alarms should clearly differ from routine information on displays. (2) *Presentation of visual alarms.* Visual alarms should flash after onset. Flashing display should change to steady display upon acknowledgement. (3) *Presentation of normal conditions (no alarm).* Alarm indicators should be designed to show no light or should not be visible on displays in normal conditions (i.e., no alarm). (4) *Flash rate.* Flashing visual alarms should be illuminated for at least 50 percent of the cycle and have a pulse frequency in the range of 0.5–1.5 Hz. (5) *Lamp testing.* Means should be provided to test the lamps. (6) *Visual alarms and night vision.* Visual alarms on the navigation bridge should not interfere with night vision. (7) *Red lighting for alarm indicators.* Alarm indicator lights should be equipped with red lights of 620 nm (nm) or higher wavelength. Low level white lights are also acceptable.

Audible Alarms

(1) *Use of audible alarms.* Audible alarms should be used simultaneously with visual alarms. (2) *Cessation of audible alarms.* Audible alarms should silence when acknowledged. (3) *Differentiation of audible alarms from other signals.* Audible alarms should be differentiated from routine signals, such as bells, buzzers and normal operation noises. (4) *Audible alarm sound characteristics.* Under normal working conditions, the alarm signals should be heard properly inside the wheelhouse and outside on the bridge wings. (5) *Sound pressure.* Audible alarm sound pressure, 1 m (39 in.) from the source should be at least 75 dB(A) and at least 10 dB(A), or preferably 20 dB(A) above ambient noise levels existing during normal operations. Audible alarm sound pressures in a space should not exceed 115 dB(A). (6) *Sound frequency.* With the exception of bells, audible alarms should have a signal frequency between 200 and 2500 Hz, with the preferable range between 500 and 1500 Hz.

Chapter 26
Procedures, Codes and Aides Memoires

Abstract This chapter discusses design considerations associated with providing navigation bridge personnel with visual and audible information that will support successful completion of tasks on the navigation bridge. This includes coding, procedures and other aides memoires.

General

This chapter discusses design considerations associated with providing navigation bridge personnel with visual and audible information that will support successful completion of tasks on the navigation bridge. This includes coding, procedures and other aides memoires.

Terminology

CIE: Commission Internationale de l'Eclairage is an international organisation that focuses on the development and implementation of safe standards in lighting.

Coding: Coding refers to a system of sensory stimuli used to represent specific information. There are various mechanisms for coding information. These include, but are not limited to:

- Audible through pitch, frequency and modulation.
- Visual through colour and shape.
- Tactile through shape and texture.

Labels: Within the context of this chapter, the terms "labels" and "labelling" refer to the descriptors used to identify controls, displays and other components on the navigation bridge and not to the internal components of displays such as meter scales or fields in a computer display.

A. A. Olsen, *Applying Physical Ergonomics to Modern Ship Design*,
https://doi.org/10.1007/978-3-031-57974-5_26

Synthesised speech: Machine-generated speech that is produced by a voice synthesiser.

Coding

The following sections provide guidance on the use of coding to support navigation bridge tasks. *Consistency.* Within any coding context (e.g., colours used on alarms, piping diagrams or simple indicator lights) coding mechanisms should be consistently applied, using the same shapes, colours or tones to convey the same meaning from console to console and space to space. *Coding simplicity.* Codes should be simple and easily recognised by limiting the number of coded levels used (e.g., use no more than seven different colours to code alarms). Bridge personnel should be able to easily sense and discriminate the code and recognise its meaning. *Flash coding.* Red flash coding should be reserved for alarms. *Colour coding.* Colour coding of indicator lights and lamps should be in accordance with Table 26.1, Colour coding recommendations for informational displays. Note that blue light is typically used for advisory indication (e.g., pumping system ready to operate) and white light indicates confirmation or selection (e.g., which pumps of a fire main in standby are online). (1) *Recommended colour coding for controls.* Table 26.2, Colour coding recommendations for controls, provides recommended colour coding for controls. (2) *Colour coding compatibility with ambient illumination.* Colour coding should be compatible with ambient illumination. Special consideration should be given to cases where low-level red illumination is used, since red backlit displays may be difficult to discern when externally illuminated by red light. (3) *Redundant colour coding.* Colour coding should be redundant with some other display feature. Add colour coding after displays have already been designed as effectively as possible in a monochrome format. (4) *Easily discriminable colours.* When selecting colours for coding discrete categories of data, those colours should be easily discriminable. (5) *Minimum colour differences.* When colour coding is used for discriminability or conspicuousness of displayed information, all colours in the set should be clearly distinct from each other.

Shape coding. When shape coding is used, codes should be based on established standards or conventional meanings. Where used, shape codes should be applied consistently. *Audible coding.* (1) *Number of audible signals.* To avoid confusion, the number of equipment-based audible signals should be minimised, ideally to communications calls, watch safety system pre-warnings, navigation alarm/warnings, machinery/cargo alarms, fire alarms; and the vessel's general alarm. (2) *Distress signals.* Distress signals should be clearly distinguishable from other calls. (3) *Distinguishing audible signals.* Differences in time patterns and/or frequencies may be used to allow bridge personnel to distinguish one signal from another. (4) *Types of audible signals.* Suggested audible signal types for navigation bridges are provided in Table 26.3, Basic types of audible signals for use on navigation bridges. Audible signals should

Table 26.1 Colour coding recommendations for informational displays

Colour	Meaning	Explanation	Typical applications
Red	Danger or alarm	Warning of potential danger or a situation that requires immediate action	Failure of pressure in a lubricating system Temperature outside specified limits Activation of a safety system
Amber/ yellow	Caution	Change or impending change of conditions	Pressure or temperature different from normal level Run/running out of time—where time is limited
Green	Safety	Indication of a safe situation or authorised to proceed	Cooling liquid circulating Automatic boiler control in operation Machine ready to be started
Blue	Instruction/ information	Any meaning not covered by the above colours	Motor ready to start, pump in standby re-circulation, RORO Ramps deployed
White	No specific meaning assigned (neutral)	Any meaning May be used where doubt exists about the application of red, green or yellow/amber Often used for confirmation	Telephone calls Synchronising lamps (for A/C bus alignment)

Table 26.2 Colour coding recommendations for controls

Colour	Meaning	Typical applications
Red	Action in case of emergency Stop or off	Emergency stop Firefighting General stop of one or more motors or open a switching device
Amber/yellow	Intervention	Intervention to suppress abnormal conditions or avoid unwanted changes
Green	Start or on	Start one or more motors or close a switching device
Blue/black/grey/white	Any meaning not covered by the above colours	Selection of station for control or standby navigator/engineer

be designed so as not to startle bridge personnel. (5) *Synthesised speech as an audible warning.* When synthesised speech is used as an audible warning, the warning should be preceded by a distinguishing signal to alert bridge personnel.

Table 26.3 Basic types of audible signals for use on navigation bridges

Type	Typical characteristics	Considerations
Buzzer	Low intensity and low frequencies	Good alerting in quiet environments without startling bridge personnel
Bell	Moderate intensity and moderate frequencies	Penetrates low frequency noise well, abrupt onset has a high alert value
Chime	Moderate intensity and moderate frequencies	Good in quiet environments, tends not to startle
Tone	Moderate intensity and limited frequency range	Convenient for intercom transmission, high alert value if intermittent

Operating and Emergency Procedures

(1) *Automatic sequence of operations.* Operations following any setting of a bridge control device, including reversing from the maximum ahead service speed in case of emergency, are to take place in an automatic sequence and with time intervals acceptable to the machinery. (2) *Emergency stopping device.* The main propulsion machinery should be provided with an emergency-stopping device on the navigation bridge, independent from the bridge control system. (3) *Availability of flooding control plan.* A plan clearly showing each deck, hold and boundaries of the watertight compartments, the openings therein with the means of closure and position of any controls thereof and the arrangements for the correction of any list due to flooding, is to be permanently exhibited or readily available on the navigation bridge. In addition, booklets containing the aforementioned information are to be made available to the officers of the vessel. *Design of procedures.* (1) *When to provide procedures.* Written procedures should be provided for complex or infrequent task sequences and for off-normal and emergency situations that do not require immediate action. ***Note:*** *immediate actions should be basic skills known by memory.* (2) *Base procedures on task requirements:*

- To the extent possible, sequence task order in procedures to reduce extent of movement required. Assign responsibility of these tasks to competent persons. Identify the purpose and scope (to whom the procedures apply) of these procedures. Summarise records and/or other requirements as the output of these procedures.
- Present data to be easily read, interpreted and to the level of precision required to perform a task.
- Refer to components in a procedure using exactly the nomenclature that identifies the associated components and settings (i.e., labels).

(3) *Postulated event base for emergency procedures.* Develop procedures based on postulated events (e.g., loss of steering, loss of thrusters, loss of navigational components such as GPS, emergency re-ballasting plans, etc.). (4) *Develop and use constrained vocabularies.* Develop and use constrained nomenclature to be used in procedures, using simple and commonly understood language. Acronyms and

Table 26.4 Examples of where consistent style should be used

Control of branching from conditional statements	Numbering systems
Fonts and paragraph styles	Stepwise numbering
Formats for information presentation	Styles and placements of graphics and tables
Identifying procedures versions, approvals and publication dates	Styles for procedural warnings and cautions
Limits and use of cross-references	Use of coding
Numbering pages	Wording style

abbreviations should be clearly and frequently defined within the procedure(s). Where procedures require external communications, they should recommend specific wording based on IMO Standard Marine Communication Phrases. Refer to Table 26.5 for the IMO SMCP Phonetic alphabet and Table 26.6 for the IMO SMCP pronunciation numbering system. (5) *Develop and use consistent style guides.* Develop and impose procedure style guides. Style guides should address the following (Table 26.4).

(6) *Wording style and sentence structure.* Wording style and sentence structure can make documents easy to read and procedures reliably followed; specifically:

- Use simple sentence structures.
- Avoid negative sentences.
- Present only one thought or action per sentence or procedure step.
- Write in a command, imperative style (e.g., "**Set autopilot to OFF**" as opposed to "**Autopilot should be OFF**").
- State procedure conditions before action steps (e.g., "**If the helmsman is at the steering wheel, then set steering control to MANUAL**" rather than "**Set steering control to MANUAL if a helmsman is at the steering wheel**").

(7) *Use of "may," shall," "will," and "should".* "Shall" and "will" are used to indicate a requirement; "should" indicates a recommended practice, preferred method or expected condition; and "may" indicates acceptable or suggested methods. It also denotes permission. (8) *Validation of content.* Content of procedures should be verified by formal review using bridge persons on watch duty as subject matter experts and by walk through of each procedure in the navigation bridge. ***Note:*** *content validation can be critical to safe operation of a ship.*

Labels, Placards and Job Performance Aids

Functional labelling. Controls and displays should be labelled clearly and unequivocally according to their function, possibly by using standardised symbols. *Label terminology.* The selection and use of terminology for labels should be consistent

Table 26.5 IMO SMCP pronunciation alphabet

Letter 2000—present	IMO SMCP pronunciations	2007—Present ITU-R pronunciations
A	Alfa	AL FAH
B	Bravo	BRAH VOH
C	Charlie	CHAR LEE or SHAR LEE
D	Delta	DELL TAH
E	Echo	ECK OH
F	Foxtrot	FOKS TROT
G	Golf	GOLF
H	Hotel	HOH TELL
I	India	IN DEE AH
J	Juliet	JEW LEE ETT
K	Kilo	KEY LOH
L	Lima	LEE MAH
M	Mike	MIKE
N	November	NO VEM BER
O	Oscar	OSS CAH
P	Papa	PAH PAH
Q	Quebec	KEH BECK
R	Romeo	ROW ME OH
S	Sierra	SEE AIR RAH
T	Tango	TANG GO
U	Uniform	YOU NEE FORM or
		OO NEE FORM
V	Victor	VIK TAH
W	Whisky	WISS KEY
X	X-ray	ECKS RAY
Y	Yankee	YANG KEY
Z	Zulu	ZOO LOO

between controls and displays. *Operating/technical manual availability.* An operating/technical manual should be placed on board the vessel that gives clear guidance to operational personnel regarding the vessel's capabilities, limitations and navigational procedures for the required bridge manning levels. This manual should include, at a minimum, the vessel name and IMO registration number, simplified diagrams, vessel navigation and manoeuvring capabilities, navigational procedures; and periodic testing procedures.

Table 26.6 IMO SMCP pronunciation numbering system

Number	2000—Present IMO SMCP pronunciations
0	ZEERO
1	WUN
2	TOO
3	TREE
4	FOWER
5	FIFE
6	SIX
7	SEVEN
8	AIT
9	NINER
.	FULL STOP/PERIOD
,	COMMA
/	SLASH
1000	TOUSAND

Chapter 27
Working Environment

General

The work environment in which tasks are performed has a significant influence on human performance. This chapter addresses task performance design considerations associated with the navigation bridge work environment including vibration; noise; lighting; device and instrument illumination; and HVAC.

Terminology

Accessibility: The ability of personnel to easily access equipment that requires maintenance, inspection, removal or replacement while wearing the appropriate clothing, including personal protective equipment and using all necessary tools and test equipment.

Anthropometrics: Data relating to physical body dimensions. It includes body characteristics such as size, breadth and the distances between anatomical landmarks, such as elbow to fingertip.

Bridge wing: Parts of the bridge on both sides of the vessel's wheelhouse extending to the sides.

Catwalk: Extension to a deck that is wide enough to allow the passage of a person.

Dark adaptation: Dark adaptation is the process by which the human eye adjusts to lower levels of illumination through increased sensitivity to light. Visual function in very dim light has the following characteristics:

- Visual detail is significantly reduced.
- The eye becomes more sensitive to blue light.

A. A. Olsen, *Applying Physical Ergonomics to Modern Ship Design*,
https://doi.org/10.1007/978-3-031-57974-5_27

- Foveal visual performance degrades significantly, while peripheral vision becomes dominant.

dB(A): Sound level in decibels, measured using A-weighting. The use of A-weighting causes the frequency response of sound level measurement devices to mimic that of the human ear, where response is maximum at about 2 kHz, less at very low or very high frequencies. A-weighted measurements correlate well with measures of speech interference and judgments of loudness.

Effective temperature: An index that combines into a single value the effect of temperature, humidity and air movement on the sensation of warmth or cold. The numerical value is that of the temperature of still, saturated air that would induce an identical sensation to the human body.

Glare: Glare is the sensation produced by luminance within the visual field that is sufficiently greater than the luminance to which the eyes are adapted causing annoyance, discomfort or loss of visual performance and visibility. Glare is experienced if windows or light sources, seen directly or by reflection in shiny surfaces, are too bright compared to the general brightness within the interior of the bridge.

Grab-rails: Hand grabs mounted to bridge surfaces, such as bulkheads that are used to help negotiate awkward accesses.

Handrails: Rails mounted along passageways, ladder ways, etc., that are used to safely negotiate spaces and work areas.

Illuminance: The amount of light falling on an object or surface. Illuminance is measured using units of Lux (lm/m^2) or foot-candles (fc; lm/ft^2). One foot-candle equals 10.76 lx.

Luminance: The amount of light emitted from a source (including that reflected from a surface). Luminance is measured in candela-per-square metre (cd/m^2).

Maintainability: The ability to carry out rapid and effective system restoration to maintain the equipment at a specified level of performance.

Maintenance: All activities necessary to keep equipment in or restore it to a specified level of performance.

Occupational safety: The science of preventing individuals from being injured or contracting diseases in the course of their work.

Whole body vibration: Mechanical vibration (or shock) transmitted to the human body as a whole. Whole body vibration is often due to the vibration of a surface supporting the body.

Vibration

Vibration levels and comfort. Uncomfortable levels of vibration should be avoided on the bridge. Vibrations should be reduced, in particular vibration, in the 0–30 Hz range, so that bridge personnel are neither hindered in their functions nor put at a health risk. *Whole body vibration limits.* Whole body vibration levels should not exceed 0.315 (preferred) to 0.4 m/s^2 (acceptable) in the frequency range of 0.5–80 Hz. *Preventing motion sickness.* Guidance to help prevent motion sickness is provided in Part 3.

Noise

The noise level and acoustic environment on a bridge should not interfere with verbal communications, mask audible alarms, be uncomfortable, contribute to fatigue of bridge personnel or reduce overall system effectiveness. Noise levels should comply with IMO Resolution A.468(X II) Code on Noise Levels on Board Ships and take into account IMO Resolution A.343(IX) Recommendation on Methods of Measuring Noise Levels at Listening Posts. *Maximum noise levels.* In general, noise levels should not exceed 65 dB(A). Note that this limit is established to facilitate task performance (i.e., communications, signal detection, etc.) and is not established as a limit for hearing protection. Specific noise levels by bridge location are discussed in Part 3. *Location of noise sources.* (1) *Exclusion of noise sources on bridges.* Noise from sources such as ventilation fans, engine intake fans and other sources should comply with IMO Resolution A.4681(XII), take into account IMO Resolution A.343(IX) and be excluded from the bridge through appropriate location and associated trunking. (2) *Location of noise sources.* Fixed sound signal apparatus should not be placed in the immediate vicinity of the bridge. The sound signal apparatus should be located as high as practicable and, if possible, forward of the bridge. *Audible alarm sound pressure levels.* Audible alarm sound pressure (one meter from the source) should be at least 75 dB(A) and at least 10 dB(A) or preferably 20 dB(A), above ambient noise levels existing during normal operations. Audible alarm sound pressures in a space should not exceed 115 dB(A).

Lighting

A satisfactory level of lighting should be provided to enable personnel to complete required navigation bridge tasks at sea and in port and under day and night conditions. Specific levels for lighting on the bridge and adjacent areas are discussed in Part 3. (1) *Task area lighting.* Individual task areas should have greater luminance than the general bridge lighting level. (2) *Glare.* Glare can significantly reduce the readability of displays, labels, placards and so on. Of particular concern is spectral glare, which is that which is reflected from smooth, highly reflective surfaces. Means to avoid glare are presented below:

- Glare and stray image reflections should be avoided in the bridge environment.
- High contrast in brightness between work area and surroundings should be avoided.
- The luminance of the task area should not be greater than three times the average luminance of the surrounding area.
- A non-reflective or matte surface should be used to keep indirect glare to a minimum.
- Lighting sources should be designed and located to avoid creating glare from working and display surfaces.
- Reflection in windows of devices, instruments and consoles and other reflective enclosures should be avoided.
- Devices should be designed and fitted to minimise glare, reflection or being obscured by strong light.

(3) *Adjustability of the lighting system.* Bridge personnel should be able to adjust the lighting, both brightness and direction as required in different areas of the bridge and by the needs of individual instruments and controls. Specific guidance for levels of bridge lighting is discussed in Part 3. (4) *Dark adaptation.* Whenever possible, red or filtered white light should be used to maintain dark adaptation in areas or on items of equipment requiring illumination in the operational mode. This should include devices in the bridge wings. Note that use of low-level red light can reduce the readability of some colour coded lights and labels, particularly those that are red, and can lead to the failure to detect the onset or offset of a red indicator light or alarm. Further guidance for low-level bridge lighting can be found in Part 3 of this book. (5) *Lighting controls.* Lighting controls should be provided at entrances and exits of enclosed workplace areas and the controls should be illuminated. (6) *Flicker avoidance.* Light sources should not have a perceptible flicker. (7) *Lighting control device.* A main workstation console lighting control device should be provided. *Indirect low-level lighting.* For operations conducted during the hours of darkness, indirect low-level red lighting or low-level white lighting should be available at deck level, especially for internal doors and staircases.

Device and Instrument Illumination

(1) *Instrument and control illumination and lighting.* All instruments and controls should have a means of illumination that is adjustable to zero with the exception of alarms and dimmer controls which should remain visible at all times. (2) *Lighting for night-time and dark operations.* Relevant equipment fitted on the bridge should include either internal or external lighting to ensure that bridge personnel can discern

them. Red light or low-level white lighting should be used to maintain dark adaptation whenever possible in areas or on items of equipment, other than the chart table, requiring illumination in the operational mode. This should include instruments in the bridge wings. Provision should be made to prevent red lights from being seen from outside the vessel. (3) *Instrument lighting*. Instruments that provide information should be presented on a background of high contrast. They should emit as little light as possible and have light text on a dark, non-reflecting background during night-time operations.

Heating, Ventilation and Air Conditioning

The bridge should be equipped with an HVAC system that allows regulation of the temperature and humidity on the bridge. Additional information on indoor climate variables can be found in Part 3. The HVAC system should maintain a temperature range from 14 to 30 °C (57 to 86 °F) with the preferred range being 18 to 26.5 °C (65 to 80 °F). If there is no mechanism to adjust the temperature, it should be 22 ± 1 °C (72 °F) as measured at the air return vent. The optimum range of effective temperature, which differs from normal temperature, for accomplishing light work while dressed appropriately for the season or climate is 21 to 27 °C (70 to 80 °F) in a warm climate or during the summer and 18 to 24 °C (65 to 75 °F) in a colder climate or during the winter. Temperature difference between any two points within the workplace should be maintained less than 5 °C (41 °F), for example the temperature of the air at floor level and at head level. Humidity should be maintained between 20 and 60% with 40 to 45% preferred. Approximately 45% relative humidity should be provided at 21 °C (70 °F). This value should decrease with rising temperatures but should remain above 20% to prevent irritation and drying of body tissues, eyes, skin and respiratory tract. Heating systems should be designed so that hot air discharge is not directed at personnel at their workstation. Air conditioning systems should be designed such that cold air discharge is not directed at personnel at their workstation. Ventilating systems should not produce air velocities exceeding 0.5 m/s. If possible, the preferred air velocity of 0.3 m/s should be used to preclude book manual pages from being turned or papers from being blown off work surfaces. The ventilation system should provide a positive pressure in the bridge. The ventilation system should supply outdoor air of at least 0.008 m^3/second for each person in the bridge.

Bridge Watch Stander Maintenance

During the course of vessel operations, some amount of maintenance may be required by bridge persons on watch duty, however, their primary responsibility is the safe and efficient conduct of vessel operations. Therefore, bridge maintenance performed by bridge personnel should be limited to that which is directly and immediately related to the safety and efficiency of operations. (1) *Minimisation of bridge maintenance requirements.* Bridge persons on watch duty and other bridge personnel maintenance requirements should be minimised and should extend to routine maintenance activities such as the following:

- Changing paper, inks, etc., in printing/plotting/trending devices.
- Replacement of simple indicator light bulbs or accessible, low voltage fuses.
- Wiping surfaces of display screens and surfaces such as radars, position plotters and charts (electronic or paper).
- Simple housekeeping.
- Maintenance such as the following should not normally be required of bridge personnel:
- Repair of window wipers or heater elements.
- Repair or replacement of main equipment or components (e.g., radio sets, circuit boards, printers).
- Cleaning, repair or replacement of furnishings.

(2) *Maintenance to be performed only within the bridge area.* Bridge persons on watch duty should not generally be charged with maintenance requiring them to leave the bridge or bridge wings. Maintenance on catwalks (e.g., to access the external surfaces of bridge windows) should be avoided. *Maintenance interface and communications.* All maintenance activities associated with bridge-managed functions (e.g., steering gear tests, taking generator sets off-line, radar calibrations, etc.) should be coordinated with the master or watch standing mate. *Bridge deck drainage.* Outside bridge decks, including the wings, should be provided with adequate drainage.

Personnel Safety on the Bridge

Occupational safety. To achieve personal safety for bridge personnel, the bridge must be designed to protect them from injury during normal conditions as well as abnormal conditions such as high seas. The following provides guidelines that will facilitate personnel safety on the bridge. (1) *Bridge physical hazards.* The bridge area should be free of physical hazards to bridge personnel. There should be no sharp edges or protuberances that could cause injury to personnel from the side or from above; and trip hazards on the bridge deck such as curled up carpet edges, loose gratings, duck boards or equipment. (2) *Securing equipment.* There should be a means for securing portable equipment. (3) *Safety equipment accessibility.* All safety equipment carried on the bridge should be clearly marked, be easily accessible and have its stowage position clearly indicated. (4) *Secure seating.* Where integrated bridge systems are provided that present a centralised position from which piloting, navigation and communications can be performed, a permanent (moveable on tracks) seat should be provided for the persons on watch duty. Where seating has been provided on the bridge, a method for securing the seating should be provided. (5) *Adequate electrical outlets.* Adequate electrical outlets should be provided on the bridge to preclude loose leads to portable equipment. (6) *Refreshment facilities.* Refreshment facilities provided on the bridge should be designed and located to preclude damage to bridge equipment or injury to personnel. (7) *Emergency lighting.* Emergency lighting should be provided for the bridge control centre, stairways and exits. Redundant lighting circuits. Bridge control centre lighting should have two separate circuits to ensure that the loss of one circuit does not leave the bridge in darkness. (8) *Handrails and grab rails.* Sufficient handrails or grab rails should be fitted to enable personnel to move or stand safely during inclement weather. Protection of stair way openings should be given special consideration. *Mechanical/electrical safety.* (1) *Electrical circuit fault isolation.* Circuits for equipment on the bridge should be designed to allow isolation of a fault without having to take other circuits out of service and to allow safe and easy replacement. (2) *Grounding of metal on the bridge.* All non-current carrying metal parts on a bridge should be effectively grounded. (3) *Component replacement safeguards.* Replaceable components on the bridge should be designed and arranged so that it would be impossible to connect or replace them incorrectly. *Bridge surfaces.* (1) *Bridge finishes.* The bridge surface finishes should be considered an integral part of the structure, layout and environmental design. (2) *Glare-free surfaces.* All prepared surfaces should be glare-free. (3) *Non-slip surfaces.* Wheelhouse, bridge wing and upper bridge decks should have non-slip standing surfaces, whether wet or dry. The surfaces of chart tables, workstations and other equipment where objects may be placed should be provided with nonslip surfaces or other means to protect objects set upon them from sliding off. (4) *Robust surfaces.* All surfaces of deck heads, bulkheads, doors and floors should be robust enough to withstand the daily wear of the at-sea environment. All surfaces should be capable of withstanding a temperature range of − 20 to + 70 °C (− 4 to + 158 °F), seawater, oils and solvents common to vessels and installations and ultra-violet light. (5) *Bridge deck drainage.* Exterior bridge decks, including the wings, should be provided with adequate drainage.

Chapter 28
Bridge Facilities

General

Human performance can be affected by the facilities provided on or adjacent to the bridge. Factors such as refreshment facilities, sanitary facilities and the interior décor should be designed to minimise human stress and distraction. The subject areas touched on in this chapter will be examined and discussed in greater detail throughout Part 3.

Terminology

Non-saturated colours: Saturation in colour usage refers to the degree that a hue differs from a grey of the same lightness. Therefore, a non-saturated colour is closer to grey and appears washed out.

Reflectance: The amount of light that is reflected (bounced back) from a surface.

Bridge Facilities and Décor

Food and refreshment. Refreshment facilities and other amenities provided for the bridge personnel should include means for preventing damage to bridge equipment and injury to personnel resulting from the use of such facilities and amenities. *Sanitary facilities.* Toilet facilities should be provided on or adjacent to the bridge. *Bridge colour choice.* Colours used in the bridge should give a calm overall impression to personnel, minimise reflections and reduce contrast between the interior of the bridge and the exterior environment. *Interior colour design.* Bright colours should not be used. Dark or mid-green colours are recommended. Alternatively, blue or brown may be used. *Reflectance range of colours.* Colours that are used on the bridge

A. A. Olsen, *Applying Physical Ergonomics to Modern Ship Design*,
https://doi.org/10.1007/978-3-031-57974-5_28

Table 28.1 Reflectance ranges of typical colour densities

Reflectance range	5–10%	11–30%	21–80%	81–90%
Typical colour densities	Dark green Blue/brown	Mid-green Blue-red	Pale green Blue-yellow	Off white Pale yellow

should minimise reflectance and glare. Table 28.1, "Reflectance ranges of typical colour densities," provides the reflectance range for some typical colour densities (Figs. 28.1 and 28.2).

Fig. 28.1 Bridge of the MV Pollux

Fig. 28.2 Bridge refreshment area, MV Pollux

Chapter 29
Evaluating Bridge Design

General

As we now know, ergonomics focuses on and represents bridge personnel needs and requirements throughout the life cycle of systems. The goal is to minimise human error, thereby maximising human and total system safety and effectiveness. This is accomplished through the application of ergonomic research and design guidance; the conduct of appropriate analyses and solicitation of bridge personnel input to derive requirements and needs; and the application of a cogent, pragmatic, ergonomic design process. This chapter presents a basic ergonomic design process that can be performed during the initial design of a new navigation bridge. It also presents a series of special interest processes that can be used to improve the interface design of existing bridges. The designers of bridges should accommodate ergonomic considerations and mitigate human error through the design of the bridge. This is accomplished through the application of an ergonomic design process for navigation bridges. One such notional process is presented later in this chapter.

Terminology

Communications: Design for clear, reliable and understandable internal-to-vessel, vessel-to-vessel and vessel-to-shore voice communications.

The original version of the chapter has been revised: Table 29.1 contains overlap in text with previously published content [1] that was inadvertently omitted. The author failed to attribute the reference [1]. The author has now obtained permission to re-use this content from CRC Press. Where [1] is Ross Jonathan M 2009 Human Factors for Naval Marine Vehicle Design and Operation, CRC Press, London. A correction to this chapter can be found at https://doi.org/10.1007/978-3-031-57974-5_53

A. A. Olsen, *Applying Physical Ergonomics to Modern Ship Design*,
https://doi.org/10.1007/978-3-031-57974-5_29

Error control: Exercising control over the probability, frequency, criticality and recoverability of human errors that may occur.

Maintenance of situation awareness: Persons on watch duty understanding of the condition of equipment, hull performance, sea and weather conditions, hazards to navigation, crew capabilities and presence and intentions of other occupants of the waterways.

Staffing levels: The number of crew who will staff the bridge during operations. This addresses bridge workload, both immediate (i.e., number of tasks that must be performed in a short period of time) and ongoing (i.e., work done over the course of a voyage). This also addresses bridge personnel fatigue.

Task-based design: Design of interfaces based on the complexity, sequence, importance and error tolerance of tasks performed on the bridge.

Work schedules: The number of continuous hours and the work shift patterns performed by bridge persons on watch duty.

Navigation Bridge Ergonomic Design Processes

New bridge design process. The bridge system design and development process are typically started in the maritime industry by the identification of service requirements for a vessel, culminating in the development of a design and performance specification that can be used by vessel builders to prepare bids and initiate detailed design. A typical process may consist of steps such as:

- Requirements determination (what the vessel must do).
- Function analysis, allocation and task analysis (generally how it will do it).
- Development of design concepts (specifically how it will do it).
- Design validation (based on evolved concepts) and generation of build specifications.

These steps generally will be performed by different entities. Requirement determination will be made by the organisation in need of a vessel for a specific service. A builder's team of naval architects will perform analyses, vessel design development and some elements of detailed design, however, some of this is left to the construction yard to determine at build time. All parties should participate in design validation, but on-site construction personnel and surveyors will perform much of this as building progresses. Throughout the bridge design process, there will be human performance design issues to be reconciled and to do so, a rigorous process is needed. Specific activities for a representative process, which ergonomic specialists would perform, are introduced below. Requirements analysis. Initial design and planning typically focuses on identifying requirements. To those commissioning the building of a vessel, fundamental requirements address considerations such as: cost of acquisition, operational and maintenance cost, performance factors (such as cargo carrying tonnage), hull characteristics (i.e., displacement, fairway access limitations due to beam, draft or cargo), crewing and so on. In terms of those considerations associated with humans and the navigation bridge, requirements may include:

- Those typically associated with the navigation bridge (such as communications, piloting, position fixing).
- Identifying the bridge requirements associated with the specific trade (e.g., tanker, container shipping, tugs, Mobile Offshore Drilling Unit, floating production unit, etc.) and possibly environments (e.g., the North Atlantic, Transpacific, the Caribbean, the Arabian Gulf, and the Gulf of Mexico, etc., including whatever Port State and Coastal State controls and requirements are imposed).
- Constraints on design (e.g., such as crew size, possibly imposed watch systems, extent of automation).
- Other factors such as the vessel's Flag State, country of vessel registry and the Classification Society to be used.

Bridge design activities focus on developing an understanding of what bridge personnel need and require for a successful system. This can be done by several methods including:

- *Interviews/questionnaires.* Where questionnaires or interviews are administered to bridge personnel to understand their needs and requirements. Bridge personnel involved in the studies should represent the full range of expected bridge personnel with respect to age gender, physical capabilities and limitations, physical size, education and aptitude and experience.
- *Focus groups.* Where a professional moderator, through discussions that address needs and requirements, leads small groups of representative bridge personnel.
- *Task observation.* Where a group of trained observers watch bridge task performance. Bridge persons on watch duty walk through operations (on a bridge or in a simulator) and the observers and masters/mates/others discuss tasks and design requirements.

These activities should be iterative in nature so that the designers gain knowledge of what bridge personnel need and require. The results of these activities provide information and data for the remainder of the activities during the initial design. Function analysis and allocation. Function analysis refers to the process whereby the functions. For example, the "action for which a person or thing is particularly fitted or employed," (e.g., jobs, hardware utilities, human tasks to be performed) are identified. Functions flow from requirements. Functions are identified from the highest level (e.g., navigate, perform surveillance, plan) to lower-level functions (e.g., plot position, plot course) and ultimately to human and machine functions and tasks (e.g., triangulate, use GPS, set vessel course and speed). With respect to the allocation of functions to humans and machines, the level of design freedom allowed bridge designers is typically constrained to the selection of devices and components (e.g., selection of existing radars sets, VHF communications devices, navigational aids, etc., that satisfy functional requirements); and the design of internal and external arrangements in the general area of the bridge. In other words, the basic design task is the selection of components (e.g., conning, navigation display, etc.) that meet functional requirements and then arranging them within the designed confines of the navigation bridge.

This being the case, function allocation involves a trade-off process whereby trade-off factors (e.g., cost, reliability, ease of human use and maintenance, extent to which requirements are satisfied, precision, compatibility) are considered and alternatively weighted and applied across the universe of already available (generally COTS) competing components. Once components are selected, human roles and tasks involved in operation/maintenance of each component are, for the most part, prescribed. To help guide consideration of ergonomics in this trade-off process, Table 29.1, Human and machine capabilities, provides a brief illustration of the types of functions/tasks that are better performed by humans or machines. Where a design objective for the navigation bridge, is to reduce persons on watch duty workload (i.e., as a control on manning, cost, fatigue control), a simple process such as that presented in Fig. 29.1, Simple function allocation model to reduce workload, can be used to support trade-offs. Task analysis. Task analysis focuses on understanding the tasks that a human will perform with the system including the inputs and outputs for those tasks in terms of information and control actions required, feedback, dependencies between tasks, sequence of performance, time and frequency of performance, accuracy requirements, criticality of the tasks; and error tolerance.

By performing a task analysis, the designer develops an understanding of the how a navigation bridge can be designed and arranged to support reliable task performance. There are various ways to perform a task analysis. Refer to Kirwan and Ainsworth (1992) for detailed guidance on task analysis. Workflow analysis. Workflow analysis focuses on understanding the flow and sequence of human activity as work is performed. It can be used to help understand how bridge personnel move between controls, displays, control panels and workstations in performing a task. Such information can be used to arrange control panels and workstations based on task sequences, frequencies and criticality. One method for performing a workflow analysis is a link analysis where physical representations of the work environment are used to draw lines and assign values for factors such as frequency and importance that illustrate workflow and how the design could be improved. Develop and explore bridge concepts. The focus here is to develop and explore various design concepts based on functions, requirements, human allocated tasks and selected equipment. Three basic steps for concept exploration of HMI design are illustrated in Fig. 29.2, Concept development and exploration, and briefly described below:

- Develop concepts for interface designs. This is typically done with the engineers on a design team. Various approaches to the design of the interface will be developed based on the output from the analyses discussed above. In some cases, an idea formation process can be employed where concepts are developed in brainstorming sessions among ergonomic professionals, masters and mates and engineering professionals.

Table 29.1 Human and machine capabilities *Source*: Ross Jonathan M 2009 Human Factors for Naval Marine Vehicle Design and Operation, CRC Press, London.

Humans excel at	Machines excel at
Detection of certain forms of very low energy levels Sensitivity to an extremely wide variety of stimuli Perceiving patterns and making generalisations about them Ability to store large amounts of information for long periods and recalling relevant facts at appropriate moments Ability to exercise judgment where events cannot be completely defined Improvising and adopting to flexible procedures Sensitivity to stimuli beyond the range of human sensitivity (infrared, radio waves, etc.) Ability to react to unexpected low probability events Applying originality in solving problems Exerting large amounts of force smoothly and precisely Ability to profit from experience and alter course of action Ability to perform fine manipulation, especially where misalignment appears unexpectedly Ability to continue to perform when overloaded Ability to reason inductively	Monitoring (both humans and machines) Performing routine, repetitive or very precise operations Responding very quickly to control signals Storing and recalling large amounts of information in short time periods Performing complex and rapid computation with high accuracy Sensitivity to stimuli beyond the range of human sensitivity (infrared, radio waves, etc.) Doing many different things at one time Insensitivity to extraneous factors Ability to repeat operations very rapidly, continuously and precisely the same way over a long period Operating in environments which are hostile to humans or beyond human tolerance Deductive processes

- Develop low-fidelity prototypes. Low-fidelity prototypes will be prepared for the various interface design concepts. These may consist of no more than paper prototypes of the interface to more complex mock-ups using foam core or computer simulations.
- Evaluate the concepts. The concepts can be evaluated through several different methods including subject matter expert (SME) assessments, bridge personnel evaluations, walk through and formal usability testing; and testing aspects of design in bridge simulators.

The results of the evaluations will be used to eliminate some of the design concepts and enhance others. This is an iterative process leading to one or more concepts to be used during concept demonstration and validation. Initiate specification development. Design specifications (even though in a preliminary stage) include functional specifications, which describe the functions and tasks being performed and how the design should support them and interface specifications that describe how bridge personnel interface will be designed to support the requirements in the functional specification. At this stage, the specifications are not fully complete and will

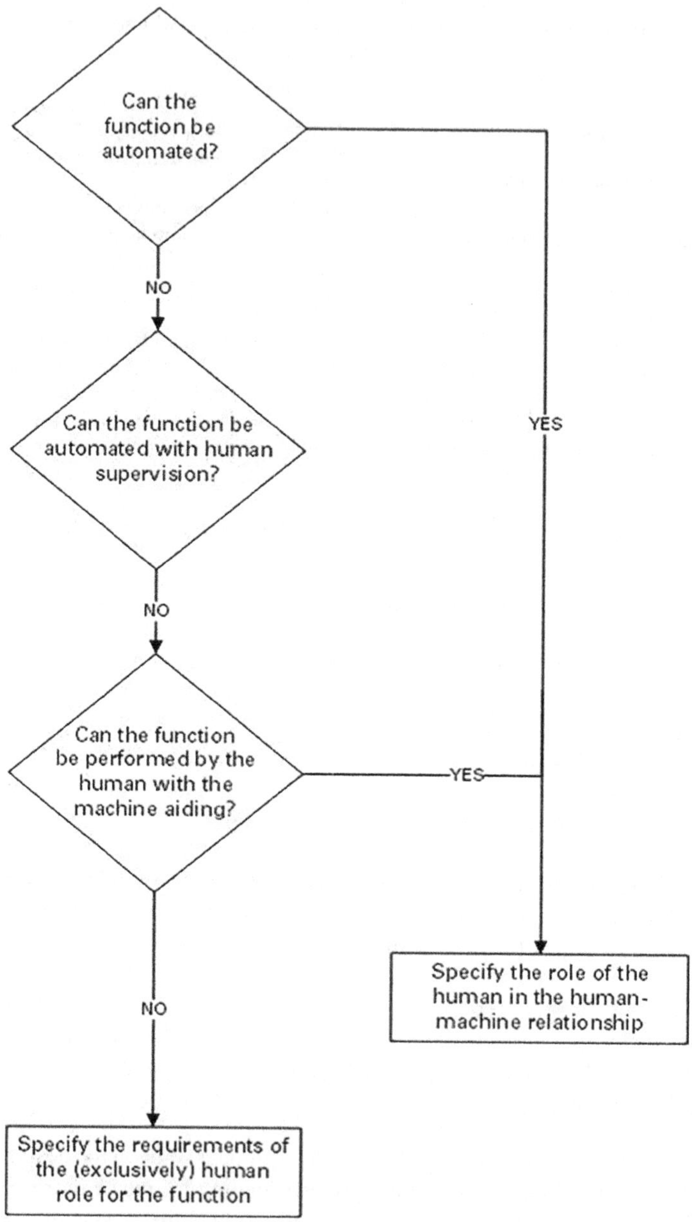

Fig. 29.1 Simple function allocation model to reduce workload

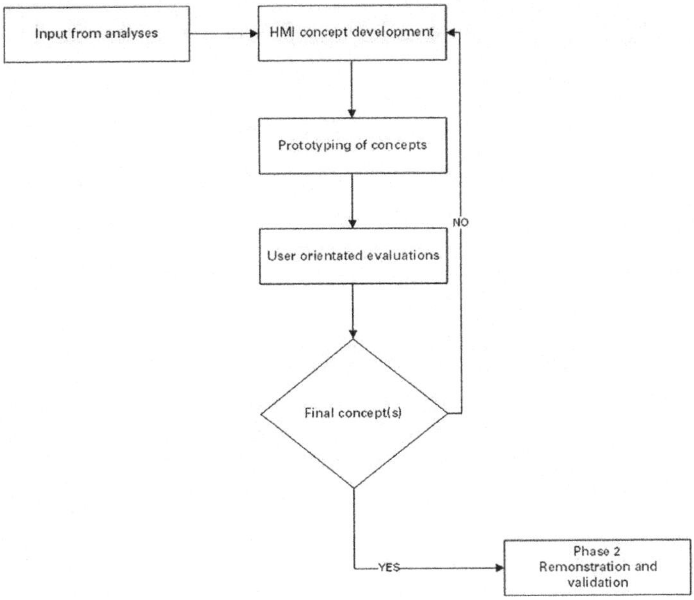

Fig. 29.2 Concept development and exploration

be fully evolved in later steps. Interface specifications should account for appropriate ergonomic standards and guidelines. Specification and design validation. The focus of specification and design validation is to verify the capabilities of the design concept(s). Typical iterative activities, illustrated in Fig. 29.3, Demonstration and validation process, include the following:

- Develop prototypes and mock-ups. The concepts emerging from the concept exploration are developed into more formal prototypes or mock-ups. These prototypes and mock-ups will be used to assess the design.
- Interface assessments. Using prototypes and mock-ups, the effectiveness of the interface is assessed with respect to human error, ease of use and bridge persons on watch duty expectations. In addition, the overall suitability of the bridge can be assessed. These assessments can be performed through various methods including assessments by professional ergonomists.
- Design and specification refinement. Based on the results of the interface assessments, the design and associated specifications are further developed. In some cases, concepts may be rejected due to lack of suitability for meeting bridge personnel requirements or other factors. Data from the interface design assessments can also be used to update function and task analysis data and refine requirements.

Detailed Design. The focus of detailed design is to take the design specifications and final concept and develop the full design of the navigation bridge. Information

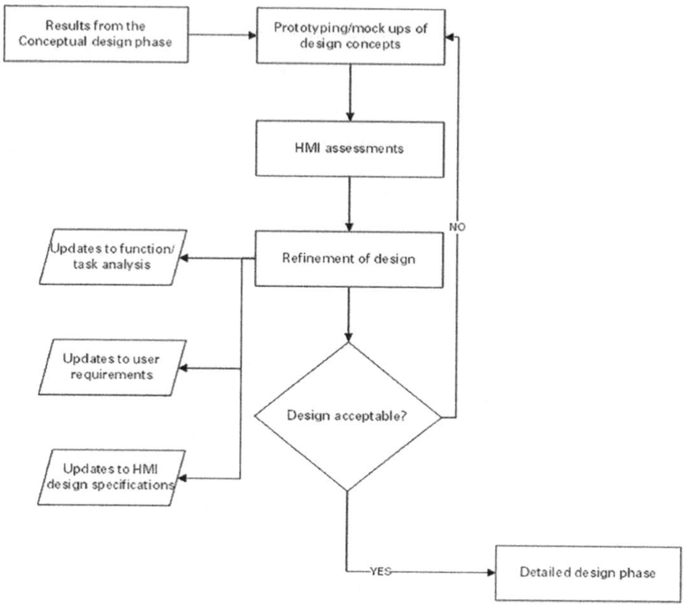

Fig. 29.3 Demonstration and validation process

gained during specification and design validation, including the updated interface design specifications, will be used to drive the detailed design of the interface. It is critical that ergonomic personnel work closely with the other design engineers to ensure that ergonomic considerations are weighed in design decisions and that the interface design is practical from an engineering perspective. Typical activities include task analysis to define procedures and training, interface specifications and task analysis to define interface design flows which illustrate bridge personnel actions and control/ display response, detailed design drawings of control panels, workstations and other areas of the vessel that have a human interface, the interface design specifications, task analysis data and other relevant documents should be updated as new requirements are learned concerning the interface, design reviews that are focused on the interface, prototypes and mock-ups, bridge personnel testing; and iterative development of detailed design. *Design implementation.* The focus of design implementation is on building the vessel. Ergonomic activities at this point will primarily be consulting on issues that arise as the design is implemented, formal acceptance testing for the interface and capturing lessons learned for future development. In particular, ergonomic personnel should develop a plan to collect operational lessons learned from bridge persons on watch duty regarding the usability of the interface design and how it could be improved. *Processes for improving design aspects of existing navigation bridges.* Presented in this section are general approaches for improving navigation bridge design (e.g., any time after construction) providing specific guidance addressing the improvement of the following ergonomic concerns:

- Ergonomic issue tracking.
- Communications.
- Labels and signs.
- Procedures and job aids.
- Information transfer and display.
- Human interfaces.
- Training.

Sample process for ergonomic issues tracking. Objectives. The objectives of this task are to identify tasks and areas with potential ergonomic issues, define standardised work environmental limits, identify safety hazards, identify design factors for improved ergonomics, define specific actions; and prepare ergonomic specifications. *Work steps: Step 1. Identify jobs and areas for design of improved work environment and safety.* A sample of tasks and areas will be selected for further analysis based on ergonomic issues. This sample will represent extremes in terms of ambient noise, temperature, humidity, illumination, safety related equipment design factors and mechanical hazards. *Step 2. Define standardised work environmental limits.* Various standards provide guidance on acceptable work environment limits. These will be used to define a set of proposed limits for work environments. *Step 3. Identify safety hazards.* Representative safety hazards will be identified for facilities and equipment that are involved in the tasks and areas from Step 1. *Step 4. Identify ergonomic design factors.* Potential solutions to be considered will include elimination of the hazard, elimination of human access, barriers and guards, redesign, warning labels, procedures and special training. For each safety hazard from Step 3, a preferred approach will be defined and documented. *Step 5. Define specific actions.* Treatments of factors that exceed the limits established in Step 2 will include design approaches such as noise attenuation at the source, noise attenuation at the ear, enhanced ventilation and cooling, protective clothing and fixed or portable additional lighting. The effectiveness of alternate solutions on environmental factors that exceed the limits from Step 2 will be assessed. *Step 6. Prepare ergonomic specification.* The rationales and results from the previous steps will be incorporated into an ergonomic specification document. This may be applicable to design of work environment and safety improvements on a current project as well as future projects. *Product.* The primary product of this task will be the ergonomic specification from Step 6.

Sample process for improvement of bridge communications. Objectives. The objectives of this activity are to analyse communications in terms of communication links and how communications are made and to recommend modifications to the communication systems. *Work steps: Step 1. Analyse communications in terms of the links, frequency, information importance and environmental conditions* under which communications are made (e.g., noise level). *Step 2. Generate a link analysis of communications requirements* among various stations (e.g., control rooms outside stations). *Step 3. Identify conditions that can potentially interfere with speech communications and speech intelligibility* (e.g., noisy equipment). *Step 4. Conduct a communication "error analysis"* to identify, for selected links, communications requirements, environmental conditions, potential errors, error criticality and causal and contributory factors. *Step 5. Develop communications criteria*, including standard message format and syntax (refer to the IMO Standard Marine Communication Phrases (SMCP), links and modes of communications among stations, to reduce error likelihood and improve communications accuracy. Step 6. Survey communications devices meeting requirements that are not in place (e.g., use of noise cancelling technology). *Step 7. Make specific recommendations concerning improved communications system design* based on assessments of the capabilities of state-of-the-art communications devices. *Step 8. Prepare ergonomic design specifications for communications* (e.g., noise levels, types of communications, location of communication devices, etc.). *Step 9. Perform speech intelligibility testing under realistic conditions. Step 10. Generate the implementation plan and a cost estimate* for implementing the communications systems specification and recommendations. *Products.* The products of this activity should be a series of recommendations and design specifications related to communications hardware and software, human-machine interfaces, message format and syntax, a list of constraints associated with communications (e.g., ambient noise); and a plan and cost estimate to implement the communications system recommendations.

Sample process for improvement of labels and signs. Objectives. The objectives are to develop and implement standards for labelling and marking of operating components, including controls, displays, communications devices, major equipment and associated operating components. *Work steps: Step 1. Survey colour coding and other conventions used on board. Step 2. Survey general bridge coding conventions. Step 3. Review applicable standards and guidelines* related to design of labels and signs and other identifying markings in industrial applications, from safety and ergonomic literature and industry conventions (see ASTM-1166, Standard Practices for Human Engineering Design for Marine Systems, Equipment and Facilities. *Step 4. Establish a specific labels and signs/placards design specification.* This should address the content, wording and nomenclature, colours, fonts and other detailed characteristics, method of affixing to components labelled, style guides for warnings and cautions, consistency with written documentation, procedures and computer display nomenclature; and method of generation of labels and markings (engraving for example). *Step 5. Develop a list of required signs and labels. Step 6. Develop a label consistent with the labelling standard,* for all components specified. *Step 7. Support development and manufacture of the labels and markings. Step 8. Monitor installation of the levels and markings. Step 9. Verify and validate.* All labels and markings installed undergo a verification and validation step. Products. The products of this activity include a platform specific "Style Guide" and standard for labels and markings, procedures for specifying and acquiring levels, markings and warning placards; and procedures for verifying and validating the content and accuracy of labels, markings and placards.

Sample process for improvement of procedures/aides memoire. Objectives. The main objective of this activity will be to specify requirements associated with generation of operating, maintenance and emergency procedures and job aids. Specific objectives include the identification of required procedures based on FMEA or similar analysis, identification of required procedures based on routine and normal operating and maintenance requirements, generation of procedures style guides; and generation of written procedures based on the style guides and requirements. *Work steps: Step 1. Identify normal operating procedures that are used in the conduct of normal and routine operations. Step 2. Identify abnormal or emergency operating procedures.* These will be identified via failure modes and effects and criticality analyses or similar. *Step 3. Survey information coding conventions.* Survey general bridge procedures and design conventions. *Step 4. Review applicable standards and guidelines related to design of procedures. Step 5. Establish procedures writing guide.* This standard should address procedural content, wording and nomenclature, colours, fonts and other detailed characteristics, information mapping and decision making, style guides for stepwise activities, style guides for embedded warnings and cautions, Consistency with labels and marking, standardised vocabulary, storage, update and accessibility; and use as part of team and individual training. *Step 6. Develop protocol for verifying and validating procedures as they are written. Step 7. Assist in writing procedures. Step 8. Verify and validate.* All procedures undergo verification and validation step. *Products.* The products of this activity include the procedures writing guide, list of applicable procedures, normal and abnormal operations; and A complete set of written procedures, verified and validated. Sample process for improvement of display design. *Objectives.* The objectives of this task are to define information requirements per critical task or operation, define information items, identify information formats, design screen layouts and human-computer interfaces per formats, define data base requirements; and prepare information display design specification. *Work steps: Step 1. Define information requirements per job/station.* Information requirements will identify the information items needed at each location for each critical task. *Step 2. Define a representative sample of information items.* Information items from Step 1 will be reviewed to establish a representative set of items for further analysis. This step cannot be performed in total independence of Step 3 because the intent will be to identify classes of information items and a representative sample of specific items per class. Steps 2 and 3 will be performed in parallel and iterated as required. *Step 3. Identify information classes and formats for workstations.* Individual information items will be categorised to identify a number of information classes. Each such class will eventually be represented as an information object in an object with associated information display requirements. Information classes/objects could include those related to critical tasks such as collision avoidance, navigation plans, port entries, tug interface, berthing and unberthing, among others. *Step 4. Prepare display specification.* Information from the above steps including object and attribute characteristics and human-computer interface functionality will be collected and documented in a display design specification. The difference between the information requirements (Step 3) and the specification will be that the former is aimed at providing information for system procurement while the latter will be aimed at developers and bridge personnel. *Step 5. Design display layouts and human-computer interfaces.*

Each information class will have unique attributes and will require a standard format for display of the attributes. *Products.* Information display designs based on critical tasks and operational information requirements. Sample process for the improvement of human-machine interfaces. *Objectives.* The objectives of this activity are to define a representative sample of existing issues, identify applicable design criteria and guidelines, identify candidate design improvements, select design improvements, design and document improvements; and develop human-machine interface specification. *Work steps: Step 1. Define a representative sample of ergonomic issues associated with interfaces.* The intent will be to identify issues and potential solutions for issues associated with human machine interfaces. *Step 2. Identify applicable design criteria and guideline.* As appropriate, design criteria and guidelines will be proposed. Step 3. Identify candidate ergonomic design improvements. Candidate improvements to interfaces will be developed for each issue from Step 1 using the applicable design guidance from Step 2. Where required, two or three candidates could be developed for each issue. Definition of candidate improvements will require bridge personnel input. *Step 4. Select ergonomic design improvement.* Candidate improvements from Step 3 will be compared on the basis of cost, expected effectiveness and breadth of application on future vessels. The set of recommended improvements will be selected using these data. *Step 5. Design and document improvements.* The selected design improvements from Step 4 will be designed in detail and documented using descriptions, specifications and graphics sufficiently that they can be implemented. *Step 6. Develop human-machine interface specification.* The applicable design criteria and guidelines will be incorporated into an interface specification document suitable for future use in vessel design. Applicable design criteria from Step 2 will be incorporated and the documentation of the specific improvements from Step 5 will be included to provide examples. *Products.* The primary products of this task will be the selected improvement concepts and designs from Step 5 and the human-machine interface specification from Step 6.

Sample process for improvement of training (individual and team). Objectives. The objectives of this task are to identify jobs and tasks for improved training methods, develop training objectives/conduct task analysis, identify performance and skill requirements, identify training media, develop training materials; and train to criteria and test skill acquisition. *Work steps: Step 1. Identify jobs and tasks for improved training methods.* Criteria for selection of jobs and tasks will include the frequency with which the tasks must be performed, the likelihood of human error and the severity of the consequences of errors. Tasks that must be performed frequently in daily operations provide opportunities for effective on-the-job training and for extensive practice. Many error modes that exist currently are readily detectable and correctable by bridge personnel and errors on these tasks have little or no consequences. It is infrequently-performed, non-detectable and non-correctable error modes that will drive the selection of tasks for improved training. Tasks that have the properties of infrequent performance and severe consequences of error are often those associated with emergency response to equipment failures or other abnormal events. The facts that performance is infrequent means that knowledge and skills associated with these tasks are "degradable" over time. Results of FMEA or other analysis will be required in this task. The target tasks would then be those necessary to recover from or to mitigate the results of credible events postulated in the FMEA. *Step 2. Develop training objectives and conduct training task analysis.* Training objectives describe observable and verifiable behaviours that bridge personnel should exhibit given necessary information. Training objectives usually involve making correct decisions and taking correct actions following receipt of information about the system, the environment, etc. As related to FMEA events, the decisions involved would be to correctly diagnose the state of the equipment and the environment. The action should be the correct one to rectify or mitigate the effects of the failure or abnormal event that has occurred. A training task analysis should be conducted to identify tasks, conditions, events and information required by the persons on watch duty, decisions and actions required of the persons on watch duty and other relevant task descriptors; and needed knowledge and skills to enable performance of the task.

This is also sometimes called a "Critical Task Analysis" in that the focus is on the highly important tasks rather than on all tasks. Critical tasks, then, do not necessarily result only from major failures but also from abnormal events that call for critical tasks to be performed. Training objectives will be defined for the tasks identified in Step 1. *Step 3. Identify performance and skill requirements.* Performance refers to measurement of trainee progress in demonstrating mastery of the training objectives. This can range from an indication that the trainee has completed a given module to performance in problem solving exercises or simulations. The intention of any training program is to impart KSA. In some cases, these elements can be treated separately in that knowledge acquisition often involves presentation of factual knowledge and testing involves assessment of acquisition. Skills and abilities usually refer to application of knowledge to problem solving. Certain skills and knowledge may be common across a number of jobs so it may be effective to split out bodies of training materials into modules. Training for a given job then consists of common modules required for the job and job specific modules. Skills, knowledge and performance measurement requirements will be identified. Specific criteria for pass/fail thresholds should be specified. *Step 4. Identify training media.* Training media can be books, simulators, embedded training incorporated into bridge applications software, desktop simulations, etc. Based on the outputs of Step 3, a preferred medium or media will be identified for each training objective and the performance measurement necessary to assess mastery of the objective. *Step 5. Develop training materials.* Training materials include the paper, slides, graphics, tapes, computer presentations and other audio and/or visual elements viewed or heard by the student. Training materials may be developed through the preparation of an outline for each module, preparation of written contents using subject matter expert inputs, development of a story board for each module, preparation of the materials including text, graphics, animations, simulations, etc.; and for computer-based training (CBT), development of courseware (using course authoring software where necessary). *Step 6. Train to criteria and test knowledge skill acquisition.* The initial application of the training content and materials from the above steps will be tested in the training course(s). Performance measures will be used to identify common errors or failures to master material on the part of students and to determine how to modify the training materials. Following any necessary modifications, the resulting course(s) will then be ready to be used to train the required audience. *Products.* The primary products of this task will be the training materials (courses, manuals, CDs, etc.). *Verification of human performance on existing systems.* Table 29.2, Potential verification measures of human performance on existing system presents a listing of potential verification measures for evaluating human performance.

Table 29.2 Potential verification measures of human performance on existing system

Evaluate the assigned role of the human in automation
Verify that each function is allocated to human or machine or a combination of the two Verify that the roles of the machine in manual tasks are defined Verify that the roles of the human in automated tasks are defined
Determine that bridge personnel workload is realistic **Evaluate training concepts**
Verify that the training analysis addresses training requirements based on job requirements Verify that the training analysis addresses all requirements for training devices, trainers and part task and full task simulators Verify that training requirements are identified in time to allow for development of any new training devices (such as simulators) Verify that the analysis addresses lessons learned from similar system training evaluations
Evaluate training concepts
Verify training requirements include specific knowledge and skills to be acquired Verify training requirements include criteria for judging skills to be learned Verify that training requirements include performance and competence measures
Evaluate human machine-interface requirements
Evaluate human-machine interface design Verify that standardisation and commonality are addressed in the design of human-machine interfaces Verify that unique human interface requirements, documentation needs and special software certifications are identified Verify that characteristics of automated decision support systems are identified Verify that human workloads and human performance requirements are assessed through human performance and task modelling, task network simulation and human-in-the-loop simulation Verify that human engineering design standards are applied to reduce human error potential
Evaluate design for usability
Have information requirements been identified? Have major deficiencies been identified that might compromise understandability or effectiveness of the proposed displays? Have bridge personnel needs been identified? Is the design of human-computer interfaces complete?

(continued)

Table 29.2 (continued)

Evaluate design for operability
Verify bridge personnel performance capability has been demonstrated
Verify that details of the design are consistent with standards
Verify that error likelihood analyses have been performed to identify types of performance errors associated with the design approach
Verify that operational procedures have been developed
Verify that control and display arrangements are based on sequence of use, priority and functional grouping
Verify that panels and consoles are designed to be maintainable
Verify that warnings are provided for hazardous operations/maintenance actions
Verify that panels are operable when bridge personnel are wearing protective clothing
Verify that work environment effects (e.g., noise, lighting, climate) have been considered in the design
Evaluate communications concepts
Verify that sufficient communication devices and systems have been provided for all communication requirements
Verify that communications system designs are based on link analyses and operational sequence analyses
Verify that speech intelligibility evaluations have been conducted
Verify that message samples, noise conditions and device fidelity are acceptable in terms of standards
Verify that messages are standardised and are based on constrained language, controlled syntax and restricted vocabulary
Verify that bridge personnel clothing conditions were considered
Verify that the range of potential environments (especially noise and vibration) was considered in design of communications
Evaluate design for habitability
Have facility human functions and associated facility requirements been identified?
Has the design effort identified access safety requirements?
Have requirements for inhabiting the facility been identified?
Verify that environmental controls are included in facilities (e.g., noise, lighting climate)
Evaluate design for maintainability
Does design for maintainability include requirements for maintenance information requirements?
Does design for maintainability include design for accessibility?
Does design for maintainability include equipment arrangement to facilitate maintenance?
Does design for maintainability include procedures-number and simplicity?
Does design for maintainability include troubleshooting diagnostics and decisions?
Does design for maintainability include built-in test and automatic test equipment?
Does design for maintainability include requirements and approaches for tools and test sets?
Does design for maintainability include requirements and approaches for equipment identification and marking?

(continued)

Table 29.2 (continued)

Evaluate design for safety
Have hazards previously identified been eliminated or the associated risks reduced to an acceptable level?
Are approaches for guarding the hazard adequate?
Are approaches for labelling the hazard adequate?
Are approaches for alarming the hazard adequate?
Are approaches for safety training/procedures adequate?
Evaluate installations
Have equipment location requirements been identified?
Has space layout for equipment installation been identified?
Have equipment configuration requirements been identified?
Have access/egress requirements been identified?
Has a design evaluation been conducted?
Evaluate inputs to engineering change proposals (ECP)
Have lessons learned been identified for the element?
Have critical tasks per function been identified?
Have human-machine interfaces been identified?
Have task requirements been identified and analysed?
Have test and evaluation been conducted to identify problems and/or validate lessons learned data?
Has the role of human versus automation been evaluated?
Have workloads been evaluated?
Have procedures been evaluated?
Have effects-of-use environments been evaluated?

Part III
Habitability in the Marine Environment

Chapter 30
General Provisions

General

This chapter is designed to provide a single source for habitability criteria suitable for ships. This chapter may be applied to vessels falling under the following categories: oil and/or chemical tankers, bulk and/or combination carriers, container vessels, multi-purpose cargo vessels, or crew areas on passenger vessels. This chapter does not apply to vessels such as offshore support vessels, tugboats, tow boats, dredgers, research vessels, drill ships, anchor handling vessels or any other vessel providing service to offshore oil and gas exploration and production. These types of vessels are addressed in Part 4 of this book. In essence, this Part 3 has been written with the objective of improving the quality of crew member performance and comfort by improving working and living environments in terms of accommodation area design and ambient environmental qualities. These habitability criteria have been chosen to provide a means to help reduce crew fatigue, improve performance and safety, and to assist with crew recruiting and retention.

Terminology

Class recognised ambient environmental testing specialist: Companies providing test or measurement services on behalf of the owner of a vessel or shipyard for the purposes of meeting any of the Class habitability notation requirements. Approval of the testing specialist for ships may also allow for testing and measurement services for other Class approved Habitability, Comfort, and MLC notations.

Accommodation areas/accommodation block: Vessel areas where the primary purpose is to rest and recreate. Accommodation spaces include cabins, medical facilities (sick bays), offices, recreation rooms, and manned spaces within the accommodation block such as the bridge or control room. For the purposes of this chapter,

A. A. Olsen, *Applying Physical Ergonomics to Modern Ship Design*,
https://doi.org/10.1007/978-3-031-57974-5_30

accommodation areas also include service spaces such as mess rooms, laundry, and storerooms.

Ambient environment: Ambient environment refers to the environmental conditions that the crew is exposed to during periods of work, leisure or rest. Specifically, this Part 3 provides criteria and limits for whole-body vibration, noise, indoor climate, and lighting.

Associated documentation: Documents referenced in this Part 3 that are needed to provide measuring techniques and further guidance.

Cargo vessel: A cargo vessel is any vessel other than a passenger vessel that is involved in commercial trade.

Crew member: Any person on board a vessel, including the master, who is not a passenger. This term is used interchangeably throughout this document with "seafarer".

Crew spaces: All areas on a vessel intended for seafarers only, such as seafarer accommodation spaces and seafarer workspaces. This term is also used interchangeably throughout this Part 3 with "seafarer spaces".

Gross tonnage: The gross tonnage calculated in accordance with the tonnage measurement regulations contained in Annex I to the International Convention on Tonnage Measurement of Ships, 1969, or any successor Convention.

Habitability: The acceptability of the conditions of a vessel in terms of whole-body vibration, noise, indoor climate, and lighting, as well as physical and spatial characteristics, according to prevailing research and standards for human efficiency and comfort.

High-speed craft: Vessels that meet the requirements of the ABS Rules for Building and Classing High-Speed Craft or alternative Class equivalent.

Manned space: For application of this Part 3, a manned space is any space where a seafarer may be present for 20 min or longer at any one time during normal, routine daily activities. Such spaces include working or living spaces. For machinery spaces, a manned space will be associated with the presence of a local work/control duty station manned for 20 min or longer at one time during normal, routine daily activities. Normal inspection, walk through (patrolling) and maintenance tasks in machinery spaces are not considered tasks contributing towards a particular space being considered manned.

Passenger: A passenger is every person other than the master and the members of the crew or other persons employed or engaged in any capacity on board a vessel for the business of that vessel.

Passenger vessel: A vessel whose primary purpose is to carry more than 12 passengers for transportation or recreational purposes. This includes cruise ships and ferries (conventional and high-speed craft).

Recreation and public spaces: Those portions of the accommodation areas that are used for halls, dining rooms, lounges, and similar spaces.

Seafarer: Any person on board a vessel, including the master, who is not a passenger. This term is used interchangeably throughout this document with "crew member".

Seafarer spaces: All areas on a vessel intended for seafarers only, such as seafarer accommodation spaces and seafarer workspaces. This term is also used interchangeably throughout this Part 3 with "crew spaces".

Shall/should: Expresses a provision that is mandatory.

Test plan: Document containing the requisite information regarding vessel design and layout, test personnel, test conditions, measurement locations, data acquisition, instruments, data analysis, and test schedule necessary for verifying the measurements for the ambient environmental aspects of habitability. This document is to be primarily prepared by the Class recognised ambient environment testing specialist with additional input provided by the designer/builder. Submission of the document for approval can be by either the testing specialist or the designer/builder.

Test report: Document containing the actual testing results from the ambient environmental tests including details of the testing conditions, measurement locations, measurement equipment, and the results of the data collected and analysed.

Workspaces: Areas allocated for work. Categories of workspaces include, but are not limited to, navigation spaces, service spaces (e.g., galley), machinery spaces, and duty/control stations.

Scope and Application

This Part 3 focuses on five habitability aspects of ship design and layout that can be controlled, measured, and assessed. These five aspects are broken into two categories: (1) accommodation areas, and (2) the ambient environment. Accommodation area criteria pertain to dimensional and outfitting aspects of spaces and open deck areas where crew members eat, sleep, recreate, and perform routine daily activities. The ambient environmental aspects of habitability pertain to the environment that the crew is exposed to during periods of work, leisure, and rest. Specifically, this Guide provides criteria, limits, and measurement methodologies for whole-body vibration (separate criteria for accommodation areas and workspaces), noise, indoor climate; and lighting. The criteria provided throughout this Part 3 are based on currently available research data and standards for the purpose of improving crew performance and providing a base level of habitability and elements of safety related to habitability.

Table 30.1 Summary of the differences between the **HAB**, **HAB+** and **HAB++** notations

	HAB	HAB+	HAB++
Accommodation areas	**HAB** requirements for accommodation areas	**HAB** and **HAB+** requirements for accommodation areas	**HAB+** and **HAB++** requirements for accommodation areas
Whole-body vibration	Level of vibration—reducing discomfort and promoting job performance	Lower level of vibration—promoting job performance and increased comfort	Lowest level of vibration—promoting job performance and increased comfort
Noise	IMO Code on Noise with modifications	**HAB** with additional requirements	**HAB+** with additional requirements
Indoor climate	IMO Code on Noise with modifications		**HAB** with additional requirements
Lighting	Typically, no differences among the notations		

Class Notations

At the vessel owner's or shipyard's request, a vessel complying with the minimum criteria for accommodation areas and the ambient environment (i.e., whole-body vibration, noise, indoor climate, and lighting) may be assigned a Class notation which signifies the vessel's compliance with the Class stipulated standards for habitability. For example, a vessel complying with the ABS **HAB** criteria and the more stringent criteria with respect to accommodation, whole-body vibration, noise, and indoor climate may be distinguished in the vessel's official Record by the notation **HAB+**. A vessel satisfying all the ABS criteria may be distinguished further in the vessel's Record by the notation **HAB++**. A summary of the differences among each of these notations is presented in Table 30.1 Summary of the differences between the Class notations **HAB**, **HAB+** and **HAB++** notations.

Data and Plans to Be Submitted to Class

General. The following general arrangement type drawings of the vessel are required to be submitted to Class for their review and approval:

- Inboard profile detailing the location of the main vertical zone boundaries, the location of the main watertight bulkheads, as well as the various deck levels.
- Plan view of each deck annotating the various spaces on each deck.

Accommodation areas. At a minimum, scaled arrangement drawings of the various accommodation spaces (elevation and plan views) and the vessel's accommodation area specifications should be submitted to class. Details of the crew accommodation area data requirements are provided in the section 'Accommodation area documentation'. *Ambient environment.* The following items are required to be submitted individually for each ambient environmental aspect:

- *Test plans.* Test plans serve as the principal means for verifying the measurements for the ambient environmental aspects of habitability. Separate test plans are required for whole body vibration, noise, indoor climate, and lighting. Specific test plan details for the various ambient environmental criteria are outlined in later sections of this Part 3:

Environmental aspect	Test plan details
Vibration	Chapter 32
Noise	Chapter 33
Indoor climate	Chapter 34
Lighting and illumination	Chapter 35

The test plans require approval by Class before any measurements are made. Class will usually notify the vessel owner or shipyard whether the test plans have been approved or require alteration. A Class approved copy of the test plan will then become part of the vessel's official documentation.

- *Test reports.* Upon completion of the ambient environmental testing, the test reports should be submitted to the Class approved surveyor for review. In addition, a copy of the test reports will become part of the vessel's official documentation. These reports contain ambient environmental information such as test results, testing details, measurement equipment details, etc. The specific report contents for the various ambient environmental criteria are outlined in later sections of this Part 3, as follows:

Environmental aspect	Test plan details
Vibration	Chapter 32
Noise	Chapter 33
Indoor climate	Chapter 34
Lighting and illumination	Chapter 35

Process for Obtaining a Class Notation

Prior to scheduling accommodation area verification or ambient environmental testing activities, the vessel owner or shipyard should certify the operational status of the vessel as being fully operational and/or inclusive of all equipment and furnishings. If the vessel is not fully operational, a complete listing of deficiencies of areas, components, equipment, etc. should be submitted to Class for review. Class will then make a determination and notify the vessel owner or shipyard as to whether accommodation area verification activities or ambient environmental testing can commence. The intent is to conclude all ambient environmental testing by the end of the vessel's sea trials. The following sections briefly describe the notation process.

Accommodation areas. Arrangement drawings, plans, and vessel design specifications for accommodation spaces should be prepared and submitted to Class for review. For new construction, the drawings should be provided to Class during the detailed design phase. For existing vessels, the arrangement drawings and plans, reflecting the current accommodation area configurations should be provided to and approved by Class, in advance of Class approved surveyor verifications. Follow-up physical verification measurements of accommodation area criteria may only be performed by a Class approved surveyor. The Class approved surveyor will then select the most appropriate verification measurement sites. The physical verification measurements of accommodation area criteria must usually be performed for all vessels in a series of vessels. The results of the Class review and actual Class approved surveyor verification are then reviewed by the Class approved surveyor during the notation confirmation process.

Ambient environment. Ambient environmental test plans for whole-body vibration, noise, indoor climate, and lighting should be individually prepared and submitted to Class for review and approval. These test plans serve as a primary means for verifying the measurement locations and measurement process, as well as specifying the testing specialist who will perform the ambient environmental testing. Testing, inspections, and data collection should be performed by the testing specialist and witnessed by a Class approved surveyor. The test reports for ambient environmental testing are to be prepared by the testing specialist and submitted to the Class approved surveyor for review. For a vessel that is expected to operate in extreme environmental conditions (e.g., Polar regions, the Persian Gulf, etc.) that are significantly different from the conditions encountered during the sea trials, the designer/shipyard may be required to submit information on the design/engineering analysis performed as part of the test plan submission verifying that the appropriate requested notation criteria can be met in the actual area of operations. If there are instances during the sea trials where the criteria was either not met, or passed by a small margin, the Class approved surveyor—in conjunction with the designer and the vessel owner—may request additional confirmatory tests to be carried out in the actual area of operation. Full ambient environmental testing as described in this Part 3 is required for the first vessel in a series of vessels. For the second and subsequent sister vessels (i.e., vessels being part of the same series of vessels that have identical structural, machinery, and outfitting designs), full scale testing for the ambient environmental aspects of the notation, is not required. For the second or all subsequent sister vessels constructed in the same shipyard, the following is applicable for all aspects of ambient environmental testing (whole-body vibration, noise, indoor climate and lighting):

Table 30.2 Threshold for measurements to be considered

Ambient environment aspect	Testing threshold
Whole body vibration	Within 10% of the criteria
Noise	Within 3 dB(A) of the criteria
Indoor climate	Within 2 °C (3.6 °F) from the lower or upper air temperature
Lighting	Within 10% of the criteria

- Spaces that were identified as "worst case" locations in the test plans for the first vessel need to be tested.
- By review of the first vessel's test reports (all ambient environment aspects), the Class approved surveyor will identify the locations which were out of compliance or close to non-compliance (refer to Table 30.2 Threshold for measurements to be considered) that need to be tested.
- Spaces that required some type of mitigation for the first vessel need to be tested.

As part of the notation assignment process, all vessels are subject to Class approved surveyor walk through. Additional testing may be required at the Class approved surveyor's discretion. For sister vessels constructed at a different shipyard, the following provisions may be applicable: full testing is required for the first vessel at each shipyard; and for the subsequent sister vessels, the procedure described previously for vessels built at the same shipyard is applicable.

Results. The Class accommodation area assessment, Class approved surveyor verification measurements, and ambient environmental test reports must be submitted to Class for review, after which Class (or their appointed Class approved surveyor) will issue a determination of notation confirmation.

Initial Requirements

The initial process for obtaining any vessel habitability notation usually comprises some form of Class engineering review, ambient environmental testing, and Class approved surveyor verifications. Testing must be carried out in accordance with the submitted test plans once they have been reviewed and approved by Class in advance of the testing regime commencing. Testing should always, where practicable, be witnessed by a Class approved surveyor. Once the criteria specified by Class (and summarised in this Part 3) have been satisfied, the appropriate notation may be confirmed.

Surveys Completed After Construction

It is intended that all surveys following the vessel's construction are aligned with Class surveys. Harmonisation of surveys may be carried out at the first available opportunity. Refer to the Class specific Rules regarding surveys completed after construction.

Alternatives

Class may consider alternative arrangements, criteria and procedures, which can be shown to satisfy the criteria directly cited or referred to in this Part 3. The demonstration of an alternative's acceptability can be made through either the presentation of satisfactory service experience or systematic analysis based on valid engineering principles. *National regulations.* Class will usually consider, for their acceptance, alternative arrangements and details, which can be shown to comply with standards recognised in the country (Flag State) in which the vessel is registered or built, provided they are deemed no less effective than those stipulated by Class. *Departures from criteria.* The criteria summarised in this Part 3 is envisioned to apply to vessels that are engaged in the usual trades and services expected of such vessels. However, it is recognised that unusual or unforeseen conditions may lead to a case where one or more of the parameters of interest in granting a notation may temporarily fall outside the range of acceptability. When a departure from criteria is identified, during either the notation's initial issuance or reconfirmation process, it must be reviewed by Class in consultation with the shipyard or the vessel owner. When the design of the accommodation areas or ambient environmental test results contain departures from the stated criteria, these will be subject to special consideration upon the receipt of details about the departure. Depending on the degree and consequences of the departure, the shipyard or vessel owner may be required to provide an assessment and corrective plan to obtain or maintain the notation. Failure to complete the agreed correction by the due date may lead to the withdrawal of the assigned notation.

Chapter 31
Accommodation Areas

General

To promote maritime safety, efficiency, and habitability, it is important that seafarers maintain appropriate levels of mental and physical fitness while on board vessels. To help accomplish this, seafarers should be provided with suitable accommodation areas. Appropriate accommodation area design helps promote reliable performance by reducing the potential for fatigue and human error. Appropriately designed and outfitted accommodation areas may also enhance crew morale, recruiting, retention, comfort, and overall quality of life at sea. Conversely, inappropriate accommodation areas can adversely impact a seafarer's ability to reliably perform assigned duties, fully relax, sleep, and recover from mentally and physically demanding work activities. This in turn can impact their ability to carry out duties on succeeding watches with the required diligence, accuracy, and attention to safety procedures. Providing an onboard environment that increases seafarer alertness and wellbeing should be of concern to vessel owners (Table 31.1). This chapter provides the measurement criteria for accommodation spaces. In particular, Tables 31.2, 31.3, 31.4, 31.5, 31.6, 31.7, 31.8, 31.9 and 31.10 encompasses criteria for access and egress, as well as crew cabins, sanitary spaces, offices, food services, recreation areas, laundry, and medical spaces. The criteria were selected to help increase crew member safety, productivity, quality of work, retention, and morale. Note: the criteria for access/egress are applicable only to the crew accommodation block and do not include crew member workspaces outside of the accommodation block. It is worth noting that compliance with the criteria outlined in this chapter are a prerequisite for the ABS Habitability (**HAB**), Habitability Plus (**HAB+**), or ship Habitability Plus Plus (**HAB++** notation. Other Classification Societies will have their own distinct and specific habitability notations, though as the criteria is derived from IMO MLC 2006, Title 3, the criteria are generally standardised across each Class.

A. A. Olsen, *Applying Physical Ergonomics to Modern Ship Design*,
https://doi.org/10.1007/978-3-031-57974-5_31

Table 31.1 Example of the ILO MLC 2006 Title 3 requirement #14

Accommodation area criteria					
General					
Requirement # = Related ILO MLC 2006, Title 3 requirement			HAB requirements	HAB+ requirements	HAB++ requirements
#	14	Accommodation area stairways have a maximum angle of inclination from the horizontal of:			
		50°	●		
		45°		●	
		40°			●

Instructions: For a **HAB** notation, the boxes with a ● in the **HAB** requirements column must be met. For a **HAB+** notation, the boxes with a ● in the **HAB+** requirements column need to be met in addition to the **HAB** requirements. For a **HAB++** notation, the boxes with a ● in the **HAB++** requirements columns need to be met in addition to the **HAB** and **HAB+** requirements

Terminology

Accommodation areas/accommodation block: Vessel areas where the primary purpose is to rest and recreate. Accommodation spaces include cabins, medical facilities (sick bays), offices, recreation rooms, and manned spaces within the accommodation block such as the bridge or control room. For the purposes of this chapter, accommodation areas also include service spaces such as mess rooms.

Accommodation Area Criteria

The accommodation area criteria are contained in Tables 31.2, 31.3, 31.4, 31.5, 31.6, 31.7, 31.8, 31.9 and 31.10. The ILO MLC, 2006 quantitative accommodation area requirements have been included in this chapter and are denoted by a "#" symbol in Tables 31.2, 31.3, 31.4, 31.5, 31.6, 31.7, 31.8, 31.9 and 31.10. The **HAB+** and **HAB++** notations have more stringent criteria than the **HAB** notation with the objective of providing enhanced living and working conditions to improve seafarer safety and comfort. This includes enhanced criteria for workspace design, crew cabins, and recreation/leisure areas. The **HAB+** notation requires meeting the accommodation area criteria of **HAB** and **HAB+**. To achieve a **HAB++** notation requires meeting the criteria for both **HAB** and **HAB+**, as well as the **HAB++** requirements. For example, Criteria #14 below provides three degrees for angles of inclination for stairways, one for **HAB**, **HAB+**, and **HAB++**. If seeking a **HAB++** notation, the 40° criteria would need to be fulfilled. By fulfilling the criteria for the **HAB++** requirements, the criteria for **HAB+** and **HAB** are met.

Table 31.2 ILO MLC 2006 Title 3 requirement (accommodation area criteria (general))

Accommodation area criteria					
General					
Requirement # = Related ILO MLC 2006, Title 3 requirement			HAB requirements	HAB+ requirements	HAB++ requirements
#	1	Headroom in all passageways, sleeping rooms, stairs, sanitary spaces, offices, food service areas, and recreational areas is at least 2030 mm (80.0 in.). This height is clear of any obstructions such as cableways, ducting, pipes, etc	●		
	2	Non-slip type deck covering is supplied where occasional water, oil or liquid on the floors is expected	●		
	3	Outside corners of bulkheads, doors, etc. are rounded	●		
	4	All edges that crew members may strike are rounded	●		
	5	Drawers and internal doors are designed: • to prevent opening and closing due to vessel motion • to be operable with one hand	●	●	
#	6	Painted wall surfaces and deck heads are light in colour	●		
	7	Interior finish materials and furnishings are designed to ease cleaning efforts and improve maintenance	●		

(continued)

Table 31.2 (continued)

Accommodation area criteria					
General					
Requirement # = Related ILO MLC 2006, Title 3 requirement			HAB requirements	HAB+ requirements	HAB++ requirements
	8	For workstations where seafarers will be working while standing for extended periods of time, a kick space of 100 mm (4 in.) high by 100 mm (4 in.) deep should be provided	●		
	9	9 Toilets are conveniently located to workplace, recreation areas, mess rooms, and shower rooms	●		
	10	A means of escape that is not readily apparent to a person from both the inside and outside of the space is adequately marked	●		
#	11	Wall surfaces and decks are washable and impervious to damp or moisture absorption	●		
#	12	For accommodation areas and recreational facilities where the floorings are made of composite materials, the joints with the sides should be profiled to avoid crevices	●		
#	13	Accommodation areas and recreational and catering facilities should be located as far as practicable from the engines, steering gear rooms, deck winches, ventilation, heating, and air-conditioning equipment, and other noisy machinery and apparatus	●		

Table 31.3 ILO MLC 2006 Title 3 requirement (accommodation area criteria (access/egress)

Accommodation area criteria					
Access/egress					
Requirement # = Related ILO MLC 2006, Title 3 requirement			HAB requirements	HAB+ requirements	HAB++ requirements
	1	Doors, hatches, or scuttles used as a means of escape are capable of being operated by one person, from either side, in both light and dark conditions	●		
	2	The method of opening a means of escape is obvious and rapid	●		
	3	Doors in accommodation spaces (with the exception of cabins), stairway, stair tower, passageway, or control spaces, open in the direction of escape, where practicable	●		
	4	Deck scuttles that serve as a means of escape are fitted with a quick acting release and a holdback device to hold the scuttle in an open position	●		
	5	High voltage electrical cabinet doors are lockable with keys provided to qualified personnel	●		
Door dimensions					
	6	Doors have the following dimensions:			
		• doors (other than emergency egress) used solely by crew members have a clear opening width of approximately 710 mm (28 in.) or ASTM size 4 doors	●		

(continued)

Table 31.3 (continued)

Accommodation area criteria					
Access/egress					
Requirement # = Related ILO MLC 2006, Title 3 requirement			HAB requirements	HAB+ requirements	HAB++ requirements
		• the distance from the deck to the top of the door is at least 1900 mm (75 in.)	●		
		• the distance from the deck to the top of the door is at least 1980 mm (78 in.)			●
Railings					
	7	Suitable storm rails/ handrails are provided in all interior passageways and at all deckhouse sides where persons on board might have normal access:			
		• storm rails/handrails are installed on both sides of passageways that are 1830 mm (72 in.) or more in width	●		
		• storm rails/handrails are 865 mm (34 in.) to 965 mm (38 in.) high	●		
		• the distance between/ or behind storm rails/ handrails and any obstruction is 75 mm (3 in.) or greater	●		
	8	Rails are installed parallel to the deck along deck edges and walkways and around open hatches, elevators, antenna platforms and along other boundaries in the following areas:			

(continued)

Table 31.3 (continued)

Accommodation area criteria					
Access/egress					
Requirement # = Related ILO MLC 2006, Title 3 requirement			HAB requirements	HAB+ requirements	HAB++ requirements
		• wherever there is danger of crew members falling to a lower level of 600 mm (23.5 in.) or more in the vessel wherever there is danger of crew members becoming enmeshed with hazardous operating machinery	●		
		• around unprotected openings with a coaming height below 760 mm (30 in.)	●		
		Note: *temporary rails can be used around unprotected openings into which a person may slip, trip or fall*			
	9	Deck/guard railings have the following design:			
		• the heights of rails or bulwarks are at least 1000 mm (39.5 in.) from the deck except where this height would interfere with the normal operation of the vessel, and toeboards which are at least 100 mm (4.0 in.) in height and have no more than a 6 mm (0.25 in.) clearance between the bottom edge of the toeboard and the walking surface	●		

(continued)

Table 31.3 (continued)

Accommodation area criteria					
Access/egress					
Requirement # = Related ILO MLC 2006, Title 3 requirement			HAB requirements	HAB+ requirements	HAB++ requirements
		• the heights of rails are at least 1070 mm (42 in.) from the deck except where this height would interfere with the normal operation of the vessel and toeboards which are at least 100 mm (4.0 in.) in height and have no more than a 6 mm (0.25 in.) clearance between the bottom edge of the toeboard and the walking surface			●
		• vertical stanchions for railings are spaced no more than 1525 mm (60 in.) apart horizontally	●		
		• at least every third vertical stanchion is supported by a bracket or stay	●		
		• chain or wire rope used as a rail is set such that the sag is not greater than 25 mm (1 in.) at the chain/rope's centre span	●		
		• maximum lengths of openings protected by wire or chain are 1830 mm (72 in.). A removable stanchion will constitute the start of a new opening	●		
		• vertical stanchions for railings are spaced no more than 1525 mm (60 in.) apart horizontally	●		

(continued)

Table 31.3 (continued)

Accommodation area criteria					
Access/egress					
Requirement # = Related ILO MLC 2006, Title 3 requirement			HAB requirements	HAB+ requirements	HAB++ requirements
		• at least every third vertical stanchion is supported by a bracket or stay	●		
		• chain or wire rope used as a rail is set such that the sag is not greater than 25 mm (1 in.) at the chain/rope's centre span	●		
		• maximum lengths of openings protected by wire or chain are 1830 mm (72 in.). A removable stanchion will constitute the start of a new opening	●		
Stairs					
	10	A clear landing at least as wide as the tread width and a minimum of 915 mm (36 in.) long is provided at the top and bottom of each stairway	●		
	11	Any change of direction in a stairway is accomplished by means of an intermediate landing at least as wide as the tread width and a minimum of 915 mm (36 in.) long	●		
	12	An intermediate landing is provided at each deck level serviced by a stair, or a maximum of every 3600 mm (144 in.) of vertical travel for stairs with a vertical rise of 6100 mm (240 in.)			●

(continued)

Table 31.3 (continued)

Accommodation area criteria					
Access/egress					
Requirement # = Related ILO MLC 2006, Title 3 requirement			HAB requirements	HAB+ requirements	HAB++ requirements
	13	Accommodation area stairways have a maximum angle of inclination from the horizontal of:			
		• 50°	●		
		• 45°		●	
		• 40°			●
	14	Stair risers and treads have the following design:			
		• the riser height is no more than 230 mm (9 in.) and the tread depth is approximately 190 mm	●		
		• (7.5 in.), with an additional 25 mm (1 in.) tread nosing		●	
		• the riser height is no more than 230 mm (9 in.) and the tread depth is approximately 230 mm (9 in.), with an additional 25 mm (1 in.) tread nosing			●
		• the riser height is no more than 230 mm (9 in.) and the tread depth is approximately 275 mm (10.75 in.), with an additional 25 mm (1 in.) tread nosing	●		

(continued)

Table 31.3 (continued)

Accommodation area criteria					
Access/egress					
Requirement # = Related ILO MLC 2006, Title 3 requirement			HAB requirements	HAB+ requirements	HAB++ requirements
		• stairs in a stairway or stair tower, the depth of the tread and the height of riser are consistent	●		
		• the minimum tread width on one-way (one person) stairs is at least 610 mm (24 in.)	●		
		• the minimum tread width on two-way (two person) stairs is at least 915 mm (36 in.)		●	
		• the minimum tread width on two-way (two person) stairs is at least 1015 mm (40 in.)			●
		• once a minimum tread width has been established at any deck, in that stair run, it is not decreased in the direction of escape	●		
		• all nosings have a non-slip surface	●		
	15	Stairway or stair towers are fitted with handrails with the following design:			
		• a handrail is provided on one side of the stair	●		
		• a handrail is provided on both sides of the stair			●

(continued)

Table 31.3 (continued)

Accommodation area criteria					
Access/egress					
Requirement # = Related ILO MLC 2006, Title 3 requirement			HAB requirements	HAB+ requirements	HAB++ requirements
		• the handrails are parallel to the pitch line of the stair flight and level at landings	●		
		• the handrail is continuous from the top to the bottom of the stair and terminates in a safe manner at both ends	●		
		• the vertical height above the tread at its nosing is at least 940 mm (37 in.) to 990 mm (39 in.)	●		
		• the distance between handrails and any obstruction is 75 mm (3 in.) or greater	●		
		• handrails on adjacent, parallel stair flights have a minimum of 100 mm (4 in.) clear distance between rails	●		
Ladders					
	16	All inclined ladders and handrails are located so as not to interfere with the opening and closing of hatches, doors, gratings, or other types of access; in all areas of the vessel other than accommodation block	●		
	17	Inclined ladders are between 50° and 60° from the horizontal	●		

(continued)

Table 31.3 (continued)

Accommodation area criteria					
Access/egress					
Requirement # = Related ILO MLC 2006, Title 3 requirement			HAB requirements	HAB+ requirements	HAB++ requirements
	18	Each vertical ladder used as a means of escape has the following design features:	●		
		• the ladder is mounted at least 180 mm (7 in.) from the nearest permanent object in the back of the ladder (distance behind each rung)	●		
		• rungs are at least 410 mm (16 in.) in width	●		
		• rungs are between 275 mm (11 in.) and 300 mm (12 in.) apart	●		
		• rungs are uniformly spaced for the length of the ladder	●		
		• at least 760 mm (30 in.) clearance in front of the ladder (climbing space)	●		
		• each ladder rung is skid/slip resistant	●		
		• grab bars are provided that extend at least 1070 mm (42 in.) above the landing platform		●	
		• Ladder landings at the top and bottom of inclined ladders:			
		• at least 760 mm (30 in.) clear in length	●		

(continued)

Table 31.3 (continued)

Accommodation area criteria					
Access/egress					
Requirement # = Related ILO MLC 2006, Title 3 requirement			HAB requirements	HAB+ requirements	HAB++ requirements
		• at least 915 mm (36 in.) clear in length			●
Ramps					
	20	Ramps are sloped < 15 degrees for inclined walking surfaces	●		
	21	Ramps have a non-skid surface	●		
	22	Ramps have a handrail on any open side of the ramp if the distance from the ramp to the nearest adjacent surface is 600 mm (23.5 in.) or more	●		
	23	Access to lifeboats is provided such that a person in a stretcher can be easily embarked into the survival craft (e.g., ramp)	●		
	24	The clear walkway width for one person in an unrestricted area, where two persons could pass is at least 710 mm (28 in	●		
	25	The clear walkway width for normal two-way traffic or any means of egress that leads to an exit or entrance is at least 915 mm (36 in.)	●		
	26	The clear walkway width for normal two-way traffic or any means of egress that leads to an exit or entrance is at least 1015 mm (40 in.)			●

Table 31.4 ILO MLC 2006 Title 3 requirement (accommodation area criteria (crew cabins)

Accommodation area criteria							
Crew cabins							
Requirement # = Related ILO MLC 2006, Title 3 requirement					HAB requirements	HAB+ requirements	HAB++ requirements
Non-passenger and non-special purpose ship floor area for various vessel sizes							
		< 3000 tonnes	3000 to 10,000 tonnes	> 10,000 tonnes			
One person cabin (seafarers not performing the duties of officers)							
#	1	4.5 m^2 (48.5 ft^2)	5.5 m^2 (60.0 ft^2)	7.0 m^2 (75.0 ft^2)	●		
		5.5 m^2 (60.0 ft^2)	6.5 m^2 (70.0 ft^2)	7.5 m^2 (80.5 ft^2)		●	
		6.5 m^2 (70.0 ft^2)	7.5 m^2 (80.5 ft^2)	8.5 m^2 (91.5 ft^2)			●
Two-person cabin (seafarers not performing the duties of officers)[1]							
#	2	7.0 m^2 (75.0 ft^2)	7.0 m^2 (75.0 ft^2)	7.0 m^2 (75.0 ft^2)	●		
		7.5 m^2 (80.5 ft^2)	8.0 m^2 (80.5 ft^2)	8.0 m^2 (86.0 ft^2)		●	
		8.0 m^2 (86.0 ft^2)	9.0 m^2 (97.0 ft^2)	9.0 m^2 (97.0 ft^2)			●
Junior officer and senior officer (where no private sitting/day room is provided)							
#	3	7.5 m^2 (80.5 ft^2)	8.5 m^2 (91.5 ft^2)	8.5 m^2 (107.5 ft^2)	●		
Passenger ship floor area for various vessel sizes							
		< 3000 tonnes	3000–10,000 tonnes	> 10,000 tonnes			
#	4	4.5 m^2 (48.5 ft^2)	5.5 m^2 (60.0 ft^2)	7.0 m^2 (75.0 ft^2)	●		
		5.5 m^2 (60.0 ft^2)	6.5 m^2 (70.0 ft^2)	7.5 m^2 (80.5 ft^2)		●	
		6.5 m^2 (70.0 ft^2)	7.5 m^2 (80.5 ft^2)	8.5 m^2 (91.5 ft^2)			●
Two-person cabin (seafarers not performing the duties of officers)							
#	5	7.0 m^2 (75.0 ft^2)	7.0 m^2 (75.0 ft^2)	7.0 m^2 (75.0 ft^2)	●		
		7.5 m^2 (80.5 ft^2)	8.0 m^2 (80.5 ft^2)	8.0 m^2 (86.0 ft^2)		●	
		8.0 m^2 (86.0 ft^2)	9.0 m^2 (97.0 ft^2)	9.0 m^2 (97.0 ft^2)			●

(continued)

Table 31.4 (continued)

Accommodation area criteria					HAB requirements	HAB+ requirements	HAB++ requirements
Crew cabins							
Requirement # = Related ILO MLC 2006, Title 3 requirement					HAB requirements	HAB+ requirements	HAB++ requirements
Three-person room (seafarers not performing the duties of officers)							
#	6	11.5 m^2 (124.0 ft^2)	11.5 m^2 (124.0 ft^2)	11.5 m^2 (124.0 ft^2)	●		
		12.5 m^2 (134.5 ft^2)	12.5 m^2 (134.5 ft^2)	12.5 m^2 (134.5 ft^2)		●	
		13.5 m^2 (145.5 ft^2)	13.5 m^2 (145.5 ft^2)	13.5 m^2 (145.5 ft^2)			●
Four-person room (seafarers not performing the duties of officers)							
#	7	14.5 m^2 (156.0 ft^2)	14.5 m^2 (156.0 ft^2)	14.5 m^2 (156.0 ft^2)	●		
		15.5 m^2 (167.0 ft^2)	15.5 m^2 (167.0 ft^2)	15.5 m^2 (167.0 ft^2)		●	
		16.5 m^2 (177.5 ft^2)	16.5 m^2 (177.5 ft^2)	16.5 m^2 (177.5 ft^2)			●
Junior officer (where no private sitting/day room is provided)[2]							
#	8	7.5 m^2 (80.5 ft^2)	7.5 m^2 (80.5 ft^2)	7.5 m^2 (80.5 ft^2)	●		
Senior officer (where no private sitting/day room is provided)3							
#	9	8.5 m^2 (91.5 ft^2)	8.5 m^2 (91.5 ft^2)	8.5 m^2 (91.5 ft^2)	●		
Occupancy level							
#	10	Individual berths are provided for each crew member			●		
#	11	Individual sleeping rooms are provided for each seafarer in ships other than passenger ships (exemption may be granted by the competent authority for ships less than 3000 GT			●		
#	12	Separate sleeping rooms should be provided for men and for women			●		
#	13	Crew members sharing rooms are in same occupation group and day persons do not share a room with watchkeepers			●		

(continued)

Table 31.4 (continued)

Accommodation area criteria					
Crew cabins					
Requirement # = Related ILO MLC 2006, Title 3 requirement			HAB requirements	HAB+ requirements	HAB++ requirements
	14	Maximum number of persons to be accommodated in any sleeping room is marked indelibly and legibly in some conveniently seen place:			
		• in the room	●		
		• outside the room	●		
Crew cabin location					
	15	Location, means of access, structure and arrangement in relation to other spaces of crew accommodation areas is to:			
		• provide adequate security	●		
		• protect against weather and sea	●		
		• insulate from heat and cold	●		
#	16	There is no direct communication into sleeping areas from cargo areas, machinery spaces, chain lockers, galleys, lamp and paint rooms or from engine, deck and other bulk storerooms, drying rooms, communal wash places or water closets	●		
	17	Crew quarters are not located:			
		• further forward in the vessel than a vertical plane located at 5 percent of the vessel's length	●		
		• abaft of the forward side of the stem at the designed summer load water line	●		
#		• immediately beneath working alleyways	●		
#		• below the loadline amidships or aft, unless satisfactory arrangements are made for lighting and ventilation and approved by the Flag State Administration	●		

(continued)

Table 31.4 (continued)

Accommodation area criteria					
Crew cabins					
Requirement # = Related ILO MLC 2006, Title 3 requirement			HAB requirements	HAB+ requirements	HAB++ requirements
#		• forward of the collision bulkhead	●		
Cabin design					
General					
#	18	Internal bulkheads are of approved material that is not likely to harbour vermin	●		
	19	Wall surfaces are not constructed from tongued and grooved boarding or any other form of construction likely to harbour vermin	●		
#	20	A sitting or day room, adjoining the sleeping room is provided for the:			
		• Master	●		
		• Chief engineer	●		
		• Chief navigating officer	●		
		• Second engineer officer		●	
		Note: *ships of less than 3000 GT may be exempted from this requirement by the vessel's Flag State*			
	21	Pipes, ventilation ducts, or other installations do not obstruct berths	●		
	22	Sleeping rooms are ventilated and heated	●		
	23	HVAC vents should not be directed at the heads of berths	●		
	24	Radiators and other heating apparatus are so placed and, where necessary, shielded as to avoid risk of fire or danger or discomfort to the occupants	●		
Berths					
	25	Multi-crew cabin berths are not side by side so that access to one berth can only be obtained over another	●		

(continued)

Table 31.4 (continued)

Accommodation area criteria					
Crew cabins					
Requirement # = Related ILO MLC 2006, Title 3 requirement			HAB requirements	HAB+ requirements	HAB++ requirements
	26	Berths are generally oriented fore and aft	●		
	27	Multi-crew cabin berths contain individually operated privacy curtains	●		
	28	Multi-crew cabin berths contain individually operated fans/blowers		●	
#	29	Multi-crew cabin arrangements are either single tier or double tiered	●		
#	30	Multi-crew cabin berths with portlights above them are only allowed in a single tier arrangement	●		
Emergency					
	31	An emergency alerting system (e.g., an audible alarm) is present for all crew cabin spaces	●		
Outfitting					
#	32	Sleeping rooms with portlights, have curtains that completely block out light	●		
#	33	Sleeping rooms are lighted with natural light and are provided with artificial light (subject to special arrangements as may be permitted in passenger vessels)	●		
	34	Grab bars and stepping surfaces are provided for access to upper berths	●		
#	35	An electric reading light is provided at the head of each berth	●		
#	36	The top berth in a tier has a dust-proof bottom of wood, canvas, or other suitable material	●		
	37	In sleeping rooms with more than one occupant, furniture (beds, lockers, drawers, etc.) are labelled identifying which furniture is dedicated to which berth	●		

(continued)

Table 31.4 (continued)

Accommodation area criteria					
Crew cabins					
Requirement # = Related ILO MLC 2006, Title 3 requirement			HAB requirements	HAB+ requirements	HAB++ requirements
	38	Each sleeping room is provided with at least the following furniture (in addition to berths and clothes lockers):			
#		• a table or desk	●		
#		• chair	●		
		• a mirror with a light	●		
#		• a small cabinet for toilet requisites for each person in the room	●		
#		• book racks	●		
#		• coat hooks	●		
#	39	Furniture is of smooth, hard material not liable to warp or corrode	●		
#	40	In vessels regularly trading to mosquito-infested ports, suitable screens are fitted to all ventilation inlets and doors to the open deck	●		
Dimensional aspects					
Aisles					
	41	Aisle widths in sleeping areas are:	●		
		• at least 610 mm (24 in.), between a single berth and the nearest obstruction	●		
		• at least 915 mm (36 in.), between facing berths			●
Berths					
#	42	The lower berth in a tier is at least 300 mm (12 in.) above the deck	●		
#	43	The upper berth is placed approximately midway between the bottom of the lower berth and the lower side of the deck head beams	●		
	44	Head clearance above each berth is:	●		
		• at least 610 mm (24 in.)	●		

(continued)

Table 31.4 (continued)

Accommodation area criteria					
Crew cabins					
Requirement # = Related ILO MLC 2006, Title 3 requirement			HAB requirements	HAB+ requirements	HAB++ requirements
		• at least 810 mm (32 in.)			●
	45	Berth inside dimensions are:			
#		• at least 1980 mm (78 in.) by 800 mm (31.5 in.)	●		
		• at least 2030 mm (80 in.) by 965 mm (38 in.)			●
#	46	The framework and leeboard of a berth is of approved material, hard, smooth and not likely to corrode or to harbour vermin	●		
#	47	Berths constructed from tubular frames are completely sealed and without perforations which would give access to vermin	●		
Storage space					
	48	Sleeping rooms provide storage space for each occupant:			
		• the clothes locker is fitted with a shelf and hasp	●		
#		• the clothes locker is at least 0.475 m3 (16.75 ft^3)	●		
#		• a drawer or equivalent space of at least 56 L; if the drawer is incorporated in the clothes-locker then the combined minimum volume of the clothes locker should be 500 L; it should be fitted with a shelf and be able to be locked by the occupant so as to provide privacy	●		
	49	Accommodation areas for the hanging of oilskins (wet weather gear) are:	●		
		• sufficient	●		
		• adequately ventilated	●		
		• provided outside but convenient to the sleeping rooms	●		

Table 31.5 ILO MLC 2006 Title 3 requirement (accommodation area criteria (sanity spaces)

Accommodation area criteria					
Sanitary spaces					
Requirement # = Related ILO MLC 2006, Title 3 requirement			HAB requirements	HAB+ requirements	HAB++ requirements
	1	Sanitary spaces are:			
		• lighted	●		
#		• heated	●		
#		• ventilated	●		
	2	Floors in sanitary spaces are:			
#		• non-slip type deck covering	●		
#		• easily cleaned	●		
#		• impervious to damp or moisture absorption	●		
#		• properly drained	●		
	3	For every six crew members who do not have a private sanitary space, there is:			
#		• one or more tubs and/or shower baths	●		
#		• one or more toilets	●		
#		• washbasins	●		
	4	Where the radio officers or operators are accommodated in an isolated position, sanitary facilities are provided near or adjacent thereto	●		
	5	Someone standing on a wet deck in sanitary spaces cannot reach light switches or electrical outlets without ground fault interrupters (GFI)	●		
	6	Common/shared spaces are gender identifiable without entering the space	●		
#	7	Separate multiple occupancy sanitary spaces are provided for men and women	●		

(continued)

Table 31.5 (continued)

Accommodation area criteria					
Sanitary spaces					
Requirement # = Related ILO MLC 2006, Title 3 requirement			HAB requirements	HAB+ requirements	HAB++ requirements
	8	Bulkheads in sanitary spaces are:			
#		• steel or other approved material	●		
#		• watertight up to 230 mm (9 in.) above deck level	●		
	9	A public sanitary facility should be situated near the ship's office(s) if the office(s) is not conveniently located near the navigation bridge[4]	●		
#	10	All sanitary spaces should have ventilation to the open air, independently of any other part of the accommodation areas	●		
	11	Free space (space available for movement without hindrance of any objects) in a common sanitary space per person is:			
		• at least 0.75 m^2 (8 ft^2)	●		
		• at least 1.1 m^2 (12 ft^2)			●
	12	Single or double occupancy sleeping rooms have a private sanitary space with a toilet, shower or tub and sink modules			●
General—based on vessel size					
	13	On vessels over 1600 GT, a toilet and washbasin having hot and cold running potable water is within easy access of the:[5]			
#		• navigation bridge deck and primarily for those on duty in the area	●		

(continued)

Table 31.5 (continued)

Accommodation area criteria					
Sanitary spaces					
Requirement # = Related ILO MLC 2006, Title 3 requirement			HAB requirements	HAB+ requirements	HAB++ requirements
#		• machinery space if not fitted near the engine room control centre	●		
		• galley/mess room(s)	●		
		• recreation room(s)	●		
		• gymnasium	●		
	14	On vessels over 1600 GT, without private or semi-private sleeping rooms or sanitary spaces, clothes changing facilities are provided for engine department personnel. These facilities are:			
#		• outside but within easy access of the machinery space and navigating bridge	●		
#		• fitted with individual lockers	●		
#		• fitted with tubs and/or showers and washbasins	●		
	15	In vessels between 5000 and 15,000 GT, at least five officers have sleeping rooms with an attached private sanitary space with a toilet, tub and/or shower and washbasin. The washbasin may be in the sleeping room	●		
	16	In vessels between 10,000 and 15,000 GT, all officers without private sanitary spaces have private intercommunicating sanitary spaces fitted with a toilet, tub and/or shower and washbasin	●		

(continued)

Table 31.5 (continued)

Accommodation area criteria					
Sanitary spaces					
Requirement # = Related ILO MLC 2006, Title 3 requirement			HAB requirements	HAB+ requirements	HAB++ requirements
	17	In vessels over 15,000 GT, all officers have sleeping rooms with an attached private sanitary space fitted with a toilet, tub and/or shower and washbasin. The washbasin may be in the sleeping room	●		
	18	In non-passenger vessels over 25,000 GT, a sanitary space is provided for every two ratings, either in an intercommunicating space between adjoining sleeping rooms or opposite the entrance of such rooms. The sanitary spaces are fitted with a toilet, tub and/or shower and washbasin	●		
#	19	Each sleeping room (both officer and ratings) is provided with a washbasin with hot and cold running potable water (except when a private sanitary space is provided)	●		
Showers/baths					
	20	Water heaters supplying showers:			
		• do not support areas that have higher water temperature requirements, such as food service area	●		
		• are provided with anti-scalding devices	●		
	21	Handholds are provided for shower and bath sanitary spaces	●		

(continued)

Table 31.5 (continued)

Accommodation area criteria					
Sanitary spaces					
Requirement # = Related ILO MLC 2006, Title 3 requirement			HAB requirements	HAB+ requirements	HAB++ requirements
	22	In sanitary spaces intended for more than one person, the shower is screened	●		
Toilets					
	23	Where there is more than one toilet in a space, they are sufficiently screened to provide privacy	●		
	24	All toilets:			
		• have flush water available at all times	●		
		• are independently controllable	●		
		• have a hand washing station	●		
	25	Toilets, intended for more than one crew member, are situated convenient to, but separate from, sleeping rooms and washrooms without direct access from:			
#		• the sleeping rooms	●		
#		• toilets to which there is no other access. (This requirement does not apply where a toilet is located in a space between two sleeping rooms having a total of not more than four persons)	●		
	26	All toilet spaces should be ventilated sufficiently to be reasonably free of disagreeable odours and condensation	●		

(continued)

Table 31.5 (continued)

Accommodation area criteria					
Sanitary spaces					
Requirement # = Related ILO MLC 2006, Title 3 requirement			HAB requirements	HAB+ requirements	HAB++ requirements
Washbasins					
	27	Water heaters supplying washbasins:			
		• do not support areas that have higher water temperature requirements	●		
		• are provided with anti-scalding devices	●		
	28	The following at a minimum are provided for each washbasin:			
		• a mirror	●		
		• toiletry shelf	●		
	29	Facilities to dry hands (e.g., towel bars, etc.) are provided at all washbasins	●		
	30	Washbasins are:			
		• constructed of approved material	●		
		• smooth surfaced	●		
		• not liable to crack, flake or corrode	●		
Urinals					
	31	Urinals have privacy partitions between units and at the end of rows if not provided by sanitary space structures	●		
	32	The dimension between the centreline of two urinals side-by-side is 690 mm (27 in.) or greater	●		

(continued)

Table 31.5 (continued)

Accommodation area criteria					
Sanitary spaces					
Requirement # = Related ILO MLC 2006, Title 3 requirement			HAB requirements	HAB+ requirements	HAB++ requirements
	33	The dimension between the centreline of a urinal and bulkhead next to the urinal is 380 mm (15 in.) or greater		●	
	34	The height of the front edge of a urinal is between 455 mm (18 in.) and 605 mm (24 in.) above the deck surface	●		

Accommodation Area Documentation

As stated earlier, accommodation area documentation must be prepared and submitted to Class for review and approval. Confirmatory verification measurements are then performed by a Class approved surveyor. The following data is typically required to be submitted to Class. Data requirements. The submitted data should serve as a means for verifying that the vessel meets the accommodation area criteria specified in Tables 31.2, 31.3, 31.4, 31.5, 31.6, 31.7, 31.8, 31.9 and 31.10.

- *Newbuilds.* For new construction, scaled arrangement drawings of the accommodation spaces (elevation and plan views), details of the accommodation area outfitting and vessel's design specification in relation to the accommodation spaces must be submitted to Class.
- *Existing vessels.* For existing vessels, appropriate arrangement drawings and plans, reflecting the current accommodation area configurations should be provided to Class together with any current vessel accommodation area design specifications.

Submission Review and Verification

Arrangement drawings, plans, and vessel specifications for the accommodation areas should be prepared and submitted for review by Class. For new construction, the drawings should be provided to Class early during the detailed design phase. For existing vessels, the arrangement drawings and plans reflecting the current accommodation area configurations may be provided to Class in advance of onboard Class approved surveyor verifications. Class will then review the submitted accommodation area documentation. Any deviations from the criteria will be highlighted to the

Table 31.6 ILO MLC 2006 Title 3 requirement (accommodation area criteria (food service areas)

Accommodation area criteria					
Food service areas					
Requirement # = Related ILO MLC 2006, Title 3 requirement			HAB requirements	HAB+ requirements	HAB++ requirements
General					
	1	Access is provided to all areas requiring cleaning, painting, or treating with pesticide/insecticide	●		
	2	There are no crevices or inaccessible voids, which might harbour vermin, cooking or food waste, or other extraneous matter	●		
	3	Places where food is prepared, served or stored are not used as sleeping areas	●		
	4	Deck drains are provided in all food service areas subject to flood type cleaning or where normal operations release or discharge water or other liquid onto the deck	●		
	5	Deck drains for food service areas are provided in number and location so that complete drainage is possible under normal conditions of list and trim	●		
	6	Where distributive systems are located in food service areas, they are installed as close as practicable to the overhead in order to provide maximum headroom	●		

(continued)

Table 31.6 (continued)

Accommodation area criteria					
Food service areas					
Requirement # = Related ILO MLC 2006, Title 3 requirement			HAB requirements	HAB+ requirements	HAB++ requirements
	7	Flow of food through vessel:			
		• is arranged in logical sequence that minimises cross-traffic or backtracking	●		
		• allows for adequate separation of clean and soiled operations	●		
Food service doors					
	8	Toilet room doors opening directly into food preparation or serving areas are self-closing and tight	●		
Food service passageways					
	9	The passage width between storage shelves is at least 710 mm (28 in.)	●		
Mess room					
General					
	10	Mess rooms are:			
#		• located apart from sleeping rooms	●		
		• adequately ventilated	●		
		• adequately insulated to prevent condensation or overheating	●		

(continued)

Table 31.6 (continued)

Accommodation area criteria					
Food service areas					
Requirement # = Related ILO MLC 2006, Title 3 requirement			HAB requirements	HAB+ requirements	HAB++ requirements
		• located as near to the galley as is practicable except where the mess room is equipped with a steam table	●		
#	11	Mess rooms have tables and seats sufficient for the number of persons likely to use them at any one time	●		
	12	Mess lines and mess rooms are protected from weather, objectionable sights (such as garbage disposal areas) and objectionable odours (such as from engines, holds, toilets, fire room, etc.)	●		
#	13	Lighting in mess rooms is provided by means of natural light and artificial light (subject to special arrangements as may be permitted in passenger vessels)	●		
	14	The following is available for crew use at all times:			
#		• refrigerator	●		
#		• facilities for hot beverages	●		

(continued)

Table 31.6 (continued)

Accommodation area criteria					
Food service areas					
Requirement # = Related ILO MLC 2006, Title 3 requirement			HAB requirements	HAB+ requirements	HAB++ requirements
#		• facilities for cool water	●		
#	15	Common or separate mess room facilities are to be provided	●		
	16	The tops of tables and seating are capable of being easily cleaned	●		
#	17	Where available pantries are not accessible to mess rooms, adequate lockers for mess utensils and proper facilities for washing utensils are provided	●		
Outfitting					
	18	A transparent sanitation shield is installed to completely shield the entire length of the food serving lines if provided with a tray slide or rail	●		
	19	Tray slides (or rails) if provided are placed along food and beverage serving lines	●		
	20	Adequate lockers for mess utensils are provided	●		

(continued)

Table 31.6 (continued)

Accommodation area criteria					
Food service areas					
Requirement # = Related ILO MLC 2006, Title 3 requirement			HAB requirements	HAB+ requirements	HAB++ requirements
Mess room dimensional aspects					
	21	Rectangular table space for each diner is:	●		
		• at least 610 mm (24 in.) wide by 380 mm (15 in.) deep	●		
		• at least 740 mm (29 in.) wide by 430 mm (17 in.) deep	●		
	22	Round Table diameter is:			
		• for seating 4 persons at least 1220 mm (48 in.)	●		
		• for seating 4 persons at least 1370 mm (54 in.)	●		
		• for seating 6 persons at least 1520 mm (60 in.)	●		
		• for seating 6 persons at least 1680 mm (66 in.)	●		
		• for seating 8 persons at least 1830 mm (72 in.)	●		
		• for seating 8 persons at least 1980 mm (78 in.)		●	
	23	Distance between tables with back-to-back seating is:			

(continued)

Table 31.6 (continued)

Accommodation area criteria					
Food service areas					
Requirement # = Related ILO MLC 2006, Title 3 requirement			HAB requirements	HAB+ requirements	HAB++ requirements
		• at least 1195 mm (47 in.)	●		
		• at least 1525 mm (60 in.)		●	
	24	Deck area requirements for planned seating capacity are:			
#		• at least 1.5 m^2 (16.2 ft^2) per person	●		
		• at least 1.7 m^2 (18.4 ft^2) per person		●	
		• at least 1.9 m^2 (20.5 ft^2) per person			●
Galley/scullery					
General					
#	25	The galley comprises space for food preparation, cooking, service and wash-up facilities. These should include areas for:			
		• baking and pastry preparation	●		
		• vegetable preparation	●		
		• meat, fish and poultry preparation	●		
		• hot services	●		
		• dish wash facilities	●		
		• pan wash facilities	●		

(continued)

Table 31.6 (continued)

Accommodation area criteria					
Food service areas					
Requirement # = Related ILO MLC 2006, Title 3 requirement			HAB requirements	HAB+ requirements	HAB++ requirements
		• equipment storage	●		
	26	Ice making facilities for at least 0.57 kg (1.25 lbs) of ice per person per day are provided	●		
	27	Hand washing stations located so that no crew member must walk more than 7.5 m (25 ft) to a station are to be provided. Slop sinks and scullery sinks are not satisfactory hand washing stations	●		
	28	Tops of food dressing tables, counters and drain boards are constructed with an integral back splash	●		
Potable water					
	29	Only potable water is piped to food storage, preparation, or service areas. The exceptions are a food waste grinder eductor (garbage disposal) line and deck washing facilities. The grinder delivery line is protected against back flow	●		

(continued)

Table 31.6 (continued)

Accommodation area criteria				
Food service areas				
Requirement # = Related ILO MLC 2006, Title 3 requirement		HAB requirements	HAB+ requirements	HAB++ requirements
30	If the vessel has a garbage room, hot and cold potable water is available for washing garbage cans	●		
31	Potable water tanks are identified with a number and the words "POTABLE WATER" adequately sized	●		
32	Potable water piping and fittings are labelled	●		
Maintenance/cleaning				
33	Drawers and bins used as food contact surfaces are readily removable and easily cleaned	●		
34	Covers, insets, or receptacles for unpackaged foods or beverages are easily removable or designed for easy cleaning in place	●		
35	Stove top or range sea rails are readily removable and easily cleanable and brackets for sea rails are easily cleanable	●		

(continued)

Table 31.6 (continued)

Accommodation area criteria				
Food service areas				
Requirement # = Related ILO MLC 2006, Title 3 requirement		HAB requirements	HAB+ requirements	HAB++ requirements
36	Ventilation baffles, vanes, louvres, dampers, filters, etc., are easily accessible or removable (without tools) for cleaning	●		
Refrigeration/freezer				
37	It is possible to open the doors of the refrigerated or cold storage rooms from the inside even though they are locked from the outside	●		
38	There is an alarm in each freezer and cold storage room to warn when the temperature is out of specification	●		
Dish washing				
39	Easy access is provided to a sink or dish washing machine	●		

vessel owner/shipyard by Class for resolution. Class will also identify any criteria that the Class approved surveyor must field verify. The surveyor will then verify that the submitted drawings match the constructed vessel and also verify any criteria that are outstanding from the Class review and document deviations from criteria.

Table 31.7 ILO MLC 2006 Title 3 requirement (accommodation area criteria (recreation)

Accommodation area criteria					
Recreation					
Requirement # = Related ILO MLC 2006, Title 3 requirement			HAB requirements	HAB+ requirements	HAB++ requirements
General					
#	1	Recreation and accommodation areas, conveniently situated and appropriately furnished are provided for the crew. (Where this is not provided separately from the mess rooms the latter is planned, furnished and equipped to give recreational facilities)	●		
#	2	Vessel has an open deck space or spaces to which the crew can have access when off duty. The space or spaces are of adequate area, having regard to the size of the vessel and the crew	●		
#	3	Recreation rooms in the crew accommodation spaces are adequately insulated to prevent condensation or overheating	●		
Recreational and exercise areas					
#	4	Furnishing for recreational facilities should as a minimum include a book-case and facilities for reading, writing and, where practicable, table and deck games	●		

(continued)

Table 31.7 (continued)

Accommodation area criteria					
Recreation					
Requirement # = Related ILO MLC 2006, Title 3 requirement			HAB requirements	HAB+ requirements	HAB++ requirements
#	5	Television and DVD/ CD player facilities are to be provided. The stock of films should be adequate for the duration of the voyage and, where necessary, changed at reasonable intervals	●		
#	6	Consideration should be given to include the following recreational facilities on board, where practicable:			
		• a smoking room	●		
		• sport equipment including exercise equipment	●		
		• facilities for swimming	●		
		• a library containing vocational and other books, the stock of which should be adequate for the duration of the voyage and changed at reasonable intervals	●		
		• facilities for recreational handicrafts	●		
		• canteen	●		
		• where appropriate, the provision of bars on board for seafarers unless these are contrary to national, religious or social customs	●		
	7	Approximately 1.0 m^2 (11 ft^2) is provided for each seat in the lounge	●		

(continued)

Table 31.7 (continued)

Accommodation area criteria					
Recreation					
Requirement # = Related ILO MLC 2006, Title 3 requirement			HAB requirements	HAB+ requirements	HAB++ requirements
	8	Approximately 1.25 m^2 (14 ft^2) is provided for each seat in the lounge			●
	9	Vessels regularly trading to or in the tropics and the Arabian Gulf, are equipped with awnings for use	●		

Table 31.8 ILO MLC 2006 Title 3 requirement (accommodation area criteria (laundry)

Accommodation area criteria					
Laundry					
Requirement # = Related ILO MLC 2006, Title 3 requirement			HAB requirements	HAB+ requirements	HAB++ requirements
General			●		
#	1	Facilities exist for washing and drying clothes on a scale appropriate to the size of the crew and the normal duration of the voyage	●		
	2	Laundry facilities are located within easy access of crew accommodation areas	●		
	3	Laundry facilities should be sufficient to allow seafarers to be provided with clean and dry underwear once per day and clean and dry outerwear and bedding once per five days	●		

(continued)

Table 31.8 (continued)

Accommodation area criteria					
Laundry					
Requirement # = Related ILO MLC 2006, Title 3 requirement			HAB requirements	HAB+ requirements	HAB++ requirements
	4	Appropriate clearance has been provided for equipment maintenance		●	
	5	Coamings and/or deck drains are provided around washing machines, water heaters, etc	●		
	6	Washer and dryer capacities are matched to assure a smooth workflow	●		
	7	Washers and dryers are placed relative to each other to facilitate the transfer of clothing from the washer to the dryer	●		
Outfitting					
	8	Tumble dryers, if provided, are exhausted directly into the weather, not into the vessel	●		
	9	Tumble dryers, if provided, are equipped with lint filters	●		
	10	The laundry facilities include at a minimum:			
#		• clothes washing machines	●		
#		• clothes drying machines (or adequately heated and ventilated drying rooms)	●		

(continued)

Table 31.8 (continued)

Accommodation area criteria					
Laundry					
Requirement # = Related ILO MLC 2006, Title 3 requirement			HAB requirements	HAB+ requirements	HAB++ requirements
#		• irons and ironing boards (or their equivalent)	●		
		• suitable sinks with an adequate supply of cold potable water and hot potable water or means of heating water	●		
		• sinks are installed in washrooms, if a separate laundry accommodation area is not reasonably practical	●		
	11	The facilities for drying clothes are:			
		• equipped with lines or other fittings for hanging clothes	●		
		• separate from sleeping rooms and mess rooms	●		
		• ventilated and heated	●		

Results

On completion of the submission review, the results of the Class review and surveyor verification will be reviewed by the Class approved surveyor against the appropriate **HAB**, **HAB+**, or **HAB++** criteria for notation award. Provided the criteria is met, the appropriate notation will then be confirmed and entered into the vessel's Record.

Table 31.9 ILO MLC 2006 Title 3 requirement (accommodation area criteria (medical)

Accommodation area criteria					
Medical					
Requirement # = Related ILO MLC 2006, Title 3 requirement			HAB requirements	HAB+ requirements	HAB++ requirements
General					
#	1	In vessels of less than 500 GT a medical chest, medical equipment and a medical guide should be carried on board as a minimum. The type of ship, the number of persons on board and the nature, destination and duration of voyages and relevant national and international recommended medical standards should be taken into account	●		
#	2	Separate medical accommodation area used exclusively for medical purposes is provided in vessels of 500 GT or more, carrying 15 or more seafarers and engaged in a voyage of more than three days' duration	●		
	3	Medical accommodation area is not used as a storage area except for medical supplies	●		
	4	Medical accommodation area is suitably located to allow for safe and efficient:			
		• access for sick or injured crew members	●		

(continued)

Table 31.9 (continued)

Accommodation area criteria					
Medical					
Requirement # = Related ILO MLC 2006, Title 3 requirement			HAB requirements	HAB+ requirements	HAB++ requirements
		• stretcher transportation from accommodation areas or work areas	●		
	5	Medical accommodation area is suitably separated from other spaces and used for the care of the sick and for no other purpose	●		
	6	Medical accommodation area is suitably designed so that the occupant can be comfortably housed and may receive proper attention in all weather	●		
	7	The arrangement of the entrance, berths, lighting, ventilation, heating and water supply is designed to provide comfort and facilitate the treatment of crew members	●		
Outfitting					
	8	If the vessel does not carry a doctor, it must have an approved medicine chest with readily understandable instructions	●		
#	9	Medical accommodation area has a toilet, washbasin and bathtub or shower conveniently situated for the use of patients	●		

(continued)

Table 31.9 (continued)

Accommodation area criteria					
Medical					
Requirement # = Related ILO MLC 2006, Title 3 requirement			HAB requirements	HAB+ requirements	HAB++ requirements
	10	The door to the toilet and washbasin should open outwards and any door latch should be capable of being opened from the outside if equipped	●		
	11	Medical accommodation area is appropriately equipped to render the necessary medical attention	●		
	12	Medical accommodation area is equipped with stretchers that meet the following:			
		• the ability to winch a sick person into a helicopter or vessel, and the capability of floating and righting itself in the water	●		
		• collapsible and suitable for use in confined spaces	●		
#	13	Toilet accommodation area is provided for the exclusive use of the occupants of the hospital accommodation area	●		
	14	Hospital and medical/ first aid accommodation area is situated and arranged so that a stretcher can be easily carried into it and placed alongside an examination table or bed		●	

(continued)

Table 31.9 (continued)

Accommodation area criteria					
Medical					
Requirement # = Related ILO MLC 2006, Title 3 requirement			HAB requirements	HAB+ requirements	HAB++ requirements
	15	Medical accommodation area has lockable storage for drugs, dressings and medical equipment	●		
	16	Medical accommodation area has an intercom or signalling system (e.g., an emergency call device) to the navigation bridge	●		

Table 31.10 ILO MLC 2006 Title 3 requirement (accommodation area criteria (ship's offices)

Accommodation area criteria					
Ship's offices					
Requirement # = Related ILO MLC 2006, Title 3 requirement			HAB requirements	HAB+ requirements	HAB++ requirements
General					
	1	1 In vessels of over 3000 GT:			
#	2	One room is provided and equipped for use as an office for the deck department	●		
#	3	One room is provided and equipped for use as an office for the engine department	●		

Notes

1. If permitted by the vessel's Flag State Administration
2. Per person
3. Per person
4. In this context, near is defined as being on the same deck or on the deck below (or above) adjacent to the stairway
5. In this context, easy access is defined as being on the same deck or on the deck below (or above) adjacent to the stairway

Chapter 32
Vibration

General

Working and/or living on board vessels imposes a series of generally low-frequency mechanical vibration, as well as single-impulse shock loads on the human body. Low-frequency vibrations are also imposed by vessel motions, which are produced by the various sea states in conjunction with vessel speed. These motions can result in motion sickness, body instability, fatigue, and increased health risk aggravated by shock loads induced by vessel slamming. Vessel slamming may be caused by dynamic impact loads being exerted on the vessel's bottom or bow flare due to vessel size, speed, and wave conditions. Higher-frequency vibration influencing comfort is often associated with rotating machinery. The imposition of higher frequency vibrations (about 1–80 Hz) induces corresponding motions and forces within the human body, creating discomfort and possibly resulting in degraded performance and health (Griffin, 1990). This chapter provides the criteria and methods for assessing whole-body vibration relating to habitability on board vessels. The criteria were selected to limit potential vibration related interference with work tasks and to improve crew comfort. Consideration of the vibration loads imposed on the body is restricted to motions transmitted from surrounding structures to the entire human body through the feet of a standing person in the frequency range 1–80 Hz. Motions transmitted to the body of a seated or recumbent person have been omitted from this chapter. Due to the provision of resilient or non-rigid surfaces on seats and beds, these surfaces will generally attenuate the transfer of vibration to levels that are lower than those experienced when standing. The motions transmitted through the feet are expected to be the highest vibration levels to which crew will be exposed. This chapter applies to manned spaces as defined in Chap. 30. Examples of such spaces include accommodation spaces (e.g., cabins, mess areas, recreation spaces, etc.) and workspaces (e.g., galley, control rooms, duty stations, etc.) both inside and, in other vessel locations, outside the accommodation block. Whole-body vibration limits defined in this chapter are based on currently available standards. Compliance with the guidance

A. A. Olsen, *Applying Physical Ergonomics to Modern Ship Design*,
https://doi.org/10.1007/978-3-031-57974-5_32

discussed in this Chapter is typically a prerequisite for the ship Habitability notation assigned by Class (e.g., **HAB**, **HAB+**, and **HAB++** for ABS Classed vessels). In the case of High-Speed Craft (as defined by Class), crew exposure to impact shock can be assessed according to the requirements expressed in the section on 'Evaluating crew exposure to shock on high-speed craft'. The results and findings of the methods presented in the section on 'Evaluating crew exposure to shock on high-speed craft' impose no requirements for acquiring a notation; rather, the results are intended to identify threats to mariner health and comfort so as to aid the vessel or craft owner in identifying means for controlling or mitigating those threats.

Terminology

Acceleration: The rate of change of velocity over time (i.e., metres-per-second, squared (m/s^2).

Calibration checks: Field calibration of a measuring instrument conducted before and after a field test, using a reference calibrated signal. Field calibrators (portable calibrators) should comply with an international or national standard.

Exposure action value: The value of vibration/repetitive shock above which controlling the whole-body vibration exposure to humans is recommended.

Exposure limit value: A value of vibration/repetitive shock above which humans should not be exposed.

Frequency: The number of complete cycles of a periodic process occurring per unit time. Frequency is expressed in Hz, which corresponds to the number of cycles observed-per-second.

Frequency weighting: A transfer function used to modify a signal according to a required dependence on vibration frequency.

- In human response to vibration, various frequency weightings have been defined in order to reflect known or hypothesised relationships between vibration frequency and human response.
- The frequency weighting used to evaluate whole-body vibration in this chapter is W_m (whole-body) for all three axes (x, y, and z), in accordance with ISO 6954.

Multi-axis acceleration value: The multi-axis acceleration value is calculated from the root sums of squares of the weighted RMS acceleration values in each axis (a_{xw}, a_{yw} and a_{zw}) at the measurement point using the following expression:

$$a_w = \sqrt{a_{xw}^2 + a_{yw}^2 + a_{zw}^2}$$

where a_{xw}, a_{yw} and a_{zw} are the weighted RMS acceleration values measured in the x-, y- and z-axes, respectively.

Multi-axis vibration: Mechanical vibration or shock acting in more than one direction simultaneously.

Reference calibration: Calibration of a measuring instrument and of a portable calibrator conducted at least every two years (or more frequently if specified by the manufacturer) by an accredited Testing and Calibration Laboratory accredited according to ISO 17025:2005, as amended, with traceability to a national or international standard.

Velocity: The rate of change of distance over time (i.e., millimetres per second, mm/s).

Vibration: The variation with time of the magnitude of a quantity which is descriptive of the motion or position of a mechanical system, when the magnitude is alternately greater and smaller than some average value.

Weighted root-mean-square acceleration value (a_w): The weighted RMS acceleration, a_w, in metres-per-second squared, is defined by the expression:

$$a_w \sqrt{\frac{1}{T}\int_0^T a_w^2(t)dt}$$

where $a_w(t)$ is the weighted acceleration as a function of time in metres-per-second squared (m/s^2) and t is the duration of the measurement in seconds.

Whole-body vibration: Mechanical vibration (or shock) transmitted to the human body as a whole. Whole-body vibration is often due to the vibration of a surface supporting the body.

Associated Documentation

The following documents provide details about test plan preparation, test measurement procedures and/or test reporting:

- ISO 6954:2000, Mechanical Vibration and Shock—Guidelines for the Measurement, Reporting and Evaluation of Vibration with regard to Habitability on Passenger and Merchant Ships.
- ISO 2631-2:2003, Mechanical Vibration and Shock—Evaluation of Human Exposure to Whole Body Vibration—Part 2, Vibration in Buildings.
- ISO 8041:2005, Human Response to Vibration—Measuring Instrumentation.
- ISO 5348:1998, Mechanical Vibration and Shock—Mechanical Mounting of Accelerometers.
- WMO: Manual on Codes, No. 306, Part A, Alphanumerical Codes (2011 Edition, Updated in 2014).

Further guidance can be found in:

- ISO 20283-2:2008, Mechanical Vibration—Measurement of Vibration on Ships—Part 2: Measurement of Structural Vibration.

Criteria

The whole-body vibration criteria for the ship Habitability notations (e.g., **HAB**, **HAB+**, or **HAB++**) are provided in Table 32.1, Maximum weighted root-mean-square acceleration level. The severity of the vibration is indicated by the weighted root-mean-square acceleration value (a_w) as defined in ISO 8041. Whole-body vibration measurements should only be taken in manned spaces. Specific locations are referred to in the section on 'Measurement locations'. The maximum vibration levels for accommodation areas and workspaces in Table 32.1, Maximum weighted root-mean-square acceleration level should not be exceeded under normal operating conditions. For instance, the **HAB** notation's maximum vibration level is primarily aimed at reducing discomfort and interference with task performance whereas the more stringent maximum level for the **HAB+** and **HAB++** notations are aimed at improving performance and comfort. In this instance, "comfort" means the ability of the crew to use a space for its intended purpose with minimal interference or annoyance from whole-body vibration. For the purposes of this chapter, the notation applies to the vibration levels occurring on the deck supporting the human body in the three translational (x-, y- and z-) axes as shown in Fig. 32.1, Measurement axes. The vibration levels are computed for each axis individually, as well as combined as a multi-axis acceleration value. Each is expressed as a frequency weighted root-mean-square (a_w) value. To meet the vibration criteria, the multi-axis a_w level must be less than or equal to the maximum level expressed in Table 32.1:

Table 32.1 Maximum weighted root-mean-square acceleration level

Notation	Frequency range (Hz)	Acceleration measurement	Maximum RMS level	
			Accommodation areas	Workspaces
HAB	1–80	a_w	178 mm/s^2 (5.0 mm/s)	214 mm/s^2 (6.0 mm/s)
HAB+	1–80	a_w	143 mm/s^2 (4.0 mm/s)	178 mm/s^2 (5.0 mm/s)
HAB++	1–80	a_w	107 mm/s^2 (3.0 mm/s)	143 mm/s^2 (4.0 mm/s)

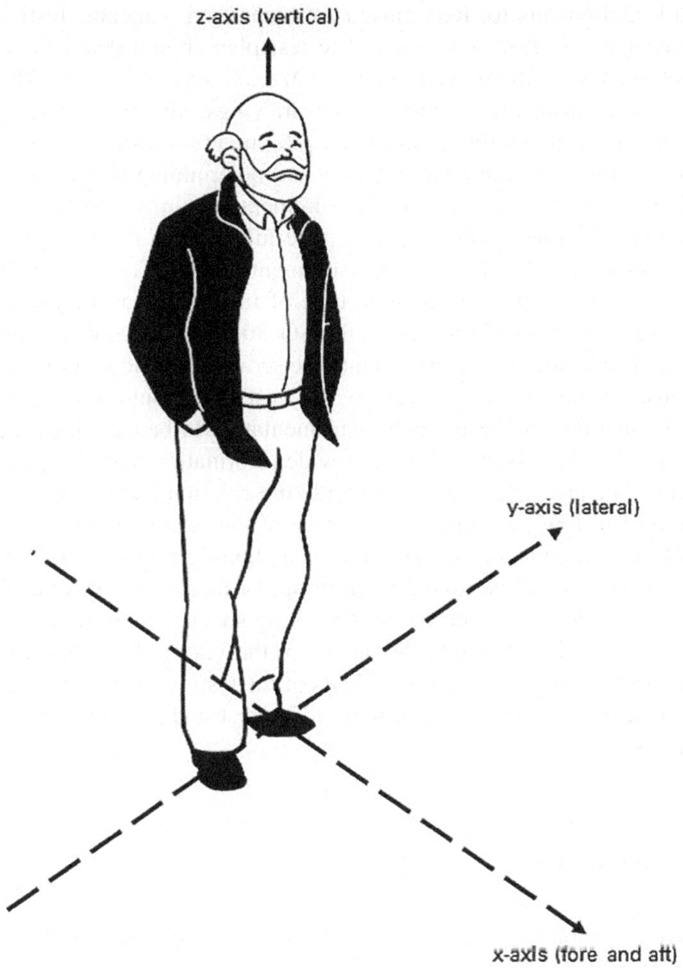

Fig. 32.1 Measurement axes

Test Plan

As stated in the section 'Test plans' in Chap. 30, a test plan should be developed to serve as the principal means for verifying the measurements to be performed to verify compliance with whole-body vibration criteria. This test plan should typically include the following sections. (1) *Documentation.* The test plan should include appropriate drawings indicating the location of all vibration sources. (2) *Test personnel.* The test plan should provide information about the testing specialist who will be conducting the test and their approval and certification in accordance with the Class specific

Procedural Requirements for Recognised Ambient Environmental Testing Specialists or equivalent. (3) *Test conditions.* The test plan should detail the conditions under which the tests will be performed. (4) *Measurement locations.* The test plan should document, in detail, on appropriate drawings, all spaces where measurements will be taken. In addition, transducer placement positions shall be indicated. Details on selecting measurement locations and determining transducer placement positions are provided in the section 'Measurement locations'. (5) *Data acquisition and instruments.* The test plan should provide information regarding the methods and instrumentation to be used for measurement and data collection. Instrumentation specification details shall include type of instruments to be used, accuracy, calibration, sensitivity, conformance with ISO 8041, and frequency range. More details on data acquisition and instruments are provided in the section 'Data acquisition and instruments'. (6) *Data analysis.* The test plan should provide information regarding the methods, software, and instrumentation to be used for data analysis. (7) *Test schedule.* The test plan should provide information regarding the proposed test schedule. The test schedule shall include information on the approximate date and duration of testing (including a summary of the scheduled measurements per day). (8) *Data collection sheets.* The test plan should provide sample data sheets (in table format) that will be used for reporting the measurement data. (9) *Conflict of interest declaration.* In the case that the testing specialist has, in any capacity or to any extent, provided input into the design of the vessel, their participation shall be indicated in the test plan including details of the testing specialist's involvement. Class may review any potential conflicts of interest and determine if any further action is necessary.

Test Requirements

Whole-body vibration measurements should be carried out in accordance with the procedures described in ISO 6954. When the requirements or procedures proscribed by Class deviate from those mentioned in ISO 6954, then the requirements or procedures proscribed by Class will take precedence unless otherwise stipulated. Data acquisition and instruments. For Class Habitability notations such as **HAB**, **HAB+**, or **HAB++**, a sample of data should be recorded for each whole-body vibration measurement position. Each whole-body vibration measurement sample should be at least 60 s in duration. For each location measured, a data sample should also be taken in accordance with the requirements set out in the section 'Test conditions'. The above measurement samples should all be taken using the appropriate Type 1 instrumentation (ISO 8041), then frequency weighted and analysed in accordance with ISO 6954. It is always desirable to employ equipment that records and stores acceleration time histories. *Test conditions.* The test conditions required for the whole-body vibration measurements should be in accordance with the following provisions: (1) *Power output.* The propulsion machinery shall run at 80% of the maximum continuous rating (MCR). Controllable pitch and Voith-Schneider propellers, if any, shall be in the normal seagoing position. For ships with special propulsion and power configurations, such as diesel-electric systems, the ship's design or operating parameters as defined in the ship's specifications (i.e., contractual service conditions) will be used (Fig. 32.2).

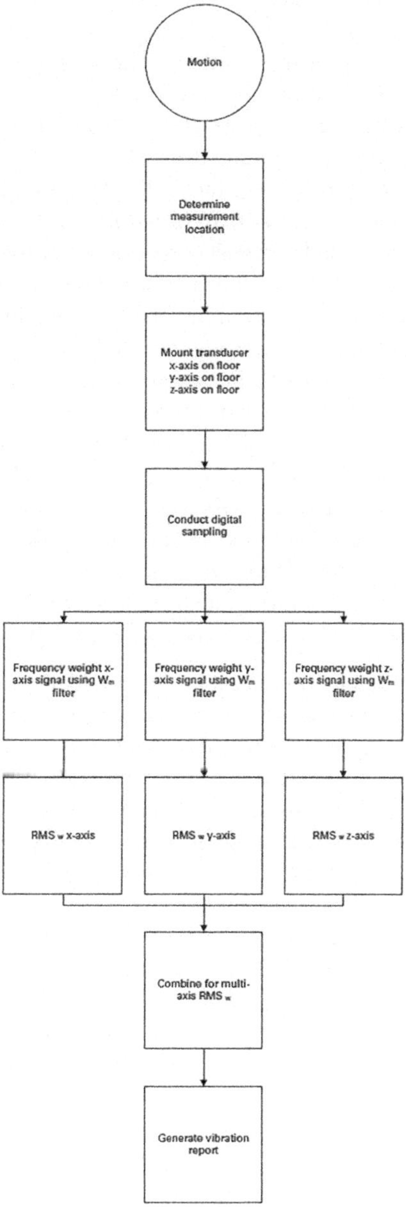

Fig. 32.2 Process for the measurement and analysis of whole-body vibration

(2) *Machinery and equipment operation.* All machinery essential for vessel operation shall operate under normal conditions throughout the measurement period. HVAC systems are to be running as for normal seagoing conditions during the whole-body vibration measurements. (3) *Course and water depth.* Whole-body vibration measurements are to be taken with the vessel in a depth of water not less than five times the draft of the vessel. For vessels that do not operate in water depths of five times draught, measurements should be taken under normal operating and steady state transit conditions. The vessel should maintain a single heading and a constant speed during the test. (4) *Rudder conditions.* Rudder action should be minimised wherever possible. (5) *Sea conditions.* Measurements are to be taken under conditions of Sea State 3 or less, as defined by the World Meteorological Organisation (WMO) Manual on Codes, No. 306, Part A, Alphanumerical Codes (2011 Edition, updated in 2019). (6) *Loading conditions.* For cargo vessels, measurements should be taken under ballast conditions. The loading condition of a passenger vessel should be as close as possible to normal operating conditions. Where this is not practicable, the loading condition may be recorded in the whole-body vibration test report. (7) *Test interference.* During the whole-body vibration measurements, vibration arising from every kind of unnecessary human activity should be avoided. For this reason, only the personnel needed for the normal operation of the equipment in the space and those carrying out the measurements may be present in the space being tested. *Measurement locations. Selection of spaces where measurements are to be conducted.* The aim when selecting vibration measurement locations is to obtain a representative sample of data that reflects the actual conditions in manned spaces. For practical reasons, it is important to select the locations such that an appropriate amount of data can be collected during the testing phase. This means the measurement locations should be selected in accordance with the following criteria:

- Select potential worst-case locations based on their proximity to vibration emitting sources such as propulsion or other rotating machinery or where vibration is likely to be transmitted to manned spaces, accommodation areas and recreation areas via the vessel's structure. Measurements should be taken in all identified worst-case locations (e.g., cabin adjacent to a machinery space).
- Where a single instance of one type of manned space exists within the vessel (e.g., the bridge, mess room, gymnasium, library, etc.), that location should be selected for measurement.
- Select a representative sample of crew cabins throughout the vessel. For vessels with less than 20 cabins, 50 percent of cabins on each deck should be selected. For vessels with greater than 20 cabins, 30 percent of cabins on each deck should be selected. These measurement locations must be selected at locations port, starboard, fore, amidships and aft. The worst-case locations are to be considered part of the representative sample for crew cabins, where applicable.
- Where multiple instances of the same types of accommodation space exist, but are not crew cabins, a representative sample of at least 50 percent of each type should be selected for measurement. The worst-case locations are to be considered part of the representative sample, where applicable.

Walk through verification inspection locations. All normally manned spaces are to be subject to a walk-through inspection by the Class approved surveyor. The number and locations of the walk-through inspections are generally determined by the surveyor. The purpose of the walk-through verification is to subjectively assess the vibration qualities. At the discretion of the surveyor, additional measurements may be required. *Transducer placement positions.* Vibration transducers (accelerometers) should be located and attached properly to the floor surface to measure the vibration at the interface between the standing crew member and the source of vibration. The mounting of accelerometers must comply with ISO 5348. When the vibration enters the human body from a non-rigid or resilient material (e.g., floor covering), secure the transducers with a suitably formed mount that does not alter the pressure distribution on the surface of the floor covering. In cabins, the vibration transducers must be placed on the deck in the centre of the space. ***Note****: this location may not provide the maximum vibration levels for this particular space.* The objective is to minimise the number of measurements yet still obtain a fair and representative sample of the exposure conditions of the person occupying the cabin. For larger spaces (public rooms, messes, recreation areas, etc.) it may be necessary to place transducers at a number of locations in order to obtain a representative sample of the whole-body vibration levels for that space. Transducer locations should be evenly distributed throughout the space. For a specific room size, the minimum number of measurement locations are to be as indicated in Table 32.2, Distribution of transducer positions within spaces.

For vessels without crew cabins, transducers should be located at standing positions normally occupied by crew near the port side, near the centreline and near the starboard side. Transducers must be evenly distributed fore and aft. Transducers located at one measurement position should be orthogonally positioned (i.e., positioned at a 90° angle) to measure whole-body vibrations in the vertical, longitudinal and transverse axes. Translational accelerometers oriented in different axes at a single measurement position should be as close together as possible.

Table 32.2 Distribution of transducer positions within spaces

Space size	Minimum number of measurement positions in room
Less than 20 m^2 (215 ft^2)	1
> 20–40 m^2 (215–431 ft^2)	2
> 40–80 m^2 (431–861 ft^2)	3
> 80–120 m^2 (861–1291 ft^2)	4
> 120–200 m^2 (1291–2150 ft^2)	5
Greater than 200 m^2 (2150 ft^2)	6

Test Report

As stated in the section on 'Test reports' in Chap. 30, a test report must be submitted to the Class approved surveyor to determine whether the vibration levels meet the whole-body vibration criteria and whether this part of the Class notation requirement has been met. The details listed in the following paragraphs are required to be provided in the whole-body vibration test report. *Test details.* The following details are required to be recorded for each period of testing:

- Loading conditions (mean draught and trim).
- Number of crew and other persons on board during tests.
- Power output.
- Vessel course and speed.
- Average water depth under keel.
- Sea state.
- Direction of swell relative to vessel heading.
- Any indications of abnormal activity during the test that might skew results.
- Machinery and equipment operated during the test.

Transducer placement positions. The actual transducer placement positions must be indicated on appropriate drawings. *Measurement equipment details.* Details of measurement and analysis equipment (e.g., manufacturer, type and serial number, accuracy and resolution), including frequency analysis parameters (e.g., resolution, averaging time, and filtering), must be provided. Copies of the relevant instrumentation reference calibration certificates, together with the results of field setup and calibration checks before and after the field tests, should also be provided with the report. *Results.* The following results per measuring location, should be provided in a tabulated format:

- Measurement position.
- Measurement period if different from requirements.
- Sample number.
- Multi-axis weighted RMS values.
- Equipment operating in proximity to the measurement position.
- Indication of pass/fail.

For all the remaining spaces that were checked through walk through verification inspection and spot check measurements, the following information should also be provided:

- Name and number of spaces.
- Walk through inspection observations.
- Measurement results if necessary.

Deviations. All deviations from the approved test plan must be reported. *Surveyor witnessing documentation.* The equipment field calibration and data collection process of vibration tests conducted at sea must be witnessed by a Class approved surveyor. The surveyor is required to sign or initial each page of the test report and prepare a witnessing document stating whether all steps of the vibration testing were completed to their satisfaction. A copy of the witnessing document is then given to the person conducting the onboard testing, for insertion into the final whole-body vibration test report. The original is usually retained by the surveyor and submitted to Class for inclusion in the vessel's Record. *Results.* On completion of the test, the whole-body vibration test report will be reviewed by the Class approved surveyor against the appropriate Class Habitation notation requirements. Where the requirements have been satisfactorily met, the vessel will be assigned the appropriate Class Habitation notation.

Evaluating Crew Exposure to Shock on High-Speed Craft

This section provides procedures and criteria for the evaluation and assessment of vessel motions (vibration and shock) for crew health (and to a lesser degree, comfort). The results of the methods and findings below impose no requirements for acquiring a notation; rather, the results are intended to identify threats to mariner health and comfort so as to aid the vessel owner in identifying means to control or mitigate those threats. Vibration and shock on high-speed craft should be evaluated using the vibration dose values (VDV) and assessed relative to a limiting exposure value of 9.1 $m/s^{1.75}$. Guidance on methods of measuring whole-body vibration at a workplace with reference to health can be found in BS EN 14253. *Measurement.* Measurements should be made using the VDV method defined in International Standard ISO 2631-1. *Location of measurements.* Measurements should be made at a distribution of locations representative of crew locations in the vessel as noted below. These should encompass the different types of seating installed on the vessel and locations both near to the centre of the craft and closest to the forward, aft, port and starboard extremities. Vibration and shock should be measured in the three orthogonal axes on the seat. For seated operations, vibration should be measured on the seat. Where suspension seats are provided, appropriate adjustments should be made for the weight of occupants. Care should be taken to avoid artifacts due to intentional movements of a seated person (e.g., when entering, leaving or moving in the seat), which can result in inflated estimates of vibration magnitudes. Inspection of acceleration time histories can be helpful to identify these events (BS EN 14253 provides additional information and examples). In locations where crew members stand on the deck, the vibration should be measured on the standing surface closely adjacent to the area of contact between the feet and that surface.

Duration of measurements. Measurements should produce vibration dose values that are representative of the shock and vibration throughout the working day of a crew member. The operating conditions during the measurements should be selected to be representative of all typical conditions (e.g., loading, speed, heading, sea state) that occur during a typical watch standing period. Measurements may be made over short periods that are representative of the different operating conditions that occur during a typical day. A predicted VDV should then be calculated from the magnitude and duration for each axis and for each condition. The partial vibration values are combined to give the overall daily VDV. *Measurement instrumentation.* Whole-body vibration measuring equipment should comply with ISO 8041, Human response to vibration—measuring instrumentation, and be in current calibration traceable to national or international standards. The accelerometers should have a flat frequency response between 0.25 and 160 Hz, and the accelerometers and measuring instrument should have a sufficient dynamic range to encompass the range of accelerations that could occur at the measurement location. High peak accelerations can occur on high- speed craft and could result in an overload unsuitable for transducers or an instrument with an inappropriate range setting. Vibration should be measured in three axes using either a tri-axial accelerometer or three single-axis accelerometers orientated in orthogonal directions. Calibration checks should be performed before and after measurements. In locations where crew members are seated, measurements should be made in three axes on seat surfaces using accelerometers moulded into a flexible disc that complies with ISO 10326-1. The centre of this disc should be positioned under the sitting bones of the seated person. In locations where the crew stands on the deck, the vibration should be measured directly on the supporting surface. The accelerometers may be mounted in the middle of a rigid plate in an area where the supporting surface is not rigid, with a person standing on the plate (refer to BS EN 14253). *Evaluation and assessment of vibration.* The vibration dose value (*VDVj*) for axis *j* (where $j = x$, y or z) is computed from:

$$VDV_j \left(\int_0^T \alpha_{wj}^4(t)dt \right)^{\frac{1}{4}}$$

Here, $\alpha_{wj}^4(t)$ is the frequency-weighted acceleration in axis j as a function of time t, and T is the duration of the exposure. When considering the risks to health from whole-body vibration and shock, ISO 2631-1 applies a weighting factor to the vibration dose values. For the x- and y-axes, the acceleration values are multiplied by 1.4. For the z-axis, the multiplying factor is 1.0. The vibration dose value to be used for the assessment of risk is the greatest of the frequency-weighted VDV in any of the three orthogonal axes after multiplication by the appropriate weighting factor:

$$VDV = \max\left[1.4\ VDV_x,\ 1.4\ VDVy,\ VDV^z\right]$$

A seafarer's daily VDV can be compared to exposure limits in the table below:

Value of vibration/repetitive shock above which controlling the whole-body vibration exposure to humans is recommended	9.1 $m/s^{-1.75}$
Value of vibration/repetitive shock above which humans should not be exposed	21 $m/s^{-1.75}$

Where exposures are found to be above these values, means to implement measures to control the risks of whole-body vibration to workers should be considered.

Chapter 33
Noise

General

A large amount of research has been performed on the effects of noise on humans. Established or commonly used criteria exist for the effects of noise on speech communication, hearing loss, sleep, concentration, and "annoyance". These have provided a basis for the guidance provided in this chapter. A detailed discussion of the effects of noise on human performance, health, and comfort may be found in Kryter (1994) The Handbook of Hearing and the Effects of Noise: Physiology, Psychology and Public Health. In this chapter, noise criteria have been selected to improve crew performance and to facilitate communication and sleep in appropriate vessel spaces. An additional goal is to enhance crew safety and comfort. In this instance, "comfort" means the ability of the crew to use a space for its intended purpose with minimal interference or annoyance from noise. The noise criteria presented in this chapter is lower than the levels commonly associated with hearing loss. Further guidance with respect to hearing conservation is provided in the IMO Resolution MSC.337(91): Code on Noise Levels Onboard Ships and should be followed for noise levels and exposure duration, particularly for areas with noise levels in excess of 85 dB(A). This chapter applies to manned spaces as defined in Chap. 30. Examples of such spaces include accommodation spaces (e.g., cabins, mess areas, recreation spaces, etc.) and workspaces (e.g., galley, control rooms, duty stations, etc.) both inside and, in other vessel locations, outside the accommodation block. Compliance with the guidance discussed in this Chapter is typically a prerequisite for the ship Habitability notation assigned by Class (e.g., **HAB**, **HAB+**, and **HAB++** for ABS Classed vessels).

A. A. Olsen, *Applying Physical Ergonomics to Modern Ship Design*,
https://doi.org/10.1007/978-3-031-57974-5_33

Terminology

A-weighted sound pressure level: The magnitude of a sound, expressed in decibels (i.e., 20 micropascals); the various frequency components are adjusted according to the A-weighted values given in IEC 61672-1 (2004) in order to account for the frequency response characteristics of the human ear. The symbol is LA; the unit is dB(A). The measurement L_{Aeq} is an equivalent continuous A-weighted sound pressure level, measured over a period of time.

Calibration checks: Field calibration of a measuring instrument conducted before and after a field test, using a reference calibrated signal or through zero calibration. Sound calibrators should comply with the standard IEC 60942:2003, as amended, type/class (1) standard and should be approved by the manufacturer of the measuring instrument used.

Equivalent continuous a-weighted sound pressure level: The A-weighted sound pressure level of a noise fluctuating over a period of time T, expressed as the amount of average energy. The symbol is L_{Aeq}; the unit is dB(A).

Reference calibration: Calibration of measuring instrument and sound calibrator, conducted at least every two (2) years (or more frequently if specified by the manufacturer) by an accredited Testing and Calibration Laboratory accredited according to ISO 17025:2005, as amended, with traceability to a national or international standard.

Associated Documentation

The following documents provide details about test plan preparation, test measurement procedures and/or test reporting:

- ISO 2923:1996, Acoustics—Measurement of Noise Onboard Vessels.
- IEC 61672-1:2013, Electroacoustics—Sound Level Meters—Part 1: Specifications.
- IEC 60942:2003, Electroacoustics—Sound Calibrators.
- IMO Resolution MSC.337(91) (2012), Code on Noise Levels Onboard Ships.
- ISO 10140-2:2010, Acoustics—Laboratory Measurement of Sound Insulation of Building Elements—Part 2: Measurements of Airborne Sound Insulation.
- ISO 717-1:2013, Acoustics—Rating of Sound Insulation in Buildings and of Building Elements—Part 1: Airborne Sound Insulation.
- WMO: Manual on Codes, No. 306, Part A, Alphanumerical Codes (2011 Edition, updated in 2019).

Criteria

Equivalent continuous a-weighted sound pressure level. The noise criteria for the Class Habitability notations (e.g., **HAB**, **HAB+**, and **HAB++**) are provided in Table 33.1, Noise criteria for ships. Noise levels are determined for the test conditions specified in the section 'Test conditions' and must not exceed the maximum acceptable L_{Aeq} levels indicated in Table 33.1, Noise criteria for ships, for each type of space or area. The maximum acceptable noise levels given in Table 33.1, Noise criteria for ships are L_{Aeq} values, determined as appropriate to the character of the noise (refer to the section on 'Measurement procedures' and recorded results in this chapter). These limit values are for sea-going conditions only and do not include any in-port activities. Acoustic insulation. The airborne sound insulation properties for bulkheads and decks within the accommodation must comply with—at a minimum—the following weighted sound reduction index (R_w) according to ISO Standard 717-1:2013:

Cabin to cabin	$R_w = 35$
Mess rooms, recreation rooms, public spaces and entertainment areas to cabins and hospitals	$R_w = 45$
Corridor to cabin	$R_w = 30$
Cabin to cabin with communicating door	$R_w = 30$

The airborne sound insulation properties should be determined by laboratory tests in accordance with ISO 10140-2:2010 as identified in the IMO Code on Noise. The manufacturer of the acoustic insulation should provide the weighted sound reduction index (R_w) value and evidence of the laboratory test.

Test Plan

As stated in Chap. 30 under the section 'Test plans', a test plan must be developed to serve as the principal means for verifying the measurements to be performed to demonstrate or confirm compliance with noise criteria. The test plan should include the following:

- *Documentation.* The test plan should include appropriate design information including noise specifications for the vessel. It should also include layout drawings indicating the locations of all noise sources and noise generating equipment. The information should be of such detail to enable the Class engineer to verify compliance with the criteria stipulated by Class. The test plan should also include the acoustic insulation plan and relevant supporting documentation for review and approval. With the test plan, the acoustic insulation test results (refer to the section on 'Acoustic insulation') should be submitted for review. The test plan must be submitted to Class Engineering for their review and approval.

Table 33.1 Noise criteria for ships

Space[a,b]	Maximum acceptable noise L_{Aeq} level dB(A)					
	HAB		HAB+		HAB++	
	Vessel size (gross tonnage)					
	< 10,000	≥ 10,000	< 10,000	≥ 10,000	< 10,000	≥ 10,000
Crew accommodation spaces and open deck recreation areas						
Cabins	60	55	55	55	50	50
Sanitary spaces (if separate from the cabin)	65	65	60	60	55	55
Dining (mess) spaces	65	60	60	55	55	50
Indoor recreation spaces	65	60	60	55	55	50
Gymnasiums	65	60	60	55	60	55
Medical and first aid centre	60	55	55	55	55	50
Open deck recreation areas	75	75	70	70	70	70
Navigation and control spaces	65	65	60	60	55	55
Wheelhouse, pilothouse, bridge	65	65	60	60	55	55
Radio room	60	60	55	55	55	55
Offices	65	60	60	55	60	55
Cargo control rooms[c]	65	65	60	60	60	60
Machinery control rooms	75	75	70	70	70	70
Service spaces						
Food preparation (e.g., galley, scullery)	75	75	70	70	65	65
Pantries	75	75	70	70	70	70
Storerooms	75	75	70	70	70	70
Laundry areas	75	75	70	70	70	70
Operating and maintenance spaces						
Machinery spaces	110	110	110	110	110	110
Workshops (other than those forming part of machinery spaces)	85	85	80	80	80	80
Non-specified workspaces[d] (other workspaces)	85	85	80	80	80	80

[a] In any manned space with noise levels above 85 dB(A), hearing protection should be worn in accordance with appropriate IMO regulations
[b] Unless this space is normally manned, no confirmatory measurements for the Class Habitability notations are required
[c] If located within machinery spaces, it is to comply with the machinery control room's criteria
[d] Examples of non-specified workspaces are open deck workspaces that are not machinery spaces, and open deck workspaces where communication is relevant

- *Test personnel.* The test plan should provide information about the testing specialist who will be conducting the test and their approval and certification in accordance with the Class specific Procedural Requirements for Recognised Ambient Environmental Testing Specialists or equivalent.
- *Test conditions.* Measurements are to be taken under conditions of Sea State 3 or less, as defined by the WMO Sea State Code.
- *Measurement locations.* The test plan should document, in detail, on appropriate drawings, all spaces where measurements will be taken. In addition, transducer placement positions should be indicated. Details on selecting measurement locations and determining transducer placement positions are provided in the section 'Measurement locations',
- *Data acquisition and instruments.* The test plan should provide information regarding the methods and instrumentation to be used for measurement and data collection. Instrumentation specification details should include the type of instruments to be used, their accuracy, calibration, and sensitivity. More details on data acquisition and instruments are provided in the section 'Data acquisition and instruments'.
- *Data analysis.* The test plan should provide information regarding the methods, software, and instrumentation to be used for data analysis.
- *Test schedule.* The test plan should provide information regarding the proposed test schedule. The test schedule should include information on the approximate date and duration of testing (including a summary of the scheduled measurements per day).
- *Data collection sheets.* The test plan should provide sample data sheets (in table format) that will be used for reporting the measurement data.
- *Conflict of interest declaration.* In the case that the testing specialist has, in any capacity or to any extent, provided input into the design of the vessel, their participation should be indicated in the test plan including details of the testing specialist's involvement. Class may review any potential conflicts of interest and determine if any further action is necessary.

Test Requirements

In general, the noise measurements should be carried out in accordance with the requirements of IMO Resolution MSC.337(91) Code on Noise Levels Onboard Ships. However, when the requirements or procedures proscribed by Class deviate from those mentioned in IMO Resolution MSC.337(91), then the requirements or procedures proscribed by Class will take precedence unless otherwise stipulated. Data acquisition and instruments. The integrating-averaging sound level meter should meet the requirements for a Type 1 instrument specified in IEC 61672-1. For each location sampled, a measurement should be taken in accordance with the requirements set forth in the section, 'Test conditions'. Test conditions. The test conditions required for the noise measurements should be in accordance with each of the following sub paragraphs, based on ISO 2923:

- *Power output.* The propulsion machinery should run at 80% of the MCR. Controllable pitch and Voith-Schneider propellers, if any, should be in the normal seagoing position. For ships with special propulsion and power configurations, such as diesel-electric systems, the actual ship's design or operating parameters as defined in the ship's specifications (i.e., contractual service conditions) should be used.
- *Machinery and equipment operation.* All machinery essential for vessel operation should operate under normal conditions throughout the measurement period. HVAC systems are to be running as for normal seagoing conditions during the noise measurements. For the galley/scullery, all equipment that runs for 20 min or more (e.g., ventilation hoods) should be running during the noise measurements. Equipment that is used only intermittently (e.g., blenders/mixers/meat cutters) do not need to be running during the noise measurements. Note any equipment that is running in the noise test report.
- *Course and water depth.* Noise measurements are to be taken with the vessel in a depth of water not less than five times the draught of the vessel. For vessels that do not operate in water depths of five times draught, measurements should be taken under normal operating and transit conditions. The vessel should maintain a single heading and a constant speed during the test.
- *Rudder* conditions. Rudder action should be minimised wherever possible.
- *Sea conditions.* Measurements are to be taken under conditions of Sea State 3 or less, as defined by the WMO Manual on Codes, No. 306, Part A, Alphanumerical Codes (2011 Edition, updated in 2019)
- *Loading conditions.* For cargo vessels, measurements should be taken under ballast conditions. The loading condition of a passenger vessel should be as close as possible to normal operating conditions. If this is not practicable, the loading condition may be recorded in the noise test report.
- *Test interference.* During the noise measurements, noise arising from every kind of unnecessary human activity must be avoided. For this reason, only the personnel needed for the normal operation of the equipment in the space and those carrying out the measurements should be present in the space being tested. Doors and windows must be closed, except where they are normally left open (such as the door on the lee side of the navigation bridge, which may normally be open). Any open doors or windows should be noted in the noise test report. Spaces should be furnished with all usual equipment, fixtures and fittings normally found in the space. Equipment should be configured to operate in its normal operating mode.

Measurement locations. Selection of spaces where measurements are to be conducted. The aim when selecting noise measurement locations is to obtain a representative sample of data that reflects the actual conditions in manned spaces. For practical reasons, it is important to select the locations such that an appropriate amount of sample data can be collected during the testing phase. The measurement locations should be selected in accordance with the following criteria:

1. Select potential worst-case locations based on their proximity to noise emitting sources such as propulsion or other rotating machinery or where noise is likely to be transmitted to manned spaces, accommodation areas and recreation areas via the vessel's structure. Measurements should be taken in all identified worst case locations (e.g., cabin adjacent to a machinery space),
2. Where a single instance of one type of manned space exists within the vessel (e.g., the bridge, mess room, gymnasium, library, etc.), that location should be selected for measurement,
3. Select a representative sample of crew cabins throughout the vessel. For vessels with less than 20 cabins 50% of cabins on each deck should be selected. For vessels with greater than 20 cabins, 30% of cabins on each deck should be selected. These measurement locations must be selected at locations port, starboard, fore, amidships and aft. The worst-case locations are to be considered part as the representative sample for crew cabins, where applicable; and/or
4. Where multiple instances of the same types of accommodation space exist, that are not crew cabins, a representative sample of at least 50% of each type should be selected for measurement. The worst-case locations are to be considered part of the representative sample, where applicable.

Walk through verification inspection locations. All normally manned spaces are to be subject to a walkthrough inspection by the Class approved surveyor. The number and locations of the walkthrough inspections are generally determined by the surveyor. The purpose of the walkthrough verification is to subjectively assess the noise qualities. At the discretion of the surveyor, additional measurements may be required. Measurement positions. The measurement positions described below are taken or adapted from ISO 2923 and IMO Resolution MSC.337(91):

> Measure at positions where persons will be seated or standing. The microphone should be at a height of approximately 1,200 mm (47 in) from the deck to represent seated persons and approximately 1,600 mm (63 in) from the deck to represent standing persons, as appropriate for the measurement position.
>
> In crew accommodation spaces, measure in the middle of the space.
>
> For all measurements, the microphone should not be closer than 500 mm (20 in) from the boundary surface (e.g., bulkhead) of a space. The measurement time should be at least fifteen (15) seconds and should be long enough to enable the measurement of the equivalent continuous A-weighted sound pressure level for any specified time interval within the stated limits of overall measurement uncertainty.
>
> If practicable, do not measure closer than 1,000 mm (39.5 in) from operating machinery, air inlets, or from decks, bulkheads or other large surfaces. Where this is not possible, measure midway between the machinery and an adjacent reflecting surface.

Measurement procedures and recorded results. (1) *Persons present during measurements.* When testing specialist personnel are conducting noise level measurements in any space, only crew members necessary for the operation of equipment or machinery in that space should be present. (2) *Sampling duration.* Equivalent continuous A-weighted sound pressure levels (L_{Aeq}) should be reported for each measurement location. The L_{Aeq} sampling duration should be sufficient to achieve a stable reading. Sampling time should be fifteen (15) seconds or longer (3) *Cyclic noise.* If the noise within a space is cyclic, the L_{Aeq} sampling duration should be sufficient to capture an integer number of complete cycles. If a long-duration sample is judged impractical, an L_{Aeq} value should be determined and reported for the high-noise portion of the cycle. (4) *Intermittent noise.* If the noise within a space is present intermittently, an L_{Aeq} value should be determined and reported for a period of high-level noise. (5) *HVAC related noise.* If HVAC system-related noise is a large contributor to the noise level in the space, a noise measurement should be made approximately 300 mm (12 in) from the vent, measured in line with the direction of airflow and recorded in the noise test report.

Test Report

As stated in Chap. 30 under the section 'Test reports', a test report must be submitted to the Class approved surveyor to determine whether the noise levels are at or below the limits and whether this part of the notation requirement has been met. The details listed below must therefore be provided in the noise test report:

Test details. The following details are required be provided for each period of testing:

- Loading conditions (mean draft and trim).
- Number of crew and number of other persons on board during testing.
- Power output.
- Vessel course and speed.
- Average water depth under keel.
- Sea state.
- Any indications of abnormal activity during the test that might skew results.
- Machinery and equipment operated during the test.

Measurement positions. Actual measurement location positions must be indicated on appropriate drawings. Measurement equipment details. Details of measurement and analysis equipment (e.g., manufacturer, type and serial number, accuracy, sampling frequency and resolution) must be provided. Copies of the relevant instrumentation reference calibration certificates, together with the results of field setup and calibration checks before and after the field tests, should also be provided with the report. Results. The following results per measuring location, should be provided in a tabulated format:

- Measurement position.
- Number of people present in the space at time of measurement.
- Measurement period if different from requirement.
- Equivalent continuous A-weighted sound pressure level (L_{Aeq}).

- Note any open doors and windows.
- Note equipment operating in proximity to the measurement position.
- Note observed direct sources of noise (such as ventilation devices) and any additional measurement data collected.
- Indication of pass/fail.

For all the remaining spaces that were checked through walk through verification inspection and spot check measurements, the following information should also be provided:

- Name and number of spaces.
- Walk through inspection observations.
- Measurement results if necessary.

Deviations. All deviations from the approved test plan must be reported. Surveyor witnessing documentation. The equipment field calibration and data collection process of vibration tests conducted at sea must be witnessed by a Class approved surveyor. The surveyor is required to sign or initial each page of the test report and prepare a witnessing document stating whether all steps of the noise testing were completed to their satisfaction. A copy of the witnessing document is then given to the person conducting the onboard testing, for insertion into the final noise test report. The original is usually retained by the surveyor and submitted to Class for inclusion in the vessel's Record.

Results

On completion of the test, the noise test report will be reviewed by the Class approved surveyor against the appropriate Class Habitation notation requirements. Where the requirements have been satisfactorily met, the vessel will be assigned the appropriate Class Habitation notation.

Chapter 34
Indoor Climate

General

Thermal comfort is defined in ISO 7730 as "... that condition of mind which expresses satisfaction with the thermal environment". The sensation of thermal comfort is therefore largely subjective and will vary from person to person. Due to differences in metabolism and expectations, there are distinct individual differences among people's perception of comfort as a function of temperature, humidity, and other atmospheric characteristics. Acclimatisation, habits, and expectations influence perceived comfort. These individual differences make it difficult to specify a single thermal environment that will be satisfactory to everyone. A thermal environment is therefore typically defined to be acceptable to at least 80% of the occupants of an interior space. Individually, the perception of thermal comfort is largely determined by the interaction of thermal environmental factors such as air temperature, air velocity, relative humidity, and factors related to activity and clothing. The thermal control or HVAC systems on a vessel should be designed to effectively control the indoor thermal environmental parameters to within acceptable limits to facilitate the thermal comfort of the occupants. To that end, this chapter provides the assessment criteria, verification and measurement methodology for indoor climate relating to habitability on ships. The criteria are based on currently available standards and were selected to provide an index of crew thermal comfort. The thermal environmental variables covered in this chapter include the ambient qualities of air temperature, air velocity, and relative humidity. Vertical thermal gradient is used to indicate uncomfortable temperature differentials between a person's head and feet. The crew cabin area horizontal gradient temperature differential between the temperature of inside bulkhead surfaces adjacent to crew cabins and the average air temperature within the space serves as an indication of potential thermal comfort or discomfort. The thermal environmental criteria provided in this chapter are for persons wearing typical indoor clothing occupied with light, primarily sedentary activity and resulting in a thermal environment acceptable to at least 80% of the occupants. This chapter applies to

A. A. Olsen, *Applying Physical Ergonomics to Modern Ship Design*,
https://doi.org/10.1007/978-3-031-57974-5_34

enclosed manned spaces as defined in Chap. 30. Examples of such spaces include crew accommodation areas, bridge, engine control room, and indoor workspaces. Compliance with the provisions outlined in this chapter are generally required to meet the Class Habitability (e.g., **HAB**, **HAB+**, and **HAB++**) notation.

Terminology

Air supply quantity: The total amount of air supplied to a specific space consisting of a percentage of recirculated air and a percentage of fresh air supply quantity, measured in litres per second (l/s).

Air temperature: The temperature of the air surrounding a person, measured with a standard thermometer.

Air velocity or movement: The rate of displacement of ambient air in a specific direction in metres-per-second (m/s) or feet-per-second (ft/s).

Comfort zone: That range of environmental conditions in which at least 80% of crew members experience thermal comfort.

Fresh air supply quantity: The amount of fresh/outdoor air supplied to a specific space, expressed in litres per second (l/s).

Horizontal gradient: The difference between the inside surface temperatures of the bulkheads bounding the berth and the average air temperature within the cabin. This is used as an indication of potential thermal discomfort due to radiant thermal sources.

HVAC zone: A space or group of spaces that is (are) independently controlled for temperature, humidity, and air distribution. A zone usually comprises common duct work fed from an air handler.

Reference calibration: Calibration of a measuring instrument, conducted at least every two years (or more frequently if specified by the manufacturer) by an accredited Testing and Calibration Laboratory accredited according to ISO 17025:2005, as amended, with traceability to a national or international standard.

Relative humidity (RH): The ratio of the amount of vapour contained in the air (absolute humidity) to the maximum amount of vapour the air can hold at a given temperature before precipitation (condensation) occurs.

Thermal comfort: Subjective index of "that condition of mind which expresses satisfaction with the thermal environment".

Ventilation: Ventilation is the process of supplying air to and removing air from any space by natural or mechanical means. From the standpoint of comfort and health, ventilation issues involve both quantity and quality.

Vertical gradient: The vertical air temperature difference within an enclosed space. The vertical gradient is used as an indication of potential local discomfort at the head and feet.

Associated Documentation

The following documents provide details about test plan preparation, test measurement procedures and/or test reporting:

- ANSI/ASHRAE 55-2013, Thermal Environmental Conditions for Human Occupancy.
- ISO 7726:1998, Ergonomics of the Thermal Environment—Instruments for Measuring Physical Quantities.
- ISO 7547:2002, Ships and Marine Technology—Air-Conditioning and Ventilation of Accommodation Spaces—Design Conditions and Basis of Calculations.
- NEBB:2005, Procedural Standards for Testing, Adjusting, Balancing of Environmental Systems.

Criteria

The indoor climate criteria for the ship Habitability notations (e.g., **HAB**, **HAB+**, and **HAB++**) are provided in Table 34.1, Summary of indoor climate requirements. Indoor climate measurements should be taken only in indoor manned spaces. Specific measurement locations are discussed in the section 'Measurement locations'. The thermal environmental comfort ranges and conditions should be achievable, under the test conditions specified in the section on 'Test conditions', for normal operating conditions, in all indoor manned spaces. For vessels with lower and median notations (e.g., **HAB** and **HAB+**), the criteria provides for a range of air temperatures maintained by a temperature controller for each zone and are primarily aimed at HVAC systems that do not make provision for individual adjustment to suit personal preferences and activities within a specific space. Vessels with an advanced notation (e.g., **HAB++**) should aim at enhancing crew comfort by making provisions for the crew to adjust indoor climate conditions, with regard to air temperature to suit personal needs within a particular space. *Air temperature.* For a **HAB** or a **HAB +** notation or equivalent, the HVAC system should be capable of providing an air temperature within the range of 20–25 °C (68–77 °F) during winter months and 23–28 °C (73.5–82.5 °F) during summer months to an HVAC zone. This temperature should be maintained by a temperature controller. Each zone must have a thermostat for reheat and dehumidification purposes. For vessels with an **HAB++** notation, or equivalent, the HVAC system should be capable of sustaining an adjustable range of air temperatures between 20 and 25 °C (68–77 °F) during winter months and 23–28 °C (73.5–82.5 °F) during summer months in indoor manned crew accommodation spaces. This temperature should be maintained by a temperature controller. Each indoor manned crew accommodation space must have its own individual controller for temperature regulation. *Relative humidity.* The HVAC system should be capable of providing and maintaining a relative humidity within a range from 30% minimum to 70% maximum. *Enclosed space vertical gradient.* The difference in temperature at 100 mm (4 in.) above the deck and 1700 mm (67 in.) above the deck should be maintained within 3 °C (6 °F). *Air velocity.* Air velocities should not exceed 30 m-per-minute or 100 ft-per-minute (0.5 m/s or 1.7 ft/s) at the measurement position in the space. *Crew cabin area horizontal gradient.* In crew cabin areas, the difference between the inside bulkhead surface temperature adjacent to crew cabins and the average air temperature within the space should be less than 10 °C (18 °F). *Air supply quantity.* The minimum quantity of fresh/outdoor air supply must be not less than 40% of the total air supplied to a specific space. The fresh/outdoor air supply quantity should not be less than 8 l/s per the number of person(s) for which the specific space is designed for (e.g., for cabins, this is the number of beds; for mess rooms, the number of seats; and for workspaces, the number of workstations, etc.). *Summary.* A summary of the indoor climate requirements is presented in Table 34.1, Summary of indoor climate requirements.

Table 34.1 Summary of indoor climate requirements

Item	Requirement or criterion	
	HAB and HAB+ or equivalent	HAB++ or equivalent
Adjustability	Non-adjustable air temperature	Adjustable range of air temperatures
Air temperature	Winter: 20–25 °C (68–77 °F) Summer: 23–28 °C (73.5–82.5 °F)	
Relative humidity	A range from 30% minimum to 70% maximum	
Vertical gradient	The acceptable range is 0–3 °C (0–6 °F)	
Air velocity	Not exceed 30 m-per-minute or 100 feet-per-minute (0.5 m/s or 1.7 ft/s)	
Horizontal gradient (crew cabin areas)	The horizontal temperature gradient in crew cabin areas should be < 10 °C (18 °F)	
Air supply quantity	The minimum quantity of fresh/outdoor air supply should be not less than 8 l/s per person and in addition will need to be at least 40% of the total air supplied to a specific space	

Test Plan

As stated in the section 'Test plans' in Chap. 30, a test plan should be developed to serve as the principal means for verifying the measurements to be performed to verify compliance with the indoor climate criteria. This test plan should typically include the following:

- *Documentation.* The test plan should include appropriate drawings indicating the location of all indoor climate systems and equipment including the HVAC system design specifications and all schematics/layout drawings of the HVAC system.
- *Test personnel.* The test plan should provide information about the testing specialist who will be conducting the test and their approval and certification in accordance with the Class specific Procedural Requirements for Recognised Ambient Environmental Testing Specialists or equivalent.
- *Test conditions.* The test plan should detail the conditions under which the tests will be performed. Details about test conditions are provided in the section on 'Test conditions'.
- *Measurement locations.* The test plan should document, in detail, on appropriate drawings, all spaces where measurements will be taken. In addition, transducer placement positions should be indicated. Details on selecting measurement locations and determining transducer placement positions are provided in the section 'Measurement locations'.
- *Data acquisition and instruments.* The test plan should provide information regarding the methods and instrumentation to be used for measurement and data collection. Instrumentation specification details should include type of instruments to be used, accuracy, calibration, sensitivity. More details on data acquisition and instruments are provided in the section 'Data acquisition and instruments'.
- *Data analysis.* The test plan should provide information regarding the methods, software, and instrumentation to be used for data analysis.
- *Test schedule.* The test plan should provide information regarding the proposed test schedule. The test schedule should include information on the approximate date and duration of testing (including a summary of the scheduled measurements per day).
- *Data collection sheets.* The test plan should provide sample data sheets (in table format) that will be used for reporting the measurement data.
- *Conflict of interest declaration.* In the event that the testing specialist has, in any capacity or to any extent, provided input into the design of the vessel, their participation should be indicated in the test plan including details of the testing specialist's involvement. Class may review any potential conflicts of interest and determine if any further action is necessary.

Test Requirements

Indoor climate measurements should be carried out in accordance with the procedures described in ANSI/ASHRAE 55-2013. When the requirements or procedures proscribed by Class deviate from those mentioned in ANSI/ASHRAE 55-2013, then the requirements or procedures proscribed by Class will take precedence unless otherwise stipulated. *Data acquisition and instruments.* The thermal measurement instrumentation used must meet or exceed the minimum characteristics of instruments for measuring physical quantities characterising an environment specified in ISO 7726. The following quantities should be measured in each of the spaces or zones identified in the test plan and the results noted in the indoor climate test report:

- Air temperature.
- Relative humidity.
- Air velocity.
- Vertical gradient.
- Horizontal gradient (in crew cabin spaces only).

The air temperature and humidity measurements should be made at least every five minutes for a minimum period of one hour. The minimum, maximum, and average values for the 1-h period should be reported for each space measured. For the **HAB++** notation, or equivalent, the air temperature measurements should be made at least every five minutes for a minimum period of one hour if the system has the capacity to reach within that hour the **HAB++** (or equivalent) requirements for lower and upper temperatures. If not, the air temperature measurements should be made at least every 10 min for a minimum period of two hours. ***Note:*** *data loggers are a type of device that can be left unattended to capture data. These type devices have been proven effective in gathering temperature and humidity values.* The measuring period for determining the average air velocity at any location should be at least three minutes. Surface temperatures for all wall surfaces that are adjacent to the head of the bed in crew cabins should be measured for determining the horizontal gradient in crew cabin spaces. The horizontal gradient is to be calculated and recorded as the difference between the wall temperature and the average air temperature at 1100 mm (43 in.) above the deck crew cabin areas. The test plan should include all supporting documentation and calculations relating to the air supply for each of the spaces or zones identified in the test plan. *Test conditions.* In order to determine the effectiveness of the HVAC system at providing the environmental conditions specified in this chapter, measurements should be made under the following conditions:

- *Testing.* Testing of identified spaces can be performed in port and at sea, provided the required test conditions stated here are in compliance at the time the measurements are made and recorded. In the case where some testing is performed in port, confirmatory testing should be performed at sea with all normally functioning equipment in its operational mode. This confirmatory testing must usually consist of a sample of the manned spaces selected as follows:

  - Spaces that were identified as "worst case" locations.
  - Spaces which were out of compliance or close to non-compliance (refer to Table 34.1. Summary of indoor climate requirements).
  - Spaces that required some type of mitigation.
  - Additional testing may be required at the Class approved surveyor's discretion.

- *Equipment operation.* The HVAC system should be operating in the normal operation or mode. A written confirmation issued by the manufacturer confirming that the HVAC system has been tested, adjusted and balanced should be provided to the surveyor before any measurement commences. The total system Testing, Adjusting, and Balancing (TAB) should be conducted in accordance with the National Environmental Balancing Bureau (NEBB) standard NEBB Procedural Standards for Testing, Adjusting, Balancing of Environmental Systems or other similar certification standard.
- *Doors and windows.* The space doors (including louvres) and windows should be closed during the evaluation period, except for routine entry and exit. Any open doors or windows should be noted in the indoor climate test report.
- *Equipment and furnishings.* Spaces should be furnished with all usual equipment, fixtures and fittings normally found in the space. Equipment must be configured to operate in its normal operating mode.
- *Weather and climatic conditions.* When thermal conditions in the manned space have a high sensitivity to time of day and weather conditions (e.g., spaces adjacent to exterior bulkheads), the measurement should be made such that the high and low extremes of the thermal parameters are determined (e.g., measurements could therefore be taken during the day and night in the same space). If possible, measurements should be taken with little or no cloud cover.
- *Test interference.* During the indoor climate measurements, any activity that might affect the indoor climatic variables in the space must be avoided. For this reason, only the personnel needed for the normal operation of the equipment in the space and those carrying out the measurements should be present in the space being tested. Doors and windows should be closed, except where they are normally left open (such as the door on the lee side of the navigation bridge, which may normally be open). Any open doors or windows should be noted in the indoor climate test report.

Measurement locations. Selection of spaces where measurements are to be conducted. The aim when selecting indoor climate measurement locations is to obtain a representative sample of data that reflects the actual conditions in manned spaces. For practical reasons, it is important to select the locations such that an appropriate amount of data can be collected during the testing phase. This means the measurement locations should be selected in accordance with the following criteria:

- Select potential problem areas where the influence of internal conditions or factors may adversely impact the quality of the indoor climate in enclosed manned spaces. Internal conditions include space proximity to equipment that radiates or absorbs heat (e.g., engine exhaust trunks, freezer spaces, galley, scullery, etc.) and surfaces with thermal differentials in excess of 10 °C (18 °F) from the ambient temperature in the space. Living and working areas at the ends of HVAC ductwork or piping runs (for heating or cooling) should be selected as potential problem locations. Measurements should be taken in all identified potential problem areas.

- Select potential problem areas where the influence of external ambient environmental conditions (e.g., sun, wind, precipitation, etc.) may adversely impact the quality of the indoor climate. These areas include manned spaces, accommodation areas and recreation areas which may be outboard or adjacent to the vessel's hull. Measurements should be taken in all identified problem areas.
- Where a single instance of one type of manned space exists within the vessel (e.g., the bridge, mess room, gymnasium, library, etc.), that location should be selected for measurement.
- Where multiple instances of one type of space exist (e.g., cabins or recreation areas), a representative sample of at least 25% of each type (e.g., one-man room, two-man room, etc.) should be selected for measurement. The worst-case locations are to be considered part of the representative sample, where applicable. All of these measurement locations must be selected at locations port, starboard, fore, amidships, and aft. The worst-case locations are to be considered part of the representative sample for crew cabins, where applicable.

Walkthrough verification inspection locations. All normally manned spaces are to be subject to a walkthrough inspection by the Class approved surveyor. The number and locations of the walkthrough inspections are generally determined by the surveyor. The purpose of the walkthrough verification is to subjectively assess the indoor climate qualities. At the discretion of the surveyor, additional measurements may be required. *Transducer placement positions.* For each space identified in the test plan, the transducer locations should be standardised as follows:

- Air temperature and relative humidity measuring instrumentation should be set up approximately in the middle of the space to measure general space temperature and humidity levels. Air temperature should be measured at approximately 100 mm (4 in.), 1100 mm (43 in.) and 1700 mm (67 in.) above the deck. Relative humidity should be measured at a height of approximately 1700 mm (67 in.) above the deck.
- Air velocity should be measured at approximately 100 mm (4 in.), 1100 mm (43 in.) and 1700 mm (67 in.) above the deck in manned spaces, as applicable (to assure air velocity is not excessive). This measurement should be taken in the centre of the space. Air velocity only needs to be measured once per testing location.
- For crew cabin spaces, inside wall surface temperatures for determining the horizontal gradient should be captured at approximately 300 mm (12 in.) above the mattress.

Test Report

As stated in the section on 'Test reports' in Chap. 30, a test report must be submitted to the Class approved surveyor to determine whether the indoor climate levels meet the criteria and whether this part of the Class notation requirement has been met. The details listed in the following paragraphs are required to be provided in the indoor climate test report. *Test details.* The following details are required to be recorded for each period of testing:

- Vessel's geographic location.
- Weather conditions and meteorological data (i.e., wind speed and direction, ambient outdoor air temperature, outdoor humidity, barometric pressure) at the onset of every data collection period and at intervals of every four hours during any data collection period. Weather conditions should also be reported at the end of each data collection period.
- Any indications of abnormal activities or conditions during the test that might skew the results.

Transducer placement positions. The actual transducer placement positions must be indicated on appropriate drawings. *Measurement equipment details.* Details of measurement and analysis equipment (e.g., manufacturer, type and serial number, accuracy and resolution must be provided. Copies of the relevant instrumentation reference calibration certificates, together with the results of field setup and calibration checks before and after the field tests, should also be provided with the report. *Results.* The following results per measuring location, should be provided in a tabulated format:

- Measurement position.
- Number of people present in the space at time of measurement.
- Measurement period if different from requirements.
- Air temperature (minimum, maximum, and average) at 100 mm (4 in.) above deck.
- Air temperature (minimum, maximum, and average) at 1100 mm (43 in.) above deck.
- Air temperature (minimum, maximum, and average) at 1700 mm (67 in.) above deck.
- Relative humidity (minimum, maximum, and average) at 1700 mm (67 in.) above deck.
- Air velocity at 100, 1100, and 1700 mm (4 in., 43 in., and 67 in.) above deck at measurement positions.
- Vertical gradient (Average air temperature at 1700 mm (67 in.) minus average air temperature at 100 mm (4 in.) above deck).
- Horizontal gradient (side wall surface temperature in crew cabin spaces minus average air temperature at 1100 mm (43 in.) above deck crew cabin areas).
- Indication of pass/fail.

For all the remaining spaces that were checked through walk through verification inspection and spot check measurements, the following information should be provided:

- Name and number of spaces.
- Walk through inspection observations.
- Measurement results if necessary.

Deviations. All deviations from the approved test plan must be reported. Surveyor witnessing documentation. The equipment field calibration and data collection process of indoor climate tests must be witnessed by a Class approved surveyor. The surveyor is required to sign or initial each page of the test report and prepare a witnessing document stating whether all steps of the indoor climate testing were completed to their satisfaction. A copy of the witnessing document is then given to the person conducting the onboard testing, for insertion into the final indoor climate test report. The original is usually retained by the surveyor and submitted to Class for inclusion in the vessel's Record.

Results

On completion of the test, the indoor climate test report will be reviewed by the Class approved surveyor against the appropriate Class Habitation notation requirements. Where the requirements have been satisfactorily met, the vessel will be assigned the appropriate Class Habitation notation.

Chapter 35
Lighting and Illumination

General

The lighting of seafarer spaces should facilitate visual task performance, facilitate movement in the space and aid in the creation of an appropriate visual environment. Lighting design involves integrating these aspects to provide adequate illumination for the safety and well-being of the crew as well as for the various tasks performed on board vessels. The selection of appropriate illumination levels for specific tasks and seafarer spaces is an important consideration in the design of lighting systems. There is a difference of opinion as to what levels of light may be considered best for visual tasks. Since illumination recommendations are generally consensus values, for any task, a range of illumination levels may apply. Since visual tasks performed within habitable spaces on board a vessel are generally similar to tasks encountered ashore, requirements for illumination on vessels generally correspond to those tasks performed in living, working, and recreation areas on shore. To that end, this chapter outlines criteria for assessing the illumination levels of general lighting and task lighting on vessels. The main objective of the assessment is to determine whether the various lighting systems comply with minimum standards to accommodate crew visual task performance and facilitate crew movements and wellbeing and safety on board vessels. Lighting criteria are based on currently available objective standards and research data. Compliance with the criteria outlined in this chapter is typically a prerequisite for attaining a Class Habitability (e.g., **HAB**, **HAB+**, and **HAB++**, or equivalent) notation.

Terminology

Disability glare: Glare which reduces the ability to perform a visual task.

A. A. Olsen, *Applying Physical Ergonomics to Modern Ship Design*,
https://doi.org/10.1007/978-3-031-57974-5_35

Discomfort glare: Glare which produces viewer discomfort, but which does not significantly interfere with visual task performance or visibility.

General lighting: Lighting designed to provide a substantially uniform level of illumination throughout an area, exclusive of any provision for special, localised tasks. Such lighting should be provided by fixed luminaires.

Glare: The discomfort or impairment of vision experienced when parts of the visual field are excessively bright in relation to the general surroundings.

Illumination: The luminous flux density at a surface (or the amount of light falling on an object or surface), i.e., the luminous flux incident-per-unit area. Illumination is measured in units of Lux (lm/m^2) or footcandles (fc; lm/ft^2). One foot-candle equals 10.76 Lux.

Lumen: The International System of Units (SI) of luminous flux, used in describing a quantity of light emitted by a source or received by a surface.

Luminaire: A complete lighting unit consisting of a lamp(s) together with the parts designed to distribute the light, to position and protect the lamp and to connect the lamp to the power supply.

Luminance: The photometric brightness of an illuminated surface (or the amount of light emitted or reflected from the surface). The SI unit of luminance is candela-per-square metre (cd/m^2).

Luminous flux: The light emitted by a source, or received by a surface and indicates the intensity of a source. Flux is expressed in lumens.

Lux: a unit of illumination, equivalent to 0.0929 foot-candle and equal to the illumination produced by luminous flux of one lumen falling perpendicularly on a surface one metre square. Also called a metre candle.

Red or low-level white illumination: Lighting provided to accommodate efficient dark adaptation in areas where seeing tasks are performed during night-time operations and in areas where people need to move from a lit interior into a dark environment and maintain good vision.

Reference calibration: Calibration of a measuring instrument, conducted at least every two years (or more frequently if specified by the manufacturer) by an accredited Testing and Calibration Laboratory accredited according to ISO 17025:2005, as amended, with traceability to a national or international standard.

Task lighting: Lighting provided to meet the illumination requirements of a specific task. Task lighting refers to the total illumination requirement that may be obtained by supplementary lighting provided in addition to the general illumination. Such lighting may be provided by fixed luminaires or via floor lamps or table lamps.

Task plane: The horizontal, vertical, or inclined plane in which the visual task lies. If no information is available, the task plane may be considered to be the horizontal

and at 750 mm (29.5 in.) above the deck for seated tasks and 1000 mm (39.5 in.) for standing tasks.

Veiling reflection (or reflected glare): A reflection of incident light that partially or totally obscures the details to be seen on a surface by reducing the contrast.

Associated Documentation

The following documents provide details about test plan preparation, test measurement procedures and/or test reporting:

- ISO 8995-1: 2002, Lighting of Workplaces—Part 1: Indoor.
- DIN EN 12464-1:2011, Light and Lighting—Lighting of Workplaces—Part 1: Indoor Lighting and Illumination Workplaces.
- DIN 5032-7:1985, Photometry; Classification of Illuminance Meters and Luminance Meters.
- CIE S 023/E:2013 (ISO/CIE 19476:2014), Characterisation of the Performance of Illuminance Meters and Luminance Meters.
- BS 667:2005, Illuminance meters. Requirements and test methods.

Criteria

The lighting criteria for Class Habitability notations are provided in:

- Table 35.1, Lighting criteria for crew accommodation spaces.
- Table 35.2, Lighting criteria for entrances and passageways.
- Table 35.3, Lighting criteria for navigation and control spaces.
- Table 35.4, Lighting criteria for service spaces
- Table 35.5, Lighting criteria for operating and maintenance spaces/areas.
- Table 35.6, Lighting criteria for red or low-level white illumination.

In this chapter, general lighting and task lighting requirements are provided for crew tasks and spaces normally encountered on vessels. The lighting levels provided in the tables are for new lamps. Emergency lighting is covered in SOLAS and IMO Resolutions and was not considered in the selection of the lighting levels provided in this chapter. The criteria in this chapter are applicable to the entire vessel, not just manned spaces. General and task lighting. The minimum maintained illumination levels stated in the Lighting Criteria Tables 35.1, 35.2, 35.3, 35.4 and 35.5 may be achieved under the test conditions specified in the section on 'Test conditions' and measured with task lighting turned on, where provided, but with external light sources (e.g., daylight, moonlight, shore light, etc.) excluded. These levels were selected from a variety of sources that are listed in the References part of this book. Red or low-level white illumination. In workspaces where red or low-level white

Table 35.1 Lighting criteria for crew accommodation spaces

Space	Illumination level in Lux	Space	Illumination level in Lux
Cabins and sanitary spaces			
General lighting	150		
Reading and writing		Sanitary spaces	
• Desk	500	• Lavatory/toilet	200
• Bunk light	200	• Bath/shower area	150
Changing room	200		
Light during sleep periods[a]	< 30		
Dining spaces			
Mess room and cafeteria	300		
Snack or coffee area	150		
Recreation spaces			
Recreation room/ lounges	200	Gymnasium	300
Library		Bulletin boards/ display areas	150
• General lighting	150	All other recreation spaces (e.g., game rooms)	200
• Reading area	500		
Multimedia resource centre (e.g., computer room)			
Medical, dental and first aid centre			
General lighting/ medical waiting areas	150	Dispensary	
Medical and dental treatment/examination room	500	• General lighting	150
Reading and writing (desk)	500	• Over the dispensary table	500
Laboratories	500	Other medical spaces	300

[a] If there is any opportunity for light to enter cabins at the times of day or night when people sleep (e.g., portlights, transoms, etc.), the maximum lighting level should be 30 Lux

illumination is provided to facilitate dark adaptation, the maintained illumination levels in Table 35.6, Lighting criteria for red or low-level white illumination, are provided for guidance only and are not generally required for a Class Habitation notation.

Table 35.2 Lighting criteria for entrances and passageways

Space	Illumination level in Lux	Space	Illumination level in Lux
Interior walkways, passageways, corridors, stairways and access ways	100	Exterior walkways, passageways, and access ways (night)	50
Elevators	100	Exterior stairways	100

Table 35.3 Lighting criteria for navigation and control spaces

Space	Illumination level in Lux	Space	Illumination level in Lux
Wheelhouse, pilothouse, bridge		Offices	
• Day	300	• General lighting	300
• Night when underway	< 30	• Reading and writing (desk)	500
Chart room		• Meeting, training rooms	500
• General lighting	150	Control rooms and control stations[a]	
• On chart table	500	• General lighting	300
Radar/gyro room	300	• Control consoles and boards, panels, instruments, etc.	300
Radio room		• Reading and writing (desk)	500

[a] At night, the wheelhouse must be kept as dark as possible to permit visibility of objects on the water, hazards to navigation, and navigational lights

Test Plan

As stated in the section ‘Test plans’ in Chap. 30, a test plan should be developed to serve as the principal means for verifying the measurements to be performed to verify compliance with the lighting and illumination criteria. This test plan should typically include the following:

- *Documentation.* The test plan should include appropriate design information and layout drawings showing the hull outline, bulkheads, access routes, location of luminaires, outlines of major furniture and equipment, and the space name and number. The drawings should be to a scale and sized to permit the scaling of measurement points (as required in the section on ‘Measurement locations’) and lighting equipment and the recording of luminance and other relevant data.

Table 35.4 Lighting criteria for service spaces

Space	Illumination level in Lux	Space	Illumination level in Lux
Food preparation		Laundries	300
• Galley	500	Chemical storage (e.g., paint store, etc.)	300
• Pantry	300	Storerooms	
• Butcher shop	500	• Large parts	200
• Thaw room	300	• Small parts	300
• Working surfaces, food preparation counter and range tops	500		
• Scullery (dishwashing) food serving lines package handling/cutting	300		
Food storage			
• Non-refrigerated	200		
• Refrigerated	100		
Mail sorting	300		

- *Test personnel.* The test plan should provide information about the testing specialist who will be conducting the test and their approval and certification in accordance with the Class specific Procedural Requirements for Recognised Ambient Environmental Testing Specialists or equivalent.
- *Test conditions.* The test plan should detail the conditions under which the tests will be performed. Details about test conditions are provided in the section on 'Test conditions'.
- *Measurement locations.* The test plan should document, in detail, on appropriate drawings, all spaces where measurements will be taken. In addition, transducer placement positions should be indicated. Details on selecting measurement locations and determining transducer placement positions are provided in the section 'Measurement locations'.
- *Data acquisition and instruments.* The test plan should provide information regarding the methods and instrumentation to be used for measurement and data collection. Instrumentation specification details should include type of instruments to be used, accuracy, calibration, sensitivity. More details on data acquisition and instruments are provided in the section 'Data acquisition and instruments'.
- *Data analysis.* The test plan should provide information regarding the methods, software, and instrumentation to be used for data analysis.

Table 35.5 Lighting criteria for operating and maintenance spaces/areas

Space	Illumination level in Lux	Space	Illumination level in Lux
Machinery spaces	200	Cargo holds (portable lighting)	
Engine room (main propulsion machinery)	300	General lighting	30
Generator, switchboard and transformer room emergency generator room	200	During cargo handling	200
Fan room/HVAC room	200	Passageways and trunks	80
Bilge/void spaces (portable lighting)	75	Storerooms	200
Pump room	200	Inspection and repair tasks	
Fire pump room	200	• Rough	300
Shaft alley	100	• Medium	500
Steering gear room	200	• Fine	750
Windlass room	200	• Extra fine	1,000
Battery room	200	Workshops	
Boiler rooms	100	• General lighting	300
		On work bench top/machine paint shop	500
Muster/embarkation area	200	Escape trunks	50
Cargo handling	200		
Weather decks	200	Crane cabin	400
Lay down area	200		
General processing area	200	Hand signalling areas between crane shack and ship deck	300
Utility area	200		
Loading ramps and bays	200		
Cargo storage and manoeuvring areas	200		

Table 35.6 Lighting criteria for red or low-level white illumination

Space	Illumination level in Lux
Where seeing is essential for charts and instruments	1–20
Interiors or spaces	5–20
Bridge areas (including chart tables, obstacles and adjacent corridors and spaces)	0–20 (continuously variable)
Stairways	5–20
Corridors	5–20
Repair work (with smaller to larger size detail)	5–55

When the space is separate from the wheelhouse

- *Test schedule.* The test plan should provide information regarding the proposed test schedule. The test schedule should include information on the approximate date and duration of testing (including a summary of the scheduled measurements per day).
- *Data collection sheets.* The test plan should provide sample data sheets (in table format) that will be used for reporting the measurement data.
- *Conflict of interest declaration.* In the case that the testing specialist has, in any capacity or to any extent, provided input into the design of the vessel, their participation should be indicated in the test plan including details of the testing specialist's involvement. Class may review any potential conflicts of interest and determine if any further action is necessary.

Test Requirements

Visual tasks encountered on vessels vary and the lighting provided can influence the ability to see and perform those tasks. Some vision and lighting considerations include task duration, visual fatigue, task criticality, veiling reflections, shadows, and abilities of the observer. The presence of glare is a concern and is often difficult to identify, measure, and assess. Glare is often transient (based on factors such as the direction of the sun to illuminated components or from the placement of lighting fixtures) and therefore difficult to anticipate. As part of lighting data collection, glare should be subjectively evaluated jointly by the testing specialist and the Class approved surveyor witnessing the collection of lighting data. These assessments should be made as part of the activities discussed in the section 'Walk through verification inspection locations'. Measurement of ambient lighting need only be performed once during sea trials or at pier side when all external light sources can be blocked out. *Data acquisition and instruments.* The illumination meter (light meter) must conform to any of the International Standards specified below (or equivalent):

- DIN 5032-7:1985 (Class A).
- CIE S 023/E:2013 (ISO/CIE 19476:2014).

- BS 667:2005 (Type F).

Note: *The measurement of red and/or low-level white illumination is not required for ship Habitability notations. Test conditions.* The test conditions required for lighting and illumination measurements must be in accordance with the following:

- *Location.* Lighting measurements may be taken in port, at sea, or both, since the measurements are not dependent on vessel operation. If testing is done in port, ship's power and not shore power should be used.
- *Spaces with windows/portlights.* In spaces with windows or portlights where the minimum lighting level is intended to be provided by artificial light sources only, lighting measurements should be taken after dark, or during the day provided all natural lighting is blocked out.
- *Spaces without windows/portlights.* Interior spaces with no windows or portlights can be measured during daylight hours.
- *Stray light.* Stray light (e.g., deck lighting and moonlight) should be capable of being blacked out as far as practicable. Where it is not possible, measurements of stray light, at appropriate positions, with all lighting turned off, should be obtained. These readings must then be deducted from readings taken at the same positions, with the lighting turned on, to determine the illumination from the lighting.
- *Light for sleep.* Lighting measurements should be taken in cabins with all cabin lights turned off and curtains, shutters, etc., closed.
- *Test interference.* During the lighting measurements, shadows on the light meter caused by any kind of human activity should be avoided. For this reason, only the personnel needed to operate any equipment during testing and those carrying out the measurements may be present in the space. Doors and windows must be closed, except where they are normally left open (such as the door on the lee side of the navigation bridge, which may normally be open). Any open doors or windows should be noted in the lighting test report. Spaces should be furnished with all usual equipment, fixtures and fittings normally found in the space.

Measurement locations. Selection of spaces where measurements are to be conducted. The aim when selecting lighting and illumination measurement locations is to obtain a representative sample of data that represents the actual conditions in the manned crew spaces listed in the Lighting Criteria Tables 35.1, 35.2, 35.3, 35.4 and 35.5 inclusive. For practical reasons, it is important to select the locations such that an appropriate amount of sample data can be collected during the testing phase. The measurement locations must be selected in accordance with the following criteria:

- Select problem areas based on the potential for excessive external illumination (daylight) into the space (e.g., bridge). Areas at the end of cable runs should be selected as potential problem locations. Measurements should be taken in all identified problem areas.

- Select potential problem areas based on the positioning of luminaires in the space as indicated on the drawings (e.g., uneven spacing of luminaires resulting in non-uniform illumination levels and dimly lit areas). Measurements must be taken in all problem areas.
- Where a single instance of one type of space exists within the vessel (e.g., the bridge, mess room, gymnasium, library, etc.), the location should be selected for measurement.
- Where multiple instances of one type of space exist (e.g., recreation areas) with the exception of cabins, a representative sample of at least 25% of each type should be selected for measurement. The problem area locations are to be considered part of the representative sample, if applicable.
- Where a number of crew cabin spaces are identical in configuration in terms of lighting systems, surface treatments, geometry, furnishings and equipment layout, only two of the spaces on each deck should be selected to determine whether the lighting requirements are met. Where cabins are not identical in configuration in terms of lighting system, surface treatments, geometry, furnishing, and equipment at least 25% should be selected on each deck for measurement.
- All of these measurement locations must be selected at locations port, starboard, fore, amidships and aft. The worst-case locations are to be considered part as the representative sample for crew cabins, if applicable.
- Where external night-time measurements are taken, spaces should be selected according to human presence in the space or task location, including exterior passageways and access ways, operating consoles, hand signalling locations, anchor handling locations, mooring locations, crew embarkations and debarkation areas, and cargo handling locations—particularly with regard to attaching and detaching cargo from crane hooks.

Walk through verification inspection locations. All normally manned spaces are to be subject to a walkthrough inspection by the Class approved surveyor. The number and locations of the walkthrough inspections are generally determined by the surveyor. The purpose of the walkthrough verification is to subjectively assess lighting and illumination qualities. The presence of discomfort or disability glare should be specifically and subjectively assessed by the Class approved surveyor and the testing specialist. Upon completion of the assessment, the result of "no glare present", "some glare present", "discomfort glare present", or "disability glare present" should be noted for each lighting test location. At the discretion of the surveyor, additional measurements may be required. General illumination measurement positions. General lighting levels should be measured with all lights turned on except supplementary task lighting, such as desk lights and berth lights. Daylight should be excluded during the measurements. Measurements should be taken on a horizontal plane approximately 750 mm (29.5 in.) above the deck. For traffic areas, readings should be taken on deck. For larger spaces (e.g., public rooms, messes, recreation areas, etc.), it may be necessary to take measurements at a number of locations in order to obtain a representative sample of the illumination for that space. Measurement locations should always be evenly distributed throughout the space.

Table 35.7 Distribution of measurement positions within spaces

Space	Minimum number of measurement positions in room
Less than 20 m^2 (215 ft^2)	1
> 20–40 m^2 (215–431 ft^2)	2
> 40–80 m^2 (431–861 ft^2)	3
> 80–120 m^2 (861–1291 ft^2)	4
> 120–200 m^2 (1,291–2150 ft^2)	5
Greater than 200 m^2 (2150 ft^2)	6

For a specific room size, the minimum number of measurement locations will be as indicated in Table 35.7, Distribution of measurement positions within spaces. The arithmetic mean of the measurement should be used for the measure of illumination of the general area. In the case of corridor lighting, measurements should be taken approximately every 3000 mm (10 ft) of corridor length. The arithmetic mean should not be used to determine compliance with corridor lighting requirements.

Task lighting measurement positions. For task area lighting (such as consoles, control stations, workbenches, computer workstations, desktops and meeting tables, medical stations, galley areas, and maintenance workstations), a representative set of readings should be taken over the task surface with the worker in the normal working position. It is best to assess the task lighting with both general and supplementary task lighting turned on. As much as possible, daylight should be excluded during the measurements. Measurements of task lighting must be made on the surface or in the plane of the task (horizontal, vertical or at an angle). For small task surfaces, i.e., smaller than 0.5 m^2 (5 ft^2), a single measurement is best taken at the centre of the task surface. For larger task surfaces (0.5 m^2 (5 ft^2) or larger), the illumination is best measured by dividing the task surface into a grid no larger than 0.5 m^2 (5 ft^2) and averaging the measurements taken at the grid intersections. For berths and bunks, measurements may be taken at a point approximately 300 mm (12 in.) above the top of the mattress and 600 mm (24 in.) from the bunk light fixture. *Red and low-level white illumination.* Guidance for the measurement of red and low-level white is provided in other national and international standards. As these measurements are not required for the Habitability notation, they are not discussed in this book.

Test Report

As stated in the section on 'Test reports' in Chap. 30, a test report must be submitted to the Class approved surveyor to determine whether the lighting and illumination levels meet the criteria and whether this part of the Class notation requirement has been met. The details listed in the following paragraphs are required to be provided in the lighting test report: test details. The following details are required to be recorded for each period of testing: (1) time of day; and external lighting conditions (e.g., were measurements taken in port or at sea and were measurements taken during day light hours or at night?). *Measurement positions.* Actual measurement locations should be indicated on appropriate drawings. *Measurement equipment details.* Details of the measuring equipment used (e.g., manufacturer, type and serial number, accuracy and resolution) should be provided. Copies of the relevant instrumentation reference calibration certificates, together with the results of field setup and calibration checks before and after the field tests, should be included with the lighting test report. *Results.* The following results, per space, should be provided in a tabulated format:

- Name and number of the space.
- Task areas (if any) in space.
- Average lighting level for general lighting.
- Average lighting level for task blighting on each task surface.
- Corridor lighting measurements.
- Lighting levels in cabins with lights turned off and all curtains, shutters, deadlights, etc. closed.
- Presence of glare at specific workstations within the space and any follow-up activity requirements.
- Indication of pass/fail.

For all the remaining spaces that were checked through walk through verification inspection and spot check measurements, the following information should be provided:

- Name and number of spaces.
- Walk through inspection observations.
- Measurement results if necessary.

Deviations. All deviations from the approved test plan must be reported. *Surveyor witnessing documentation.* The equipment field calibration and data collection process of indoor climate tests must be witnessed by a Class approved surveyor. The surveyor is required to sign or initial each page of the test report and prepare a witnessing document stating whether all steps of the indoor climate testing were completed to their satisfaction. A copy of the witnessing document is then given to the person conducting the onboard testing, for insertion into the final indoor climate test report. The original is usually retained by the surveyor and submitted to Class for inclusion in the vessel's Record.

Results

On completion of the test, the lighting test report will be reviewed by the Class approved surveyor against the appropriate Class Habitation notation requirements. Where the requirements have been satisfactorily met, the vessel will be assigned the appropriate Class Habitation notation.

Chapter 36
Requirements for Class Recognised Test Specialists

General

The objective of this chapter is to outline the basic standards for qualifying and certifying testing specialists performing ambient environmental testing and evaluation. These standards apply to the approval of testing specialists that provide whole-body vibration measurements and analysis, noise measurements and analysis, indoor climate measurement and analysis; and lighting measurement and analysis. The general requirements concerning testing specialists are provided in the section on 'General requirements' whereas the specific requirements for the test services listed above given in the section titled 'Detailed requirements by ambient environmental aspect'.

Procedure for Approval and Certification

Documentation requirements. The following documents are to be submitted to Class for review and approval:

- An outline of the company (e.g., organisation and management structure) including subsidiaries or subcontractors to be included in the approval/certification.
- A list of company experience in the specific ambient environmental aspect.
- A list of test personnel documenting training and experience in conducting tests within the relevant ambient environmental aspect and qualifications according to recognised national, international or industry standards, as applicable.
- Description of equipment used for the measurement and analysis of the particular ambient environmental aspect for which approval is sought (e.g., calibration, accuracy, etc.).

A. A. Olsen, *Applying Physical Ergonomics to Modern Ship Design*,
https://doi.org/10.1007/978-3-031-57974-5_36

The following documents are typically reviewed during the initial approval audit (refer to the section on 'Auditing of the testing specialist'):

- A guide for operators of such equipment.
- Training programmes for test personnel.
- Draft check lists and data recording sheets for recording results of the services provided.
- Quality Manual and/or documented procedures covering the requirements.
- Documented procedures for communication with the crew prior to commencing work in order to provide a safe system of work in place as necessary.
- Evidence of approval/acceptance by certifying bodies, if any.
- Information about other activities which may present a conflict of interest (i.e., in the case that the testing specialist provides, in any capacity or to any extent, input on the design of in part or whole of a vessel).
- Record of customer claims and of corrective actions requested by certification bodies for the past year.
- Where relevant, list and documentation of licenses granted by equipment's manufacturer.
- Example test plan(s) for the ambient environmental quality for which approval is requested.
- Example test reports.

General Requirements

Extent of approval. The testing specialist is expected to demonstrate, as discussed in the section 'Training of personnel through to reporting', that they have the competence, quality control, and quality assurance needed to perform the test and analysis services for which approval is sought. *Training of personnel.* The testing specialist is responsible for the qualification and training of its personnel to a recognised national, international, or industry standard as applicable. Where such standards do not exist, the testing specialist may be required to define the standards for the training and qualification of its personnel relevant to the functions each is authorised to perform. The personnel should also have adequate experience and be familiar with the operation of any necessary equipment. In most cases, personnel are required to have had a minimum of one year tutored on-the-job training. Where it is not possible to perform internal training, a programme of external training may be considered as acceptable. *Supervision.* The testing specialist usually provides supervision for all services provided. The responsible supervisor is typically required to have had a minimum of one year of experience in supervising tests and at least two years of performing tests in the ambient environmental aspect for which the testing specialist is recognised. *Personnel records.* The testing specialist is required to maintain accurate records of the recognised test personnel. The records should contain information about formal education, training, and experience for the ambient environmental test services for which they are recognised. *Equipment and facilities.* The testing specialist is expected to have the necessary equipment and facilities for the ambient environmental aspect to be tested. A record of the equipment used for ambient environmental testing must be accurately kept. This record contains information about the equipment maintenance and calibration.

Control of data. When computers are used for the acquisition, processing, recording, reporting, storage, measurement assessment and monitoring of data, the ability of computer software to satisfy the intended application should be adequately documented and confirmed by the testing specialist. *Procedures.* The testing specialist must have documented work procedures covering all ambient environmental test services supplied. *Subcontractors.* The testing specialist is typically required to provide information of any agreements and arrangements should any parts of the services provided be subcontracted. Particular emphasis may be given to the quality management by the testing specialist in following-up of such subcontracts. Subcontractors providing anything other than subcontracted personnel or equipment are jointly required to meet the requirements outlined in the sections on 'General requirements' and 'Quality assurance system'. *Verification.* The testing specialist is required to verify that the services provided are carried out in accordance with approved procedures. *Reporting.* The report issued by the testing specialist is must typically be prepared in a form acceptable to Class and include a copy of the Certificate of approval.

Auditing of the testing specialist. Upon reviewing the submitted documents with satisfactory result, testing specialists are usually audited for the initial certification process and then every three years in order to ascertain that the testing specialist is duly organised and managed in accordance with the submitted documents and that it is considered capable of conducting the test services for which approval/certification is sought. *Certification.* Certification is conditional on a practical demonstration to Class or its agent of the specific ambient environmental test service performance as well as a sample of a satisfactory report. *Quality assurance system.* The testing specialist is expected to have a documented system covering, as a minimum, the following:

- Operating instructions for the test equipment.
- Maintenance and reference calibration of equipment.
- Training programs for test personnel.
- Supervision and verification to verify compliance with test procedures.
- Recording and reporting of information.
- Quality management of subsidiaries and subcontractors.
- Job preparation.
- Periodic reviews of test process procedures, complaints, corrective actions and issuance maintenance and control of documents.

Code of conduct for the relevant activity. A documented quality assurance system complying with the applicable ISO 9000 standard or equivalent and including the above items is usually considered acceptable to the majority of Class'. *Testing specialist relations with the test equipment manufacturer.* A company, which works as a service station and conducts reference calibrations of equipment for a manufacturer (and is a testing specialist in this field) should be assessed by the manufacturer(s) and nominated as their agent. The manufacturer should verify that appropriate instruction manuals, material, etc., are available for the agent, as well as verifying proper training of the agent's technicians has occurred. Such testing specialists are typically recognised either on a case-by-case basis or as follows: if a manufacturer of equipment (and the testing specialist) applies for inclusion of its nominated agents and/or subsidiaries in the approval, then the manufacturer must have implemented a quality assurance system certified in accordance with the relevant ISO 9000 standard or equivalent. The manufacturer must have effective controls of its agents and/or subsidiaries, and these agents/subsidiaries must have an equally effective quality control system complying with the relevant ISO 9000 or equivalent. Such approvals are based upon an evaluation of the quality assurance system implemented by the applicable company ISO 9000 or equivalent. Class will usually follow-up the adherence to this quality assurance system by performing audits on such agents or subsidiaries against the relevant ISO 9000 standard or equivalent.

Certificate of Approval

Upon satisfactory completion of both the audit of the testing specialist and practical demonstration, Class will issue a Certificate of approval stating that the testing specialist's test and analysis service operation system has been found to be satisfactory and that the results of test and analysis services performed in accordance with that system may be accepted and utilised by Class in making decisions affecting optional Habitability/Comfort classification notations. The Certificate will clearly state the type and scope of services and any limitations or restrictions imposed. The testing specialist will also then be included in Class's records of recognised testing specialists. Where several ambient environmental aspect measurements are conducted by a given company, each aspect is to be assessed and recognised, except as specified above in the section on 'Testing specialist relations with the test equipment manufacturer'. *Renewal.* The Certificate of approval is subject to renewal or endorsement at intervals not exceeding three years per service supplier procedure. The renewal or endorsement is typically accomplished by verification through audits to ensure that approved conditions and standards are maintained. *Alterations.* When any alteration to the certified test and analysis service operation system of the testing specialist is made, Class is to be immediately notified. A reaudit may be required when deemed necessary by Class.

Cancellation of Approval

Class approval of the test specialist may be cancelled or annulled in any of the following cases:

- Where services are improperly carried out or the results are improperly reported.
- Where deficiencies are found in the recognised services of the testing specialist and appropriate corrective action is not taken.
- Where the testing specialist fails to inform Class of any alteration, as per the section on 'Alterations'.
- Where a renewal audit, if requested as per Class's typical Renewal regime, has not been carried out.
- Where wilful acts or omissions are ascertained.

Class will always reserve the right to cancel their approval in the event of evidence coming to light of any or all of the above improprieties. A testing specialist whose approval is cancelled may apply for re-approval provided the non-conformities, which resulted in the cancellation of their Class approval have been corrected and that Class is able to confirm, to their satisfaction, that the corrective action has been effectively implemented.

Detailed Requirements by Ambient Environmental Aspect

Whole-Body vibration

1. *Extent of engagement.* Whole-body vibration measurement testing specialists are engaged to conduct vibration measurements and analyses on board vessels. In almost all cases, the following minimum requirements apply.
2. *Supervisors.* The supervisor should be qualified to Level II according to a recognised national or international Non-Destructive Testing (NDT) standard (ANSI/ASNT CP-189, ISO 9712 or EN 473); or have a documented history of at least one year supervising and of at least two years performing vibration testing on board marine vessels. The supervisor should have sufficient knowledge of vessel structures and equipment, measurement and analysis of whole-body vibration according to ISO 6954, to verify that test procedures are compliant with the required test conditions.
3. *Test Personnel.* The test personnel carrying out the measurements should be qualified to Level I according to a recognised national or international NDT standard (ANSI/ASNT CP-189, ISO 9712 or EN 473); or have a documented history of one year experience in performing vibration environmental testing on board marine vessels; and have knowledge in the field of vibration, vibration measurements and handling of measurement equipment. In addition, be able to document theoretical and practical training on board in using a vibration transducer; and have adequate knowledge of vessel structures and equipment.

4. *Equipment*. Class will ordinarily verify that the equipment to be used is in accordance with the applicable measurement standard. It should be demonstrated to Class's satisfaction that the equipment to be used is fit for the intended purpose. Whole-body vibration measurements are generally taken using an instrument manufactured to ISO 8041:2005, as amended, type/class (1) standard; then frequency weighted and analysed in accordance with ISO 6954:2000, as amended. It is usually desirable to employ equipment that records and stores acceleration time histories. Portable calibrators, where used, should comply with an international or national standard. Calibration of the vibration transducer and of the portable calibrator, is to be conducted at least every two years (or more frequently if specified by the manufacturer) by an accredited testing and calibration laboratory accredited according to ISO 17025:2005, as amended.
5. *Procedures*. Documented work procedures or test instructions are to contain, at a minimum, information about the following:
   - Test preparation.
   - Selection and identification of measurement locations.
   - Surface preparation.
   - Calibration checks.
   - Testing methods.
   - Equipment handling.
   - Report preparation and content.
   - Method for handling previous results if subsequent calibration shows instruments to be out of tolerance.
   - Reporting.
6. *Access to documents*. The Testing specialist is expected to have access to the following document(s):
   - ISO 6954:2000, Mechanical Vibration and Shock—Guidelines for the Measurement, Reporting and Evaluation of Vibration with regard to Habitability on Passenger and Merchant Ships.

Noise

1. *Extent of engagement*. Noise measurement testing specialists are engaged to conduct noise measurements and analyses on board vessels. In almost all cases, the following minimum requirements apply.
2. *Supervisor*. The supervisor should be qualified to Level II according to a recognised national or international NDT standard (ANSI/ASNT CP-189, ISO 9712 or EN 473); or have a documented history of at least one year supervising and of at least two years performing noise testing on board marine vessels. The supervisor is expected to have sufficient knowledge of vessel structures, measuring equipment, ISO 2923, IEC 61672-1, and IMO Resolution MSC.337(91) to verify that test procedures are compliant with the required test conditions.
3. *Test Personnel*. The test personnel carrying out the measurements are required to be qualified to Level I according to a recognised national or international NDT

standard (ANSI/ASNT CP-189, ISO 9712 or EN 473); or have a documented history of one year experience in performing noise testing on board marine vessels including participation in a minimum of five measurement campaigns as an assistant test personnel; and have knowledge in the field of noise, sound measurements and handling of measurement equipment. In addition, be able to document theoretical and practical training on board in using a sound level meter; and have adequate knowledge of the applicable international requirements (SOLAS Regulation II-1/3-12, as amended, and IMO Code on noise levels on board Ships, as amended) and the requisite Class crew habitability guides; and have training concerning the procedures specified in the IMO Code on Noise Level; and have adequate knowledge of vessel structures and measuring equipment.

4. *Equipment.* Class will ordinarily verify that the equipment to be used is in accordance with the applicable measurement standard. It should be demonstrated to Class's satisfaction that the equipment to be used is fit for the intended purpose. Noise measurements are taken using an integrating-averaging sound level meter manufactured to IEC 61672-1:2013, as amended, type/class (1) standard. When used alone, or in conjunction with a sound level meter, as appropriate, an octave filter set should conform to IEC 61260-1:2014, as amended, standard. Sound calibrators should comply with the standard IEC 60942:2003, as amended, type/class (1) standard and should be approved by the manufacturer of the measuring instrument used. Calibration of the sound level meter and of the sound calibrator, is to be conducted at least every two years by an accredited Testing and Calibration Laboratory accredited according to ISO 17025:2005, as amended. It is strongly advised to use a microphone wind screen when taking readings outside or on deck, and below deck where there is any substantial air movement. The windscreen should not affect the measurement level of similar sounds by more than 0.5 dB(A) in "no wind" conditions.
5. *Procedures.* Documented work procedures or test instructions are to contain, at a minimum, information about the following:

   - Test preparation
   - Selection and identification of measurement locations
   - Surface preparation
   - Calibration checks
   - Testing methods
   - Equipment handling.

6. *Report preparation and content.* The report should contain details on the methods employed for handling previous results if subsequent calibration shows instruments to be out of tolerance.
7. *Access to documents.* The testing specialist is expected to have access to the following document(s):

   - SOLAS 1974, as amended (Reg.II-1/3-12).
   - Resolution A.468(XII) and IMO Resolution MSC.337(91) code on noise levels on board ships.

- Resolution A.343(IX) Recommendation on methods of measuring noise levels at listening posts.
- Class crew habitability guides.

Indoor climate

1. *Extent of engagement*. Indoor climate measurement Testing specialists are engaged to conduct indoor climate measurements and analyses on board vessels. In almost all cases, the following minimum requirements apply.
2. *Supervisor*. The supervisor should be qualified to Level II according to a recognised national or international NDT standard (ANSI/ASNT CP-189, ISO 9712 or EN 473); or have a documented history of at least one year supervising and of at least two years performing ambient environmental testing on board marine vessels. The supervisor should have sufficient knowledge of vessel structures, measuring equipment, ANSI/ ASHRAE 55-2013 as well as ISO 7726 and ISO 7547, to verify that test procedures are compliant with the required test conditions.
3. *Test personnel*. The test personnel carrying out the measurements should be qualified to Level I according to a recognised national or international NDT standard (ANSI/ASNT CP-189, ISO 9712 or EN 473); or have a documented history of one year experience in performing ambient environmental testing on board marine vessels; and have knowledge in the field of indoor climate, indoor climate measurements and handling of measurement equipment; and have adequate knowledge of vessel structures and equipment.
4. *Equipment*. Class will ordinarily verify that the equipment to be used is in accordance with the applicable measurement standard. It should be demonstrated to Class's satisfaction that the equipment to be used is fit for the intended purpose. Indoor Climate measurement should be taken using an instrument manufactured to meet or exceed the minimum characteristics of instruments for measuring physical quantities characterising an environment specified in ISO 7726:1998, as amended, or those of an equivalent standard. Calibration of the measuring instrument is to be conducted at least every two years (or more frequently if specified by the manufacturer) by an accredited Testing and Calibration Laboratory accredited according to ISO 17025:2005, as amended.
5. *Procedures*. Documented work procedures or test instructions are to contain, at a minimum, information about the following:
   - Test preparation.
   - Selection and identification of measurement locations.
   - Surface preparation, if applicable.
   - Testing methods.
   - Equipment handling.
   - Report preparation and content.
   - Methods for handling previous results if subsequent calibration shows instruments to be out of tolerance.
6. *Access to documents*. The *testing* specialist should have access to the following document(s):

- ANSI/ASHRAE 55-2013, Thermal Environmental Conditions for Human Occupancy.
- ISO 7726:1998, Ergonomics of the Thermal Environment—Instruments for Measuring Physical Quantities

Lighting and illumination

1. *Extent of engagement.* Lighting measurement testing specialists are engaged to conduct illuminance measurements and analyses on *board* vessels. In almost all cases, the following minimum requirements apply.
2. *Supervisor.* The supervisor should be qualified to Level II according to a recognised national or international NDT standard (ANSI/ASNT CP-189, ISO 9712 or EN 473); or have a documented history of at least one year supervising and of at least two years performing ambient environmental testing on board marine vessels. The supervisor should have sufficient knowledge of vessel structures, measuring equipment, ISO 8995:2002 and DIN EN 12464-1:2011 to verify that test procedures are compliant with the required test conditions.
3. *Test personnel.* The test personnel carrying out the measurements should be qualified to Level I according to a recognised *national* or international NDT standard (ANSI/ASNT CP-189, ISO 9712 or EN 473) or have a documented history of one year experience in performing ambient environmental testing on board marine vessels; and have knowledge in the field of lighting, illumination level measurements and handling of measurement equipment; and have adequate knowledge of vessel structures and equipment.
4. *Equipment.* Class will ordinarily verify that the equipment to be used is in accordance with the applicable measurement standard. It should be demonstrated to Class's satisfaction that the equipment to be used is fit for the intended *purpose.* Lighting measurement should be taken using an illumination meter conforming to any of the International Standards specified below (or equivalent):

   - DIN 5032-7:1985 (Class A).
   - CIE S 023/E:2013 (ISO/CIE 19476:2014).
   - BS 667:2005 (Type F).

Calibration of the illumination meter is to be conducted at least every two years (or more frequently if specified by the manufacturer) by a Testing and Calibration Laboratory accredited according to ISO 17025:2005 standard, as amended.

5. *Procedures. Documented* work procedures or test instructions should contain, at a minimum, information about the following:

   - Test preparation.
   - Selection and identification of measurement locations.
   - Surface preparation.
   - Calibration checks.
   - Testing methods.
   - Equipment handling.
   - Report preparation and content.

- Methods for handling previous results if subsequent calibration shows instruments to be out of tolerance.

6. *Access to documents*. The testing *specialist* should have access to the following document(s):

   - Class specific crew habitability guides.

Part IV
Habitability on Workboats

Chapter 37
General Provisions for Workboats

General

This chapter is designed to provide a single source for habitability criteria suitable for workboats. This chapter may be applied to vessels falling under the following categories: offshore support vessels, tugboats, tow boats, dredgers, research vessels, anchor handling vessels, or other vessels providing service to the offshore oil and gas exploration and production industries. This chapter does not apply to vessels such as oil and chemical tankers, bulk or combination carriers, container carriers, multipurpose cargo vessels, or MODU. These types of vessels are addressed in Part 3 of this book. In essence, this Part 4 has been written with the objective of improving the quality of crew member performance and comfort by improving working and living environments in terms of accommodation area design and ambient environmental qualities. These habitability criteria have been chosen to provide a means to help reduce crew fatigue, improve performance and safety, and to assist with crew recruiting and retention.

Terminology

Accommodation areas/accommodation block: Vessel areas where the primary purpose is to rest and recreate. Accommodation spaces include cabins, medical facilities (sick bays), offices, recreation rooms, and manned spaces within the accommodation block such as the bridge or control room. For the purposes of this chapter, accommodation areas also include service spaces such as mess rooms, laundry, and storerooms.

Ambient environment: Ambient environment refers to the environmental conditions that the crew is exposed to during periods of work, leisure or rest. Specifically, this

A. A. Olsen, *Applying Physical Ergonomics to Modern Ship Design*,
https://doi.org/10.1007/978-3-031-57974-5_37

Part 4 provides criteria and limits for whole-body vibration, noise, indoor climate, and lighting.

Associated documentation: Documents referenced in this Part 4 that are needed to provide measuring techniques and further guidance.

Class recognised ambient environmental testing specialist: Companies providing test or measurement services on behalf of the owner of a vessel or shipyard for the purposes of meeting any of the Class habitability notation requirements. Approval of the testing specialist for ships may also allow for testing and measurement services for other Class approved Habitability, Comfort, and MLC notations.

Crew member: Any person on board a vessel, including the master, who is not a passenger. This term is used interchangeably throughout this document with "seafarer".

Crew spaces: All areas on a vessel intended for seafarers only, such as seafarer accommodation spaces and seafarer workspaces. This term is also used interchangeably throughout this Part 4 with "seafarer spaces".

Dynamic positioning: A system to automatically maintain a workboat's position and heading by controlling propellers and/or thrusters. Dynamic positioning can maintain a position to a fixed point over the bottom, or in relation to a moving object (such as another vessel). It can also be used to position the vessel at a favourable angle towards wind, waves, and current.

Habitability: The acceptability of the conditions of a vessel in terms of whole-body vibration, noise, indoor climate, and lighting, as well as physical and spatial characteristics, according to prevailing research and standards for human efficiency and comfort.

Manned space: For application of this Part 4, a manned space is any space where a seafarer may be present for 20 min or longer at any one time during normal, routine daily activities. Such spaces include working or living spaces. For machinery spaces, a manned space will be associated with the presence of a local work/control duty station manned for 20 min or longer at one time during normal, routine daily activities. Normal inspection, walk through (patrolling) and maintenance tasks in machinery spaces are not considered tasks contributing towards a particular space being considered manned.

Passenger: A passenger is every person other than the master and the members of the crew or other persons employed or engaged in any capacity on board a vessel for the business of that vessel.

Recreation and public spaces: Those portions of the accommodation areas that are used for halls, dining rooms, lounges, and similar spaces.

Seafarer spaces: All areas on a vessel intended for seafarers only, such as seafarer accommodation spaces and seafarer workspaces. This term is also used interchangeably throughout this Part 4 with "crew spaces".

Seafarer: Any person on board a vessel, including the master, who is not a passenger. This term is used interchangeably throughout this document with "crew member".

Should: Expresses a provision that is mandatory.

Test plan: Document containing the requisite information regarding vessel design and layout, test personnel, test conditions, measurement locations, data acquisition, instruments, data analysis, and test schedule necessary for verifying the measurements for the ambient environmental aspects of habitability. This document is to be primarily prepared by the Class recognised ambient environment testing specialist with additional input provided by the designer/builder. Submission of the document for approval can be by either the testing specialist or the designer/builder.

Transit conditions: Those conditions where the vessel is transitioning (moving) from one location to another.

Workboat: Vessels falling under the categories of offshore support vessels, tugboats, tow boats, dredgers, or research vessels and some special service vessels.

Workspaces: Areas allocated for work. Categories of workspaces include, but are not limited to, navigation spaces, service spaces (e.g., galley), machinery spaces, and duty/control stations.

Scope and Application

This Part 4 focuses on five habitability aspects of ship design and layout that can be controlled, measured, and assessed. These five aspects are broken into two categories: (1) accommodation areas, and (2) the ambient environment. Accommodation area criteria pertain to dimensional and outfitting aspects of spaces and open deck areas where crew members eat, sleep, recreate, and perform routine daily activities. The ambient environmental aspects of habitability pertain to the environment that the crew is exposed to during periods of work, leisure, and rest. Specifically, this chapter provides criteria, limits, and measurement methodologies for the following:

- Whole-body vibration (separate criteria for accommodation areas and workspaces).
- Noise.
- Indoor climate.
- Lighting.

The criteria provided throughout this Part 4 are based on currently available research data and standards for the purpose of improving crew performance and providing a base level of habitability and elements of safety related to habitability.

Table 37.1 Summary of the differences between the HAB(WB), HAB+(WB) and HAB++(WB) notations

	HAB(WB)	HAB+(WB)	HAB++(WB)
Accommodation areas	**HAB(WB)** requirements for accommodation areas	**HAB(WB)** and **HAB+(WB)** requirements for accommodation areas	**HAB+(WB)** and **HAB++(WB)** requirements for accommodation areas
Whole-bod vibration	Level of vibration—reducing discomfort and promoting job performance	Lower level of vibration—promoting job performance and increased comfort	Lowest level of vibration—promoting job performance and increased comfort
Noise	IMO Code on noise with modifications	**HAB(WB)** with additional requirements	**HAB+(WB)** with additional requirements
Indoor climate	No provision for individual temperature adjustment		Aimed at enhancing crew comfort by making provisions for individual adjustments of indoor climate temperature
Lighting	No differences among the notations		

Class Notations

At the vessel owner's or shipyard's request, a vessel complying with the minimum criteria for accommodation areas and the ambient environment (i.e., whole-body vibration, noise, indoor climate, and lighting) may be assigned a Class notation which signifies the vessel's compliance with the Class stipulated standards for habitability. For example, a vessel complying with the ABS **HAB(WB)** criteria and the more stringent criteria with respect to accommodation, whole body vibration, noise, and indoor climate may be distinguished in the Record by the notation **HAB+(WB)**. A vessel satisfying all the ABS criteria may be distinguished further in the Record by the notation **HAB++(WB)**. A summary of the differences among each of these notations is presented below (Table 37.1).

Plans and Data to Be Submitted to Class

General. The following general arrangement type drawings of the vessel are required to be submitted to Class for their review and approval:

- Inboard profile detailing the location of the main vertical zone boundaries, the location of the main watertight bulkheads, as well as the various deck levels.
- Plan view of each deck annotating the various spaces on each deck.

Accommodation areas. At a minimum, scaled arrangement drawings of the various accommodation spaces (elevation and plan views) and the vessel's accommodation area specifications should be submitted to class. Details of the crew accommodation area data requirements are provided in the section 'Accommodation area documentation'. Ambient environment. The following items are required to be submitted individually for each ambient environmental aspect:

- *Test plans.* Test plans serve as the principal means for verifying the measurements for the ambient environmental aspects of habitability. Separate test plans are required for whole body vibration, noise, indoor climate, and lighting. Specific test plan details for the various ambient environmental criteria are outlined in later sections of this Part 4:

  The test plans require approval by Class before any measurements are made. Class will usually notify the vessel owner or shipyard whether the test plans have been approved or require alteration. A Class approved copy of the test plan will then become part of the vessel's official documentation.
- *Test reports.* Upon completion of the ambient environmental testing, the test reports should be submitted to the Class approved surveyor for review. In addition, a copy of the test reports will become part of the vessel's official documentation. These reports contain ambient environmental information such as test results, testing details, measurement equipment details, etc. The specific report contents for the various ambient environmental criteria are outlined in later sections of this Part 4, as follows.

Process for Obtaining a Class Notation

Prior to scheduling accommodation area verification or ambient environmental testing activities, the vessel owner or shipyard should certify the operational status of the vessel as being fully operational and/or inclusive of all equipment and furnishings. If the vessel is not fully operational, a complete listing of deficiencies of areas, components, equipment, etc. should be submitted to Class for review. Class will then make a determination and notify the vessel owner or shipyard as to whether accommodation area verification activities or ambient environmental testing can commence. The intent is to conclude all ambient environmental testing by the end of the vessel's sea trials. Figure 37.1 Process for obtaining a habitability notation, charts the process for obtaining an ABS **HAB(WB)**, **HAB+(WB)**, or a **HAB++(WB)** notation. The following sections briefly describe the notation process. Accommodation areas. Arrangement drawings, plans, and vessel design specifications for accommodation spaces should be prepared and submitted to Class for review. For new construction, the drawings should be provided to Class during the detailed design phase. For existing vessels, the arrangement drawings and plans, reflecting the current accommodation area configurations should be provided to and approved by Class, in advance of Class approved surveyor verifications. Follow-up physical verification measurements of accommodation area criteria may only be performed by a Class approved surveyor. The Class approved surveyor will then select the most appropriate

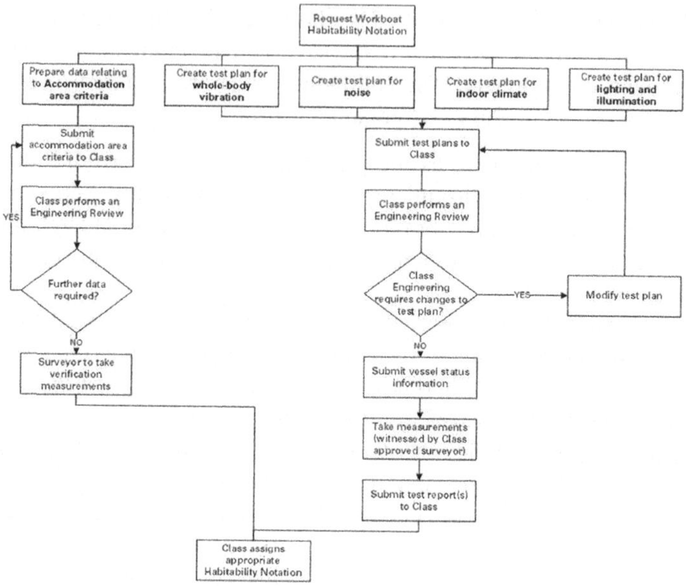

Fig. 37.1 Process for obtaining a class habitability notation

verification measurement sites. The physical verification measurements of accommodation area criteria must usually be performed for all vessels in a series of vessels. The results of the Class review and actual Class approved surveyor verification are then reviewed by the Class approved surveyor during the notation confirmation process.

- *Ambient environment.* Ambient environmental test plans for whole-body vibration, noise, indoor climate, and lighting should be individually prepared and submitted to Class for review and approval. These test plans should serve as a primary means for verifying the measurement locations and measurement process, as well as specifying the testing specialist who will perform the ambient environmental testing. Testing, inspections, and data collection should be performed by the testing specialist and witnessed by a Class approved surveyor. The test reports for ambient environmental testing are to be prepared by the testing specialist and submitted to the Class approved surveyor for review. For a vessel that is expected to operate in extreme environmental conditions (e.g., Polar regions, the Persian Gulf, etc.) that are significantly different from the conditions encountered during the sea trials, the designer/shipyard may be required to submit information on the design/engineering analysis performed as part of the test plan submission verifying that the appropriate requested notation criteria can be met in the actual area of operations. If there are instances during the sea trials where the criteria were either not met, or passed by a small margin, the Class approved surveyor—in conjunction with the designer and the vessel owner—may request additional confirmatory tests to be carried out in the actual area of operation. Full ambient environmental testing as described in this Part 4 is required for the first vessel in a series of vessels. For the second and subsequent sister vessels (i.e., vessels being part of the same series of vessels that have identical structural, machinery, and outfitting designs), full scale testing for the ambient environmental aspects of the notation, is not required. For the second or all subsequent sister vessels constructed in the same shipyard, the following is applicable for all aspects of ambient environmental testing (whole-body vibration, noise, indoor climate and lighting):

  - Spaces that were identified as "worst case" locations in the test plans for the first vessel need to be tested.
  - By review of the first vessel's test reports (all ambient environment aspects), the Class approved surveyor will identify the locations which were out of compliance or close to non-compliance that need to be tested.
  - Spaces that required some type of mitigation for the first vessel need to be tested.

- As part of the notation assignment process, all vessels are subject to Class approved surveyor walk through. Additional testing may be required at the Class approved surveyor's discretion. For sister vessels constructed at a different shipyard, the following provisions may be applicable:

  - Full testing is required for the first vessel at each shipyard.
  - For the subsequent sister vessels, the procedure described previously for vessels built at the same shipyard is applicable.

Results. The Class accommodation area assessment, Class approved surveyor verification measurements, and ambient environmental test reports must be submitted to Class for review, after which Class (or their appointed Class approved surveyor) will issue a determination of notation confirmation.

Initial Requirements

The initial process for obtaining any vessel habitability notation usually comprises some form of Class engineering review, ambient environmental testing, and Class approved surveyor verifications. Testing must be carried out in accordance with the submitted test plans once they have been reviewed and approved by Class in advance of the testing regime commencing. Testing should always, where practicable, be witnessed by a Class approved surveyor. Once the criteria specified by Class (and summarised in this Part 4) have been satisfied, the appropriate notation may be confirmed.

Surveys Completed After Construction

It is intended that all surveys following the vessel's construction are aligned with Class surveys. Harmonisation of surveys may be carried out at the first available opportunity. In order to maintain the **HAB(WB)**, **HAB+(WB)**, or **HAB++(WB)**, or equivalent, notation, an annual survey must be carried out within three months before or after each annual anniversary date of the crediting of the initial survey or the previous special periodical survey. The following information is typically reviewed by the attending Class approved surveyor for issues that could affect the Workboat Habitability notation:

- Collision and grounding reports since previous initial, annual or special periodical survey.
- Fire, repair, and damage reports since previous initial, annual or special periodical survey.
- A list of all structural or mechanical modifications to the vessel since previous initial, annual or special periodical survey.
- Verification that equipment and facilities continue to be fit for purpose and are operating in accordance with accommodation area criteria.

During the attending surveyor's review of the submitted information, a determination will be made as to whether changes or alterations have taken place that could affect the Workboat Habitability notation. As a result, the vessel may be subject to the review, ambient environmental testing, and inspection requirements outlined in this Part 4. Special periodical surveys. In order to maintain the **HAB(WB)**, **HAB+(WB)**, or **HAB++(WB)**, or equivalent, notation, a special periodical survey must be completed within five years after the date of build or after the crediting date of the previous special periodical survey. A special periodical survey will be credited as of the completion date of the survey but not later than five years from date of build or from the date recorded for the previous special periodical survey. If the special periodical survey is completed within three months prior to the due date, the special periodical survey will be credited to agree with the effective due date. The special periodical survey may be commenced fifteen months prior to the due date and be continued with completion by the due date. Survey requirements. The survey typically comprises a Class Engineering review, Class approved surveyor verification, and ambient environmental testing. Most surveys cover all five habitability aspects. In preparation of the survey, the following documentation should be submitted to Class at least three months prior to the ambient environmental testing being conducted:

- Fire, repair, or damage reports since previous annual survey.
- A list of all structural or mechanical modifications to the vessel since previous annual survey.
- Drawings/arrangements of seafarer spaces, HVAC, electrical, etc., affected by alterations.
- Copy of approved initial test plans and test reports.
- Test plans and test reports resulting from annual surveys.

- Previous special periodical survey test plans and reports, if applicable.
- Proposed special periodical survey test plans for the current survey.

The special periodical survey data submission serves three purposes. The first is to perform a Class Engineering review of seafarer spaces against any alterations to the vessel since the initial survey, with measurements verified by the Class approved surveyor. The second purpose is to provide a history of ambient environmental testing, as well as the special periodical survey ambient environmental test plans for review and approval. The third is to allow scheduling of measurement verifications and ambient environmental testing. A special periodical survey test plan for each ambient environmental aspect of Habitability should be submitted in accordance with the criteria stated below. The approved initial test plans should be used as a basis for creating the special periodical survey test plans. For creation of the special periodical survey test plans, the sections titled 'Test plan and Test requirements' may be referenced as they outline the requirements for each ambient environmental aspect. For specifying measurement locations for the special periodical survey test plans, the following should be followed:

- Measurements are taken in all areas affected by vessel alterations. Measurements are limited to the ambient environmental aspect affected by the alteration. For example, structural changes require both whole-body vibration and noise measurements. Structural changes do not necessarily require indoor climate or lighting measurements. Changes to luminaires require lighting measurements but not whole-body vibration, noise, or indoor climate measurements.
- For all ambient environmental aspects, measurements are taken in all worst case or problem area locations based on the requirements outlined in the section 'Selection of spaces where measurements are to be conducted'.
- For all ambient environmental aspects, measurements are taken in 25% of crew cabins and staterooms identified in the initial test plans. The cabin locations must be representative of locations port, starboard, fore, amidships, and aft. Any worst-case locations can be considered part of the representative sample for crew cabins and staterooms, if applicable.
- For all ambient environmental aspects, measurements are taken where a single instance of one type of a manned space exists within the vessel (e.g., the bridge, radio room, officer's mess, gymnasium, library, etc.). The worst-case locations can be considered part of the single instance representative sample, if applicable; and
- Where multiple instances of the same type space exist, a representative sample of at least 25% of each type are selected for measurement for all ambient environmental aspects.

The worst-case locations are to be considered part of the representative sample, if applicable. For all ambient environmental conditions, visual/walk-through inspections should be conducted in accordance with the appropriate guidance provided throughout this Part 4. *Requirements for vessel alterations.* No alterations which affect or may affect the Workboat Habitability notation awarded, including alterations to the structure, machinery, electrical systems, piping, furnishings or lighting systems, are to be made to the vessel unless plans of the proposed alterations are submitted to and approved by Class before the work of alteration is commenced. Where Class determines that the alteration will affect the Workboat Habitability notation, the altered vessel may be subject to the review, verification, and ambient environmental testing requirements.

Alternatives

Class may consider alternative arrangements, criteria and procedures, which can be shown to satisfy the criteria directly cited or referred to in this Part 4. The demonstration of an alternative's acceptability can be made through either the presentation of satisfactory service experience or systematic analysis based on valid engineering principles. *National regulations.* Class will usually consider, for their acceptance, alternative arrangements and details, which can be shown to comply with standards recognised in the country (Flag State) in which the vessel is registered or built, provided they are deemed no less effective than those stipulated by Class. *Departures from criteria.* The criteria summarised in this Part 4 is envisioned to apply to vessels that are engaged in the usual trades and services expected of such vessels. However, it is recognised that unusual or unforeseen conditions may lead to a case where one or more of the parameters of interest in granting a notation may temporarily fall outside the range of acceptability. When a departure from criteria is identified, during either the notation's initial issuance or reconfirmation process, it must be reviewed by Class in consultation with the shipyard or the vessel owner. When the design of the accommodation areas or ambient environmental test results contain departures from the stated criteria, these will be subject to special consideration upon the receipt of details about the departure. Depending on the degree and consequences of the departure, the shipyard or vessel owner may be required to provide an assessment and corrective plan to obtain or maintain the notation. Failure to complete the agreed correction by the due date may lead to the withdrawal of the assigned notation.

Chapter 38
Accommodation Areas on Workboats

General

To promote maritime safety, efficiency, and habitability, it is important that seafarers maintain appropriate levels of mental and physical fitness while on board vessels. To help accomplish this, seafarers should be provided with suitable accommodation areas. Appropriate accommodation area design helps promote reliable performance by reducing the potential for fatigue and human error. Appropriately designed and outfitted accommodation areas may also enhance crew morale, recruiting, retention, comfort, and overall quality of life at sea. Conversely, inappropriate accommodation areas can adversely impact a seafarer's ability to reliably perform assigned duties, fully relax, sleep, and recover from mentally and physically demanding work activities. This in turn can impact their ability to carry out duties on succeeding watches with the required diligence, accuracy, and attention to safety procedures. Providing an onboard environment that increases seafarer alertness and wellbeing should be of concern to vessel owners. This chapter provides the measurement criteria for accommodation spaces. In particular, Tables 38.1, 38.2, 38.3, 38.4, 38.5, 38.6, 38.7, 38.8, 38.9 and 38.10 encompasses criteria for access and egress, as well as crew cabins, sanitary spaces, offices, food services, recreation areas, laundry, and medical spaces. The criteria were selected to help increase crew member safety, productivity, quality of work, retention, and morale. *Note The criteria for access/egress are applicable only to the crew accommodation block and do not include crew member workspaces outside of the accommodation block.* It is worth noting that compliance with the criteria outlined in this chapter are a prerequisite for the ABS Workboat Habitability (**HAB(WB)**), Workboat Habitability Plus (**HAB+(WB)**), or Workboat Habitability Plus Plus (**HAB++(WB)**) notation. Other classification societies will have their own distinct and specific habitability notations, though as the criteria is derived from IMO MLC 2006, Title 3, the criteria are generally standardised across each Class.

A. A. Olsen, *Applying Physical Ergonomics to Modern Ship Design*,
https://doi.org/10.1007/978-3-031-57974-5_38

Table 38.1 Example of the ILO MLC 2006 Title 3 requirement #14

Accommodation area criteria					
General					
Requirement # = Related ILO MLC 2006, Title 3 requirement			HAB(WB) requirements	HAB+(WB) requirements	HAB++(WB) requirements
#	14	Accommodation area stairways have a maximum angle of inclination from the horizontal of			
		50°	•		
		45°		•	
		40°			•

Terminology

Accommodation areas/accommodation block: Vessel areas where the primary purpose is to rest and recreate. Accommodation spaces include cabins, medical facilities (sick bays), offices, recreation rooms, and manned spaces within the accommodation block such as the bridge or control room. For the purposes of this chapter, accommodation areas also include service spaces such as mess rooms,

Accommodation Area Criteria

The accommodation area criteria are contained in Tables 38.2, 38.3, 38.4, 38.5, 38.6, 38.7, 38.8, 38.9 and 38.10. The ILO MLC, 2006 quantitative accommodation area requirements have been included in this chapter and are denoted by a "#" symbol in Tables 38.2, 38.3, 38.4, 38.5, 38.6, 38.7, 38.8, 38.9 and 38.10. The **HAB+(WB)** and **HAB++(WB)** notations have more stringent criteria than the **HAB(WB)** notation with the objective of providing enhanced living and working conditions to improve seafarer safety and comfort. This includes enhanced criteria for workspace design, crew cabins, and recreation/leisure areas. The **HAB+(WB)** notation requires meeting the accommodation area criteria of **HAB(WB)** and **HAB+(WB)**. To achieve a **HAB++(WB)** notation requires meeting the criteria for both **HAB(WB)** and **HAB+(WB)**, as well as the **HAB++(WB)** requirements. For example, Criteria #14 below provides three degrees for angles of inclination for stairways, one for **HAB(WB)**, **HAB+(WB)**, and **HAB++(WB)**. If seeking a **HAB++(WB)** notation, the 40° criteria would need to be fulfilled. By fulfilling the criteria for the **HAB++(WB)** requirements, the criteria for **HAB+(WB)** and **HAB(WB)** are met.

Instructions: For a **HAB(WB)** notation, the boxes with a • in the **HAB(WB)** requirements column must be met. For a **HAB+(WB)** notation, the boxes with a • in

Table 38.2 ILO MLC 2006 Title 3 requirement (accommodation area criteria (general)

Accommodation area criteria					
General					
Requirement # = Related ILO MLC 2006, Title 3 requirement			HAB(WB) requirements	HAB+(WB) requirements	HAB++(WB) requirements
#	1	Headroom in all passageways, sleeping rooms, stairs, sanitary spaces, offices, food service areas, and recreational areas is at least 2030 mm (80.0 in.). This height is clear of any obstructions such as cableways, ducting, pipes, etc.	•		
	2	Non-slip type deck covering is supplied where occasional water, oil or liquid on the floors is expected	•		
	3	Outside corners of bulkheads, doors, etc. are rounded	•		
	4	All edges that crew members may strike are rounded	•		
	5	Drawers and internal doors are designed			
		• To prevent opening and closing due to vessel motion	•		
		• To be operable with one hand		•	
#	6	Painted wall surfaces and deck heads are light in colour	•		
	7	Interior finish materials and furnishings are designed to ease cleaning efforts and improve maintenance	•		

(continued)

Table 38.2 (continued)

Accommodation area criteria					
General					
Requirement # = Related ILO MLC 2006, Title 3 requirement			HAB(WB) requirements	HAB+(WB) requirements	HAB++(WB) requirements
	8	For workstations where seafarers will be working while standing for extended periods of time, a kick space of 100 mm (4 in.) high by 100 mm (4 in.) deep should be provided	•		
	9	9 Toilets are conveniently located to workplace, recreation areas, mess rooms, and shower rooms	•		
	10	A means of escape that is not readily apparent to a person from both the inside and outside of the space is adequately marked	•		
#	11	Wall surfaces and decks are washable and impervious to damp or moisture absorption	•		
#	12	For accommodation areas and recreational facilities where the floorings are made of composite materials, the joints with the sides should be profiled to avoid crevices	•		

(continued)

Table 38.2 (continued)

Accommodation area criteria					
General					
Requirement # = Related ILO MLC 2006, Title 3 requirement			HAB(WB) requirements	HAB+(WB) requirements	HAB++(WB) requirements
#	13	Accommodation areas and recreational and catering facilities should be located as far as practicable from the engines, steering gear rooms, deck winches, ventilation, heating, and air-conditioning equipment, and other noisy machinery and apparatus	•		

the **HAB+(WB)** requirements column need to be met in addition to the **HAB(WB)** requirements. For a **HAB++(WB)** notation, the boxes with a • in the **HAB++(WB)** requirements columns need to be met in addition to the **HAB(WB)** and **HAB+(WB)** requirements.

Accommodation Area Documentation

As stated earlier, accommodation area documentation must be prepared and submitted to Class for review and approval. Confirmatory verification measurements are then performed by a Class approved surveyor. The following data is typically required to be submitted to Class. Data requirements. The submitted data shall serve as a means for verifying that the vessel meets the accommodation area criteria specified in Tables 38.2, 38.3, 38.4, 38.5, 38.6, 38.7, 38.8, 38.9 and 38.10.

- *Newbuilds.* For new construction, scaled arrangement drawings of the accommodation spaces (elevation and plan views), details of the accommodation area outfitting and vessel's design specification in relation to the accommodation spaces must be submitted to Class.
- *Existing vessels.* For existing vessels, appropriate arrangement drawings and plans, reflecting the current accommodation area configurations shall be provided to Class together with any current vessel accommodation area design specifications.

Table 38.3 ILO MLC 2006 Title 3 requirement (accommodation area criteria (access/egress)

Accommodation area criteria				
Access/egress				
Requirement # = Related ILO MLC 2006, Title 3 requirement		HAB(WB) requirements	HAB+(WB) requirements	HAB++(WB) requirements
1	Doors, hatches, or scuttles used as a means of escape are capable of being operated by one person, from either side, in both light and dark conditions	•		
2	The method of opening a means of escape is obvious and rapid	•		
3	Doors in accommodation spaces (with the exception of cabins), stairway, stair tower, passageway, or control spaces, open in the direction of escape, where practicable	•		
4	Deck scuttles that serve as a means of escape are fitted with a quick acting release and a holdback device to hold the scuttle in an open position	•		
5	High voltage electrical cabinet doors are lockable with keys provided to qualified personnel	•		
Door dimensions				
6	Doors have the following dimensions			
	• Doors (other than emergency egress) used solely by crew members have a clear opening width of approximately 710 mm (28 in.) or ASTM size 4 doors	•		
	• The distance from the deck to the top of the door is at least 1900 mm (75 in.)	•		

(continued)

Table 38.3 (continued)

Accommodation area criteria				
Access/egress				
Requirement # = Related ILO MLC 2006, Title 3 requirement		HAB(WB) requirements	HAB+(WB) requirements	HAB++(WB) requirements
	• The distance from the deck to the top of the door is at least 1980 mm (78 in.)			•
Railings				
7	Suitable storm rails/handrails are provided in all interior passageways and at all deckhouse sides where persons on board might have normal access			
	• Storm rails/handrails are installed on both sides of passageways that are 1830 mm (72 in.) or more in width	•		
	• Storm rails/handrails are 865 mm (34 in.) to 965 mm (38 in.) high	•		
	• The distance between/or behind storm rails/handrails and any obstruction is 75 mm (3 in.) or greater	•		
8	Rails are installed parallel to the deck along deck edges and walkways and around open hatches, elevators, antenna platforms and along other boundaries in the following areas			
	• Wherever there is danger of crew members falling to a lower level of 600 mm (23.5 in.) or more in the vessel wherever there is danger of crew members becoming enmeshed with hazardous operating machinery	•		

(continued)

Table 38.3 (continued)

Accommodation area criteria				
Access/egress				
Requirement # = Related ILO MLC 2006, Title 3 requirement		HAB(WB) requirements	HAB+(WB) requirements	HAB++(WB) requirements
	• Around unprotected openings with a coaming height below 760 mm (30 in.)	•		
	Note Temporary rails can be used around unprotected openings into which a person may slip, trip or fall			
9	Deck/guard railings have the following design			
	• The heights of rails or bulwarks are at least 1000 mm (39.5 in.) from the deck except where this height would interfere with the normal operation of the vessel, and toeboards which are at least 100 mm (4.0 in.) in height and have no more than a 6 mm (0.25 in.) clearance between the bottom edge of the toeboard and the walking surface	•		
	• The heights of rails are at least 1070 mm (42 in.) from the deck except where this height would interfere with the normal operation of the vessel and toeboards which are at least 100 mm (4.0 in.) in height and have no more than a 6 mm (0.25 in.) clearance between the bottom edge of the toeboard and the walking surface			•

(continued)

Table 38.3 (continued)

Accommodation area criteria				
Access/egress				
Requirement # = Related ILO MLC 2006, Title 3 requirement		HAB(WB) requirements	HAB+(WB) requirements	HAB++(WB) requirements
	• Vertical stanchions for railings are spaced no more than 1525 mm (60 in.) apart horizontally	•		
	• At least every third vertical stanchion is supported by a bracket or stay	•		
	• Chain or wire rope used as a rail is set such that the sag is not greater than 25 mm (1 in.) at the chain/rope's centre span	•		
	• Maximum lengths of openings protected by wire or chain are 1830 mm (72 in.). A removable stanchion will constitute the start of a new opening	•		
	• Vertical stanchions for railings are spaced no more than 1525 mm (60 in.) apart horizontally	•		
	• At least every third vertical stanchion is supported by a bracket or stay	•		
	• Chain or wire rope used as a rail is set such that the sag is not greater than 25 mm (1 in.) at the chain/rope's centre span	•		
	• Maximum lengths of openings protected by wire or chain are 1830 mm (72 in.). A removable stanchion will constitute the start of a new opening	•		

(continued)

Table 38.3 (continued)

Accommodation area criteria				
Access/egress				
Requirement # = Related ILO MLC 2006, Title 3 requirement		HAB(WB) requirements	HAB+(WB) requirements	HAB++(WB) requirements
Stairs				
10	A clear landing at least as wide as the tread width and a minimum of 915 mm (36 in.) long is provided at the top and bottom of each stairway	•		
11	Any change of direction in a stairway is accomplished by means of an intermediate landing at least as wide as the tread width and a minimum of 915 mm (36 in.) long	•		
12	An intermediate landing is provided at each deck level serviced by a stair, or a maximum of every 3600 mm (144 in.) of vertical travel for stairs with a vertical rise of 6100 mm (240 in.)			•
13	Accommodation area stairways have a maximum angle of inclination from the horizontal of			
	• 50°	•		
	• 45°		•	
	• 40°			•
14	Stair risers and treads have the following design			
	• The riser height is no more than 230 mm (9 in.) and the tread depth is approximately 190 mm	•		
	• (7.5 in.), with an additional 25 mm (1 in.) tread nosing		•	

(continued)

Table 38.3 (continued)

Accommodation area criteria				
Access/egress				
Requirement # = Related ILO MLC 2006, Title 3 requirement		HAB(WB) requirements	HAB+(WB) requirements	HAB++(WB) requirements
	• The riser height is no more than 230 mm (9 in.) and the tread depth is approximately 230 mm (9 in.), with an additional 25 mm (1 in.) tread nosing			•
	• The riser height is no more than 230 mm (9 in.) and the tread depth is approximately 275 mm (10.75 in.), with an additional 25 mm (1 in.) tread nosing	•		
	• Stairs in a stairway or stair tower, the depth of the tread and the height of riser are consistent	•		
	• The minimum tread width on one-way (one person) stairs is at least 610 mm (24 in.)	•		
	• The minimum tread width on two-way (two person) stairs is at least 915 mm (36 in.)		•	
	• The minimum tread width on two-way (two person) stairs is at least 1015 mm (40 in.)			•
	• Once a minimum tread width has been established at any deck, in that stair run, it is not decreased in the direction of escape	•		
	• All nosings have a non-slip surface	•		
15	Stairway or stair towers are fitted with handrails with the following design			
	• A handrail is provided on one side of the stair	•		

(continued)

Table 38.3 (continued)

Accommodation area criteria				
Access/egress				
Requirement # = Related ILO MLC 2006, Title 3 requirement		HAB(WB) requirements	HAB+(WB) requirements	HAB++(WB) requirements
	• A handrail is provided on both sides of the stair			•
	• The handrails are parallel to the pitch line of the stair flight and level at landings	•		
	• The handrail is continuous from the top to the bottom of the stair and terminates in a safe manner at both ends	•		
	• The vertical height above the tread at its nosing is at least 940 mm (37 in.) to 990 mm (39 in.)	•		
	• The distance between handrails and any obstruction is 75 mm (3 in.) or greater	•		
	• Handrails on adjacent, parallel stair flights have a minimum of 100 mm (4 in.) clear distance between rails	•		
Ladders				
16	All inclined ladders and handrails are located so as not to interfere with the opening and closing of hatches, doors, gratings, or other types of access; in all areas of the vessel other than accommodation block	•		
17	Inclined ladders are between 50° and 60° from the horizontal	•		
18	Each vertical ladder used as a means of escape has the following design features	•		

(continued)

Table 38.3 (continued)

Accommodation area criteria				
Access/egress				
Requirement # = Related ILO MLC 2006, Title 3 requirement		HAB(WB) requirements	HAB+(WB) requirements	HAB++(WB) requirements
	• The ladder is mounted at least 180 mm (7 in.) from the nearest permanent object in the back of the ladder (distance behind each rung)	•		
	• Rungs are at least 410 mm (16 in.) in width	•		
	• Rungs are between 275 mm (11 in.) and 300 mm (12 in.) apart	•		
	• Rungs are uniformly spaced for the length of the ladder	•		
	• At least 760 mm (30 in.) clearance in front of the ladder (climbing space)	•		
	• Each ladder rung is skid/slip resistant	•		
	• Grab bars are provided that extend at least 1070 mm (42 in.) above the landing platform		•	
	• Ladder landings at the top and bottom of inclined ladders			
	• At least 760 mm (30 in.) clear in length	•		
	• At least 915 mm (36 in.) clear in length			•
Ramps				
20	Ramps are sloped < 15 degrees for inclined walking surfaces	•		
21	Ramps have a non-skid surface	•		

(continued)

Table 38.3 (continued)

Accommodation area criteria				
Access/egress				
Requirement # = Related ILO MLC 2006, Title 3 requirement		HAB(WB) requirements	HAB+(WB) requirements	HAB++(WB) requirements
22	Ramps have a handrail on any open side of the ramp if the distance from the ramp to the nearest adjacent surface is 600 mm (23.5 in.) or more	•		
23	Access to lifeboats is provided such that a person in a stretcher can be easily embarked into the survival craft (e.g., ramp)	•		
24	The clear walkway width for one person in an unrestricted area, where two persons could pass is at least 710 mm (28 in.)	•		
25	The clear walkway width for normal two-way traffic or any means of egress that leads to an exit or entrance is at least 915 mm (36 in.)	•		
26	The clear walkway width for normal two-way traffic or any means of egress that leads to an exit or entrance is at least 1015 mm (40 in.)			•

Submission Review and Verification

Arrangement drawings, plans, and vessel specifications for the accommodation areas should be prepared and submitted for review by Class. For new construction, the drawings should be provided to Class early during the detailed design phase. For existing vessels, the arrangement drawings and plans reflecting the current accommodation area configurations may be provided to Class in advance of onboard Class approved surveyor verifications. Class will then review the submitted accommodation area documentation. Any deviations from the criteria will be highlighted to the vessel owner/shipyard by Class for resolution. Class will also identify any criteria that the Class approved surveyor must field verify. The surveyor will then verify that

Table 38.4 ILO MLC 2006 Title 3 requirement (accommodation area criteria (crew cabins)

Accommodation area criteria							
Crew cabins							
Requirement # = Related ILO MLC 2006, Title 3 requirement					HAB(WB) requirements	HAB+(WB) requirements	HAB++(WB) requirements
Non-passenger and non-special purpose ship floor area for various vessel sizes							
		< 3000 tonnes	3000 to 10,000 tonnes	> 10,000 tonnes			
One person cabin (seafarers not performing the duties of officers)							
#	1	4.5 m^2 (48.5 ft^2)	5.5 m^2 (60.0 ft^2)	7.0 m^2 (75.0 ft^2)	•		
		5.5 m^2 (60.0 ft^2)	6.5 m^2 (70.0 ft^2)	7.5 m^2 (80.5 ft^2)		•	
		6.5 m^2 (70.0 ft^2)	7.5 m^2 (80.5 ft^2)	8.5 m^2 (91.5 ft^2)			•
Two-person cabin (seafarers not performing the duties of officers)							
#	2	7.0 m^2 (75.0 ft^2)	7.0 m^2 (75.0 ft^2)	7.0 m^2 (75.0 ft^2)	•		
		7.5 m^2 (80.5 ft^2)	8.0 m^2 (80.5 ft^2)	8.0 m^2 (86.0 ft^2)		•	
		8.0 m^2 (86.0 ft^2)	9.0 m^2 (97.0 ft^2)	9.0 m^2 (97.0 ft^2)			•
Junior officer and senior officer (where no private sitting/day room is provided)							
#	3	7.5 m^2 (80.5 ft^2)	8.5 m^2 (91.5 ft^2)	8.5 m^2 (107.5 ft^2)	•		
Passenger ship floor area for various vessel sizes							
		< 3000 tonnes	3000 to 10,000 tonnes	> 10,000 tonnes			
#	4	4.5 m^2 (48.5 ft^2)	5.5 m^2 (60.0 ft^2)	7.0 m^2 (75.0 ft^2)	•		
		5.5 m^2 (60.0 ft^2)	6.5 m^2 (70.0 ft^2)	7.5 m^2 (80.5 ft^2)		•	
		6.5 m^2 (70.0 ft^2)	7.5 m^2 (80.5 ft^2)	8.5 m^2 (91.5 ft^2)			•
Two-person cabin (seafarers not performing the duties of officers)							
#	5	7.0 m^2 (75.0 ft^2)	7.0 m^2 (75.0 ft^2)	7.0 m^2 (75.0 ft^2)	•		
		7.5 m^2 (80.5 ft^2)	8.0 m^2 (80.5 ft^2)	8.0 m^2 (86.0 ft^2)		•	
		8.0 m^2 (86.0 ft^2)	9.0 m^2 (97.0 ft^2)	9.0 m^2 (97.0 ft^2)			•

(continued)

Table 38.4 (continued)

Accommodation area criteria							
Crew cabins							
Requirement # = Related ILO MLC 2006, Title 3 requirement					HAB(WB) requirements	HAB+(WB) requirements	HAB++(WB) requirements
Three-person room (seafarers not performing the duties of officers)							
#	6	11.5 m^2 (124.0 ft^2)	11.5 m^2 (124.0 ft^2)	11.5 m^2 (124.0 ft^2)	•		
		12.5 m^2 (134.5 ft^2)	12.5 m^2 (134.5 ft^2)	12.5 m^2 (134.5 ft^2)		•	
		13.5 m^2 (145.5 ft^2)	13.5 m^2 (145.5 ft^2)	13.5 m^2 (145.5 ft^2)			•
Four-person room (seafarers not performing the duties of officers)							
#	7	14.5 m^2 (156.0 ft^2)	14.5 m^2 (156.0 ft^2)	14.5 m^2 (156.0 ft^2)	•		
		15.5 m^2 (167.0 ft^2)	15.5 m^2 (167.0 ft^2)	15.5 m^2 (167.0 ft^2)		•	
		16.5 m^2 (177.5 ft^2)	16.5 m^2 (177.5 ft^2)	16.5 m^2 (177.5 ft^2)			•
Junior officer (where no private sitting/day room is provided)							
#	8	7.5 m^2 (80.5 ft^2)	7.5 m^2 (80.5 ft^2)	7.5 m^2 (80.5 ft^2)	•		
Senior officer (where no private sitting/day room is provided)3							
#	9	8.5 m^2 (91.5 ft^2)	8.5 m^2 (91.5 ft^2)	8.5 m^2 (91.5 ft^2)	•		
Occupancy level							
#	10	Individual berths are provided for each crew member			•		
#	11	Individual sleeping rooms are provided for each seafarer in ships other than passenger ships (exemption may be granted by the competent authority for ships less than 3000 GT			•		
#	12	Separate sleeping rooms should be provided for men and for women			•		
#	13	Crew members sharing rooms are in same occupation group and day persons do not share a room with watchkeepers			•		

(continued)

Table 38.4 (continued)

Accommodation area criteria					
Crew cabins					
Requirement # = Related ILO MLC 2006, Title 3 requirement			HAB(WB) requirements	HAB+(WB) requirements	HAB++(WB) requirements
	14	Maximum number of persons to be accommodated in any sleeping room is marked indelibly and legibly in some conveniently seen place			
		• In the room	•		
		• Outside the room	•		
Crew cabin location					
	15	Location, means of access, structure and arrangement in relation to other spaces of crew accommodation areas is to			
		• Provide adequate security	•		
		• Protect against weather and sea	•		
		• Insulate from heat and cold	•		
#	16	There is no direct communication into sleeping areas from cargo areas, machinery spaces, chain lockers, galleys, lamp and paint rooms or from engine, deck and other bulk storerooms, drying rooms, communal wash places or water closets	•		
	17	Crew quarters are not located			
		• Further forward in the vessel than a vertical plane located at 5 percent of the vessel's length	•		
		• Abaft of the forward side of the stem at the designed summer load water line	•		
#		• Immediately beneath working alleyways	•		
#		• Below the loadline amidships or aft, unless satisfactory arrangements are made for lighting and ventilation and approved by the Flag State Administration	•		

(continued)

Table 38.4 (continued)

Accommodation area criteria					
Crew cabins					
Requirement # = Related ILO MLC 2006, Title 3 requirement			HAB(WB) requirements	HAB+(WB) requirements	HAB++(WB) requirements
#		• Forward of the collision bulkhead	•		
Cabin design					
General					
#	18	Internal bulkheads are of approved material that is not likely to harbour vermin	•		
	19	Wall surfaces are not constructed from tongued and grooved boarding or any other form of construction likely to harbour vermin	•		
#	20	A sitting or day room, adjoining the sleeping room is provided for the			
		• Master	•		
		• Chief engineer	•		
		• Chief navigating officer	•		
		• Second engineer officer		•	
		Note Ships of less than 3000 GT may be exempted from this requirement by the vessel's Flag State			
	21	Pipes, ventilation ducts, or other installations do not obstruct berths	•		
	22	Sleeping rooms are ventilated and heated	•		
	23	HVAC vents should not be directed at the heads of berths	•		
	24	Radiators and other heating apparatus are so placed and, where necessary, shielded as to avoid risk of fire or danger or discomfort to the occupants	•		
Berths					
	25	Multi-crew cabin berths are not side by side so that access to one berth can only be obtained over another	•		

(continued)

Table 38.4 (continued)

Accommodation area criteria					
Crew cabins					
Requirement # = Related ILO MLC 2006, Title 3 requirement			HAB(WB) requirements	HAB+(WB) requirements	HAB++(WB) requirements
	26	Berths are generally oriented fore and aft	•		
	27	Multi-crew cabin berths contain individually operated privacy curtains	•		
	28	Multi-crew cabin berths contain individually operated fans/blowers		•	
#	29	Multi-crew cabin arrangements are either single tier or double tiered	•		
#	30	Multi-crew cabin berths with portlights above them are only allowed in a single tier arrangement	•		
Emergency					
	31	An emergency alerting system (e.g., an audible alarm) is present for all crew cabin spaces	•		
Outfitting					
#	32	Sleeping rooms with portlights, have curtains that completely block out light	•		
#	33	Sleeping rooms are lighted with natural light and are provided with artificial light (subject to special arrangements as may be permitted in passenger vessels)	•		
	34	Grab bars and stepping surfaces are provided for access to upper berths	•		
#	35	An electric reading light is provided at the head of each berth	•		
#	36	The top berth in a tier has a dust-proof bottom of wood, canvas, or other suitable material	•		
	37	In sleeping rooms with more than one occupant, furniture (beds, lockers, drawers, etc.) are labelled identifying which furniture is dedicated to which berth	•		

(continued)

Table 38.4 (continued)

Accommodation area criteria					
Crew cabins					
Requirement # = Related ILO MLC 2006, Title 3 requirement			HAB(WB) requirements	HAB+(WB) requirements	HAB++(WB) requirements
	38	Each sleeping room is provided with at least the following furniture (in addition to berths and clothes lockers)			
#		• A table or desk	•		
#		• Chair	•		
		• A mirror with a light	•		
#		• A small cabinet for toilet requisites for each person in the room	•		
#		• Book racks	•		
#		• Coat hooks	•		
#	39	Furniture is of smooth, hard material not liable to warp or corrode	•		
#	40	In vessels regularly trading to mosquito-infested ports, suitable screens are fitted to all ventilation inlets and doors to the open deck	•		
Dimensional aspects					
Aisles					
	41	Aisle widths in sleeping areas are	•		
		• At least 610 mm (24 in.), between a single berth and the nearest obstruction	•		
		• At least 915 mm (36 in.), between facing berths			•
Berths					
#	42	The lower berth in a tier is at least 300 mm (12 in.) above the deck	•		
#	43	The upper berth is placed approximately midway between the bottom of the lower berth and the lower side of the deck head beams	•		
	44	Head clearance above each berth is	•		
		• At least 610 mm (24 in.)	•		

(continued)

Table 38.4 (continued)

Accommodation area criteria					
Crew cabins					
Requirement # = Related ILO MLC 2006, Title 3 requirement			HAB(WB) requirements	HAB+(WB) requirements	HAB++(WB) requirements
		• At least 810 mm (32 in.)			•
	45	Berth inside dimensions are			
#		• At least 1980 mm (78 in.) by 800 mm (31.5 in.)	•		
		• At least 2030 mm (80 in.) by 965 mm (38 in.)			•
#	46	The framework and leeboard of a berth is of approved material, hard, smooth and not likely to corrode or to harbour vermin	•		
#	47	Berths constructed from tubular frames are completely sealed and without perforations which would give access to vermin	•		
Storage space					
	48	Sleeping rooms provide storage space for each occupant			
		• The clothes locker is fitted with a shelf and hasp	•		
#		• The clothes locker is at least 0.475 m^3 (16.75 ft^3)	•		
#		• A drawer or equivalent space of at least 56 l; if the drawer is incorporated in the clothes-locker then the combined minimum volume of the clothes locker should be 500 litres; it should be fitted with a shelf and be able to be locked by the occupant so as to provide privacy	•		
	49	Accommodation areas for the hanging of oilskins (wet weather gear) are	•		
		• Sufficient	•		
		• Adequately ventilated	•		
		• Provided outside but convenient to the sleeping rooms	•		

Table 38.5 ILO MLC 2006 Title 3 requirement (accommodation area criteria (sanity spaces)

Accommodation area criteria					
Sanitary spaces					
Requirement # = Related ILO MLC 2006, Title 3 requirement			HAB(WB) requirements	HAB+(WB) requirements	HAB++(WB) requirements
	1	Sanitary spaces are			
		• Lighted	•		
#		• Heated	•		
#		• Ventilated	•		
	2	Floors in sanitary spaces are			
#		• Non-slip type deck covering	•		
#		• Easily cleaned	•		
#		• Impervious to damp or moisture absorption	•		
#		• Properly drained	•		
	3	For every six crew members who do not have a private sanitary space, there is			
#		• One or more tubs and/or shower baths	•		
#		• One or more toilets	•		
#		• Washbasins	•		
	4	Where the radio officers or operators are accommodated in an isolated position, sanitary facilities are provided near or adjacent thereto	•		
	5	Someone standing on a wet deck in sanitary spaces cannot reach light switches or electrical outlets without ground fault interrupters (GFI)	•		
	6	Common/shared spaces are gender identifiable without entering the space	•		
#	7	Separate multiple occupancy sanitary spaces are provided for men and women	•		

(continued)

Table 38.5 (continued)

Accommodation area criteria					
Sanitary spaces					
Requirement # = Related ILO MLC 2006, Title 3 requirement			HAB(WB) requirements	HAB+(WB) requirements	HAB++(WB) requirements
	8	Bulkheads in sanitary spaces are			
#		• Steel or other approved material	•		
#		• Watertight up to 230 mm (9 in.) above deck level	•		
	9	A public sanitary facility should be situated near the ship's office(s) if the office(s) is not conveniently located near the navigation bridge	•		
#	10	All sanitary spaces should have ventilation to the open air, independently of any other part of the accommodation areas	•		
	11	Free space (space available for movement without hindrance of any objects) in a common sanitary space per person is			
		• At least 0.75 m^2 (8 ft^2)	•		
		• At least 1.1 m^2 (12 ft^2)			•
	12	Single or double occupancy sleeping rooms have a private sanitary space with a toilet, shower or tub and sink modules			•
General-based on vessel size					
	13	On vessels over 1600 GT, a toilet and washbasin having hot and cold running potable water is within easy access of the			

(continued)

Table 38.5 (continued)

Accommodation area criteria					
Sanitary spaces					
Requirement # = Related ILO MLC 2006, Title 3 requirement			HAB(WB) requirements	HAB+(WB) requirements	HAB++(WB) requirements
#		• Navigation bridge deck and primarily for those on duty in the area	•		
#		• Machinery space if not fitted near the engine room control centre	•		
		• Galley/mess room(s)	•		
		• Recreation room(s)	•		
		• Gymnasium	•		
	14	On vessels over 1600 GT, without private or semi-private sleeping rooms or sanitary spaces, clothes changing facilities are provided for engine department personnel. These facilities are			
#		• Outside but within easy access of the machinery space and navigating bridge	•		
#		• Fitted with individual lockers	•		
#		• Fitted with tubs and/or showers and washbasins	•		
	15	In vessels between 5000 and 15,000 GT, at least five officers have sleeping rooms with an attached private sanitary space with a toilet, tub and/or shower and washbasin. The washbasin may be in the sleeping room	•		

(continued)

Table 38.5 (continued)

Accommodation area criteria					
Sanitary spaces					
Requirement # = Related ILO MLC 2006, Title 3 requirement			HAB(WB) requirements	HAB+(WB) requirements	HAB++(WB) requirements
	16	In vessels between 10,000 and 15,000 GT, all officers without private sanitary spaces have private intercommunicating sanitary spaces fitted with a toilet, tub and/or shower and washbasin	•		
	17	In vessels over 15,000 GT, all officers have sleeping rooms with an attached private sanitary space fitted with a toilet, tub and/or shower and washbasin. The washbasin may be in the sleeping room	•		
	18	In non-passenger vessels over 25,000 GT, a sanitary space is provided for every two ratings, either in an intercommunicating space between adjoining sleeping rooms or opposite the entrance of such rooms. The sanitary spaces are fitted with a toilet, tub and/or shower and washbasin	•		
#	19	Each sleeping room (both officer and ratings) is provided with a washbasin with hot and cold running potable water (except when a private sanitary space is provided)	•		

(continued)

Table 38.5 (continued)

Accommodation area criteria					
Sanitary spaces					
Requirement # = Related ILO MLC 2006, Title 3 requirement			HAB(WB) requirements	HAB+(WB) requirements	HAB++(WB) requirements
Showers/baths					
	20	Water heaters supplying showers			
		• Do not support areas that have higher water temperature requirements, such as food service area	•		
		• Are provided with anti-scalding devices	•		
	21	Handholds are provided for shower and bath sanitary spaces	•		
	22	In sanitary spaces intended for more than one person, the shower is screened	•		
Toilets					
	23	Where there is more than one toilet in a space, they are sufficiently screened to provide privacy	•		
	24	All toilets			
		• Have flush water available at all times	•		
		• Are independently controllable	•		
		• Have a hand washing station	•		
	25	Toilets, intended for more than one crew member, are situated convenient to, but separate from, sleeping rooms and washrooms without direct access from			
#		• The sleeping rooms	•		

(continued)

Table 38.5 (continued)

Accommodation area criteria					
Sanitary spaces					
Requirement # = Related ILO MLC 2006, Title 3 requirement			HAB(WB) requirements	HAB+(WB) requirements	HAB++(WB) requirements
#		• Toilets to which there is no other access. (This requirement does not apply where a toilet is located in a space between two sleeping rooms having a total of not more than four persons)	•		
	26	All toilet spaces should be ventilated sufficiently to be reasonably free of disagreeable odours and condensation	•		
Washbasins					
	27	Water heaters supplying washbasins			
		• Do not support areas that have higher water temperature requirements	•		
		• Are provided with anti-scalding devices	•		
	28	The following at a minimum are provided for each washbasin			
		• A mirror	•		
		• Toiletry shelf	•		
	29	Facilities to dry hands (e.g., towel bars, etc.) are provided at all washbasins	•		
	30	Washbasins are			
		• Constructed of approved material	•		
		• Smooth surfaced	•		
		• Not liable to crack, flake or corrode	•		

(continued)

Table 38.5 (continued)

Accommodation area criteria					
Sanitary spaces					
Requirement # = Related ILO MLC 2006, Title 3 requirement			HAB(WB) requirements	HAB+(WB) requirements	HAB++(WB) requirements
Urinals					
	31	Urinals have privacy partitions between units and at the end of rows if not provided by sanitary space structures	•		
	32	The dimension between the centreline of two urinals side-by-side is 690 mm (27 in.) or greater	•		
	33	The dimension between the centreline of a urinal and bulkhead next to the urinal is 380 mm (15 in.) or greater		•	
	34	The height of the front edge of a urinal is between 455 mm (18 in.) and 605 mm (24 in.) above the deck surface	•		

the submitted drawings match the constructed vessel and also verify any criteria that are outstanding from the Class review and document deviations from criteria.

Results

On completion of the submission review, the results of the Class review and surveyor verification will be reviewed by the Class approved surveyor against the appropriate **HAB(WB)**, **HAB+(WB)**, or **HAB++(WB)** criteria for notation award. Provided the criteria is met, the appropriate notation will then be confirmed and entered into the vessel's Record.

Table 38.6 ILO MLC 2006 Title 3 requirement (accommodation area criteria (food service areas)

Accommodation area criteria					
Food service areas					
Requirement # = Related ILO MLC 2006, Title 3 requirement			HAB(WB) requirements	HAB+(WB) requirements	HAB++(WB) requirements
General					
	1	Access is provided to all areas requiring cleaning, painting, or treating with pesticide/insecticide	•		
	2	There are no crevices or inaccessible voids, which might harbour vermin, cooking or food waste, or other extraneous matter	•		
	3	Places where food is prepared, served or stored are not used as sleeping areas	•		
	4	Deck drains are provided in all food service areas subject to flood type cleaning or where normal operations release or discharge water or other liquid onto the deck	•		
	5	Deck drains for food service areas are provided in number and location so that complete drainage is possible under normal conditions of list and trim	•		
	6	Where distributive systems are located in food service areas, they are installed as close as practicable to the overhead in order to provide maximum headroom	•		
	7	Flow of food through vessel			

(continued)

Table 38.6 (continued)

Accommodation area criteria					
Food service areas					
Requirement # = Related ILO MLC 2006, Title 3 requirement			HAB(WB) requirements	HAB+(WB) requirements	HAB++(WB) requirements
		• Is arranged in logical sequence that minimises cross-traffic or backtracking	•		
		• Allows for adequate separation of clean and soiled operations	•		
Food service doors					
	8	Toilet room doors opening directly into food preparation or serving areas are self-closing and tight	•		
Food service passageways					
	9	The passage width between storage shelves is at least 710 mm (28 in.)	•		
Mess room					
General					
	10	Mess rooms are			
#		• Located apart from sleeping rooms	•		
		• Adequately ventilated	•		
		• Adequately insulated to prevent condensation or overheating	•		

(continued)

Table 38.6 (continued)

Accommodation area criteria					
Food service areas					
Requirement # = Related ILO MLC 2006, Title 3 requirement			HAB(WB) requirements	HAB+(WB) requirements	HAB++(WB) requirements
		• Located as near to the galley as is practicable except where the mess room is equipped with a steam table	•		
#	11	Mess rooms have tables and seats sufficient for the number of persons likely to use them at any one time	•		
	12	Mess lines and mess rooms are protected from weather, objectionable sights (such as garbage disposal areas) and objectionable odours (such as from engines, holds, toilets, fire room, etc.)	•		
#	13	Lighting in mess rooms is provided by means of natural light and artificial light (subject to special arrangements as may be permitted in passenger vessels)	•		
	14	The following is available for crew use at all times			
#		• Refrigerator	•		
#		• Facilities for hot beverages	•		

(continued)

Table 38.6 (continued)

Accommodation area criteria					
Food service areas					
Requirement # = Related ILO MLC 2006, Title 3 requirement			HAB(WB) requirements	HAB+(WB) requirements	HAB++(WB) requirements
#		• Facilities for cool water	•		
#	15	Common or separate mess room facilities are to be provided	•		
	16	The tops of tables and seating are capable of being easily cleaned	•		
#	17	Where available pantries are not accessible to mess rooms, adequate lockers for mess utensils and proper facilities for washing utensils are provided	•		
Outfitting					
	18	A transparent sanitation shield is installed to completely shield the entire length of the food serving lines if provided with a tray slide or rail	•		
	19	Tray slides (or rails) if provided are placed along food and beverage serving lines	•		
	20	Adequate lockers for mess utensils are provided	•		
Mess room dimensional aspects					
	21	Rectangular table space for each diner is	•		

(continued)

Table 38.6 (continued)

Accommodation area criteria					
Food service areas					
Requirement # = Related ILO MLC 2006, Title 3 requirement			HAB(WB) requirements	HAB+(WB) requirements	HAB++(WB) requirements
		• At least 610 mm (24 in.) wide by 380 mm (15 in.) deep	•		
		• At least 740 mm (29 in.) wide by 430 mm (17 in.) deep	•		
	22	Round Table diameter is			
		• For seating 4 persons at least 1220 mm (48 in.)	•		
		• For seating 4 persons at least 1370 mm (54 in.)	•		
		• For seating 6 persons at least 1520 mm (60 in.)	•		
		• For seating 6 persons at least 1680 mm (66 in.)	•		
		• For seating 8 persons at least 1830 mm (72 in.)	•		
		• For seating 8 persons at least 1980 mm (78 in.)		•	
	23	Distance between tables with back-to-back seating is			
		• At least 1195 mm (47 in.)	•		
		• At least 1525 mm (60 in.)		•	

(continued)

Table 38.6 (continued)

Accommodation area criteria					
Food service areas					
Requirement # = Related ILO MLC 2006, Title 3 requirement			HAB(WB) requirements	HAB+(WB) requirements	HAB++(WB) requirements
	24	Deck area requirements for planned seating capacity are			
#		• At least 1.5 m^2 (16.2 ft^2) per person	•		
		• At least 1.7 m^2 (18.4 ft^2) per person		•	
		• At least 1.9 m^2 (20.5 ft^2) per person			•
Galley/scullery					
General					
#	25	The galley comprises space for food preparation, cooking, service and wash-up facilities. These should include areas for			
		• Baking and pastry preparation	•		
		• Vegetable preparation	•		
		• Meat, fish and poultry preparation	•		
		• Hot services	•		
		• Dish wash facilities	•		
		• Pan wash facilities	•		
		• Equipment storage	•		

(continued)

Table 38.6 (continued)

Accommodation area criteria				
Food service areas				
Requirement # = Related ILO MLC 2006, Title 3 requirement		HAB(WB) requirements	HAB+(WB) requirements	HAB++(WB) requirements
26	Ice making facilities for at least 0.57 kg (1.25 lbs) of ice per person per day are provided	•		
27	Hand washing stations located so that no crew member must walk more than 7.5 m (25 ft) to a station are to be provided. Slop sinks and scullery sinks are not satisfactory hand washing stations	•		
28	Tops of food dressing tables, counters and drain boards are constructed with an integral back splash	•		
Potable water				
29	Only potable water is piped to food storage, preparation, or service areas. The exceptions are a food waste grinder eductor (garbage disposal) line and deck washing facilities. The grinder delivery line is protected against back flow	•		

(continued)

Table 38.6 (continued)

Accommodation area criteria					
Food service areas					
Requirement # = Related ILO MLC 2006, Title 3 requirement			HAB(WB) requirements	HAB+(WB) requirements	HAB++(WB) requirements
	30	If the vessel has a garbage room, hot and cold potable water is available for washing garbage cans	•		
	31	Potable water tanks are identified with a number and the words "POTABLE WATER" adequately sized	•		
	32	Potable water piping and fittings are labelled	•		
Maintenance/cleaning					
	33	Drawers and bins used as food contact surfaces are readily removable and easily cleaned	•		
	34	Covers, insets, or receptacles for unpackaged foods or beverages are easily removable or designed for easy cleaning in place	•		
	35	Stove top or range sea rails are readily removable and easily cleanable and brackets for sea rails are easily cleanable	•		

(continued)

Table 38.6 (continued)

Accommodation area criteria					
Food service areas					
Requirement # = Related ILO MLC 2006, Title 3 requirement			HAB(WB) requirements	HAB+(WB) requirements	HAB++(WB) requirements
	36	Ventilation baffles, vanes, louvres, dampers, filters, etc., are easily accessible or removable (without tools) for cleaning	•		
Refrigeration/freezer					
	37	It is possible to open the doors of the refrigerated or cold storage rooms from the inside even though they are locked from the outside	•		
	38	There is an alarm in each freezer and cold storage room to warn when the temperature is out of specification	•		
Dish washing					
	39	Easy access is provided to a sink or dish washing machine	•		

Table 38.7 ILO MLC 2006 Title 3 requirement (accommodation area criteria (recreation)

Accommodation area criteria					
Recreation					
Requirement # = Related ILO MLC 2006, Title 3 requirement			HAB(WB) requirements	HAB+(WB) requirements	HAB++(WB) requirements
General					
#	1	Recreation and accommodation areas, conveniently situated and appropriately furnished are provided for the crew. (Where this is not provided separately from the mess rooms the latter is planned, furnished and equipped to give recreational facilities)	•		
#	2	Vessel has an open deck space or spaces to which the crew can have access when off duty. The space or spaces are of adequate area, having regard to the size of the vessel and the crew	•		
#	3	Recreation rooms in the crew accommodation spaces are adequately insulated to prevent condensation or overheating	•		
Recreational and exercise areas					
#	4	Furnishing for recreational facilities should as a minimum include a book-case and facilities for reading, writing and, where practicable, table and deck games	•		

(continued)

Table 38.7 (continued)

Accommodation area criteria					
Recreation					
Requirement # = Related ILO MLC 2006, Title 3 requirement			HAB(WB) requirements	HAB+(WB) requirements	HAB++(WB) requirements
#	5	Television and DVD/CD player facilities are to be provided. The stock of films should be adequate for the duration of the voyage and, where necessary, changed at reasonable intervals	•		
#	6	Consideration should be given to include the following recreational facilities on board, where practicable			
		• A smoking room	•		
		• Sport equipment including exercise equipment	•		
		• Facilities for swimming	•		
		• A library containing vocational and other books, the stock of which should be adequate for the duration of the voyage and changed at reasonable intervals	•		

(continued)

Table 38.7 (continued)

Accommodation area criteria					
Recreation					
Requirement # = Related ILO MLC 2006, Title 3 requirement			HAB(WB) requirements	HAB+(WB) requirements	HAB++(WB) requirements
		• Facilities for recreational handicrafts	•		
		• Canteen	•		
		• Where appropriate, the provision of bars on board for seafarers unless these are contrary to national, religious or social customs	•		
	7	Approximately 1.0 m^2 (11 ft^2) is provided for each seat in the lounge	•		
	8	Approximately 1.25 m^2 (14 ft^2) is provided for each seat in the lounge			•
	9	Vessels regularly trading to or in the tropics and the Arabian Gulf, are equipped with awnings for use	•		

Table 38.8 ILO MLC 2006 Title 3 requirement (accommodation area criteria (laundry)

Accommodation area criteria					
Laundry					
Requirement # = Related ILO MLC 2006, Title 3 requirement			HAB(WB) requirements	HAB+(WB) requirements	HAB++(WB) requirements
General			•		
#	1	Facilities exist for washing and drying clothes on a scale appropriate to the size of the crew and the normal duration of the voyage	•		
	2	Laundry facilities are located within easy access of crew accommodation areas	•		
	3	Laundry facilities should be sufficient to allow seafarers to be provided with clean and dry underwear once per day and clean and dry outerwear and bedding once per five days	•		
	4	Appropriate clearance has been provided for equipment maintenance		•	
	5	Coamings and/or deck drains are provided around washing machines, water heaters, etc.	•		
	6	Washer and dryer capacities are matched to assure a smooth workflow	•		
	7	Washers and dryers are placed relative to each other to facilitate the transfer of clothing from the washer to the dryer	•		

(continued)

Table 38.8 (continued)

Accommodation area criteria					
Laundry					
Requirement # = Related ILO MLC 2006, Title 3 requirement			HAB(WB) requirements	HAB+(WB) requirements	HAB++(WB) requirements
Outfitting					
	8	Tumble dryers, if provided, are exhausted directly into the weather, not into the vessel	•		
	9	Tumble dryers, if provided, are equipped with lint filters	•		
	10	The laundry facilities include at a minimum			
#		• Clothes washing machines	•		
#		• Clothes drying machines (or adequately heated and ventilated drying rooms)	•		
#		• Irons and ironing boards (or their equivalent)	•		
		• Suitable sinks with an adequate supply of cold potable water and hot potable water or means of heating water	•		
		• Sinks are installed in washrooms, if a separate laundry accommodation area is not reasonably practical	•		
	11	The facilities for drying clothes are			
		• Equipped with lines or other fittings for hanging clothes	•		

(continued)

Table 38.8 (continued)

Accommodation area criteria					
Laundry					
Requirement # = Related ILO MLC 2006, Title 3 requirement			HAB(WB) requirements	HAB+(WB) requirements	HAB++(WB) requirements
		• Separate from sleeping rooms and mess rooms	•		
		• Ventilated and heated	•		

Table 38.9 ILO MLC 2006 Title 3 requirement (accommodation area criteria (medical)

Accommodation area criteria					
Medical					
Requirement # = Related ILO MLC 2006, Title 3 requirement			HAB(WB) requirements	HAB+(WB) requirements	HAB++(WB) requirements
General					
#	1	In vessels of less than 500 GT a medical chest, medical equipment and a medical guide should be carried on board as a minimum. The type of ship, the number of persons on board and the nature, destination and duration of voyages and relevant national and international recommended medical standards should be taken into account	•		
#	2	Separate medical accommodation area used exclusively for medical purposes is provided in vessels of 500 GT or more, carrying 15 or more seafarers and engaged in a voyage of more than three days' duration	•		

(continued)

Table 38.9 (continued)

Accommodation area criteria					
Medical					
Requirement # = Related ILO MLC 2006, Title 3 requirement			HAB(WB) requirements	HAB+(WB) requirements	HAB++(WB) requirements
	3	Medical accommodation area is not used as a storage area except for medical supplies	•		
	4	Medical accommodation area is suitably located to allow for safe and efficient			
		• Access for sick or injured crew members	•		
		• Stretcher transportation from accommodation areas or work areas	•		
	5	Medical accommodation area is suitably separated from other spaces and used for the care of the sick and for no other purpose	•		
	6	Medical accommodation area is suitably designed so that the occupant can be comfortably housed and may receive proper attention in all weather	•		
	7	The arrangement of the entrance, berths, lighting, ventilation, heating and water supply is designed to provide comfort and facilitate the treatment of crew members	•		

(continued)

Table 38.9 (continued)

Accommodation area criteria					
Medical					
Requirement # = Related ILO MLC 2006, Title 3 requirement			HAB(WB) requirements	HAB+(WB) requirements	HAB++(WB) requirements
Outfitting					
	8	If the vessel does not carry a doctor, it must have an approved medicine chest with readily understandable instructions	•		
#	9	Medical accommodation area has a toilet, washbasin and bathtub or shower conveniently situated for the use of patients	•		
	10	The door to the toilet and washbasin should open outwards and any door latch should be capable of being opened from the outside if equipped	•		
	11	Medical accommodation area is appropriately equipped to render the necessary medical attention	•		
	12	Medical accommodation area is equipped with stretchers that meet the following			
		• The ability to winch a sick person into a helicopter or vessel, and the capability of floating and righting itself in the water	•		

(continued)

Table 38.9 (continued)

Accommodation area criteria					
Medical					
Requirement # = Related ILO MLC 2006, Title 3 requirement			HAB(WB) requirements	HAB+(WB) requirements	HAB++(WB) requirements
		• Collapsible and suitable for use in confined spaces	•		
#	13	Toilet accommodation area is provided for the exclusive use of the occupants of the hospital accommodation area	•		
	14	Hospital and medical/ first aid accommodation area is situated and arranged so that a stretcher can be easily carried into it and placed alongside an examination table or bed		•	
	15	Medical accommodation area has lockable storage for drugs, dressings and medical equipment	•		
	16	Medical accommodation area has an intercom or signalling system (e.g., an emergency call device) to the navigation bridge	•		

Table 38.10 ILO MLC 2006 Title 3 requirement (accommodation area criteria (ship's offices)

Accommodation area criteria					
Ship's offices					
Requirement # = Related ILO MLC 2006, Title 3 requirement			HAB(WB) requirements	HAB+(WB) requirements	HAB++(WB) requirements
General					
	1	1 In vessels of over 3000 GT			
#	2	One room is provided and equipped for use as an office for the deck department	•		
#	3	One room is provided and equipped for use as an office for the engine department	•		

Chapter 39
Vibration on Workboats

General

Working and/or living on board vessels imposes a series of generally low-frequency mechanical vibration, as well as single-impulse shock loads on the human body. Low-frequency vibrations are also imposed by vessel motions, which are produced by the various sea states in conjunction with vessel speed. These motions can result in motion sickness, body instability, fatigue, and increased health risk aggravated by shock loads induced by vessel slamming. Vessel slamming may be caused by dynamic impact loads being exerted on the vessel's bottom or bow flare due to vessel size, speed, and wave conditions. Higher-frequency vibration influencing comfort is often associated with rotating machinery. The imposition of higher frequency vibrations (about 1–80 Hz) induces corresponding motions and forces within the human body, creating discomfort and possibly resulting in degraded performance and health (Griffin 1990). This chapter provides the criteria and methods for assessing whole-body vibration relating to habitability on board vessels. The criteria were selected to limit potential vibration related interference with work tasks and to improve crew comfort. Consideration of the vibration loads imposed on the body is restricted to motions transmitted from surrounding structures to the entire human body through the feet of a standing person in the frequency range 1–80 Hz. Motions transmitted to the body of a seated or recumbent person have been omitted from this guidance. Due to the provision of resilient or non-rigid surfaces on seats and beds, these surfaces will generally attenuate the transfer of vibration to levels that are lower than those experienced when standing. The motions transmitted through the feet are expected to be the highest vibration levels to which crew will be exposed. This chapter applies to manned spaces as defined in Chap. 30. Examples of such spaces include accommodation spaces (e.g., cabins, mess areas, recreation spaces, etc.) and workspaces (e.g., galley, control rooms, duty stations, etc.) both inside and, in other vessel locations, outside the accommodation block. Whole-body vibration limits defined in this chapter are based on currently available standards. Compliance with the guidance

A. A. Olsen, *Applying Physical Ergonomics to Modern Ship Design*,
https://doi.org/10.1007/978-3-031-57974-5_39

discussed in this chapter is typically a prerequisite for the Workboat Habitability notation assigned by Class (e.g., **HAB(WB)**, **HAB+(WB)**, and **HAB++(WB)** for ABS Classed vessels).

Terminology

Acceleration: The rate of change of velocity over time (i.e., metres-per-second, squared (m/s^2).

Dynamic positioning: A system to automatically maintain a workboat's position and heading by controlling propellers and/or thrusters. Dynamic positioning can maintain a position to a fixed point over the bottom, or in relation to a moving object (such as another vessel). It can also be used to position the vessel at a favourable angle towards wind, waves, and current.

Frequency: The number of complete cycles of a periodic process occurring per unit time. Frequency is expressed in Hz which corresponds to the number of cycles observed-per-second.

Frequency weighting: A transfer function used to modify a signal according to a required dependence on vibration frequency.

- In human response to vibration, various frequency weightings have been defined in order to reflect known or hypothesised relationships between vibration frequency and human response.
- The frequency weighting used to evaluate whole-body vibration in this chapter is W_m (whole-body) for all three axes (x, y, and z), in accordance with ISO 6954.

Manned space: Any space where a seafarer may be present for 20 min or longer at one time during normal, routine daily activities. Such spaces would include working or living spaces.

Multi-axis acceleration value: The Multi-Axis Acceleration Value is calculated from the root-sums-of squares of the weighted RMS acceleration values in each axis (a_{xw}, a_{yw} and a_{zw}) at the measurement point using the following expression:

$$a_w = \sqrt{a_{xw}^2 + a_{yw}^2 + a_{zw}^2}$$

where a_{xw}, a_{yw} and a_{zw} are the weighted RMS acceleration values measured in the x-, y- and z-axes, respectively.

Multi-axis vibration: Mechanical vibration or shock acting in more than one direction simultaneously.

Reference calibration: Calibration of a measuring instrument and of a portable calibrator conducted at least every two years (or more frequently if specified by the manufacturer) by an accredited Testing and Calibration Laboratory accredited according to ISO 17025:2005, as amended, with traceability to a national or international standard.

Transit conditions: Those conditions where the vessel is transitioning (moving) from one location to another.

Vibration: The variation with time of the magnitude of a quantity which is descriptive of the motion or position of a mechanical system, when the magnitude is alternately greater and smaller than some average value.

Weighted root-mean-square acceleration value (a_w): The weighted RMS acceleration, a_w, in metres-per-second squared, is defined by the expression:

$$a_w \sqrt{\frac{1}{T}\int_0^T a_w^2(t)dt}$$

where $a_{w(t)}$ is the weighted acceleration as a function of time in metres-per-second squared (m/s^2) and t is the duration of the measurement in seconds.

Whole-body vibration: Mechanical vibration (or shock) transmitted to the human body as a whole. Whole-body vibration is often due to the vibration of a surface supporting the body.

Associated Documentation

The following documents provide details about test plan preparation, test measurement procedures and/or test reporting:

- ISO 6954:2000, Mechanical Vibration and Shock—Guidelines for the Measurement, Reporting and Evaluation of Vibration with regard to Habitability on Passenger and Merchant Ships.
- ISO 2631-2:2003, Mechanical Vibration and Shock—Evaluation of Human Exposure to Whole Body Vibration—Part 2, Vibration in Buildings.
- ISO 8041:2005, Human Response to Vibration—Measuring Instrumentation.
- ISO 5348:1998, Mechanical Vibration and Shock—Mechanical Mounting of Accelerometers.
- WMO: Manual on Codes, No. 306, Part A, Alphanumerical Codes (2011 Edition, Updated in 2014).

Further guidance can be found in:

- ISO 2923: 1996, Acoustics—Measurement of noise onboard vessels.

- ISO 20283-2:2008, Mechanical Vibration—Measurement of Vibration on Ships—Part 2: Measurement of Structural Vibration.

Criteria

The whole-body vibration criteria for the ship Habitability notations (e.g., **HAB(WB)**, **HAB+(WB)**, or **HAB++(WB)**) are provided in Table 39.1, Maximum weighted root-mean-square acceleration level (workboats less than or equal to 60 m in length) and Table 39.2, Maximum weighted root-mean-square acceleration level (workboats greater than 60 m in length). Maximum whole-body vibration levels are provided for transit and dynamic positioning conditions. The severity of the vibration is indicated by the weighted root-mean-square acceleration value (aw) as defined in ISO 8041. Whole-body vibration measurements should only be taken in manned spaces. A space is considered "manned" if it is occupied by a crew member for 20 min or longer at a time for normal, routine daily activities. Specific locations are referred to in the section on 'Measurement locations'. The maximum vibration levels for accommodation areas and workspaces in Table 39.1, Maximum weighted root-mean-square acceleration level (workboats less than or equal to 60 m in length) and Table 39.2, Maximum weighted RMS acceleration level (workboats greater than 60 m in length) should not be exceeded under normal operating conditions, either in transit or dynamic positioning. For instance, the **HAB(WB)** notation's maximum vibration level is primarily aimed at reducing discomfort and interference with task performance whereas the more stringent maximum level for the **HAB+(WB)** and **HAB++(WB)** notations are aimed at improving performance and comfort. In this instance, "comfort" means the ability of the crew to use a space for its intended purpose with minimal interference or annoyance from whole-body vibration. For the purposes of this chapter, the notation applies to the vibration levels occurring on the deck supporting the human body in the three (3) translational (x-, y- and z-) axes as shown in Fig. 39.1, Measurement axes. The vibration levels are computed for each axis individually, as well as combined as a multi-axis acceleration value. Each is expressed as a frequency weighted root-mean-square (a_w) value. To meet the vibration criteria, the multi-axis a_w level must be less than or equal to the maximum level expressed in Table 39.1, Maximum weighted RMS acceleration level (workboats less than or equal to 60 m in length) and Table 39.2, Maximum weighted root-mean-square acceleration level (workboats greater than 60 m in length):

Table 39.1 Maximum weighted root-mean-square acceleration level (workboats less than or equal to 60 m in length)

Notation	Frequency range (Hz)	Acceleration measurement	Maximum RMS level			
			Transit conditions		Dynamic positioning conditions	
			Accommodation areas	Workspaces	Accommodation areas	Workspaces
HAB(WB)	1–80	a_w	178 mm/s^2 (5.0 mm/s)	214 mm/s^2 (6.0 mm/s)	178 mm/s^2 (5.0 mm/s)	214 mm/s^2 (6.0 mm/s)
HAB+(WB)	1–80	a_w	161 mm/s^2 (4.5 mm/s)	196 mm/s^2 (5.5 mm/s)	161 mm/s^2 (4.5 mm/s)	196 mm/s^2 (5.5 mm/s)
HAB++(WB)	1–80	a_w	143 mm/s^2 (4.0 mm/s)	178 mm/s2 (5.0 mm/s)	143 mm/s^2 (4.0 mm/s)	178 mm/s^2 (5.0 mm/s)

Table 39.2 Maximum weighted root-mean-square acceleration level (workboats greater than 60 m in length)

Notation	Frequency range (Hz)	Acceleration measurement	Maximum RMS level			
			Transit conditions		Dynamic positioning conditions	
			Accommodation areas	Workspaces	Accommodation areas	Workspaces
HAB(WB)	1–80	a_w	161 mm/s^2 (4.5 mm/s)	178 mm/s^2 (5.0 mm/s)	178 mm/s^2 (5.0 mm/s)	196 mm/s^2 (5.5 mm/s)
HAB+(WB)	1–80	a_w	143 mm/s^2 (4.0 mm/s)	161 mm/s^2 (4.5 mm/s)	161 mm/s^2 (4.5 mm/s)	178 mm/s^2 (5.0 mm/s)
HAB++(WB)	1–80	a_w	125 mm/s^2 (3.5 mm/s)	143 mm/s^2 (4.0 mm/s)	143 mm/s^2 (4.0 mm/s)	161 mm/s^2 (4.5 mm/s)

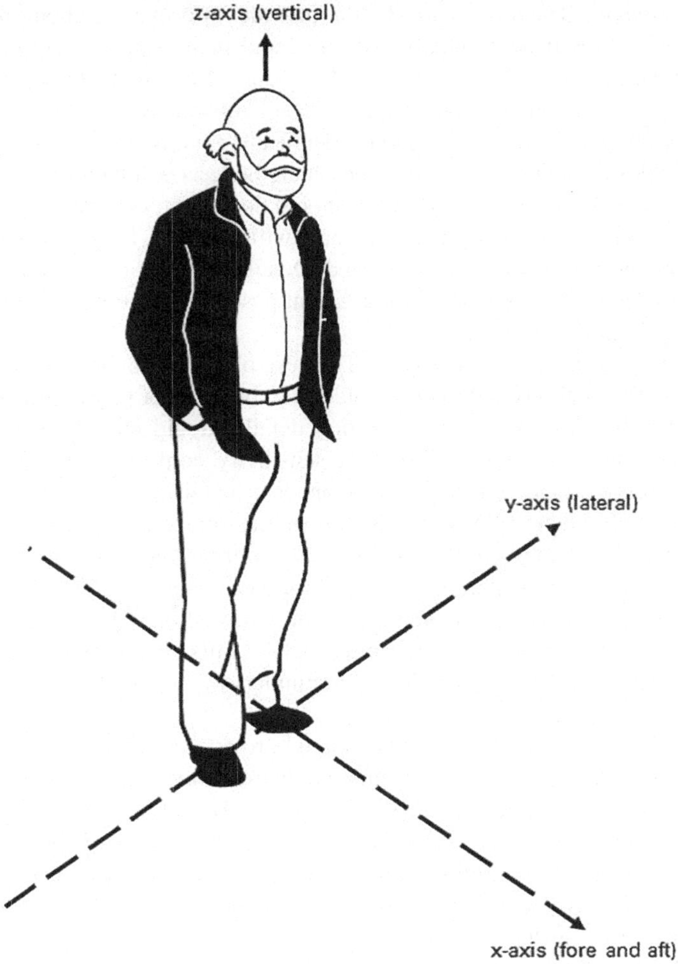

Fig. 39.1 Measurement axes

Test Plan

As stated in the section 'Test plans' in Chap. 37, a test plan should be developed to serve as the principal means for verifying the measurements to be performed to verify compliance with whole-body vibration criteria. This test plan should typically include the following:

- *Documentation.* The test plan should include appropriate drawings indicating the location of all vibration sources.

- *Test personnel.* The test plan should provide information about the testing specialist who will be conducting the test and their approval and certification in accordance with the Class specific Procedural Requirements for Recognised Ambient Environmental Testing Specialists or equivalent.
- *Test conditions.* The test plan should detail the conditions for both transit and dynamic positioning conditions under which the tests will be performed.
- *Measurement locations.* The test plan should document, in detail, on appropriate drawings, all spaces where measurements will be taken. In addition, transducer placement positions should be indicated. Details on selecting measurement locations and determining transducer placement positions are provided in the section 'Measurement locations'.
- *Data acquisition and instruments.* The test plan should provide information regarding the methods and instrumentation to be used for measurement and data collection. Instrumentation specification details should include type of instruments to be used, accuracy, calibration, sensitivity, conformance with ISO 8041, and frequency range. Further details on data acquisition and instruments are provided in the section 'Data acquisition and instruments'.
- *Data analysis.* The test plan should provide information regarding the methods, software, and instrumentation to be used for data analysis.
- *Test schedule.* The test plan should provide information regarding the proposed test schedule. The test schedule should include information on the approximate date and duration of testing (including a summary of the scheduled measurements per day).
- *Data collection sheets.* The test plan should provide sample data sheets (in table format) that will be used for reporting the measurement data.
- *Conflict of interest declaration.* In the case that the testing specialist has, in any capacity or to any extent, provided input into the design of the vessel, their participation should be indicated in the test plan including details of the testing specialist's involvement. Class may review any potential conflicts of interest and determine if any further action is necessary.

Test Requirements

Whole-body vibration measurements should be carried out in accordance with the procedures described in ISO 6954. When the requirements or procedures proscribed by Class deviate from those mentioned in ISO 6954, then the requirements or procedures proscribed by Class will take precedence unless otherwise stipulated. Data acquisition and instruments. For Class Habitability notations such as **HAB(WB)**, **HAB+(WB)**, or **HAB++(WB)**, a sample of data should be recorded for each whole-body vibration measurement position. Each whole-body vibration measurement sample should be at least 60 s in duration. For each location measured, a data sample should also be taken in accordance with the requirements set out in the section 'Test conditions'. The above measurement samples should all be taken using the appropriate Type 1 instrumentation (ISO 8041), then frequency weighted and analysed in accordance with ISO 6954. It is always desirable to employ equipment that records and stores acceleration time histories. *Test conditions.* The test conditions required for the whole-body vibration measurements should be in accordance with the following:

- *Power output*:
  1. *Transit*: The propulsion machinery should run at contractual service conditions.
  2. *Dynamic positioning*: Devices such as azimuth or tunnel thrusters automatic or manual should run at contractual service conditions.
- *Equipment operation.* As appropriate for the mode of operation (transit or dynamic positioning), all machinery essential for vessel operation should be operated under normal conditions throughout the measurement period. HVAC systems are to be running as for normal seagoing conditions during the whole-body vibration measurements.
- *Course and water depth.* Whole-body vibration measurements are to be taken with the vessel in a depth of water not less than five times the draft of the vessel. For vessels that do not operate in water depths of five times draught, measurements should be taken under normal operating and steady state transit conditions. The vessel should maintain a single heading and a constant speed during the test. Measurements during dynamic positioning should be taken while the vessel is maintaining a position relative to another vessel or fixed point (e.g., offshore installation).
- *Rudder conditions.* Rudder action should be minimised wherever possible. During dynamic positioning, measurements should be taken while the vessel is maintaining a position relative to another vessel or fixed point (e.g., offshore installation).
- *Sea conditions.* Measurements are to be taken under conditions of Sea State 3 or less, as defined by the WMO Manual on Codes, No. 306, Part A, Alphanumerical Codes (2011 Edition, updated in 2019).
- *Loading conditions.* For cargo vessels, measurements should be taken under normal operating conditions.
- *Test interference.* During the whole-body vibration measurements, vibration arising from every kind of unnecessary human activity should be avoided. For this reason, only the personnel needed for the normal operation of the equipment in the space and those carrying out the measurements may be present in the space being tested.

Measurement locations. Selection of spaces where measurements are to be conducted. The aim when selecting vibration measurement locations is to obtain a representative sample of data that reflects the actual conditions in manned spaces. For practical reasons, it is important to select the locations such that an appropriate amount of data can be collected during the testing phase. This means the measurement locations should be selected in accordance with the following criteria:

- Select potential worst-case locations based on their proximity to vibration emitting sources such as propulsion or other rotating machinery or where vibration is likely to be transmitted to manned spaces, accommodation areas and recreation areas via the vessel's structure. Measurements should be taken in all identified worst-case locations (e.g., cabin adjacent to a machinery space).
- Where a single instance of one type of manned space exists within the vessel (e.g., the bridge, mess room, gymnasium, library, etc.), that location should be selected for measurement.
- Select a representative sample of crew cabins throughout the vessel. For vessels with less than 20 cabins, 50% of cabins on each deck should be selected. For vessels with greater than 20 cabins, 30% of cabins on each deck should be selected. These measurement locations must be selected at locations port, starboard, fore, amidships and aft. The worst-case locations are to be considered part of the representative sample for crew cabins, where applicable.
- Where multiple instances of the same types of accommodation space exist, but are not crew cabins, a representative sample of at least 50% of each type should be selected for measurement. The worst-case locations are to be considered part of the representative sample, where applicable.

Walk through verification inspection locations. All normally manned spaces are to be subject to a walkthrough inspection by the Class approved surveyor. The number and locations of the walkthrough inspections are generally determined by the surveyor. The purpose of the walkthrough verification is to subjectively assess the vibration qualities. At the discretion of the surveyor, additional measurements may be required. *Transducer placement positions.* Vibration transducers (accelerometers) should be located and attached properly to the floor surface to measure the vibration at the interface between the standing crew member and the source of vibration. The mounting of accelerometers must comply with ISO 5348. When the vibration enters the human body from a non-rigid or resilient material (e.g., floor covering), secure the transducers with a suitably formed mount that does not alter the pressure distribution on the surface of the floor covering. In cabins, the vibration transducers must be placed on the deck in the centre of the space. ***Note****: this location may not provide the maximum vibration levels for this particular space. The objective is to minimise the number of measurements yet still obtain a fair and representative sample of the exposure conditions of the person occupying the cabin.* For larger spaces (public rooms, messes, recreation areas, etc.) it may be necessary to place transducers at a number of locations in order to obtain a representative sample of the whole-body vibration levels for that space. Transducer locations should be evenly distributed throughout the space. For a specific room size, the minimum number of measurement locations are to be as indicated in Table 39.2, Distribution of transducer positions within spaces (Fig. 39.2; Table 39.3).

For vessels without crew cabins, transducers should be located at standing positions normally occupied by crew near the port side, near the centreline and near the starboard side. Transducers must be evenly distributed fore and aft. Transducers

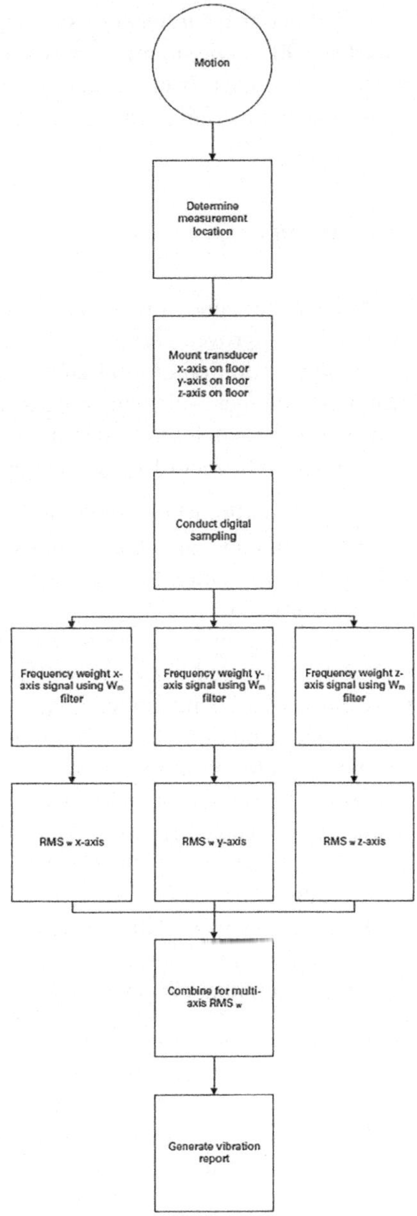

Fig. 39.2 Process for the measurement and analysis of whole-body vibration

Table 39.3 Distribution of transducer positions within spaces

Space size	Minimum number of measurement positions in room
Less than 40 m^2 (431 ft^2)	1
Greater than 40 m^2 (431 ft^2)	2

located at one measurement position should be orthogonally positioned (i.e., positioned at a 90° angle) to measure whole-body vibrations in the vertical, longitudinal and transverse axes. Translational accelerometers oriented in different axes at a single measurement position should be as close together as possible.

Test Report

As stated in the section on 'Test reports' in Chap. 37, a test report must be submitted to the Class approved surveyor to determine whether the vibration levels meet the whole-body vibration criteria and whether this part of the Class notation requirement has been met. The details listed in the following paragraphs are required to be provided in the whole-body vibration test report. Test details. The following details are required to be recorded for each period of testing:

- Loading conditions (mean draught and trim).
- Number of crew and other persons on board during tests.
- Bridge confirmation that operating conditions complied with contractual service conditions.
- What class of dynamic positioning (e.g., DP-0, DP-1, etc.).
- Vessel course and speed, as well as latitude and longitude coordinates of tests.
- Average water depth under keel.
- Weather conditions and meteorological data (i.e., wind speed and direction, ambient outdoor air temperature, outdoor humidity, barometric pressure) at the onset of every data collection period and at intervals of every four hours (if needed) during any data collection period. Weather conditions should also be reported at the end of each data collection period.
- Sea state.
- Direction of swell relative to vessel heading.
- Any indications of abnormal activity during the test that might skew results.
- Vessel equipment operated during the test.

Transducer placement positions. The actual transducer placement positions within the measured spaces (for both transit and dynamically positioned operations) must be indicated on appropriate drawings. *Measurement equipment details.* Details of measurement and analysis equipment (e.g., manufacturer, type and serial number, accuracy and resolution), including frequency analysis parameters (e.g., resolution, averaging time, and filtering), must be provided. Copies of the relevant instrumentation reference calibration certificates, together with the results of field setup and calibration checks before and after the field tests, should also be provided with the report. *Results.* The following results per measuring location, should be provided in a tabulated format (for both transit and dynamically positioned operations):

- Measurement position (i.e., space and location within space).
- Measurement period if different from requirements.

- Sample number.
- Multi-axis weighted RMS values.
- Equipment operating in proximity to the measurement position.
- Indication of pass/fail.

For all the remaining spaces that were checked through walk through verification inspection and spot check measurements, the following information should also be provided:

- Name and number of spaces.
- Walk through inspection observations.
- Measurement results if necessary.

Deviations. All deviations from the approved test plan must be reported. Surveyor witnessing documentation. The equipment field calibration and data collection process of vibration tests conducted at sea must be witnessed by a Class approved surveyor. The surveyor is required to sign or initial each page of the test report and prepare a witnessing document stating whether all steps of the vibration testing were completed to their satisfaction. A copy of the witnessing document is then given to the person conducting the onboard testing, for insertion into the final whole-body vibration test report. The original is usually retained by the surveyor and submitted to Class for inclusion in the vessel's Record.

Results

On completion of the test, the whole-body vibration test report will be reviewed by the Class approved surveyor against the appropriate Class Habitation notation requirements. Where the requirements have been satisfactorily met, the vessel will be assigned the appropriate Class Habitation notation.

Chapter 40
Noise on Workboats

General

A large amount of research has been performed on the effects of noise on humans. Established or commonly used criteria exist for the effects of noise on speech communication, hearing loss, sleep, concentration, and "annoyance". These have provided a basis for the guidance provided in this chapter. A detailed discussion of the effects of noise on human performance, health, and comfort may be found in Kryter (1994): The Handbook of Hearing and the Effects of Noise: Physiology, Psychology and Public Health. In this chapter, noise criteria have been selected to improve crew performance and to facilitate communication and sleep in appropriate vessel spaces. An additional goal is to enhance crew safety and comfort. In this instance, "comfort" means the ability of the crew to use a space for its intended purpose with minimal interference or annoyance from noise. The noise criteria presented in this chapter is lower than the levels commonly associated with hearing loss. Further guidance with respect to hearing conservation is provided in the IMO Resolution MSC.337(91): Code on Noise Levels Onboard Ships and should be followed for noise levels and exposure duration, particularly for areas with noise levels in excess of 85 dB(A). This chapter applies to manned spaces and other areas occupied by seafarers for 20 min or longer at any one time during normal, routine daily activities. Examples of such spaces include cabins, staterooms, workspaces (duty stations), mess areas, and recreation spaces, both inside and in other vessel locations outside the accommodation block. Compliance with the guidance discussed in this chapter is typically a prerequisite for the Workboat Habitability notation assigned by Class (e.g., **HAB(WB)**, **HAB+(WB)**, and **HAB++(WB)** for ABS Classed vessels).

A. A. Olsen, *Applying Physical Ergonomics to Modern Ship Design*,
https://doi.org/10.1007/978-3-031-57974-5_40

Terminology

A-weighted sound pressure level: The magnitude of a sound, expressed in decibels (i.e., 20 micropascals); the various frequency components are adjusted according to the A-weighted values given in IEC 61672-1 (2004) in order to account for the frequency response characteristics of the human ear. The symbol is LA; the unit is dB(A). The measurement L_{Aeq} is an equivalent continuous A-weighted sound pressure level, measured over a period of time.

Calibration checks: Field calibration of a measuring instrument conducted before and after a field test, using a reference calibrated signal or through zero calibration. Sound calibrators should comply with the standard IEC 60942:2003, as amended, type/class (1) standard and should be approved by the manufacturer of the measuring instrument used.

Dynamic positioning: A system to automatically maintain a workboat's position and heading by controlling propellers and/or thrusters. Dynamic positioning can maintain a position to a fixed point over the bottom, or in relation to a moving object (such as another vessel). It can also be used to position the vessel at a favourable angle towards wind, waves, and current.

Equivalent continuous a-weighted sound pressure level: The A-weighted sound pressure level of a noise fluctuating over a period of time T, expressed as the amount of average energy. The symbol is L_{Aeq}; the unit is dB(A).

Reference calibration: Calibration of measuring instrument and sound calibrator, conducted at least every two years (or more frequently if specified by the manufacturer) by an accredited Testing and Calibration Laboratory accredited according to ISO 17025:2005, as amended, with traceability to a national or international standard.

Transit conditions: Those conditions where the vessel is transitioning (moving) from one location to another.

Associated Documentation

The following documents provide details about Test plan preparation, test measurement procedures and/or test reporting:

- ISO 2923:1996, Acoustics—Measurement of Noise Onboard Vessels.
- IEC 61672-1:2013, Electroacoustics—Sound Level Meters—Part 1: Specifications.
- IEC 60942:2003, Electroacoustics—Sound Calibrators.
- IMO Resolution MSC.337(91) (2012), Code on Noise Levels Onboard Ships.
- ISO 10140-2:2010, Acoustics—Laboratory Measurement of Sound Insulation of Building Elements—Part 2: Measurements of Airborne Sound Insulation.

- ISO 717-1:2013, Acoustics—Rating of Sound Insulation in Buildings and of Building Elements—Part 1: Airborne Sound Insulation.
- WMO: Manual on Codes, No. 306, Part A, Alphanumerical Codes (2011 Edition, updated in 2019).

Criteria

Equivalent continuous a-weighted sound pressure level. The noise criteria for the Class Habitability notations (e.g., **HAB(WB)**, **HAB+(WB)**, and **HAB++(WB)**) are provided in Table 40.1, Noise criteria for workboats. Noise levels are determined for the test conditions specified in the section 'Test conditions' and must not exceed the maximum acceptable L_{Aeq} levels indicated in Table 40.1, Noise criteria for workboats, for each type of space or area. The maximum acceptable noise levels given in Table 40.1, Noise criteria for workboats are L_{Aeq} values, determined as appropriate to the character of the noise (refer to the section on 'Measurement procedures' and recorded results in this chapter). These limit values are for transit and dynamic positioning conditions only and do not include any in-port activities.

Test Plan

As stated in Chap. 37 under the section 'Test plans', a test plan must be developed to serve as the principal means for verifying the measurements to be performed to demonstrate or confirm compliance with noise criteria. The test plan should include the following:

- *Documentation.* The test plan should include appropriate design information including noise specifications for the vessel. It should also include layout drawings indicating the locations of all noise sources and noise generating equipment. The information should be of such detail to enable the Class engineer to verify compliance with the criteria stipulated by Class. The test plan should also include the acoustic insulation plan and relevant supporting documentation for review and approval. The test plan must be submitted to Class Engineering for their review and approval.
- *Test personnel.* The test plan should provide information about the testing specialist who will be conducting the test and their approval and certification in accordance with the Class specific Procedural Requirements for Recognised Ambient Environmental Testing Specialists or equivalent.
- *Test conditions.* Measurements are to be taken under conditions of Sea State 3 or less, as defined by the WMO Sea State Code.

Table 40.1 Noise criteria for workboats

Space[1,2]	Maximum acceptable noise L_{Aeq} level dB(A)					
	HAB(WB)		HAB+(WB)		HAB++(WB)	
	Transit conditions			Dynamic positioning conditions		
	HAB(WB)	HAB+(WB)	HAB++(WB)	HAB(WB)	HAB+(WB)	HAB++(WB)
Crew accommodation spaces and open deck recreation areas						
Cabins	60	55	50	60	55	55
Sanitary spaces (if separate from the cabin)	65	60	55	65	65	60
Dining (mess) spaces	65	60	55	65	60	60
Indoor recreation spaces	65	60	55	65	60	60
Gymnasiums	65	60	60	65	60	60
Medical and first aid centre	60	55	55	60	60	60
Open deck recreation areas	75	70	70	75	75	70
Navigation and control spaces						
Wheelhouse, pilothouse, bridge	65	60	55	65	60	60
Radio room	60	55	55	60	55	55
Offices	65	60	60	65	60	55
Cargo control rooms	65	60	60	65	60	60
Machinery control rooms	75	70	70	75	70	65
Service spaces						
Food preparation (e.g., galley, scullery)	75	70	65	75	70	70
Pantries	75	70	70	75	70	70
Storerooms	75	70	70	75	70	70
Laundry areas	75	70	70	75	70	70
Operating and maintenance spaces						
Continuously manned machinery spaces	90	90	90	90	90	90
Not continuously manned machinery spaces	110	110	110	110	110	110
Fan and generator rooms	85	80	80	85	85	80

[1] In any manned space with noise levels above 85 dB(A), hearing protection should be worn in accordance with appropriate IMO regulations.

[2] Unless this space is normally manned, no confirmatory measurements for the Class Habitability notations are required.

- *Measurement locations.* The test plan should document, in detail, on appropriate drawings, all spaces where measurements will be taken for both transit and dynamic positioning operations. In addition, transducer placement positions should be indicated. Details on selecting measurement locations and determining transducer placement positions are provided in the section 'Measurement locations'.
- *Data acquisition and instruments.* The integrating-averaging sound level meter must meet the requirements for a Type 1 instrument as specified in IEC 61672-1. For each location sampled, a measurement should be taken for transit and dynamic positioning operations in accordance with the requirements outlined under test conditions. It is recommended that dynamic positioning measurements are noted as to whether positioning was maintained dynamically or manually.
- *Data analysis.* The test plan should provide information regarding the methods, software, and instrumentation to be used for data analysis.
- *Test schedule.* The test plan should provide information regarding the proposed test schedule.

Test Requirements

In general, the noise measurements should be carried out in accordance with the requirements of IMO Resolution MSC.337(91) Code on Noise Levels Onboard Ships. However, when the requirements or procedures proscribed by Class deviate from those mentioned in IMO Resolution MSC.337(91), then the requirements or procedures proscribed by Class will take precedence unless otherwise stipulated. *Data acquisition and instruments. The* integrating-averaging sound level meter must meet the requirements for a Type 1 instrument as specified in IEC 61672-1. For each location sampled, a measurement must be taken for transit and dynamic positioning operations in accordance with the requirements outlined in the section, 'Test conditions'. It is recommended that dynamic positioning measurements be noted as to whether positioning was maintained dynamically or manually. *Test conditions.* The test conditions required for the noise measurements should be in accordance with each of the following sub paragraphs, based on ISO 2923.

- *Power output:*

  1. *Transit:* The propulsion machinery should run at contractual service conditions.
  2. *Dynamic positioning:* Devices such as azimuth or tunnel thrusters automatic or manual should run at contractual service conditions.

- *Equipment operation.* As appropriate for the mode of operation (transit or dynamic positioning), all machinery essential for vessel operation must be operated under normal conditions throughout the measurement period. HVAC systems are to be running as for normal seagoing conditions during the noise measurements. For the galley/scullery, all equipment that runs for 20 min or more (e.g., ventilation hoods) should be running during the noise measurements. Equipment that is used only intermittently (e.g., blender/mixer/meat cutter) need not be running during the noise measurements. Note the equipment that is running in the noise test report.

- *Course and water depth.* Noise measurements are to be taken with the vessel in a depth of water not less than five times the draught of the vessel. For vessels that do not operate in water depths of five times draught, measurements should be taken under normal operating and transit conditions. The vessel should maintain a single heading and a constant speed during the test.
- *Rudder conditions.* Rudder action should be minimised wherever possible. During dynamic positioning, measurements should be taken while the vessel is maintaining a position relative to another vessel or fixed point (e.g., offshore installation).
- *Sea conditions.* Measurements are to be taken under conditions of Sea State 3 or less, as defined by the WMO Manual on Codes, No. 306, Part A, Alphanumerical Codes (2011 Edition, updated in 2019).
- *Loading conditions.* The loading condition of the workboat should be as close as possible to normal operating conditions.
- *Test interference.* During the noise measurements, noise arising from every kind of unnecessary human activity must be avoided. For this reason, only the personnel needed for the normal operation of the equipment in the space and those carrying out the measurements should be present in the space being tested. Doors and windows must be closed, except where they are normally left open (such as the door on the lee side of the navigation bridge, which may normally be open). Any open doors or windows should be noted in the noise test report. Spaces should be furnished with all usual equipment, fixtures and fittings normally found in the space. Equipment should be configured to operate in its normal operating mode.

Measurement locations. Selection of spaces where measurements are to be conducted. The aim when selecting noise measurement locations is to obtain a representative sample of data that reflects the actual conditions in manned spaces. For practical reasons, it is important to select the locations such that an appropriate amount of sample data can be collected during the testing phase. The measurement locations should be selected in accordance with the following criteria:

- Select potential worst-case locations based on their proximity to noise emitting sources such as propulsion or other rotating machinery or where noise is likely to be transmitted to manned spaces, accommodation areas and recreation areas via the vessel's structure. Measurements should be taken in all identified worst-case locations (e.g., cabin adjacent to a machinery space).
- Where a single instance of one type of manned space exists within the vessel (e.g., the bridge, mess room, gymnasium, library, etc.), that location should be selected for measurement.
- Select a representative sample of crew cabins throughout the vessel. For vessels with less than 20 cabins 50% of cabins on each deck should be selected. For vessels with greater than 20 cabins, 30% of cabins on each deck should be

selected. These measurement locations must be selected at locations port, starboard, fore, amidships and aft. The worst-case locations are to be considered part as the representative sample for crew cabins, where applicable.

- Where multiple instances of the same types of accommodation space exist, that are not crew cabins, a representative sample of at least 50% of each type should be selected for measurement. The worst-case locations are to be considered part of the representative sample, where applicable.

Walk through verification inspection locations. All normally manned spaces are to be subject to a walkthrough inspection by the Class approved surveyor. The number and locations of the walkthrough inspections are generally determined by the surveyor. The purpose of the walkthrough verification is to subjectively assess the noise qualities. At the discretion of the surveyor, additional measurements may be required. *Measurement positions.* The measurement positions described below are taken or adapted from ISO 2923 and IMO Resolution MSC.337(91):

> Measure at positions where persons will be seated or standing. The microphone should be at a height of approximately 1200 mm (47 in.) from the deck to represent seated persons and approximately 1600 mm (63 in.) from the deck to represent standing persons, as appropriate for the measurement position.
>
> In crew accommodation spaces, measure in the middle of the space.
>
> For all measurements, the microphone should not be closer than 500 mm (20 in.) from the boundary surface (e.g., bulkhead) of a space. The measurement time should be at least 15 seconds and should be long enough to enable the measurement of the equivalent continuous A-weighted sound pressure level for any specified time interval within the stated limits of overall measurement uncertainty.
>
> If practicable, do not measure closer than 1000 mm (39.5 in.) from operating machinery, air inlets, or from decks, bulkheads or other large surfaces. Where this is not possible, measure midway between the machinery and an adjacent reflecting surface.

Measurement procedures and recorded results. Persons present during measurements. When testing specialist personnel are conducting noise level measurements in any space, only crew members necessary for the operation of equipment or machinery in that space should be present. *Sampling duration.* Equivalent continuous A-weighted sound pressure levels (L_{Aeq}) should be reported for each measurement location. The L_{Aeq} sampling duration should be sufficient to achieve a stable reading. Sampling time should be 15 s or longer. *Cyclic noise.* If the noise within a space is cyclic, the L_{Aeq} sampling duration should be sufficient to capture an integer number of complete cycles. If a long-duration sample is judged impractical, an L_{Aeq} value should be determined and reported for the high-noise portion of the cycle. Intermittent noise. If the noise within a space is present intermittently, an L_{Aeq} value should be determined and reported for a period of high-level noise. *HVAC related noise.* If HVAC system-related noise is a large contributor to the noise level in the space, a noise measurement should be made approximately 300 mm (12 in.) from the vent, measured in line with the direction of airflow, and recorded in the noise test report.

Test Report

As stated in Chap. 37 under the section 'Test reports', a test report must be submitted to the Class approved surveyor to determine whether the noise levels are at or below the limits and whether this part of the notation requirement has been met. The details listed below must therefore be provided in the noise test report. *Test details.* The following details are required be provided for each period of testing:

- Loading conditions (mean draught and trim).
- Number of crew and number of other persons on board during testing.
- Bridge confirmation that operating conditions complied with contractual service conditions.
- Vessel course and speed.
- Average water depth under keel.
- Weather conditions and meteorological data (i.e., wind speed and direction, ambient outdoor air temperature, outdoor humidity, barometric pressure) at the onset of every data collection period and at intervals of every four hours (if needed) during any data collection period. Weather conditions should also be reported at the end of each data collection period.
- Sea state.
- Any indications of abnormal activity during the test that might skew results.
- Equipment operated during the test.

Measurement positions. For transit and dynamic positioning measurements, actual measurement location positions must be indicated on appropriate drawings. *Measurement equipment details.* Details of measurement and analysis equipment (e.g., manufacturer, type and serial number, accuracy, sampling frequency and resolution) must be provided. Copies of the relevant instrumentation reference calibration certificates, together with the results of field setup and calibration checks before and after the field tests, should also be provided with the report. *Results.* The following results per measuring location, should be provided in a tabulated format:

- Measurement position (i.e., space and location within space).
- Measurement conditions (transit or dynamic positioning, power output and pitch angles, sea states).
- Number of people present in the space at time of measurement.
- Measurement period.
- Time at start and finish of measurement.
- Measurement period if different from requirement.
- Equivalent continuous A-weighted sound pressure level (L_{Aeq}).
- Note any open doors and windows.
- Note equipment operating in proximity to the measurement position.
- Note observed direct sources of noise (such as ventilation devices) and any additional measurement data collected.
- Indication of pass/fail.

For all the remaining spaces that were checked through walk through verification inspection and spot check measurements, the following information should also be provided:

- Name and number of spaces.
- Walk through inspection observations.
- Measurement results if necessary.

Deviations. All deviations from the approved test plan must be reported. *Surveyor witnessing documentation.* The equipment field calibration and data collection process of noise tests conducted at sea must be witnessed by a Class approved surveyor. The surveyor is required to sign or initial each page of the test report and prepare a witnessing document stating whether all steps of the noise testing were completed to their satisfaction. A copy of the witnessing document is then given to the person conducting the onboard testing, for insertion into the final noise test report. The original is usually retained by the surveyor and submitted to Class for inclusion in the vessel's Record.

Results

On completion of the test, the noise test report will be reviewed by the Class approved surveyor against the appropriate Class Habitation notation requirements. Where the requirements have been satisfactorily met, the vessel will be assigned the appropriate Class Habitation notation.

Chapter 41
Indoor Climate on Workboats

General

Thermal comfort is defined in ISO 7730 as "… that condition of mind which expresses satisfaction with the thermal environment". The sensation of thermal comfort is therefore largely subjective and will vary from person to person. Due to differences in metabolism and expectations, there are distinct individual differences among people's perception of comfort as a function of temperature, humidity, and other atmospheric characteristics. Acclimatisation, habits, and expectations influence perceived comfort. These individual differences make it difficult to specify a single thermal environment that will be satisfactory to everyone. A thermal environment is therefore typically defined to be acceptable to at least 80% of the occupants of an interior space. Individually, the perception of thermal comfort is largely determined by the interaction of thermal environmental factors such as air temperature, air velocity, relative humidity, and factors related to activity and clothing. The thermal control or HVAC systems on a vessel should be designed to effectively control the indoor thermal environmental parameters to within acceptable limits to facilitate the thermal comfort of the occupants. To that end, this chapter provides the assessment criteria, verification and measurement methodology for indoor climate relating to habitability on ships. The criteria are based on currently available standards and were selected to provide an index of crew thermal comfort. The thermal environmental variables covered in this chapter include the ambient qualities of air temperature, air velocity, and relative humidity. Vertical thermal gradient is used to indicate uncomfortable temperature differentials between a person's head and feet. The crew cabin area horizontal gradient temperature differential between the temperature of inside bulkhead surfaces adjacent to crew cabins and the average air temperature within the space serves as an indication of potential thermal comfort or discomfort. The thermal environmental criteria provided in this chapter are for persons wearing typical indoor clothing occupied with light, primarily sedentary activity and resulting in a thermal environment acceptable to at least 80% of the occupants. This chapter applies to

A. A. Olsen, *Applying Physical Ergonomics to Modern Ship Design*,
https://doi.org/10.1007/978-3-031-57974-5_41

enclosed manned spaces occupied by crew members for 20 min or longer at any one time for normal, routine daily activities. Examples of seafarer spaces include crew accommodation areas, bridge, engine control room, and indoor workspaces. Compliance with the provisions outlined in this chapter are generally required to meet the Class Habitability (e.g., Workboat Habitability (**HAB(WB)**), Workboat Habitability Plus (**HAB+(WB)**), or Workboat Habitability Plus Plus (**HAB++(WB)**) notations.

Terminology

Air temperature: The temperature of the air surrounding a person, measured with a standard thermometer.

Air velocity or movement: The rate of displacement of ambient air in a specific direction in metres-per-second (m/s) or feet-per-second (ft/s).

Comfort zone: That range of environmental conditions in which at least 80% of crew members experience thermal comfort.

Fresh air supply quantity: The amount of fresh/outdoor air supplied to a specific space, expressed in litres per second (l/s).

Horizontal gradient: The difference between the inside surface temperatures of the bulkheads bounding the berth and the average air temperature within the cabin. This is used as an indication of potential thermal discomfort due to radiant thermal sources.

HVAC zone: A space or group of spaces that is (are) independently controlled for temperature, humidity, and air distribution. A zone usually comprises common duct work fed from an air handler.

Reference calibration: Calibration of a measuring instrument, conducted at least every two years (or more frequently if specified by the manufacturer) by an accredited Testing and Calibration Laboratory accredited according to ISO 17025:2005, as amended, with traceability to a national or international standard.

Relative humidity (RH): The ratio of the amount of vapour contained in the air (absolute humidity) to the maximum amount of vapour the air can hold at a given temperature before precipitation (condensation) occurs.

Thermal comfort: Subjective index of "that condition of mind which expresses satisfaction with the thermal environment".

Ventilation: Ventilation is the process of supplying air to and removing air from any space by natural or mechanical means. From the standpoint of comfort and health, ventilation issues involve both quantity and quality.

Vertical gradient: The vertical air temperature difference within an enclosed space. The vertical gradient is used as an indication of potential local discomfort at the head and feet.

Associated Documentation

The following documents provide details about test plan preparation, test measurement procedures and/or test reporting:

- ANSI/ASHRAE 55-2013, Thermal Environmental Conditions for Human Occupancy.
- ANSI/ASHRAE (15) (2010). Practices for Measuring, Testing, Adjusting, and Balancing Shipboard HVAC&R Systems.
- ISO 7726:1998, Ergonomics of the Thermal Environment—Instruments for Measuring Physical Quantities.
- ISO 7547:2002, Ships and Marine Technology—Air-Conditioning and Ventilation of Accommodation Spaces—Design Conditions and Basis of Calculations.
- NEBB:2005, Procedural Standards for Testing, Adjusting, Balancing of Environmental Systems.

Criteria

The indoor climate criteria for the ship Habitability notations (e.g., **HAB(WB)**, **HAB+(WB)**, and **HAB++(WB)**) are provided in Table 41.1, Summary of indoor climate requirements. Indoor climate measurements should be taken only in indoor manned spaces. Specific measurement locations are discussed in the section 'Measurement locations'. The thermal environmental comfort ranges and conditions should be achievable, under the test conditions specified in the section on 'Test conditions', for normal operating conditions, in all indoor manned spaces. For vessels with lower and median notations (e.g., **HAB(WB)** and **HAB+(WB)**), the criteria provides for a range of air temperatures maintained by a temperature controller for each zone and are primarily aimed at HVAC systems that do not make provision for individual adjustment to suit personal preferences and activities within a specific space. Vessels with an advanced notation (e.g., **HAB++(WB)**) should aim at enhancing crew comfort by making provisions for the crew to adjust indoor climate conditions, with regard to air temperature to suit personal needs within a particular space.

Air temperature. The HVAC system should be capable of providing a pre-set return air temperature of 20–25 °C (68–77°F) during winter months and 22–27 °C (71.5–80°F) during summer months to an HVAC zone for a set of habitable spaces. This temperature should be maintained by a temperature controller. Each zone should have a thermostat for reheat and dehumidification purposes. The HVAC system should

Table 41.1 Summary of indoor climate requirements

Item	Requirement or criterion	
	HAB(WB) and HAB+(WB) or equivalent	HAB++(WB) or equivalent
Adjustability	Non-adjustable air temperature	Adjustable range of air temperatures
Air temperature	Winter: 20–25 °C (68–77°F) Summer: 23–28 °C (73.5–82.5°F)	
Relative humidity	A range from 30% minimum to 70% maximum	
Vertical gradient	The acceptable range is 0–3 °C (0–6°F)	
Air velocity	Not exceed 30 m-per-minute or 100 feet-per-minute (0.5 m/s or 1.7 ft/s)	
Horizontal gradient (crew cabin areas)	The horizontal temperature gradient in crew cabin areas should be < 10 °C (18°F)	
Air supply quantity	The minimum quantity of fresh/outdoor air supply should be not less than 8 l/s per person and in addition will need to be at least 40% of the total air supplied to a specific space	

be capable of sustaining an adjustable range of air temperatures between 20 and 25 °C (68–77°F) inclusive during winter months and 23–28 °C (73.5–82.5°F) during summer months in all indoor manned spaces. This temperature should be maintained by a temperature controller. Each manned space should have its own individual controller for temperature regulation. *Relative humidity.* The HVAC system should be capable of providing and maintaining a relative humidity within a range from 30% minimum to 70% maximum. *Enclosed space vertical gradient.* The difference in temperature at 100 mm (4 in.) above the deck and 1700 mm (67 in.) above the deck should be maintained within 3 °C (6°F). *Air velocity.* Air velocities should not exceed 30 m-per-minute or 100 feet-per-minute (0.5 m/s or 1.7 ft/s) at the measurement position in the space. *Crew cabin area horizontal gradient.* In crew cabin areas, the difference between the inside bulkhead surface temperature adjacent to crew cabins and the average air temperature within the space should be less than 10 °C (18°F). *Air exchange rate.* The rate of air exchange for enclosed spaces should be at least six complete changes-per-hour. *Test plan.* As stated in the section 'Test plans' in Chap. 37, a test plan should be developed to serve as the principal means for verifying the measurements to be performed to verify compliance with the indoor climate criteria. This test plan should typically include the following:

- *Documentation.* The test plan should include appropriate drawings indicating the location of all indoor climate systems and equipment including the HVAC system design specifications and all schematics/layout drawings of the HVAC system.
- *Test personnel.* The test plan should provide information about the testing specialist who will be conducting the test and their approval and certification in accordance with the Class specific Procedural Requirements for Recognised Ambient Environmental Testing Specialists or equivalent.

- *Test conditions.* The test plan should detail the conditions under which the tests will be performed. Details about test conditions are provided in the section on 'Test conditions'.
- *Measurement locations.* The test plan should document, in detail, on appropriate drawings, all spaces where measurements will be taken. In addition, transducer placement positions should be indicated. Details on selecting measurement locations and determining transducer placement positions are provided in the section 'Measurement locations'.
- *Data acquisition and instruments.* The test plan should provide information regarding the methods and instrumentation to be used for measurement and data collection. Instrumentation specification details should include type of instruments to be used, accuracy, calibration, sensitivity. More details on data acquisition and instruments are provided in the section 'Data acquisition and instruments'.
- *Data analysis.* The test plan should provide information regarding the methods, software, and instrumentation to be used for data analysis.
- *Test schedule.* The test plan should provide information regarding the proposed test schedule. The test schedule should include information on the approximate date and duration of testing (including a summary of the scheduled measurements per day).
- *Data collection sheets.* The test plan should provide sample data sheets (in table format) that will be used for reporting the measurement data.
- *Conflict of interest declaration.* In the case that the testing specialist has, in any capacity or to any extent, provided input into the design of the vessel, their participation should be indicated in the test plan including details of the testing specialist's involvement. Class may review any potential conflicts of interest and determine if any further action is necessary.

Test Requirements

Indoor climate measurements should be carried out in accordance with the procedures described in ANSI/ASHRAE 55-2013 and ANSI/ASHRAE 15. When the requirements or procedures proscribed by Class deviate from those mentioned in ANSI/ASHRAE 55-2013 and ANSI/ASHRAE 15, then the requirements or procedures proscribed by Class will take precedence unless otherwise stipulated. *Data acquisition and instruments.* The thermal measurement instrumentation used must meet or exceed the minimum characteristics of instruments for measuring physical quantities characterising an environment specified in ISO 7726. The following quantities should be measured in each of the spaces or zones identified in the test plan and the results noted in the indoor climate test report:

- Air temperature.
- Relative humidity.
- Air velocity.
- Vertical gradient.

- Horizontal gradient (in crew cabin spaces only).

The air temperature and humidity measurements should be made at least every five minutes for a minimum period of one hour. The minimum, maximum, and average values for the 1-h period should be reported for each space measured. For the **HAB+(WB)** and **HAB++(WB)** notations, the temperature control for a specific space should be set to the lower requirement level of 20 °C (68°F) for the first half-hour and to 27 °C (80°F) for the second half-hour of measurement. ***Note***: *Data loggers are a type of device that can be left unattended to capture data. These type devices have been proven effective in gathering temperature and humidity values.* The measuring period for determining the average air velocity at any location should be three minutes. Surface temperatures for all wall surfaces that are adjacent to crew cabins should be measured for determining the horizontal gradient in crew cabin spaces. *Test conditions.* In order to determine the effectiveness of the HVAC system at providing the environmental conditions specified in this chapter measurements should be made under the following conditions:

- *Testing.* Testing of identified spaces can be performed in port and at sea, provided the required test conditions stated here are in compliance at the time the measurements are made and recorded. In the case where some testing is performed in port, confirmatory testing should be performed at sea with all normally functioning equipment in its operational mode. This confirmatory testing must usually consist of a sample of the manned spaces selected as follows:
  - Spaces that were identified as “worst case” locations.
  - Spaces which were out of compliance or close to non-compliance (refer to Table 41.1, Summary of indoor climate requirements).
  - Spaces that required some type of mitigation.
  - Additional testing may be required at the Class approved surveyor’s discretion.
- *Equipment operation.* The HVAC system should be operating in the normal operation or mode. A written confirmation issued by the manufacturer confirming that the HVAC system has been tested, adjusted and balanced should be provided to the surveyor before any measurement commences. The total system TAB should be conducted in accordance with the NNEBB standard NEBB Procedural Standards for Testing, Adjusting, Balancing of Environmental Systems or other similar certification standard.
- *Doors and windows.* The space doors (including louvres) and windows should be closed during the evaluation period, except for routine entry and exit. Any open doors or windows should be noted in the indoor climate test report.
- *Equipment and furnishings.* Spaces should be furnished with all usual equipment, fixtures and fittings normally found in the space. Equipment must be configured to operate in its normal operating mode.
- *Weather and climatic conditions.* When thermal conditions in the manned space have a high sensitivity to time of day and weather conditions (e.g., spaces adjacent to exterior bulkheads), the measurement should be made such that the high and low extremes of the thermal parameters are determined (e.g., measurements

could therefore be taken during the day and night in the same space). If possible, measurements should be taken with little or no cloud cover.

- *Test interference.* During the indoor climate measurements, any activity that might affect the indoor climatic variables in the space must be avoided. For this reason, only the personnel needed for the normal operation of the equipment in the space and those carrying out the measurements should be present in the space being tested. Doors and windows should be closed, except where they are normally left open (such as the door on the lee side of the navigation bridge, which may normally be open). Any open doors or windows should be noted in the indoor climate test report.

Measurement locations. Selection of spaces where measurements are to be conducted. The aim when selecting indoor climate measurement locations is to obtain a representative sample of data that reflects the actual conditions in manned spaces. For practical reasons, it is important to select the locations such that an appropriate amount of data can be collected during the testing phase. This means the measurement locations should be selected in accordance with the following criteria:

- Select potential problem areas where the influence of internal conditions or factors may adversely impact the quality of the indoor climate in enclosed manned spaces. Internal conditions include space proximity to equipment that radiates or absorbs heat (e.g., engine exhaust trunks, freezer spaces, galley, scullery, etc.) and surfaces with thermal differentials in excess of 10 °C (18°F) from the ambient temperature in the space. Living and working areas at the ends of HVAC ductwork or piping runs (for heating or cooling) should be selected as potential problem locations. Measurements should be taken in all identified potential problem areas.
- Select potential problem areas where the influence of external ambient environmental conditions (e.g., sun, wind, precipitation, etc.) may adversely impact the quality of the indoor climate. These areas include manned spaces, accommodation areas and recreation areas which may be outboard or adjacent to the vessel's hull. Measurements should be taken in all identified problem areas.
- Where a single instance of one type of manned space exists within the vessel (e.g., the bridge, mess room, gymnasium, library, etc.), that location should be selected for measurement.
- Where multiple instances of one type of space exist (e.g., cabins or recreation areas), a representative sample of at least 25% of each type (e.g., one-man room, two-man room, etc.) should be selected for measurement. The worst-case locations are to be considered part of the representative sample, where applicable. All of these measurement locations must be selected at locations port, starboard, fore, amidships, and aft. The worst-case locations are to be considered part of the representative sample for crew cabins, where applicable. *Walk through verification inspection locations.* All normally manned spaces are to be subject to a walkthrough inspection by the Class approved surveyor. The number and locations of the walkthrough inspections are generally determined by the surveyor. The purpose of the walkthrough verification is to subjectively assess the indoor climate qualities. At the discretion of the surveyor, additional measurements may

be required. *Transducer placement positions.* For each space identified in the test plan, the transducer locations should be standardised as follows.

- Air temperature and relative humidity measuring instrumentation should be set up approximately in the middle of the space to measure general space temperature and humidity levels. Air temperature should be measured at approximately 100 mm (4 in.), 1100 mm (43 in.) and 1700 mm (67 in.) above the deck. Relative humidity should be measured at a height of approximately 1700 mm (67 in.) above the deck.
- Air velocity should be measured at approximately 100 mm (4 in.), 1100 mm (43 in.) and 1700 mm (67 in.) above the deck in manned spaces, as applicable (to assure air velocity is not excessive). This measurement should be taken in the centre of the space. Air velocity only needs to be measured once per testing location.
- For crew cabin spaces, inside wall surface temperatures for determining the horizontal gradient should be captured at approximately 300 mm (12 in.) above the mattress.

Test Report

As stated in the section on 'Test reports' in Chap. 37, a test report must be submitted to the Class approved surveyor to determine whether the indoor climate levels meet the criteria and whether this part of the Class notation requirement has been met. The details listed in the following paragraphs are required to be provided in the indoor climate test report. *Test details.* The following details are required to be recorded for each period of testing:

- Vessel's latitude and longitude coordinates during testing.
- Weather conditions and meteorological data (i.e., wind speed and direction, ambient outdoor air temperature, outdoor humidity, barometric pressure) at the onset of every data collection period and at intervals of every four (4) hours during any data collection period. Weather conditions should also be reported at the end of each data collection period.
- Any indications of abnormal activities or conditions during the test that might skew the results.

Transducer placement positions. The actual transducer placement positions must be indicated on appropriate drawings. *Measurement equipment details.* Details of measurement and analysis equipment (e.g., manufacturer, type and serial number, accuracy and resolution must be provided. Copies of the relevant instrumentation reference calibration certificates, together with the results of field setup and calibration checks before and after the field tests, should also be provided with the report. *Results.* The following results per measuring location, should be provided in a tabulated format:

- Measurement position.
- Number of people present in the space at time of measurement.
- Measurement period if different from requirements.
- Air temperature (minimum, maximum, and average) at 100 mm (4 in.) above deck.
- Air temperature (minimum, maximum, and average) at 1100 mm (43 in.) above deck.
- Air temperature (minimum, maximum, and average) at 1700 mm (67 in.) above deck.
- Relative humidity (minimum, maximum, and average) at 1700 mm (67 in.) above deck.
- Air velocity at 100, 1100, and 1700 mm (4, 43, and 67 in.) above deck at measurement positions.
- Vertical gradient (Average air temperature at 1700 mm (67 in.) minus average air temperature at 100 mm (4 in.) above deck).
- Horizontal gradient (side wall surface temperature in crew cabin spaces minus average air temperature at 1100 mm (43 in.) above deck crew cabin areas).
- Indication of pass/fail.

For all the remaining spaces that were checked through walk through verification inspection and spot check measurements, the following information should be provided:

- Name and number of spaces.
- Walk through inspection observations.
- Measurement results if necessary.

Deviations. All deviations from the approved test plan must be reported. *Surveyor witnessing documentation.* The equipment field calibration and data collection process of indoor climate tests must be witnessed by a Class approved surveyor. The surveyor is required to sign or initial each page of the test report and prepare a witnessing document stating whether all steps of the indoor climate testing were completed to their satisfaction. A copy of the witnessing document is then given to the person conducting the onboard testing, for insertion into the final indoor climate test report. The original is usually retained by the surveyor and submitted to Class for inclusion in the vessel's Record.

Results

On completion of the test, the indoor climate test report will be reviewed by the Class approved surveyor against the appropriate Class Habitation notation requirements. Where the requirements have been satisfactorily met, the vessel will be assigned the appropriate Class Habitation notation.

Chapter 42
Lighting and Illumination on Workboats

General

The lighting of seafarer spaces should facilitate visual task performance, facilitate movement in the space and aid in the creation of an appropriate visual environment. Lighting design involves integrating these aspects to provide adequate illumination for the safety and wellbeing of the crew as well as for the various tasks performed on board vessels. The selection of appropriate illumination levels for specific tasks and seafarer spaces is an important consideration in the design of lighting systems. There is a difference of opinion as to what levels of light may be considered best for visual tasks. Since illumination recommendations are generally consensus values, for any task, a range of illumination levels may apply. Since visual tasks performed within habitable spaces on board a vessel are generally similar to tasks encountered ashore, requirements for illumination on vessels generally correspond to those tasks performed in living, working, and recreation areas on shore. To that end, this chapter outlines criteria for assessing the illumination levels of general lighting and task lighting on vessels. The main objective of the assessment is to determine whether the various lighting systems comply with minimum standards to accommodate crew visual task performance and facilitate crew movements and wellbeing and safety on board vessels. Lighting criteria are based on currently available objective standards and research data. Compliance with the criteria outlined in this chapter is typically a prerequisite for attaining a Class Habitability (e.g., **HAB(WB)**, **HAB+(WB)**, and **HAB++(WB)** or equivalent) notation.

Terminology

Disability glare: Glare which reduces the ability to perform a visual task.

A. A. Olsen, *Applying Physical Ergonomics to Modern Ship Design*,
https://doi.org/10.1007/978-3-031-57974-5_42

Discomfort glare: Glare which produces viewer discomfort, but which does not significantly interfere with visual task performance or visibility.

General lighting: Lighting designed to provide a substantially uniform level of illumination throughout an area, exclusive of any provision for special, localised tasks. Such lighting should be provided by fixed luminaires.

Glare: The discomfort or impairment of vision experienced when parts of the visual field are excessively bright in relation to the general surroundings.

Illumination: The luminous flux density at a surface (or the amount of light falling on an object or surface), i.e., the luminous flux incident-per-unit area. Illumination is measured in units of Lux (lm/m^2) or footcandles (fc; lm/ft^2). One foot-candle equals 10.76 Lux.

Lumen: The SI of luminous flux, used in describing a quantity of light emitted by a source or received by a surface.

Luminaire: A complete lighting unit consisting of a lamp(s) together with the parts designed to distribute the light, to position and protect the lamp and to connect the lamp to the power supply.

Luminance: The photometric brightness of an illuminated surface (or the amount of light emitted or reflected from the surface). The SI unit of luminance is candela-per-square metre (cd/m^2).

Luminous flux: The light emitted by a source, or received by a surface and indicates the intensity of a source. Flux is expressed in lumens.

Lux: a unit of illumination, equivalent to 0.0929 foot-candle and equal to the illumination produced by luminous flux of one lumen falling perpendicularly on a surface one metre square. Also called a metre candle.

Red or low-level white illumination: Lighting provided to accommodate efficient dark adaptation in areas where seeing tasks are performed during night-time operations and in areas where people need to move from a lit interior into a dark environment and maintain good vision.

Reference calibration: Calibration of a measuring instrument, conducted at least every two years (or more frequently if specified by the manufacturer) by an accredited Testing and Calibration Laboratory accredited according to ISO 17025:2005, as amended, with traceability to a national or international standard.

Task lighting: Lighting provided to meet the illumination requirements of a specific task. Task lighting refers to the total illumination requirement that may be obtained by supplementary lighting provided in addition to the general illumination. Such lighting may be provided by fixed luminaires or via floor lamps or table lamps.

Task plane: The horizontal, vertical, or inclined plane in which the visual task lies. If no information is available, the task plane may be considered to be the horizontal

and at 750 mm (29.5 in.) above the deck for seated tasks and 1000 mm (39.5 in.) for standing tasks.

Veiling reflection (or reflected glare): A reflection of incident light that partially or totally obscures the details to be seen on a surface by reducing the contrast.

Associated Documentation

The following documents provide details about test plan preparation, test measurement procedures and/or test reporting.

- ISO 8995-1: 2002, Lighting of Workplaces—Part 1: Indoor.
- DIN EN 12464-1:2011, Light and Lighting—Lighting of Workplaces—Part 1: Indoor Workplaces.
- DIN 5032-7:1985, Photometry; Classification of Illuminance Meters and Luminance Meters.
- CIE S 023/E:2013 (ISO/CIE 19476:2014), Characterisation of the Performance of Illuminance Meters and Luminance Meters.
- BS 667:2005, Illuminance meters. Requirements and test methods.

Criteria

The lighting criteria for Class Habitability notations are provided in:

- Table 42.1, Lighting criteria for crew accommodation spaces.
- Table 42.2, Lighting criteria for navigation and control spaces.
- Table 42.3, Lighting criteria for service spaces.
- Table 42.4, Lighting criteria for operating and maintenance spaces/areas.
- Table 42.5, Lighting criteria for red or low-level white illumination.

For lighting, the criteria for **HAB(WB)**, **HAB+(WB)**, and **HAB++(WB)** notations are the same. In this chapter, general lighting and task lighting requirements are provided for crew tasks and spaces normally encountered on vessels. The lighting levels provided in the tables are for new lamps. Emergency lighting is covered in SOLAS and IMO Resolutions and was not considered in the selection of the lighting levels provided in this chapter. The criteria in this chapter are applicable to the entire vessel, not just manned spaces. *General and task lighting*. The minimum maintained illumination levels stated in the Lighting criteria Tables 42.1 throubh 42.4 may be achieved under the test conditions specified in the section on 'Test conditions' and measured with task lighting turned on, where provided, but with external light sources (e.g., daylight, moonlight, shore light, etc.) excluded. These levels were selected from a variety of sources that are listed in the References part of this book. *Red or low-level white illumination*. In workspaces where red or low-level white

Table 42.1 Lighting criteria for crew accommodation spaces

Space	Illumination level in Lux	Space	Illumination level in Lux
Cabins and sanitary spaces			
General lighting	150		
Reading and writing		Sanitary spaces	
• Desk	500	• Lavatory/toilet	200
• Bunk light	200	• Bath/shower area	150
Changing room	200		
Light during sleep periods[1]	<30		
Dining spaces			
Mess room and cafeteria	300		
Snack or coffee area	150		
Recreation spaces			
Recreation room/lounges	200	Gymnasium	300
Library		Bulletin boards/ display areas	150
• General lighting	150	All other recreation spaces (e.g., game rooms)	200
• Reading area	500		
Multimedia resource centre (e.g., computer room)			
Medical, dental and first aid centre			
General lighting/medical waiting areas	150	Dispensary	
Medical and dental treatment/examination room	500	• General lighting	500
Reading and writing (desk)	500	• Over the dispensary table	500
Laboratories	500	Other medical spaces	200

illumination is provided to facilitate dark adaptation, the maintained illumination levels in Table 42.5, Lighting criteria for red or low-level white illumination, are provided for guidance only and are not generally required for a Class Habitation notation.

[1] If there is any opportunity for light to enter cabins at the times of day or night when people sleep (e.g., portlights, transoms, etc.), the maximum lighting level should be 30 Lux.

Table 42.2 Lighting criteria for entrances and passageways

Space	Illumination level in Lux	Space	Illumination level in Lux
Interior walkways, passageways, corridors, stairways and access ways	100	Exterior walkways, passageways, and access ways (night)	50
Elevators	100	Exterior stairways	100

Table 42.3 Lighting criteria for navigation and control spaces

Space	Illumination level in Lux	Space	Illumination level in Lux
Wheelhouse, pilothouse, bridge		Offices	
• Day	300	• General lighting	300
• Night when underway	<30	• Reading and writing (desk)	500
Chart room		• Meeting, training rooms	500
• General lighting	150	Control rooms and control stations[2]	
• On chart table	500	• General lighting	300
Radar/gyro room[2]	300	• Control consoles and boards, panels, instruments, etc.	300
Radio room		• Reading and writing (desk)	500

Test Plan

As stated in the section 'Test plans' in Chap. 37, a test plan should be developed to serve as the principal means for verifying the measurements to be performed to verify compliance with the lighting and illumination criteria. This test plan should typically include the following:

- *Documentation.* The test plan should include appropriate design information and layout drawings showing the hull outline, bulkheads, access routes, location of luminaires, outlines of major furniture and equipment, and the space name and number. The drawings should be to a scale and sized to permit the scaling of measurement points (as required in the section on 'Measurement locations') and lighting equipment and the recording of luminance and other relevant data.

[2] At night, the wheelhouse must be kept as dark as possible to permit visibility of objects on the water, hazards to navigation, and navigational lights.

Table 42.4 Lighting criteria for service spaces

Space	Illumination level in Lux	Space	Illumination level in Lux
Food preparation		Laundries	300
• Galley	500	Chemical storage (e.g., paint store, etc.)	300
• Pantry	300	Storerooms	
• Butcher shop	500	• Large parts	200
• Thaw room	300	• Small parts	300
• Working surfaces, food preparation counter and range tops	500		
• Scullery (dishwashing) food serving lines package handling/cutting	300		
Food storage			
• Non-refrigerated	200		
• Refrigerated	100		
Mail sorting	300		

- *Test personnel.* The test plan should provide information about the testing specialist who will be conducting the test and their approval and certification in accordance with the Class specific Procedural Requirements for Recognised Ambient Environmental Testing Specialists or equivalent.
- *Test conditions.* The test plan should detail the conditions under which the tests will be performed. Details about test conditions are provided in the section on 'Test conditions'.
- *Measurement locations.* The test plan should document, in detail, on appropriate drawings, all spaces where measurements will be taken. In addition, transducer placement positions should be indicated. Details on selecting measurement locations and determining transducer placement positions are provided in the section 'Measurement locations'.
- *Data acquisition and instruments.* The test plan should provide information regarding the methods and instrumentation to be used for measurement and data collection. Instrumentation specification details should include type of instruments to be used, accuracy, calibration, sensitivity. More details on data acquisition and instruments are provided in the section 'Data acquisition and instruments'.
- *Data analysis.* The test plan should provide information regarding the methods, software, and instrumentation to be used for data analysis.

Table 42.5 Lighting criteria for operating and maintenance spaces/areas

Space	Illumination level in Lux	Space	Illumination level in Lux
Machinery spaces	200	Cargo holds (portable lighting)	
Engine room (main propulsion machinery)	300	General lighting[2]	30
Generator, switchboard and transformer room emergency generator room	200	During cargo handling[3]	200
Fan room/HVAC room	200	Passageways and trunks	80
Bilge/void spaces (portable lighting)	75	Storerooms	200
Pump room	200	Inspection and repair tasks	
Fire pump room	200	• Rough	300
Shaft alley	100	• Medium	500
Steering gear room	200	• Fine	750
Windlass room	200	• Extra fine	1000
Battery room	200	Workshops	
Boiler rooms	100	• General lighting	300
		• On work bench top/ machine paint shop	500
Muster/embarkation area	200	Escape trunks	50
Cargo handling[3]	200		
Weather decks	200	Crane cabin	400
Lay down area	200		
General processing area	200	Hand signalling areas between crane shack and ship deck	300
Utility area	200		
Loading ramps and bays	200		
Cargo storage and manoeuvring areas	200		

- *Test schedule.* The test plan should provide information regarding the proposed test schedule. The Test schedule should include information on the approximate date and duration of testing (including a summary of the scheduled measurements per day).
- *Data collection sheets.* The test plan should provide sample data sheets (in table format) that will be used for reporting the measurement data.
- *Conflict of interest declaration.* In the case that the testing specialist has, in any capacity or to any extent, provided input into the design of the vessel, their participation should be indicated in the test plan including details of the testing specialist's involvement. Class may review any potential conflicts of interest and determine if any further action is necessary.

Test Requirements

Visual tasks encountered on vessels vary and the lighting provided can influence the ability to see and perform those tasks. Some vision and lighting considerations include task duration, visual fatigue, task criticality, veiling reflections, shadows, and abilities of the observer. The presence of glare is a concern and is often difficult to identify, measure, and assess. Glare is often transient (based on factors such as the direction of the sun to illuminated components or from the placement of lighting fixtures) and therefore difficult to anticipate. As part of lighting data collection, glare should be subjectively evaluated jointly by the testing specialist and the Class approved surveyor witnessing the collection of lighting data. These assessments should be made as part of the activities discussed in the section 'Walk through verification inspection locations'. Measurement of ambient lighting need only be performed once during sea trials or at pier side when all external light sources can be blocked out. *Data acquisition and instruments.* The illuminance meter (light meter) must conform to any of the International Standards specified below (or equivalent):

- The illuminance meter (light meter) should meet the requirements specified in Section B.5 of IESNA RP-12-97. This defines the instrumentation requirements for measuring white light.
- Instruments used to measure red illuminance should meet the requirements specified in Section B.8.4 of IESNA RP-12-97.
- *Test conditions.* The test conditions required for lighting and illumination measurements must be in accordance with the following.
- *Location.* Lighting measurements may be taken in port, at sea, or both, since the measurements are not dependent on vessel operation. If testing is done in port, ship's power and not shore power should be used.
- *Spaces with windows/portlights.* In spaces with windows or portlights where the minimum lighting level is intended to be provided by artificial light sources only, lighting measurements should be taken after dark, or during the day provided all natural lighting is blocked out.

- *Spaces without windows/portlights.* Interior spaces with no windows or portlights can be measured during daylight hours.
- *Stray light.* Stray light (e.g., deck lighting and moonlight) should be capable of being blacked out as far as practicable. Where it is not possible, measurements of stray light, at appropriate positions, with all lighting turned off, should be obtained. These readings must then be deducted from readings taken at the same positions, with the lighting turned on, to determine the illumination from the lighting.
- *Light for sleep.* Lighting measurements should be taken in cabins with all cabin lights turned off and curtains, shutters, etc., closed.
- *Test interference.* During the lighting measurements, shadows on the light meter caused by any kind of human activity should be avoided. For this reason, only the personnel needed to operate any equipment during testing and those carrying out the measurements may be present in the space. Doors and windows must be closed, except where they are normally left open (such as the door on the lee side of the navigation bridge, which may normally be open). Any open doors or windows should be noted in the lighting test report. Spaces should be furnished with all usual equipment, fixtures and fittings normally found in the space.
- *Measurement locations.* Selection of spaces where measurements are to be conducted. The aim when selecting lighting and illumination measurement locations is to obtain a representative sample of data that represents the actual conditions in the manned crew spaces listed in the Lighting criteria Tables 42.1, 42.2, 42.3 and 42.4 inclusive. For practical reasons, it is important to select the locations such that an appropriate amount of sample data can be collected during the testing phase. The measurement locations must be selected in accordance with the following criteria:

  - Select problem areas based on the potential for excessive external illumination (daylight) into the space (e.g., bridge). Areas at the end of cable runs should be selected as potential problem locations. Measurements should be taken in all identified problem areas.
    Select potential problem areas based on the positioning of luminaires in the space as indicated on the drawings (e.g., uneven spacing of luminaires resulting in non-uniform illumination levels and dimly lit areas). Measurements must be taken in all problem areas.
  - Where a single instance of one type of space exists within the vessel (e.g., the bridge, mess room, gymnasium, library, etc.), the location should be selected for measurement.
  - Where multiple instances of one type of space exist (e.g., recreation areas) with the exception of cabins, a representative sample of at least 25% of each type should be selected for measurement. The problem area locations are to be considered part of the representative sample, if applicable.
  - Where a number of crew cabin spaces are identical in configuration in terms of lighting systems, surface treatments, geometry, furnishings and equipment layout, only two of the spaces on each deck should be selected to determine whether the lighting requirements are met. Where cabins are not identical

in configuration in terms of lighting system, surface treatments, geometry, furnishing, and equipment at least 25% should be selected on each deck for measurement.

- All of these measurement locations must be selected at locations port, starboard, fore, amidships and aft. The worst-case locations are to be considered part as the representative sample for crew cabins, if applicable.
- Where external night-time measurements are taken, spaces should be selected according to human presence in the space or task location, including exterior passageways and access ways, operating consoles, hand signalling locations, anchor handling locations, mooring locations, crew embarkations and debarkation areas, and cargo handling locations—particularly with regard to attaching and detaching cargo from crane hooks.

Walk through verification inspection locations. All normally manned spaces are to be subject to a walkthrough inspection by the Class approved surveyor. The number and locations of the walkthrough inspections are generally determined by the surveyor. The purpose of the walkthrough verification is to subjectively assess lighting and illumination qualities. The presence of discomfort or disability glare should be specifically and subjectively assessed by the Class approved surveyor and the testing specialist. Upon completion of the assessment, the result of "no glare present", "some glare present", "discomfort glare present", or "disability glare present" should be noted for each lighting test location. At the discretion of the surveyor, additional measurements may be required. General illumination measurement positions. General lighting levels should be measured with all lights turned on except supplementary task lighting, such as desk lights and berth lights. Daylight should be excluded during the measurements. Measurements should be taken on a horizontal plane approximately 750 mm (29.5 in.) above the deck. For traffic areas, readings should be taken on deck. For larger spaces (e.g., public rooms, messes, recreation areas, etc.), it may be necessary to take measurements at a number of locations in order to obtain a representative sample of the illumination for that space. Measurement locations should always be evenly distributed throughout the space. For a specific room size, the minimum number of measurement locations will be as indicated in Table 42.6, Distribution of measurement positions within spaces.

The arithmetic mean of the measurement should be used for the measure of illumination of the general area. In the case of corridor lighting, measurements should

Table 42.6 Distribution of measurement positions within spaces

Space	Minimum number of measurement positions in room
Less than 40 m^2 (431 ft^2)	1
Less than 80 m^2 (861 ft^2)	2
Less than 200 m^2 (2142 ft^2)	3
Greater than or equal to 200 m^2 (2142 ft^2)	4

be taken approximately every 3000 mm (10 ft) of corridor length. The arithmetic mean should not be used to determine compliance with corridor lighting requirements. *Task lighting measurement positions.* For task area lighting (such as consoles, control stations, workbenches, computer workstations, desktops and meeting tables, medical stations, galley areas, and maintenance workstations), a representative set of readings should be taken over the task surface with the worker in the normal working position. It is best to assess the task lighting with both general and supplementary task lighting turned on. As much as possible, daylight should be excluded during the measurements. Measurements of task lighting must be made on the surface or in the plane of the task (horizontal, vertical or at an angle). For small task surfaces, i.e., smaller than 0.5 m^2 (5 ft^2), a single measurement is best taken at the centre of the task surface. For larger task surfaces (0.5 m^2 (5 ft^2) or larger), the illumination is best measured by dividing the task surface into a grid no larger than 0.5 m^2 (5 ft^2) and averaging the measurements taken at the grid intersections. Measurements for mirrors should be taken approximately 400 mm (16 in.) away from the surface of the mirror. For berths and bunks, measurements may be taken at a point approximately 300 mm (12 in.) above the top of the mattress and 600 mm (24 in.) from the bunk light fixture. *Red and low-level white illumination.* Guidance for the measurement of red and low-level white is provided in B.8.4 in IESNA RP-12-97. These measurements are not required for the Habitability notation.

Test Report

As stated in the section on 'Test reports' in Chap. 37, a test report must be submitted to the Class approved surveyor to determine whether the lighting and illumination levels meet the criteria and whether this part of the Class notation requirement has been met. The details listed in the following paragraphs are required to be provided in the lighting test report. *Test details.* The following details are required to be recorded for each period of testing:

- Time of day.
- External lighting conditions (e.g., were measurements taken in port or at sea and were measurements taken during day light hours or at night?).

Measurement positions. Actual measurement locations should be indicated on appropriate drawings. *Measurement equipment details.* Details of the measuring equipment used (e.g., manufacturer, type and serial number, accuracy and resolution) should be provided. Copies of the relevant instrumentation reference calibration certificates, together with the results of field setup and calibration checks before and after the field tests, should be included with the lighting test report. *Results.* The following results, per space, should be provided in a tabulated format:

- Name and number of space(s).
- Task areas (if any) in space.

- Average lighting level for general lighting.
- Average lighting level for task blighting on each task surface.
- Corridor lighting measurements.
- Lighting levels in cabins with lights turned off and all curtains, shutters, deadlights, etc., closed.
- Presence of glare at specific workstations within the space and any follow-up activity requirements
- Indication of pass/fail.

For all the remaining spaces that were checked through walk through verification inspection and spot check measurements, the following information should be provided:

- Name and number of spaces.
- Walk through inspection observations.
- Measurement results if necessary.

Deviations. All deviations from the approved test plan must be reported. *Surveyor witnessing documentation.* The equipment field calibration and data collection process of indoor climate tests must be witnessed by a Class approved surveyor. The surveyor is required to sign or initial each page of the test report and prepare a witnessing document stating whether all steps of the indoor climate testing were completed to their satisfaction. A copy of the witnessing document is then given to the person conducting the onboard testing, for insertion into the final indoor climate test report. The original is usually retained by the surveyor and submitted to Class for inclusion in the vessel's Record.

Results

On completion of the test, the lighting test report will be reviewed by the Class approved surveyor against the appropriate Class Habitation notation requirements. Where the requirements have been satisfactorily met, the vessel will be assigned the appropriate Class Habitation notation.

Part V
Habitability on Offshore Installations

Chapter 43
General Provisions for Offshore Installations

General

This chapter is designed to provide a single source of guidance and information relating to the habitability criteria suitable for new and existing offshore installations. This chapter may be applied to installations falling under the categories of tension leg platforms (TLP), floating production, storage and offloading (FPSO), floating, storage and offloading (FSO), spars, fixed platforms, or any other buoyant or non-buoyant structure supported by or attached to the seafloor. It does not apply to offshore service vessels, dredgers, or mobile offshore drilling units such as self-elevating drilling units (SEDU), column-stabilised drilling units, or drillships as these are discussed in Part 4 of this book. For installation owners/operators seeking to achieve the optional Habitability notation (for instance, the **HAB(OS)**, the Habitability Offshore Plus (**HAB+(OS)** notation, or the Habitability Plus Plus (**HAB++(OS)**) provided by ABS, this chapter provides the essential criteria required by Class to review the accommodation area design for compliance with ILO MLC, 2006 Title 3 requirements, as relates to the MLC-ACCOM notation. As with the previous chapters which reference the notation criteria for vessels and workboats, this Part 5 focuses on the five habitability aspects of offshore installation design and layout that can be controlled, measured and assessed. These five aspects are divided into two categories: personnel accommodation areas and the ambient environment. Accommodation area criteria pertain to dimensional and outfitting aspects of spaces and open deck areas where personnel eat, sleep, recreate and perform routine daily activities. The ambient environmental aspects of habitability pertain to the environment that personnel are exposed to during periods of work, leisure and rest. Specifically, this Part 5 provides criteria, limits and measurement methodologies for whole-body vibration (separate criteria for personnel accommodation areas and workspaces), noise, indoor climate; and lighting. The criteria provided in this Part 5 are based on currently available research data and standards for the purpose of improving personnel performance and providing a base level of habitability and elements of safety related to habitability.

A. A. Olsen, *Applying Physical Ergonomics to Modern Ship Design*,
https://doi.org/10.1007/978-3-031-57974-5_43

Terminology

Accommodation areas/accommodation block: Offshore installation areas where the primary purpose is to rest and recreate. Accommodation spaces include cabins, medical facilities (sick bays), offices, recreation rooms, and manned spaces within the accommodation block such as the bridge or control room. For the purposes of this chapter, accommodation areas also include service spaces such as mess rooms, laundry, and storerooms.

Ambient environment: Ambient environment refers to the environmental conditions that personnel are exposed to during periods of work, leisure or rest. Specifically, this Part 5 provides criteria and limits for whole-body vibration, noise, indoor climate, and lighting.

Associated documentation: Documents referenced in this Part 5 that are needed to provide measuring techniques and further guidance.

Class recognised ambient environmental testing specialist: Companies providing test or measurement services on behalf of the owner/operator of the offshore installation or builder for the purposes of meeting any of the Class habitability notation requirements. Approval of the testing specialist for ships may also allow for testing and measurement services for other Class approved Habitability, Comfort, and MLC notations.

Dynamic positioning: A system to automatically maintain a workboat's position and heading by controlling propellers and/or thrusters. Dynamic positioning can maintain a position to a fixed point over the bottom, or in relation to a moving object (such as another offshore installation or vessel). It can also be used to position the installation at a favourable angle towards wind, waves, and current.

Fixed installation: Bottom-founded offshore installation permanently affixed to the sea floor. The term includes, but is not limited to, fixed platforms, guyed towers, and converted fixed installations.

Floating installation: Offshore installation designed to provide hydrocarbon processing and/or hydrocarbon storage and offload hydrocarbons. The term floating installation is used to generically identify a buoyant facility that is site-specific. This installation is securely and substantially moored so that it cannot be moved without a special effort. The term includes, but not limited to TLP, spar buoy, semi-submersibles and permanently moored shipshape hulls (FPSO, FSO, FPO).

Gross tonnage: The gross tonnage calculated in accordance with the tonnage measurement regulations contained in Annex I to the International Convention on Tonnage Measurement of Ships, 1969, or any successor Convention.

Habitability: The acceptability of the conditions of an installation in terms of whole-body vibration, noise, indoor climate, and lighting, as well as physical and spatial characteristics, according to prevailing research and standards for human efficiency and comfort.

Manned facility: A facility with permanent, occupied living accommodation areas, or one that requires continuous presence of personnel for more than 12 h in successive 24-h periods.

Manned space: For application of this Part 5, a manned space is any space where a seafarer may be present for 20 min or longer at any one time during normal, routine daily activities. Such spaces include working or living spaces. For machinery spaces, a manned space will be associated with the presence of a local work/control duty station manned for 20 min or longer at one time during normal, routine daily activities. Normal inspection, walk through (patrolling) and maintenance tasks in machinery spaces are not considered tasks contributing towards a particular space being considered manned.

Offshore installation: A buoyant or non-buoyant structure, supported by or attached to the sea floor, whose design is based on foundation and long-term environmental conditions at a particular site where it is intended to remain.

Personnel accommodation areas/personnel accommodation block: Areas where the primary purpose is to rest and recreate. Personnel accommodation spaces include cabins, medical facilities (sick bays), offices, recreation rooms, and manned spaces within the personnel accommodation block such as the bridge or control room. For the purposes of this book, personnel accommodation areas also include service spaces such as mess rooms, laundry, and storerooms.

Personnel spaces: All areas on an offshore installation intended for personnel only, such as personnel accommodation spaces and workspaces.

Personnel: Any person on board an offshore installation including mariners, industrial personnel, seafarers, and offshore support personnel.

Recreation and public spaces: Those portions of the personnel accommodation areas that are used for halls, dining rooms, lounges, and similar spaces.

Service spaces: Those spaces used for galleys, pantries containing cooking appliances, lockers and store-rooms, workshops other than those forming part of the machinery spaces, and similar spaces and trunks to such spaces.

Should: Expresses a provision that is mandatory.

Test plan: Document containing the requisite information regarding installation design and layout, test personnel, test conditions, measurement locations, data acquisition, instruments, data analysis and test schedule necessary for verifying the measurements for the ambient environmental aspects of habitability. This document is to be primarily prepared by the class recognised ambient environment testing specialist with additional input provided by the designer/builder. Submission of the document for approval can be by either the testing specialist or the designer/builder.

Test report: Document containing the actual testing results from the ambient environmental tests including details of the testing conditions, measurement locations, measurement equipment, and the results of the data collected and analysed.

Workspaces: Areas allocated for work. Categories of workspaces include, but are not limited to control rooms, service spaces (e.g., galley), production areas, machinery spaces, and duty/control stations.

Class Notations

At the offshore installation owner/operator or builders request, an installation complying with the minimum criteria for personnel accommodation areas and the ambient environment (i.e., whole-body vibration, noise, indoor climate, and lighting) may be assigned a Class notation which signifies the installation's compliance with the Class stipulated standards for habitability. For example, an installation complying with the **HAB(OS)** criteria and the more stringent criteria with respect to personnel accommodation, whole-body vibration, noise, and indoor climate may be distinguished in the Record by the notation **HAB+(OS)**. An installation satisfying all the ABS criteria may be distinguished further in the Record by the notation **HAB++(OS)**. A summary of the differences among each of these notations is presented below in Table 43.1.

Table 43.1 Summary of the differences between the **HAB(OS)**, **HAB+(OS)** and **HAB++(OS)** notations

	HAB(OS)	**HAB+(OS)**	**HAB++(OS)**
Personnel accommodation areas	**HAB(OS)** requirements for accommodation areas	**HAB(OS)** and **HAB+(OS)** requirements for accommodation areas. There is no difference between **HAB+(OS)** and **HAB++(OS)** notations	
Whole-body vibration	Level of vibration – reducing discomfort and promoting job performance	Lower level of vibration – promoting job performance and increased comfort	Lowest level of vibration – promoting job performance and increased comfort
Noise	IMO Code on Noise with Modifications	**HAB(OS)** with additional requirements	**HAB+(OS)** with additional requirements
Indoor climate	No provision for individual temperature adjustment		Aimed at enhancing crew comfort by making provisions for individual adjustments of indoor climate temperature
Lighting	No differences among the notations		

Plans and Data to be Submitted to Class

General. The following general arrangement type drawings of the installation are required to be submitted to Class for their review and approval: outboard and inboard profiles that show the elevations of all major decks above a specified baseline; and plans of each deck showing the watertight and non-tight bulkhead locations, structural and non- structural bulkhead locations, layout of major machinery, etc. Personnel accommodation areas. At a minimum, scaled arrangement drawings of the various personnel accommodation spaces (elevation and plan views) and the installation's accommodation area specifications should be submitted to class. Details of the personnel accommodation area data requirements are provided in the section 'Personnel accommodation area documentation'. *Ambient environment.* The following items are required to be submitted individually for each ambient environmental aspect:

- *Test plans.* Test plans serve as the principal means for verifying the measurements for the ambient environmental aspects of habitability. Separate test plans are required for wholebody vibration, noise, indoor climate, and lighting. The test plans require approval by Class before any measurements are made. Class will usually notify the installation owner, operator or builder whether the test plans have been approved or require alteration. A Class approved copy of the test plan will then become part of the installation's official documentation.
- *Test reports.* Upon completion of the ambient environmental testing, the test reports should be submitted to the Class approved surveyor for review. In addition, a copy of the test reports will become part of the installation's official documentation. These reports contain ambient environmental information such as test results, testing details, measurement equipment details, etc.

Process for Obtaining a Class Notation

Prior to scheduling accommodation area verification or ambient environmental testing activities, the installation owner, operator or builder should certify the operational status of the installation as being fully operational and/or inclusive of all equipment and furnishings. If the installation is not fully operational, a complete listing of deficiencies of areas, components, equipment, etc. should be submitted to Class for review. Class will then make a determination and notify the installation owner, operator or builder as to whether accommodation area verification activities or ambient environmental testing can commence. The intent is to conclude all ambient environmental testing during installation and commissioning in the field. Figure 43.1 Habitability notation process with initial testing only, charts the process for obtaining an ABS **HAB(OS)**, **HAB+(OS)**, or a **HAB++(OS)** notation. The following sections briefly describe the notation process. *Personnel accommodation areas.* Arrangement drawings, plans, and installation design specifications for accommodation spaces

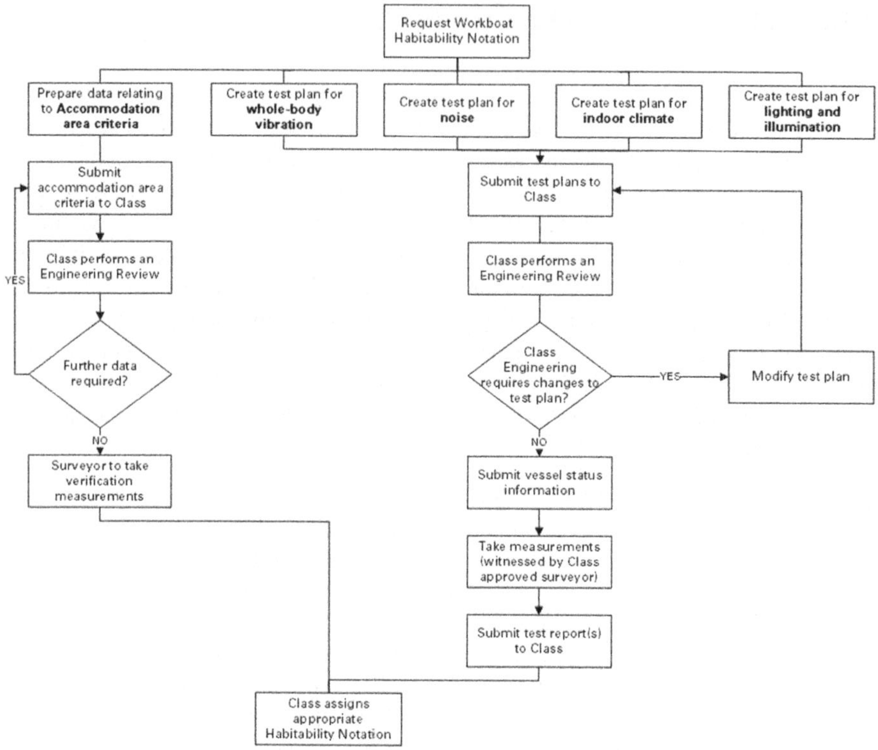

Fig. 43.1 Process for obtaining a Class habitability notation

should be prepared and submitted to Class for review. For new construction, the drawings should be provided to Class during the detailed design phase. For existing installations, the arrangement drawings and plans, reflecting the current personnel accommodation area configurations should be provided to and approved by Class, in advance of Class approved surveyor verifications. Follow-up physical verification measurements of the personnel accommodation area criteria may only be performed by a Class approved surveyor. The Class approved surveyor will then select the most appropriate verification measurement sites. The physical verification measurements of accommodation area criteria must usually be performed for all installations in a series of installations. The results of the Class review and actual Class approved surveyor verification are then reviewed by the Class approved surveyor during the notation confirmation process. *Ambient environment.* Ambient environmental test plans for whole-body vibration, noise, indoor climate, and lighting should be individually prepared and submitted to Class for review and approval. These test plans should serve as a primary means for verifying the measurement locations and measurement process, as well as specifying the testing specialist who will perform the ambient environmental testing. Testing, inspections, and data collection should be performed by the testing specialist and witnessed by a Class approved surveyor. The test reports for ambient environmental testing are to be prepared by the testing specialist and submitted to the Class approved surveyor for review.

For an installation that is expected to operate in extreme environmental conditions (e.g., Polar regions, the Persian Gulf, etc.) that are significantly different from the conditions encountered during the field trials, the installation designer/builder may be required to submit information on the design/engineering analysis performed as part of the test plan submission verifying that the appropriate requested notation criteria can be met in the actual area of operations. If there are instances during the field trials where the criteria were either not met, or passed by a small margin, the Class approved surveyor—in conjunction with the designer/builder and the installation owner/operator—may request additional confirmatory tests to be carried out in the actual area of operation. Full ambient environmental testing as described in this Part 5 is required for the first installation constructed as part of a series of installations. For the second and subsequent sister installations (i.e., installations being part of the same series of installations having identical structural, machinery, and outfitting designs), full scale testing for the ambient environmental aspects of the notation, is not required. For the second or all subsequent sister installations constructed by the same builder, the following is applicable for all aspects of ambient environmental testing (whole-body vibration, noise, indoor climate and lighting):

- Spaces that were identified as "worst case" locations in the test plans for the first installation need to be tested.
- By review of the first installation's test reports (all ambient environment aspects), the Class approved surveyor will identify the locations which were out of compliance or close to non-compliance that need to be tested.
- Spaces that required some type of mitigation for the first installation need to be tested.

As part of the notation assignment process, all installations are subject to Class approved surveyor walk through. Additional testing may be required at the Class approved surveyor's discretion. For sister installations constructed by a different builder, the following provisions may be applicable: full testing is required for the first installation from each builder; and for the subsequent sister installation's, the procedure described previously for installations built by the same builder is applicable. *Results.* The Class accommodation area assessment, Class approved surveyor verification measurements, and ambient environmental test reports must be submitted to Class for review, after which Class (or their appointed Class approved surveyor) will issue a determination of notation confirmation.

Initial Requirements

The initial process for obtaining any installation habitability notation usually comprises some form of Class engineering review, ambient environmental testing, and Class approved surveyor verifications. Testing must be carried out in accordance with the submitted test plans once they have been reviewed and approved by Class in advance of the testing regime commencing. Testing should always, where practicable, be witnessed by a Class approved surveyor. Once the criteria specified by Class (and summarised in this Part 5) have been satisfied, the appropriate notation may be confirmed.

Surveys Completed After Construction

It is intended that all surveys following the installation's construction are aligned with Class surveys. Harmonisation of surveys may be carried out at the first available opportunity.

Alternatives

Class may consider alternative arrangements, criteria and procedures, which can be shown to satisfy the criteria directly cited or referred to in this Part 5. The demonstration of an alternative's acceptability can be made through either the presentation of satisfactory service experience or systematic analysis based on valid engineering principles. *National regulations.* Class will usually consider, for their acceptance, alternative arrangements and details, which can be shown to comply with standards recognised in the country (Flag State) in which the installation is registered or built, provided they are deemed no less effective than those stipulated by Class. *Departures from criteria.* The criteria summarised in this Part 5 is envisioned to apply to installations that are engaged in the usual activities and services expected of such installations. However, it is recognised that unusual or unforeseen conditions may lead to a case where one or more of the parameters of interest in granting a notation may temporarily fall outside the range of acceptability. When a departure from criteria is identified, during either the notation's initial issuance or reconfirmation process, it must be reviewed by Class in consultation with the installation's builder and owner/operator. When the design of the personnel accommodation areas or ambient environmental test results contain departures from the stated criteria, these will be subject to special consideration upon the receipt of details about the departure. Depending on the degree and consequences of the departure, the installation's builder or owner/operator may be required to provide an assessment and corrective plan to obtain or maintain the notation. Failure to complete the agreed correction by the due date may lead to the withdrawal of the assigned notation.

Chapter 44
Accommodation Areas on Offshore Installations

General

To promote maritime safety, efficiency, and habitability, it is important that personnel maintain appropriate levels of mental and physical fitness while on board installations. To help accomplish this, personnel employed on board offshore installations should be provided with suitable accommodation areas. Appropriate personnel accommodation area design helps promote reliable performance by reducing the potential for fatigue and human error. Appropriately designed and outfitted personnel accommodation areas may also enhance personnel morale, recruiting, retention, comfort, and overall quality of life at sea. Conversely, inappropriate personnel accommodation areas can adversely impact the personnel's ability to reliably perform assigned duties, fully relax, sleep, and recover from mentally and physically demanding work activities. This in turn can impact their ability to carry out their duties with the required diligence, accuracy, and attention to safety procedures. Providing an on board environment that increases seafarer alertness and well being should be of concern to installation owners/operators. This chapter provides the measurement criteria for personnel accommodation spaces. In particular, Tables 44.2 through 44.10 encompass the criteria for access and egress, as well as personnel cabins, sanitary spaces, offices, food services, recreation areas, laundry, and medical spaces. The criteria were selected to help increase personnel safety, productivity, quality of work, retention, and morale. Note: the criteria for access/egress are applicable only to the personnel accommodation block and do not include personnel workspaces outside of the accommodation block. It is worth noting that compliance with the criteria outlined in this chapter are a prerequisite for the ABS offshore installation Habitability (**HAB(OS)**), offshore installation Habitability Plus (**HAB+(OS)**), or offshore installation Habitability Plus Plus (**HAB++(OS)**) notation. Other classification societies will have their own distinct and specific habitability notations, though as the criteria is derived from IMO MLC 2006, Title 3, the criteria are generally standardised across each Class Table 44.1.

A. A. Olsen, *Applying Physical Ergonomics to Modern Ship Design*,
https://doi.org/10.1007/978-3-031-57974-5_44

Table 44.1 Example of the ILO MLC 2006 Title 3 requirement #17

Accommodation area criteria				
General				
Requirement # = related ILO MLC 2006, title 3 requirement			HAB(OS) requirements	HAB+(OS)/ HAB++(OS) requirements
#	17	Accommodation area stairways have a maximum angle of inclination from the horizontal of		
		45°	•	
		40°		•

Instructions: For a **HAB(OS)** notation, the boxes with a • in the **HAB(OS)** requirements column must be met. For a **HAB+(OS)** notation, the boxes with a • in the **HAB+(OS)** requirements column need to be met in addition to the **HAB(OS)** requirements. For a **HAB++(OS)** notation, the boxes with a • in the **HAB++(OS)** requirements columns need to be met in addition to the **HAB(OS)** and **HAB+(OS)** requirements

Terminology

Personnel accommodation areas/personnel accommodation block: Areas where the primary purpose is to rest and recreate. Personnel accommodation spaces include cabins, medical facilities (sick bays), offices, recreation rooms, and manned spaces within the personnel accommodation block such as the bridge or control room. For the purposes of this chapter, personnel accommodation areas also include service spaces such as mess rooms, laundry, and storerooms.

Service spaces: Those used for galleys, pantries containing cooking appliances, lockers and storerooms, workshops other than those forming part of the machinery spaces, and similar spaces and trunks to such spaces.

Associated Documentation

- International Labour Organisation (ILO) Conventions 92 and 133.
- ILO MLC, Title 3, 2006.

Personnel Accommodation Area Criteria

The accommodation area criteria are contained in Tables 44.2 through 44.10. The ILO MLC, 2006 quantitative accommodation area requirements have been included in this chapter and are denoted by a "#" symbol in Tables 44.2 through 44.10.

Table 44.2 ILO MLC 2006 Title 3 requirement (accommodation area criteria (general)

Accommodation area criteria					
General					
Requirement # = related ILO MLC 2006, Title 3 requirement			HAB(OS) requirements	HAB+(OS) requirements	HAB++(OS) requirements
#	1	Headroom in all passageways, sleeping rooms, stairs, sanitary spaces, offices, food service areas, and recreational areas is at least 2030 mm (80.0 in.). This height is clear of any obstructions such as cableways, ducting, pipes, etc.	•		
	2	Non-slip type deck covering is supplied where occasional water, oil or liquid on the floors is expected	•		
	3	Outside corners of bulkheads, doors, etc. are rounded	•		
	4	All edges that crew members may strike are rounded	•		
	5	Drawers and internal doors are designed			
		• To prevent opening and closing due to vessel motion	•		
		• To be operable with one hand	•		
#	6	Painted wall surfaces and deck heads are light in colour	•		
	7	Interior finish materials and furnishings are designed to ease cleaning efforts and improve maintenance	•		

(continued)

Table 44.2 (continued)

Accommodation area criteria					
General					
Requirement # = related ILO MLC 2006, Title 3 requirement			HAB(OS) requirements	HAB+(OS) requirements	HAB++(OS) requirements
	8	For workstations where seafarers will be working while standing for extended periods of time, a kick space of 100 mm (4 in.) high by 100 mm (4 in.) deep should be provided	•		
	9	9 Toilets are conveniently located to workplace, recreation areas, mess rooms, and shower rooms	•		
	10	A means of escape that is not readily apparent to a person from both the inside and outside of the space is adequately marked	•		
#	11	Wall surfaces and decks are washable and impervious to damp or moisture absorption	•		
#	12	For accommodation areas and recreational facilities where the floorings are made of composite materials, the joints with the sides should be profiled to avoid crevices	•		
#	13	Accommodation areas and recreational and catering facilities should be located as far as practicable from the engines, steering gear rooms, deck winches, ventilation, heating, and air-conditioning equipment, and other noisy machinery and apparatus	•		

(continued)

Table 44.2 (continued)

Accommodation area criteria					
General					
Requirement # = related ILO MLC 2006, Title 3 requirement			HAB(OS) requirements	HAB+(OS) requirements	HAB++(OS) requirements
	14	Transparent or translucent surfaces in doors, gates, walls and partitions should be of a safety material or be adequately protected against breakage in the following cases			
		• In doors and gates	•		
		• Door and gate side panels	•		
		• where any part of the transparent or translucent surface is at shoulder level or below	•		
		• In windows, walls and partitions, where any part of the transparent or translucent surface is at waist level or below	•		

The **HAB+(OS)** and **HAB++(OS)** notations have more stringent criteria than the **HAB(OS)** notation with the objective of providing enhanced living and working conditions to improve seafarer safety and comfort. This includes enhanced criteria for workspace design, personnel cabins, and recreational/leisure areas (Table 44.1).

Accommodation Area Documentation

As stated earlier, personnel accommodation area documentation must be prepared and submitted to Class for review and approval. Confirmatory verification measurements are then performed by a (Class approved) surveyor. The following data is typically required to be submitted to Class. *Data requirements.* The submitted data serves as a means for verifying that the installation meets the personnel accommodation area criteria specified in Tables 44.2 through 44.10.

Table 44.3 ILO MLC 2006 Title 3 requirement (Accommodation area criteria (access/egress)

Accommodation area criteria					
Access/egress					
Requirement # = related ILO MLC 2006, title 3 requirement			HAB(OS) requirements	HAB+(OS) requirements	HAB++(OS) requirements
	1	Doors, hatches, or scuttles used as a means of escape are capable of being operated by one person, from either side, in both light and dark conditions	•		
	2	The method of opening a means of escape is obvious and rapid	•		
	3	Doors in accommodation spaces (with the exception of cabins), stairway, stair tower, passageway, or control spaces, open in the direction of escape, where practicable	•		
	4	Deck scuttles that serve as a means of escape are fitted with a quick acting release and a holdback device to hold the scuttle in an open position	•		
	5	Handles for sliding doors are designed to avoid accidental crushing of fingers	•		
	6	Sliding doors should have a stop or other effective means to prevent the door coming off the end of the track. They should also have a retaining rail to prevent the door from falling should the suspension system fail or the rollers leave the track			

(continued)

Table 44.3 (continued)

Accommodation area criteria					
Access/egress					
Requirement # = related ILO MLC 2006, title 3 requirement			HAB(OS) requirements	HAB+(OS) requirements	HAB++(OS) requirements
	7	High voltage electrical cabinet doors are lockable with keys provided to qualified personnel			
Door dimensions					
	8	Doors have the following dimensions			
		• Doors (other than emergency egress) used solely by crew members have a clear opening width of approximately 710 mm (28 in.) or ASTM size 4 doors	•		
Railings					
	9	For installations where severe motions are possible, suitable storm rails/handrails are provided in all interior passageways and at all accommodations block sides where persons onboard might have normal access			
		• Storm rails/handrails are installed on both sides of passageways that are 1830 mm (72 in.) or more in width	•		
		• Storm rails/handrails are 865 mm (34 in.) to 965 mm (38 in.) high	•		

(continued)

Table 44.3 (continued)

Accommodation area criteria					
Access/egress					
Requirement # = related ILO MLC 2006, title 3 requirement			HAB(OS) requirements	HAB+(OS) requirements	HAB++(OS) requirements
		• The distance between/ or behind storm rails/ handrails and any obstruction is 75 mm (3 in.) or greater	•		
	10	Rails are installed parallel to the deck along deck edges and walkways and around open hatches, elevators, antenna platforms and along other boundaries in the following areas			
		• Wherever there is danger of crew members falling to a lower level of 600 mm (23.5 in.) or more in the vessel wherever there is danger of crew members becoming enmeshed with hazardous operating machinery	•		
		• Around unprotected openings with a coaming height below 760 mm (30 in.)	•		
		Note: *temporary rails can be used around unprotected openings into which a person may slip, trip or fall*			
	11	Deck/guard railings have the following design			

(continued)

Table 44.3 (continued)

Accommodation area criteria					
Access/egress					
Requirement # = related ILO MLC 2006, title 3 requirement			HAB(OS) requirements	HAB+(OS) requirements	HAB++(OS) requirements
		• The heights of rails or bulwarks are at least 1000 mm (39.5 in.) from the deck except where this height would interfere with the normal operation of the vessel, and toeboards which are at least 100 mm (4.0 in.) in height and have no more than a 6 mm (0.25 in.) clearance between the bottom edge of the toeboard and the walking surface	•		
		• The heights of rails are at least 1070 mm (42 in.) from the deck except where this height would interfere with the normal operation of the vessel and toeboards which are at least 100 mm (4.0 in.) in height and have no more than a 6 mm (0.25 in.) clearance between the bottom edge of the toeboard and the walking surface		•	•
		• Vertical stanchions for railings are spaced no more than 1525 mm (60 in.) apart horizontally	•		
		• At least every third vertical stanchion is supported by a bracket or stay	•		

(continued)

Table 44.3 (continued)

Accommodation area criteria					
Access/egress					
Requirement # = related ILO MLC 2006, title 3 requirement			HAB(OS) requirements	HAB+(OS) requirements	HAB++(OS) requirements
		• Chain or wire rope used as a rail is set such that the sag is not greater than 25 mm (1 in.) at the chain/rope's centre span	•		
		• Maximum lengths of openings protected by wire or chain are 1830 mm (72 in.). A removable stanchion will constitute the start of a new opening	•		
		• Vertical stanchions for railings are spaced no more than 1525 mm (60 in.) apart horizontally	•		
		• At least every third vertical stanchion is supported by a bracket or stay	•		
		• Chain or wire rope used as a rail is set such that the sag is not greater than 25 mm (1 in.) at the chain/rope's centre span	•		
		• Maximum lengths of openings protected by wire or chain are 1830 mm (72 in.). A removable stanchion will constitute the start of a new opening	•		

(continued)

Table 44.3 (continued)

Accommodation area criteria					
Access/egress					
Requirement # = related ILO MLC 2006, title 3 requirement			HAB(OS) requirements	HAB+(OS) requirements	HAB++(OS) requirements
Stairs					
	12	A clear landing at least as wide as the tread width and a minimum of 915 mm (36 in.) long is provided at the top and bottom of each stairway	•		
	13	Any change of direction in a stairway is accomplished by means of an intermediate landing at least as wide as the tread width and a minimum of 915 mm (36 in.) long	•		
	14	An intermediate landing is provided at each deck level serviced by a stair, or a maximum of every 3600 mm (144 in.) of vertical travel for stairs with a vertical rise of 6100 mm (240 in.)		•	•
	15	Accommodation area stairways have a maximum angle of inclination from the horizontal of			
		• 50°	•		
		• 45°		•	•
		• 40°		•	•
	16	Stair risers and treads have the following design			
		• The riser height is no more than 230 mm (9 in.) and the tread depth is approximately 190 mm	•		

(continued)

Table 44.3 (continued)

Accommodation area criteria					
Access/egress					
Requirement # = related ILO MLC 2006, title 3 requirement			HAB(OS) requirements	HAB+(OS) requirements	HAB++(OS) requirements
		• (7.5 in.), with an additional 25 mm (1 in.) tread nosing		•	•
		• The riser height is no more than 230 mm (9 in.) and the tread depth is approximately 230 mm (9 in.), with an additional 25 mm (1 in.) tread nosing		•	•
		• The riser height is no more than 230 mm (9 in.) and the tread depth is approximately 275 mm (10.75 in.), with an additional 25 mm (1 in.) tread nosing		•	•
		• Stairs in a stairway or stair tower, the depth of the tread and the height of riser are consistent	•		
		• The minimum tread width on one-way (one person) stairs is at least 610 mm (24 in.)	•		
		• The minimum tread width on two-way (two person) stairs is at least 915 mm (36 in.)			
		• The minimum tread width on two-way (two person) stairs is at least 1015 mm (40 in.)		•	•

(continued)

Table 44.3 (continued)

Accommodation area criteria					
Access/egress					
Requirement # = related ILO MLC 2006, title 3 requirement			HAB(OS) requirements	HAB+(OS) requirements	HAB++(OS) requirements
		• Once a minimum tread width has been established at any deck, in that stair run, it is not decreased in the direction of escape	•		
		• All nosings have a non-slip surface	•		
	17	Stairway or stair towers are fitted with handrails with the following design			
		• A handrail is provided on one side of the stair	•		
		• A handrail is provided on both sides of the stair	•		
		• The handrails are parallel to the pitch line of the stair flight and level at landings		•	•
		• The handrail is continuous from the top to the bottom of the stair and terminates in a safe manner at both ends	•		
		• The vertical height above the tread at its nosing is at least 940 mm (37 in.) to 990 mm (39 in.)	•		
		• The distance between handrails and any obstruction is 75 mm (3 in.) or greater	•		

(continued)

Table 44.3 (continued)

Accommodation area criteria					
Access/egress					
Requirement # = related ILO MLC 2006, title 3 requirement			HAB(OS) requirements	HAB+(OS) requirements	HAB++(OS) requirements
		• Handrails on adjacent, parallel stair flights have a minimum of 100 mm (4 in.) clear distance between rails	•		
Ladders					
	18	All inclined ladders and handrails are located so as not to interfere with the opening and closing of hatches, doors, gratings, or other types of access; in all areas of the vessel other than accommodation block	•		
	19	Inclined ladders are between 50° and 60° from the horizontal	•		
	20	Each vertical ladder used as a means of escape has the following design features	•		
		• The ladder is mounted at least 180 mm (7 in.) from the nearest permanent object in the back of the ladder (distance behind each rung)	•		
		• Rungs are at least 410 mm (16 in.) in width	•		
		• Rungs are between 275 mm (11 in.) and 300 mm (12 in.) apart	•		
		• Rungs are uniformly spaced for the length of the ladder	•		

(continued)

Table 44.3 (continued)

Accommodation area criteria					
Access/egress					
Requirement # = related ILO MLC 2006, title 3 requirement			HAB(OS) requirements	HAB+(OS) requirements	HAB++(OS) requirements
		• At least 760 mm (30 in.) clearance in front of the ladder (climbing space)	•		
		• Each ladder rung is skid/slip resistant	•		
		• Grab bars are provided that extend at least 1070 mm (42 in.) above the landing platform		•	•
		• Ladder landings at the top and bottom of inclined ladders			
		• At least 760 mm (30 in.) clear in length	•		
		• At least 915 mm (36 in.) clear in length		•	•
Ramps					
	21	Ramps are sloped < 15° for inclined walking surfaces	•		
	22	Ramps have a non-skid surface	•		
	23	Ramps have a handrail on any open side of the ramp if the distance from the ramp to the nearest adjacent surface is 600 mm (23.5 in.) or more	•		
	24	Access to lifeboats is provided such that a person in a stretcher can be easily embarked into the survival craft (e.g., ramp)	•		

(continued)

Table 44.3 (continued)

Accommodation area criteria					
Access/egress					
Requirement # = related ILO MLC 2006, title 3 requirement			HAB(OS) requirements	HAB+(OS) requirements	HAB++(OS) requirements
	25	The clear walkway width for one person in an unrestricted area, where two persons could pass is at least 710 mm (28 in.)	•		
	26	The clear walkway width for normal two-way traffic or any means of egress that leads to an exit or entrance is at least 915 mm (36 in.)	•		
	27	The clear walkway width for normal two-way traffic or any means of egress that leads to an exit or entrance is at least 1015 mm (40 in.)	•		

- *Newbuilds.* For new construction, scaled arrangement drawings of the personnel accommodation spaces (elevation and plan views), details of the personnel accommodation area outfitting and installation's design specification in relation to the accommodation spaces must be submitted to Class.
- *Existing installations.* For existing offshore installations, appropriate arrangement drawings and plans, reflecting the current personnel accommodation area configurations should be provided to Class together with any current installation accommodation area design specifications.

Submission Review and Verification

Arrangement drawings, plans, and installation specifications for the personnel accommodation areas should be prepared and submitted for review by Class. For new construction, the drawings should be provided to Class early during the detailed design phase. For existing installations, the arrangement drawings and plans reflecting the current personnel accommodation area configurations may be provided to Class in advance of onboard surveyor verifications. Class will then review the submitted accommodation area documentation. Any deviations from the criteria will

Table 44.4 ILO MLC 2006 Title 3 requirement (Accommodation area criteria (personnel cabins)

Accommodation area criteria					
Crew cabins					
Requirement # = related ILO MLC 2006, title 3 requirement			HAB(OS) requirements	HAB+(OS) requirements	HAB++(OS) requirements
Room size			•		
#	1	Cabin floor area for single occupancy rooms is at least 7.0 m^2 (75 ft^2)			•
	2	Cabin floor area for single occupancy rooms is at least 8.5 m^2 (91.5 ft^2)	•		
#	3	Cabin floor area for double occupancy rooms is at least 8.0 m^2 (86 ft^2)			•
	4	Cabin floor area for double occupancy rooms is at least 10.0 m^2 (107.5 ft^2)	•		
Occupancy level			•		
#	5	Sleeping rooms are not occupied by more than two persons	•		
		• The OIM	•		
		• The production supervisor	•	•	•
		• The maintenance supervisor	•		
		• The facilities engineer	•		
		• The marine supervisor	•		
		• The logistics/ transport coordinator	•		
		• Marine officers (e.g., master, chief engineer) if applicable	•		
	6	No more than two senior operational personnel (e.g., managers) or two senior marine personnel (e.g., petty officers) occupy a single sleeping room	•		

(continued)

Table 44.4 (continued)

Accommodation area criteria					
Crew cabins					
Requirement # = related ILO MLC 2006, title 3 requirement			HAB(OS) requirements	HAB+(OS) requirements	HAB++(OS) requirements
#	7	Personnel sharing rooms are in same occupation group and day persons do not share a room with watchkeepers	•		
	8	Maximum number of persons to be accommodated in any sleeping room is marked indelibly and legibly in some conveniently seen place	•		
		• In the room	•		
#		• Outside the room	•		
#		• Separate sleeping rooms are provided for men and for women	•		
		• Storm rails/handrails are 865 mm (34 in.) to 965 mm (38 in.) high	•		
		• The distance between or behind storm rails/ handrails and any obstruction is 75 mm (3 in.) or greater	•		
Cabin location			•		•
#	9	There is no direct access into sleeping areas from cargo areas, machinery spaces, chain lockers, galleys, lamp and paint rooms or from engine, deck and other bulk storerooms, drying rooms, communal wash places or toilets	•		

(continued)

Table 44.4 (continued)

Accommodation area criteria				
Crew cabins				
Requirement # = related ILO MLC 2006, title 3 requirement		HAB(OS) requirements	HAB+(OS) requirements	HAB++(OS) requirements
10	A high risk/low risk location grading concept should be adopted such that the accommodation areas are located as far as possible from high-risk areas such as gas compression, process and wellhead areas. Low risk facilities such as water injection and utilities should be positioned between the high-risk areas and the accommodation areas	•		
11	Personnel accommodation areas are not located immediately beneath high-traffic alleyways or other noisy areas	•		
Cabin design		•		
General		•		
12	There are a sufficient number of beds for the maximum number of people required to sleep on the installation without using the same beds for successive shifts	•		
13	Wall surfaces are not constructed from tongued and grooved boarding or any other form of construction likely to harbour vermin	•		

(continued)

Table 44.4 (continued)

Accommodation area criteria					
Crew cabins					
Requirement # = related ILO MLC 2006, title 3 requirement			HAB(OS) requirements	HAB+(OS) requirements	HAB++(OS) requirements
#	14	Internal bulkheads are of approved material that is not likely to harbour vermin	•		
#	15	That part of a bulkhead separating such places from sleeping rooms and external bulkheads are efficiently constructed of steel or other approved substance and be watertight and gas tight	•		
	16	Pipes, ventilation ducts, or other installations do not obstruct berths	•		
	17	Sleeping rooms are ventilated and heated	•		
	18	HVAC vents are not directed at the heads of berths	•		
	18	Radiators and other heating apparatus are so placed and, where necessary, shielded as to avoid risk of fire or danger or discomfort to the occupants	•		
Berths			•		
	19	Multi-personnel cabin berths are not side by side so that access to one berth can only be obtained over another	•		
	20	Multi-personnel cabin berths contain individually operated privacy curtains	•		

(continued)

Table 44.4 (continued)

Accommodation area criteria					
Crew cabins					
Requirement # = related ILO MLC 2006, title 3 requirement			HAB(OS) requirements	HAB+(OS) requirements	HAB++(OS) requirements
	21	Multi-personnel cabin berths contain individually operated fans/blowers	•		
	22	Multi-personnel cabin arrangements are either single tier or double tiered	•		
	23	Multi- personnel cabin berths with portlights above them are only allowed in a single tier arrangement	•		
Emergencies			•		•
	24	An emergency alerting system (e.g., an audible alarm) is present for all personnel cabin spaces	•		
Outfitting			•		
#	25	Sleeping rooms, with portlights or windows, have curtains that completely block out light	•		
#	26	Sleeping rooms are lighted with natural light and are provided with artificial light (subject to special arrangements as may be permitted by the Flag State)	•		
	27	Grab bars and stepping surfaces are provided for access to upper berths	•		

(continued)

Table 44.4 (continued)

Accommodation area criteria					
Crew cabins					
Requirement # = related ILO MLC 2006, title 3 requirement			HAB(OS) requirements	HAB+(OS) requirements	HAB++(OS) requirements
#	28	An electric reading light is provided at the head of each berth	•		
#	29	The top berth in a tier has a dust-proof bottom of wood, canvas or other suitable material	•		
	30	In sleeping rooms with more than one occupant, furniture (beds, lockers, drawers, etc.) are labelled identifying which furniture is dedicated to which berth	•		
	31	Each sleeping room is provided with at least the following furniture (in addition to berths and clothes lockers)	•		
#		• A table or desk	•		
#		• A chair	•		
#		• A mirror with a light	•		
#		• A small cabinet for toilet requisites for each person in the room	•		
#		• A book rack	•		
#		• Coat hooks	•		
#	32	Furniture is of smooth, hard material, not liable to warp or corrode	•		
Dimensional aspects					
Aisles					
	33	Aisle widths in sleeping areas are			

(continued)

Table 44.4 (continued)

Accommodation area criteria					
Crew cabins					
Requirement # = related ILO MLC 2006, title 3 requirement			HAB(OS) requirements	HAB+(OS) requirements	HAB++(OS) requirements
		• At least 610 mm (24 in.), between a single berth and the nearest obstruction	•		
		• At least 915 mm (36 in.), between facing berths	•		
Berths					
#	34	The lower berth in a tier is at least 300 mm (12 in.) above the deck	•		
#	35	The upper berth is placed approximately midway between the bottom of the lower berth and the lower side of the deck head beams	•		
	36	Head clearance above each berth is			
		• At least 610 mm (24 in.)	•		
		• At least 810 mm (32 in.)		•	•
	37	Berth inside dimensions are			
#		• At least 1980 mm (78 in.) by 800 mm (31.5 in.)	•		
		• At least 2030 mm (80 in.) by 965 mm (38 in.)		•	•
#	38	The framework and lee-board of a berth is of approved material, hard, smooth and not likely to corrode or to harbour vermin	•		

(continued)

Table 44.4 (continued)

Accommodation area criteria					
Crew cabins					
Requirement # = related ILO MLC 2006, title 3 requirement			HAB(OS) requirements	HAB+(OS) requirements	HAB++(OS) requirements
#	39	Berths constructed from tubular frames are completely sealed and without perforations which would give access to vermin	•		
Storage space					
	40	Sleeping rooms provide storage space for each occupant			
		• The clothes locker is fitted with a shelf and hasp	•		
#		• A drawer or equivalent space of at least 56 litres; if the drawer is incorporated in the clothes locker then the combined minimum volume of the clothes locker is 500 litres; it is fitted with a shelf and be able to be locked by the occupant so as to provide privacy	•		
#		• A drawer or equivalent space of at least 56 litres; if the drawer is incorporated in the clothes locker then the combined minimum volume of the clothes locker is 500 litres; it is fitted with a shelf and be able to be locked by the occupant so as to provide privacy	•		

(continued)

Table 44.4 (continued)

Accommodation area criteria					
Crew cabins					
Requirement # = related ILO MLC 2006, title 3 requirement			HAB(OS) requirements	HAB+(OS) requirements	HAB++(OS) requirements
	41	Accommodation areas for the hanging of oilskins/rain gear are			
		• Sufficient	•		
		• Adequately ventilated	•		
		• Provided outside but convenient to the sleeping rooms	•		

be highlighted to the installation's builder and/or owner/operator by Class for resolution. Class will also identify any criteria that the surveyor must field verify. The surveyor will then verify that the submitted drawings match the constructed installation and also verify any criteria that are outstanding from the Class review and document deviations from criteria.

Results

On completion of the submission review, the results of the Class review and surveyor verification will be reviewed by the surveyor against the appropriate Class notation criteria for the installation. Provided the criteria is satisfactorily met, the appropriate notation will then be confirmed and entered into the installation's Record (Tables 44.3, 44.4, 44.5, 44.6, 44.7, 44.8 and 44.9).

Table 44.5 ILO MLC 2006 Title 3 requirement (Accommodation area criteria (sanity spaces)

Accommodation area criteria					
Sanitary spaces					
Requirement # = related ILO MLC 2006, title 3 requirement			HAB(OS) requirements	HAB+(OS) requirements	HAB++(OS) requirements
	1	Sanitary spaces are			
		• Lighted	•		
#		• Heated	•		
#		• Ventilated	•		
	2	Floors in sanitary spaces are			
#		• Non-slip type deck covering	•		
#		• Easily cleaned	•		
#		• Impervious to damp or moisture absorption	•		
#		• Properly drained	•		
	3	For every six crew members who do not have a private sanitary space, there is			
#		• One or more tubs and/or shower baths	•		
#		• One or more toilets	•		
#		• Washbasins	•		
	4	Where personnel are working in an isolated area, sanitary spaces are provided near or adjacent thereto	•		
	5	Someone standing on a wet deck in sanitary spaces cannot reach light switches or electrical outlets without ground fault interrupters (GFI)	•		
	6	Common/shared spaces are gender identifiable without entering the space	•		
#	7	Separate multiple occupancy sanitary spaces are provided for men and women	•		

(continued)

Table 44.5 (continued)

Accommodation area criteria					
Sanitary spaces					
Requirement # = related ILO MLC 2006, title 3 requirement			HAB(OS) requirements	HAB+(OS) requirements	HAB++(OS) requirements
	8	Separate multiple occupancy sanitary spaces are provided for men and women	•		
#		Bulkheads in sanitary spaces are made of steel or other approved material	•		
	9	Bulkheads are watertight up to 230 mm (9 in.) above deck level	•		
	10	A public sanitary facility should be situated near the ship's office(s) if the office(s) is not conveniently located near the navigation bridge[4]	•		
		Note: *in this context, near is defined as being on the same deck or on the deck below (or above) adjacent to the stairway*			
	11	Free space (space available for movement without hindrance of any objects) in a common sanitary space per person is			
		• At least 0.75 m^2 (8 ft^2)	•		
		• At least 1.1 m^2 (12 ft^2)		•	•
	12	Sleeping rooms have a private sanitary space with a toilet, shower or tub, and sink modules	•		
	13	A toilet and washbasin having hot and cold running potable water is within easy access of the (where applicable)			
#		• Navigation bridge deck and primarily for those on duty in the area	•		

(continued)

Table 44.5 (continued)

Accommodation area criteria					
Sanitary spaces					
Requirement # = related ILO MLC 2006, title 3 requirement			HAB(OS) requirements	HAB+(OS) requirements	HAB++(OS) requirements
#		• Machinery space if not fitted near the engine room control centre	•		
		• Galley/mess room(s)	•		
		• Recreation room(s)	•		
		• Gymnasium	•		
		Note: *in this context, easy access is defined as being on the same deck or on the deck below (or above) adjacent to the stairway*			
	14	Each sleeping room is provided with a washbasin with hot and cold running potable water	•		
Showers/baths					
	15	Water heaters supplying showers			
		• Do not support areas that have higher water temperature requirements, such as food service area	•		
		• Are provided with anti-scalding devices	•		
	16	Handholds are provided for shower and bath sanitary spaces	•		
	17	In sanitary spaces intended for more than one person, the shower is screened	•		
	18	A locker room is provided for personnel to change from and store their working clothes, boots and helmets before entering the clean area of the accommodation areas	•		

(continued)

Table 44.5 (continued)

Accommodation area criteria				
Sanitary spaces				
Requirement # = related ILO MLC 2006, title 3 requirement		HAB(OS) requirements	HAB+(OS) requirements	HAB++(OS) requirements
19	The locker room is located so as to act as a natural transit between the working and personnel living spaces with separate but direct access between these spaces	•		
20	Toilet and washing facilities are located adjacent to the locker room and available to personnel before changing	•		
21	The locker room accommodates lockers for 150% of the total number of beds	•		
22	Lockers are two tiered and arranged in banks, with adequate space between banks for personnel to change. Island bench seats and/or seat adjacent to the lockers are provided	•		
23	Individual locker dimensions are at least 1000 mm (39.5 in.) high × 300 mm (12 in.) wide × 500 mm (20 in.) deep	•		
Toilets				
24	Where there is more than one toilet in a space, they are sufficiently screened to provide privacy	•		
25	All toilets			
	• Have flush water available at all times	•		
	• Are independently controllable	•		
	• Have a hand washing station	•		

(continued)

Table 44.5 (continued)

Accommodation area criteria					
Sanitary spaces					
Requirement # = related ILO MLC 2006, title 3 requirement			HAB(OS) requirements	HAB+(OS) requirements	HAB++(OS) requirements
	26	Toilets, intended for more than one crew member, are situated convenient to, but separate from, sleeping rooms and washrooms without direct access from			
#		• The sleeping rooms	•		
#		• Toilets to which there is no other access. (This requirement does not apply where a toilet is located in a space between two sleeping rooms having a total of not more than four persons)	•		
		Note: *this requirement does not apply where a toilet is located in a space between two sleeping rooms having a total of not more than four persons*			
	27	All toilet spaces should be ventilated sufficiently to be reasonably free of disagreeable odours and condensation	•		
Washbasins					
	28	Water heaters supplying washbasins			
		• Do not support areas that have higher water temperature requirements	•		
		• Are provided with anti-scalding devices	•		

(continued)

Table 44.5 (continued)

Accommodation area criteria					
Sanitary spaces					
Requirement # = related ILO MLC 2006, title 3 requirement			HAB(OS) requirements	HAB+(OS) requirements	HAB++(OS) requirements
	29	The following at a minimum are provided for each washbasin			
		• A mirror	•		
		• Toiletry shelf	•		
	30	Facilities to dry hands (e.g., towel bars, etc.) are provided at all washbasins	•		
	31	Washbasins are			
		• Constructed of approved material	•		
		• Smooth surfaced	•		
		• Not liable to crack, flake or corrode	•		
Urinals					
	32	Urinals have privacy partitions between units and at the end of rows if not provided by sanitary space structures	•		
	33	The dimension between the centreline of two urinals side-by-side is 690 mm (27 in.) or greater	•		
	34	The dimension between the centreline of a urinal and bulkhead next to the urinal is 380 mm (15 in.) or greater	•		
	35	The height of the front edge of a urinal is between 455 mm (18 in.) and 605 mm (24 in.) above the deck surface	•		

Table 44.6 ILO MLC 2006 Title 3 requirement (Accommodation area criteria (food service areas)

Requirement # = related ILO MLC 2006, title 3 requirement			HAB(OS) requirements	HAB+(OS) requirements	HAB++(OS) requirements
Accommodation area criteria					
Food service areas					
General					
	1	A separate office is provided for food service/catering management within the food services area	•		
	2	Access is provided to all areas requiring cleaning, painting or treating with pesticide/insecticide	•		
	3	There are no crevices or inaccessible voids, which might harbour vermin, cooking or food waste, or other extraneous matter	•		
	4	Places where food is prepared, served or stored are not used as sleeping areas	•		
	5	Deck drains are provided in all food service areas subject to flood type cleaning or where normal operations release or discharge water or other liquid onto the deck	•		
	6	Deck drains for food service areas are provided in number and location so that complete drainage is possible under normal conditions of list and trim	•		

(continued)

Table 44.6 (continued)

Accommodation area criteria					
Food service areas					
Requirement # = related ILO MLC 2006, title 3 requirement			HAB(OS) requirements	HAB+(OS) requirements	HAB++(OS) requirements
	7	Where distributive systems are located in food service areas, they are installed as close as practicable to the overhead in order to provide maximum headroom	•		
	8	Flow of food through the installation			
		• Is arranged in logical sequence that minimises cross-traffic or backtracking	•		
		• Allows for adequate separation of clean and soiled operations	•		
Food service doors					
	9	Toilet room doors opening directly into food preparation or serving areas are self-closing and tight	•		
Food service passageways					
	10	The passage width between storage shelves is at least 710 mm (28 in.)	•		
Mess room					
General					
	11	Mess rooms are			
#		• Located apart from sleeping rooms	•		

(continued)

Table 44.6 (continued)

Accommodation area criteria					
Food service areas					
Requirement # = related ILO MLC 2006, title 3 requirement			HAB(OS) requirements	HAB+(OS) requirements	HAB++(OS) requirements
		• Adequately ventilated	•		
		• Adequately insulated to prevent condensation or overheating	•		
		• Located as near to the galley as is practicable except where the mess room is equipped with a steam table	•		
#	12	Mess rooms have tables and seats sufficient for the number of persons likely to use them at any one time	•		
	13	Mess lines and mess rooms are protected from weather, objectionable sights (such as garbage disposal areas) and objectionable odours (such as from engines, holds, toilets, fire room, etc.)	•		
#	14	Lighting in mess rooms is provided by means of natural light and artificial light (subject to special arrangements as may be permitted in passenger vessels)	•		

(continued)

Table 44.6 (continued)

Accommodation area criteria					
Food service areas					
Requirement # = related ILO MLC 2006, title 3 requirement			HAB(OS) requirements	HAB+(OS) requirements	HAB++(OS) requirements
	15	The following is available for crew use at all times			
#		• Refrigerator	•		
#		• Facilities for hot beverages	•		
#		• Facilities for cool water	•		
#	16	Common or separate mess room facilities are to be provided	•		
	17	The tops of tables and seating are capable of being easily cleaned	•		
#	18	Where available pantries are not accessible to mess rooms, adequate lockers for mess utensils and proper facilities for washing utensils are provided	•		
Outfitting					
	19	A transparent sanitation shield is installed to completely shield the entire length of the food serving lines if provided with a tray slide or rail		•	•
	20	Tray slides (or rails) if provided are placed along food and beverage serving lines	•		

(continued)

Table 44.6 (continued)

Accommodation area criteria					
Food service areas					
Requirement # = related ILO MLC 2006, title 3 requirement			HAB(OS) requirements	HAB+(OS) requirements	HAB++(OS) requirements
	21	Adequate lockers for mess utensils are provided	•		
Mess room dimensional aspects					
	22	Rectangular table space for each diner is	•		
		• At least 610 mm (24 in.) wide by 380 mm (15 in.) deep	•		
		• At least 740 mm (29 in.) wide by 430 mm (17 in.) deep		•	•
	23	Round Table diameter is			
		• For seating 4 persons at least 1220 mm (48 in.)	•		
		• For seating 4 persons at least 1370 mm (54 in.)		•	•
		• For seating 6 persons at least 1520 mm (60 in.)	•		
		• For seating 6 persons at least 1680 mm (66 in.)		•	•
		• For seating 8 persons at least 1830 mm (72 in.)	•		
		• For seating 8 persons at least 1980 mm (78 in.)		•	•

(continued)

Table 44.6 (continued)

Accommodation area criteria					
Food service areas					
Requirement # = related ILO MLC 2006, title 3 requirement			HAB(OS) requirements	HAB+(OS) requirements	HAB++(OS) requirements
	24	Distance between tables with back-to-back seating is			
		• At least 1195 mm (47 in.)	•		
		• At least 1525 mm (60 in.)		•	
	25	Deck area requirements for planned seating capacity are			
#		• At least 1.5 m^2 (16.2 ft^2) per person	•		
		• At least 1.7 m^2 (18.4 ft^2) per person		•	
		• At least 1.9 m^2 (20.5 ft^2) per person			•
Galley/scullery					
General					
#	26	The galley comprises space for food preparation, cooking, service and wash-up facilities. These should include areas for			
		• Baking and pastry preparation	•		
		• Vegetable preparation	•		

(continued)

Table 44.6 (continued)

Accommodation area criteria					
Food service areas					
Requirement # = related ILO MLC 2006, title 3 requirement			HAB(OS) requirements	HAB+(OS) requirements	HAB++(OS) requirements
		• Meat, fish and poultry preparation	•		
		• Hot services	•		
		• Dish wash facilities	•		
		• Pan wash facilities	•		
		• Equipment storage	•		
	27	Ice making facilities for at least 0.57 kg (1.25 lbs) of ice per person per day are provided	•		
	28	Hand washing stations located so that no crew member must walk more than 7.5 m (25 ft) to a station are to be provided. Slop sinks and scullery sinks are not satisfactory hand washing stations	•		
	29	Tops of food dressing tables, counters and drain boards are constructed with an integral back splash	•		
Potable water					
	30	Only potable water is piped to food storage, preparation, or service areas. The exceptions are a food waste grinder eductor (garbage disposal) line and deck washing facilities. The grinder delivery line is protected against back flow	•		

(continued)

Table 44.6 (continued)

Accommodation area criteria				
Food service areas				
Requirement # = related ILO MLC 2006, title 3 requirement		HAB(OS) requirements	HAB+(OS) requirements	HAB++(OS) requirements
31	If the vessel has a garbage room, hot and cold potable water is available for washing garbage cans	•		
32	Potable water tanks are identified with a number and the words "POTABLE WATER" adequately sized	•		
33	Potable water piping and fittings are labelled	•		
Maintenance/cleaning				
34	Drawers and bins used as food contact surfaces are readily removable and easily cleaned	•		
35	Covers, insets, or receptacles for unpackaged foods or beverages are easily removable or designed for easy cleaning in place	•		
36	Stove top or range sea rails are readily removable and easily cleanable and brackets for sea rails are easily cleanable	•		

(continued)

Table 44.6 (continued)

Accommodation area criteria					
Food service areas					
Requirement # = related ILO MLC 2006, title 3 requirement			HAB(OS) requirements	HAB+(OS) requirements	HAB++(OS) requirements
	37	Ventilation baffles, vanes, louvres, dampers, filters, etc., are easily accessible or removable (without tools) for cleaning	•		
Refrigeration/freezer					
	38	It is possible to open the doors of the refrigerated or cold storage rooms from the inside even though they are locked from the outside	•		
	39	There is an alarm in each freezer and cold storage room to warn when the temperature is out of specification		•	•
Dish washing					
	40	Easy access is provided to a sink or dish washing machine	•		

Table 44.7 ILO MLC 2006 Title 3 requirement (Accommodation area criteria (recreation)

Accommodation area criteria					
Recreation					
Requirement # = related ILO MLC 2006, title 3 requirement			HAB(OS) requirements	HAB+(OS) requirements	HAB++(OS) requirements
General					
#	1	Recreation and accommodation areas, conveniently situated and appropriately furnished are provided for the crew. (Where this is not provided separately from the mess rooms the latter is planned, furnished and equipped to give recreational facilities)	•		
	2	Rooms provided for the various activities within the recreation area should be located and arranged with movable partitions so that areas can be enlarged or reduced to accommodate particular requirements, where practicable	•		
#	3	Installation has an open deck space or spaces to which personnel can have access when off duty	•		
	4	Where space permits, the facilities should be located adjacent to the dining room, and positioned so that a logical flow sequence is established between the recreation facilities and the dining room. The route between the recreation facilities, the stairway and the dining room should be designed to afford easy circulation and avoid congestion			

(continued)

Table 44.7 (continued)

Accommodation area criteria				
Recreation				
Requirement # = related ILO MLC 2006, title 3 requirement		HAB(OS) requirements	HAB+(OS) requirements	HAB++(OS) requirements
5	Recreation rooms in the personnel accommodation spaces are adequately insulated to prevent condensation or overheating			
Recreational and exercise areas				
6	Areas outfitted with sport/exercise equipment should be provided	•		
7	The exercise area(s) should be located adjacent to recreation rooms but remote from quiet areas to avoid nuisance noise	•		
8	Consideration should be given to provide facilities for swimming			
Recreational and quiet areas				
9	Furnishing for recreational facilities should as a minimum include a book-case and facilities for reading, writing and, where practicable, table and deck games			
10	Television and DVD/CD player facilities are to be provided. The stock of films should be adequate for the duration of the voyage and, where necessary, changed at reasonable intervals			

(continued)

Table 44.7 (continued)

Accommodation area criteria					
Recreation					
Requirement # = related ILO MLC 2006, title 3 requirement			HAB(OS) requirements	HAB+(OS) requirements	HAB++(OS) requirements
	11	Consideration should be given to include the following recreational facilities on board, where practicable			
		• A smoking room	•		
		• Sport equipment including exercise equipment	•		
		• Facilities for swimming	•		
		• A library containing vocational and other books, the stock of which should be adequate for the duration of the voyage and changed at reasonable intervals	•		
		• Facilities for recreational handicrafts	•		
		• Canteen	•		
		• Where appropriate, the provision of bars on board for seafarers unless these are contrary to national, religious or social customs	•		

(continued)

Table 44.7 (continued)

Accommodation area criteria					
Recreation					
Requirement # = related ILO MLC 2006, title 3 requirement			HAB(OS) requirements	HAB+(OS) requirements	HAB++(OS) requirements
	12	One or more quiet lounge(s) should be designed to provide seating for a minimum of 10% of the number of beds with both chairs with cushioning and stackable chair seating to the same ratio as for the games room (minimum 6 chairs) including the following facilities	•		
		• Coffee tables.	•		
		• Two folding card tables with associated stackable chairs	•		
		• Two folding card tables with associated stackable chairs	•		
		• A range of storage cabinets and shelving for books, magazines, etc. with one lockable section	•		
		• Approximately 1.0 m^2 (11 ft^2) is provided for each seat in the lounge	•		
		• Approximately 1.25 m^2 (14 ft^2) is provided for each seat in the lounge		•	•

Table 44.8 ILO MLC 2006 Title 3 requirement (Accommodation area criteria (laundry)

Accommodation area criteria					
Laundry					
Requirement # = related ILO MLC 2006, title 3 requirement			HAB(OS) requirements	HAB+(OS) requirements	HAB++(OS) requirements
General			•		
#	1	Facilities exist for washing and drying clothes on a scale appropriate to the size of the crew and the normal duration of the voyage	•		
	2	Laundry facilities are located within easy access of crew accommodation areas	•		
	3	Laundry facilities should be sufficient to allow seafarers to be provided with clean and dry underwear once per day and clean and dry outerwear and bedding once per five days	•		
	4	Appropriate clearance has been provided for equipment maintenance		•	
	5	Coamings and/or deck drains are provided around washing machines, water heaters, etc.	•		
	6	Washer and dryer capacities are matched to assure a smooth workflow	•		
	7	Washers and dryers are placed relative to each other to facilitate the transfer of clothing from the washer to the dryer	•		

(continued)

Table 44.8 (continued)

Accommodation area criteria					
Laundry					
Requirement # = related ILO MLC 2006, title 3 requirement			HAB(OS) requirements	HAB+(OS) requirements	HAB++(OS) requirements
Outfitting					
	8	Where self-service laundries are provided, tumble dryers are exhausted directly into the weather, not into the installation	•		
	9	Where self-service laundries are provided, tumble dryers are equipped with lint filters	•		
	10	Where self-service laundries are provided, the laundry facilities include at a minimum			
#		• Clothes washing machines	•		
#		• Clothes drying machines (or adequately heated and ventilated drying rooms)	•		
#		• Irons and ironing boards (or their equivalent)	•		
		• Suitable sinks with an adequate supply of cold potable water and hot potable water or means of heating water	•		
		• Sinks are installed in washrooms, if a separate laundry accommodation area is not reasonably practical	•		

Table 44.9 ILO MLC 2006 Title 3 requirement (Accommodation area criteria (medical)

Accommodation area criteria					
Medical					
Requirement # = related ILO MLC 2006, title 3 requirement			HAB(OS) requirements	HAB+(OS) requirements	HAB++(OS) requirements
General					
#	1	The personnel accommodation area contains a dedicated medical facility	•		
#	2	The medical accommodation area is fitted with an examination couch and at least one berth, preferably. Single occupancy rooms can be utilised for medical needs	•		
	3	If double tier beds are provided, the upper tier is either hinged or removable	•		
	4	The medical accommodation area is not used as a storage area except for medical supplies	•		
	5	The medical accommodation area is suitably located to allow for safe and efficient			
		• Access for sick or injured crew members	•		
		• Stretcher transportation from accommodation areas or work areas	•		

(continued)

Table 44.9 (continued)

Accommodation area criteria					
Medical					
Requirement # = related ILO MLC 2006, title 3 requirement			HAB(OS) requirements	HAB+(OS) requirements	HAB++(OS) requirements
	6	Medical accommodation area is suitably separated from other spaces and used for the care of the sick and for no other purpose	•		
	7	Medical accommodation area is suitably designed so that the occupant can be comfortably housed and may receive proper attention in all weather	•		
	8	The arrangement of the entrance, berths, lighting, ventilation, heating and water supply is designed to provide comfort and facilitate the treatment of crew members	•		
Outfitting					
	9	If the vessel does not carry a doctor, it must have an approved medicine chest with readily understandable instructions	•		
	10	Medical accommodation area has a toilet, washbasin and bathtub or shower conveniently situated for the use of patients	•		
	11	If the installation does not carry a doctor, it must have an approved medicine chest with readily understandable instructions	•		

(continued)

Table 44.9 (continued)

Accommodation area criteria					
Medical					
Requirement # = related ILO MLC 2006, title 3 requirement			HAB(OS) requirements	HAB+(OS) requirements	HAB++(OS) requirements
#	12	The medical accommodation area has a toilet, washbasin and bathtub or shower conveniently situated for the use of patients	•		
	13	The door to the toilet and washbasin opens outwards and any door latch is capable of being opened from the outside	•		
	14	The medical accommodation area is equipped with stretchers that meet the following			
		• The ability to winch a sick person into a helicopter or vessel, and the capability of floating and righting itself in the water	•		
		• Collapsible and suitable for use in confined spaces	•		
	15	Medical accommodation area is appropriately equipped to render the necessary medical attention	•		
	16	Toilet accommodation area is provided for the exclusive use of the occupants of the hospital accommodation area	•		

(continued)

Table 44.9 (continued)

Accommodation area criteria					
Medical					
Requirement # = related ILO MLC 2006, title 3 requirement			HAB(OS) requirements	HAB+(OS) requirements	HAB++(OS) requirements
	17	Hospital and medical/first aid accommodation area is situated and arranged so that a stretcher can be easily carried into it and placed alongside an examination table or bed	•		
	18	The medical accommodation area has lockable storage for drugs, dressings and medical equipment	•		
	19	Medical accommodation area has an intercom or signalling system (e.g., an emergency call device) to the navigation bridge	•		

Table 44.10 ILO MLC 2006 Title 3 requirement (Accommodation area criteria (offshore installation offices)

Accommodation area criteria					
Offshore installation's offices					
Requirement # = related ILO MLC 2006, title 3 requirement			HAB(OS) requirements	HAB+(OS) requirements	HAB++(OS) requirements
General					
	1	Toilets and washbasins are in proximity to offices	•		
#	2	All installations are provided with separate offices or a common office for use appropriate personnel	•		

Chapter 45
Vibration on Offshore Installations

General

Working and/or living on board offshore installations imposes a series of generally low-frequency mechanical vibration, as well as single-impulse shock loads on the human body. Low-frequency vibrations are also imposed by the installation's motions, which are produced by the various sea states. These motions can result in motion sickness, body instability, fatigue, and increased health risk aggravated by shock loads Some shock loads may be produced by wave "slamming". Floating installations with unusual hull forms may be particularly susceptible to certain wave frequency motions. Higher-frequency vibration influencing comfort is often associated with rotating machinery. The imposition of higher frequency vibrations (about 1–80 Hz) induces corresponding motions and forces within the human body, creating discomfort and possibly resulting in degraded performance and health (Griffin, 1990). This chapter provides the criteria and methods for assessing whole-body vibration relating to habitability on board offshore installations. The criteria were selected to limit potential vibration related interference with work tasks and to improve personnel comfort. Consideration of the vibration loads imposed on the body is restricted to motions transmitted from surrounding structures to the entire human body through the feet of a standing person in the frequency range 1–80 Hz. Motions transmitted to the body of a seated or recumbent person have been omitted from this chapter. Due to the provision of resilient or non-rigid surfaces on seats and beds, these surfaces will generally attenuate the transfer of vibration to levels that are lower than those experienced when standing. The motions transmitted through the feet are expected to be the highest vibration levels to which personnel will be exposed. This chapter applies to manned spaces as defined in Chap. 43. Examples of such spaces include accommodation spaces (e.g., cabins, mess areas, recreation spaces, etc.) and workspaces (e.g., galley, control rooms, duty stations, etc.) both inside and, in other installation locations, outside the accommodation block. Whole-body vibration limits defined in this chapter are based on currently available standards. Compliance with the guidance

A. A. Olsen, *Applying Physical Ergonomics to Modern Ship Design*,
https://doi.org/10.1007/978-3-031-57974-5_45

discussed in this chapter are typically a prerequisite for the offshore installation Habitability notation assigned by Class (e.g., **HAB(OS)**, **HAB+(OS)**, and **HAB++(OS)** for ABS Classed offshore installations or other Class equivalents).

Terminology

Acceleration: The rate of change of velocity over time (i.e., metres-per-second, squared (m/s^2).

Calibration checks: Field calibration of a measuring instrument conducted before and after a field test, using a reference calibrated signal. Field calibrators (portable calibrators) should comply with an international or national standard.

Dynamic positioning: A system to automatically maintain a workboat's position and heading by controlling propellers and/or thrusters. Dynamic positioning can maintain a position to a fixed point over the bottom, or in relation to a moving object (such as another offshore installation or vessel). It can also be used to position the installation at a favourable angle towards wind, waves, and current.

Frequency: The number of complete cycles of a periodic process occurring per unit time. Frequency is expressed in Hz which corresponds to the number of cycles observed-per-second.

Frequency weighting: A transfer function used to modify a signal according to a required dependence on vibration frequency.

- In human response to vibration, various frequency weightings have been defined in order to reflect known or hypothesised relationships between vibration frequency and human response.
- The frequency weighting used to evaluate whole-body vibration in this chapter is W_m (whole-body) for all three axes (x, y, and z), in accordance with ISO 6954.

Manned space: Any space where a seafarer may be present for 20 min or longer at one time during normal, routine daily activities. Such spaces would include working or living spaces.

Multi-axis acceleration value: The Multi-Axis Acceleration Value is calculated from the root-sums of squares of the weighted RMS acceleration values in each axis (a_{xw}, a_{yw} and a_{zw}) at the measurement point using the following expression:

$$a_w = \sqrt{a_{xw}^2 + a_{yw}^2 + a_{zw}^2}$$

where a_{xw}, a_{yw} and a_{zw} are the weighted RMS acceleration values measured in the x-, y- and z-axes, respectively.

Multi-axis vibration: Mechanical vibration or shock acting in more than one direction simultaneously.

Reference calibration: Calibration of a measuring instrument and of a portable calibrator conducted at least every two years (or more frequently if specified by the manufacturer) by an accredited Testing and Calibration Laboratory accredited according to ISO 17025:2005, as amended, with traceability to a national or international standard.

Velocity: The rate of change of distance over time (i.e., millimetres per second, mm/s).

Vibration: The variation with time of the magnitude of a quantity which is descriptive of the motion or position of a mechanical system, when the magnitude is alternately greater and smaller than some average value.

Weighted root-mean-square acceleration value (aw): The weighted root-mean-square (RMS) acceleration, a_w, in metres-per-second squared, is defined by the expression:

$$a_w \sqrt{\frac{1}{T}\int_0^T a_w^2(t)dt}$$

where $a_{w(t)}$ is the weighted acceleration as a function of time in meters-per-second squared (m/s^2) and t is the duration of the measurement in seconds.

Whole-body vibration: Mechanical vibration (or shock) transmitted to the human body as a whole. Whole-body vibration is often due to the vibration of a surface supporting the body.

Associated Documentation

The following documents provide details about test plan preparation, test measurement procedures and/or test reporting:

- ISO 6954:2000, Mechanical Vibration and Shock—Guidelines for the Measurement, Reporting and Evaluation of Vibration with regard to Habitability on Passenger and Merchant Ships.
- ISO 2631-2:2003, Mechanical Vibration and Shock—Evaluation of Human Exposure to Whole Body Vibration—Part 2, Vibration in Buildings.
- ISO 8041:2005, Human Response to Vibration—Measuring Instrumentation.
- ISO 5348:1998, Mechanical Vibration and Shock—Mechanical Mounting of Accelerometers.
- WMO: Manual on Codes, No. 306, Part A, Alphanumerical Codes (2011 Edition, Updated in 2014).

Further guidance can be found in:

- ISO 20283-2:2008, Mechanical Vibration—Measurement of Vibration on Ships—Part 2: Measurement of Structural Vibration.

Criteria

The whole-body vibration criteria for offshore installation Habitability notations (e.g., **HAB(OS)**, **HAB+(OS)**, or **HAB++(OS)**) are provided in Table 45.1, Maximum weighted root-mean-square acceleration level. Maximum whole-body vibration levels are provided for transit and dynamic positioning conditions. The severity of the vibration is indicated by the weighted root-mean square acceleration value (aw) as defined in ISO 8041. Whole-body vibration measurements should only be taken in manned spaces. A space is considered "manned" if it is occupied by personnel for 20 min or longer at a time for normal, routine daily activities. Specific locations are referred to in the section on 'Measurement locations'. The maximum vibration levels for accommodation areas and workspaces in Table 45.1, Maximum weighted root-mean-square acceleration level should not be exceeded under normal operating conditions, either in transit or dynamic positioning. For instance, the **HAB(OS)** notation's maximum vibration level is primarily aimed at reducing discomfort and interference with task performance whereas the more stringent maximum level for the **HAB+(OS)** and **HAB++(OS)** notations are aimed at improving performance and comfort. In this instance, "comfort" means the ability of personnel to use a space for its intended purpose with minimal interference or annoyance from whole-body vibration. For the purposes of this chapter, the notation applies to the vibration levels occurring on the deck supporting the human body in the three translational ($x-$, $y-$ and $z-$) axes as shown in Fig. 45.1, Measurement axes. The vibration levels are computed for each axis individually, as well as combined as a multi-axis acceleration value. Each is expressed as a frequency weighted root mean-square (a_w) value. To meet the vibration criteria, the multi-axis aw level must be less than or equal to the maximum level expressed in Table 45.1, Maximum weighted root-mean-square acceleration level.

Table 45.1 Maximum weighted root-mean-square acceleration level

Notation	Frequency range (Hz)	Acceleration measurement	Maximum RMS level	
			Accommodation spaces	Workspaces
HAB(OS)	1–80	a_w	161 mm/s^2 (4.5 mm/s)	178 mm/s^2 (5.0 mm/s)
HAB+(OS)	1–80	a_w	143 mm/s^2 (4.0 mm/s)	161 mm/s^2 (4.5 mm/s)
HAB++(OS)	1–80	a_w	125 mm/s^2 (3.5 mm/s)	143 mm/s^2 (4.0 mm/s)

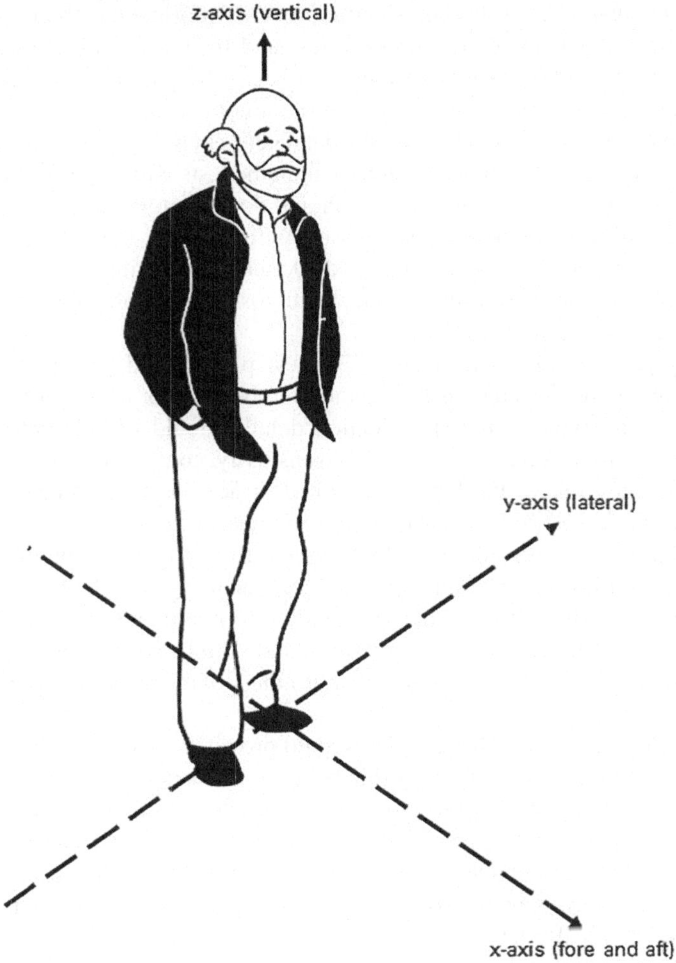

Fig. 45.1 Measurement axes

Test Plan

As stated in the section 'Test plans' in Chap. 43, a test plan should be developed to serve as the principal means for verifying the measurements to be performed to verify compliance with whole-body vibration criteria. This test plan should typically include the following:

- *Documentation.* The test plan should include appropriate drawings indicating the location of all vibration sources.

- *Test personnel.* The test plan should provide information about the testing specialist who will be conducting the test and their approval and certification in accordance with the Class specific Procedural Requirements for Recognised Ambient Environmental Testing Specialists or equivalent.
- *Test conditions.* The test plan should detail the conditions for both transit and dynamic positioning conditions under which the tests will be performed.
- *Measurement locations.* The test plan should document, in detail, on appropriate drawings, all spaces where measurements will be taken. In addition, transducer placement positions should be indicated. Details on selecting measurement locations and determining transducer placement positions are provided in the section 'Measurement locations'.
- *Data acquisition and instruments.* The test plan should provide information regarding the methods and instrumentation to be used for measurement and data collection. Instrumentation specification details should include type of instruments to be used, accuracy, calibration, sensitivity, conformance with ISO 8041, and frequency range. Further details on data acquisition and instruments are provided in the section 'Data acquisition and instruments'.
- *Data analysis.* The test plan should provide information regarding the methods, software, and instrumentation to be used for data analysis.
- *Test schedule.* The test plan should provide information regarding the proposed test schedule. The test schedule should include information on the approximate date and duration of testing (including a summary of the scheduled measurements per day).
- *Data collection sheets.* The test plan should provide sample data sheets (in table format) that will be used for reporting the measurement data.
- *Conflict of interest declaration.* In the case that the testing specialist has, in any capacity or to any extent, provided input into the design of the installation, their participation should be indicated in the test plan including details of the testing specialist's involvement. Class may review any potential conflicts of interest and determine if any further action is necessary.

Test Requirements

Whole-body vibration measurements should be carried out in accordance with the procedures described in ISO 6954. When the requirements or procedures proscribed by Class deviate from those mentioned in ISO 6954, then the requirements or procedures proscribed by Class will take precedence unless otherwise stipulated. Data acquisition and instruments. For Class Habitability notations such as **HAB(OS)**, **HAB+(OS)**, or **HAB++(OS)**, or other Class equivalents, a sample of data should be recorded for each whole-body vibration measurement position. Each whole-body vibration measurement sample should be at least 60 s in duration. For each location measured, a data sample should also be taken in accordance with the requirements set out in the section 'Test conditions'. The above measurement samples should all be taken using the appropriate Type 1 instrumentation (ISO 8041), then frequency weighted and analysed in accordance with ISO 6954. It is always desirable to employ equipment that records and stores acceleration time histories. *Test conditions.* The test conditions required for the whole-body vibration measurements should be in accordance with the following:

- *Location.* The installation should be on location, properly moored or fixed, and in normal operating mode,
- *Operating conditions.* The typical loading condition (e.g., variable load, ballast, produced hydrocarbon, produced water, etc.) of the installation should be as close as possible to normal operating conditions. If this is not practicable, the loading condition should be highlighted in the whole-body vibration test report,
- *Machinery and equipment operation.* HVAC systems are to be running during the vibration measurements. All machinery essential for operation (fluid, turbo generators, and additional machinery) should operate under normal conditions throughout the measurement period. Due to the nature of offshore operations, a log should be maintained with respect to equipment and facility operations during testing periods (i.e., flaring, compressors starts, automatic starts and stops of equipment, etc.),
- *Heading control.* Thrusters used to maintain or assist the heading of an installation should be operating as required under normal operating environmental conditions, and
- *Test interference.* During the whole-body vibration measurements, vibration arising from every kind of unnecessary human activity should be avoided. For this reason, only the personnel needed for the normal operation of the equipment in the space and those carrying out the measurements may be present in the space being tested.

Measurement locations. Selection of spaces where measurements are to be conducted. The aim when selecting vibration measurement locations is to obtain a representative sample of data that reflects the actual conditions in manned spaces. For practical reasons, it is important to select the locations such that an appropriate amount of data can be collected during the testing phase. This means the measurement locations should be selected in accordance with the following criteria:

- Select potential worst-case locations based on their proximity to vibration emitting sources such as propulsion or other rotating machinery or where vibration is likely to be transmitted to manned spaces, accommodation areas and recreation areas via the installation's structure. Measurements should be taken in all identified worst-case locations (e.g., cabin adjacent to a machinery space).
- Where a single instance of one type of manned space exists within the installation (e.g., control rooms, mess room, gymnasium, library, etc.), that location should be selected for measurement.
- Select a representative sample of personnel cabins throughout the installation. For installations with less than 20 cabins, 50% of cabins on each deck should

be selected. For installations with greater than 20 cabins, 30% of cabins on each deck should be selected.

- Where multiple instances of the same types of accommodation space exist, but are not personnel cabins, a representative sample of at least 50% of each type should be selected for measurement. The worst-case locations are to be considered part of the representative sample, where applicable.

Walk through verification inspection locations. All normally manned spaces are to be subject to a walkthrough inspection by the Class approved surveyor. The number and locations of the walkthrough inspections are generally determined by the surveyor. The purpose of the walkthrough verification is to subjectively assess the vibration qualities. At the discretion of the surveyor, additional measurements may be required. *Transducer placement positions.* Vibration transducers (accelerometers) should be located and attached properly to the floor surface to measure the vibration at the interface between the standing personnel and the source of vibration. The mounting of accelerometers must comply with ISO 5348. When the vibration enters the human body from a non-rigid or resilient material (e.g., floor covering), secure the transducers with a suitably formed mount that does not alter the pressure distribution on the surface of the floor covering. In cabins, the vibration transducers must be placed on the deck in the centre of the space. ***Note***: *this location may not provide the maximum vibration levels for this particular space.* The objective is to minimise the number of measurements yet still obtain a fair and representative sample of the exposure conditions of the person occupying the cabin. For larger spaces (public rooms, messes, recreation areas, etc.) it may be necessary to place transducers at a number of locations in order to obtain a representative sample of the whole-body vibration levels for that space. Transducer locations should be evenly distributed throughout the space. For a specific room size, the minimum number of measurement locations are to be as indicated in Table 45.2, Distribution of transducer positions within spaces.

Transducers located at one measurement position should be orthogonally positioned (i.e., positioned at a 90° angle) to measure whole-body vibrations in the vertical, longitudinal and transverse axes. Translational accelerometers oriented in different axes at a single measurement position should be as close together as possible. *Test report.* As stated in the section on ‘Test reports’ in Chap. 43, a test report must be submitted to the Class approved surveyor to determine whether the vibration levels meet the whole-body vibration criteria and whether this part of the Class notation requirement has been met. The details listed below are required to be provided in the whole-body vibration test report. *Test details.* The following details are required to be recorded for each period of testing:

Table 45.2 Distribution of transducer positions within spaces

Space size	Minimum number of measurement positions in room
Less than 40 m^2 (431 ft^2)	1
Greater than 40 m^2 (431 ft^2)	2

- Loading location including latitude and longitude coordinates and appropriate block identification.
- Machinery and equipment operated during the test.
- Number of personnel and total number of persons on board during tests.
- Any indication of abnormal activities during the test that might skew results.

Transducer placement positions. The actual transducer placement positions within the measured spaces must be indicated on appropriate drawings. *Measurement equipment details.* Details of measurement and analysis equipment (e.g., manufacturer, type and serial number, accuracy and resolution), including frequency analysis parameters (e.g., resolution, averaging time, and filtering), must be provided. Copies of the relevant instrumentation reference calibration certificates, together with the results of field setup and calibration checks before and after the field tests, should also be provided with the report. *Results.* The following results per measuring location, should be provided in a tabulated format (for both transit and dynamically positioned operations):

- Measurement position.
- Measurement period if different from requirements.
- Sample number.
- Multi-axis weighted RMS values.
- Equipment operating in proximity to the measurement position.
- Indication of pass/fail.

For all the remaining spaces that were checked through walk through verification inspection and spot check measurements, the following information should also be provided:

- Name and number of spaces.
- Walk through inspection observations.
- Measurement results if necessary.

Deviations. All deviations from the approved test plan must be reported. Surveyor witnessing documentation. The equipment field calibration and data collection process of vibration tests conducted at sea must be witnessed by a Class approved surveyor. The surveyor is required to sign or initial each page of the test report and prepare a witnessing document stating whether all steps of the vibration testing were completed to their satisfaction. A copy of the witnessing document is then given to the person conducting the onboard testing, for insertion into the final whole-body vibration test report. The original is usually retained by the surveyor and submitted to Class for inclusion in the installation's Record. *Results.* On completion of the test, the whole-body vibration test report will be reviewed by the Class approved surveyor against the appropriate Class Habitation notation requirements. Where the requirements have been satisfactorily met, the installation will be assigned the appropriate Class Habitation notation.

Chapter 46
Noise on Offshore Installations

General

A large amount of research has been performed on the effects of noise on humans. Established or commonly used criteria exist for the effects of noise on speech communication, hearing loss, sleep, concentration, and "annoyance". These have provided a basis for the guidance provided in this chapter. A detailed discussion of the effects of noise on human performance, health, and comfort may be found in Kryter (1994): The Handbook of Hearing and the Effects of Noise: Physiology, Psychology and Public Health. In this chapter, noise criteria have been selected to improve personnel performance and to facilitate communication and sleep in appropriate offshore installation spaces. An additional goal is to enhance personnel safety and comfort. In this instance, "comfort" means the ability of the personnel to use a space for its intended purpose with minimal interference or annoyance from noise. The noise criteria presented in this chapter is lower than the levels commonly associated with hearing loss. Further guidance with respect to hearing conservation is provided in the IMO Resolution MSC.337(91): Code on Noise Levels Onboard Ships and should be followed for noise levels and exposure duration, particularly for areas with noise levels in excess of 85 dB(A). This chapter applies to manned spaces such as cabins, workspaces (duty stations), mess areas, and recreation spaces, both inside and in other installation locations outside the accommodation block. Compliance with the guidance discussed in this chapter is typically a prerequisite for the Workboat Habitability notation assigned by Class (e.g., **HAB(OS)**, **HAB+(OS)**, and **HAB++(OS)** for ABS Classed installations).

A. A. Olsen, *Applying Physical Ergonomics to Modern Ship Design*,
https://doi.org/10.1007/978-3-031-57974-5_46

Terminology

A-weighted sound pressure level: The magnitude of a sound, expressed in decibels (i.e., 20 micropascals); the various frequency components are adjusted according to the A-weighted values given in IEC 61672-1 (2004) in order to account for the frequency response characteristics of the human ear. The symbol is L_A; the unit is dB(A). The measurement L_{Aeq} is an equivalent continuous A-weighted sound pressure level, measured over a period of time.

Calibration checks: Field calibration of a measuring instrument conducted before and after a field test, using a reference calibrated signal or through zero calibration. Sound calibrators should comply with the standard IEC 60942:2003, as amended, type/class (1) standard and should be approved by the manufacturer of the measuring instrument used.

Dynamic positioning: A system to automatically maintain a workboat's position and heading by controlling propellers and/or thrusters. Dynamic positioning can maintain a position to a fixed point over the bottom, or in relation to a moving object (such as another installation or vessel). It can also be used to position the installation at a favourable angle towards wind, waves, and current.

Equivalent continuous a-weighted sound pressure level: The A-weighted sound pressure level of a noise fluctuating over a period of time T, expressed as the amount of average energy. The symbol is L_{Aeq}; the unit is dB(A).

Reference calibration: Calibration of measuring instrument and sound calibrator, conducted at least every two years (or more frequently if specified by the manufacturer) by an accredited Testing and Calibration Laboratory accredited according to ISO 17025:2005, as amended, with traceability to a national or international standard.

Associated Documentation

The following documents provide details about test plan preparation, test measurement procedures and/or test reporting.

- ISO 2923:1996, Acoustics—Measurement of Noise Onboard Vessels.
- IEC 61672-1:2013, Electroacoustics—Sound Level Meters—Part 1: Specifications.
- IEC 60942:2003, Electroacoustics—Sound Calibrators.
- IMO Resolution MSC.337(91) (2012), Code on Noise Levels Onboard Ships.
- ISO 10140-2:2010, Acoustics—Laboratory Measurement of Sound Insulation of Building Elements—Part 2: Measurements of Airborne Sound Insulation.
- ISO 717-1:2013, Acoustics—Rating of Sound Insulation in Buildings and of Building Elements—Part 1: Airborne Sound Insulation.

- WMO: Manual on Codes, No. 306, Part A, Alphanumerical Codes (2011 Edition, updated in 2019).

Criteria

Equivalent continuous a-weighted sound pressure level. The noise criteria for the Class Habitability notations (e.g., **HAB(OS)**, **HAB+(OS)**, and **HAB++(OS)**) are provided in Table 46.1, Noise criteria for offshore installations. Noise levels are determined for the test conditions specified in the section 'Test conditions' and must not exceed the maximum acceptable L_{Aeq} levels indicated in Table 46.1, Noise criteria for offshore installations, for each type of space or area. The maximum acceptable noise levels given in Table 46.1, Noise criteria for offshore installations are L_{Aeq} values, determined as appropriate to the character of the noise (refer to the section on 'Measurement procedures' and recorded results in this chapter). These limit values are for normal operating conditions with an installation on location and properly moored or fixed.

Acoustic Insulation

The airborne sound insulation properties for bulkheads and decks within the accommodation should comply at least with the following weighted sound reduction index (R_w) according to ISO Standard 717-1:2013.

Cabin to cabin	$R_w = 35$
Mess rooms, recreation rooms, public spaces and entertainment areas to cabins and hospitals	$R_w = 45$
Corridor to cabin	$R_w = 30$
Cabin to cabin with communicating door	$R_w = 30$

The airborne sound insulation properties should be determined by laboratory tests in accordance with ISO 10140-2:2010 as identified in the IMO Code on Noise. The manufacturer of the acoustic insulation should provide the weighted sound reduction index (R_w) value and evidence of the laboratory test.

Table 46.1 Noise criteria for offshore installations

Space[1,2]	Maximum acceptable noise L_{Aeq} level dB(A)		
	HAB(OS)	HAB+(OS)	HAB++(OS
Personnel accommodation spaces and open deck recreation areas			
Cabins	55	55	50
Sanitary spaces (if separate from the cabin)	60	55	50
Dining (mess) spaces	60	55	50
Indoor recreation spaces	60	55	55
Gymnasiums	60	55	50
Medical and first aid centre	55	50	65
Open deck recreation areas	70	65	60
Navigation and control spaces			
Wheelhouse, pilothouse, bridge	60	55	55
Radio room	60	55	55
Offices	60	55	55
Cargo control rooms[3]	60	60	55
Machinery control rooms	70	65	65
Emergency command centre	60	55	55
Service spaces			
Food preparation (e.g., galley, scullery)	70	70	65
Pantries	75	70	70
Storerooms	75	70	70
Laundry areas	75	70	70
Operating and maintenance spaces			
Machinery spaces	110	110	110
Workshops (other than those forming part of machinery spaces)	85	80	80
Non-specified workspaces[4] (other workspaces)	85	80	80

[1] In any manned space with noise levels above 85 dB(A), hearing protection should be worn in accordance with appropriate IMO regulations.

[2] Unless this space is normally manned, no confirmatory measurements for the Class Habitability notations are required.

[3] If located within machinery spaces, it is to comply with the machinery control room's criteria.

[4] Examples of non-specified workspaces are open deck workspaces that are not machinery spaces, and open deck workspaces where communication is relevant.

Test Plan

As stated in Chap. 43 under the section 'Test plans', a test plan must be developed to serve as the principal means for verifying the measurements to be performed to demonstrate or confirm compliance with noise criteria. The test plan should include the following.

- *Documentation.* The test plan should include appropriate design information including noise specifications for the installation. It should also include layout drawings indicating the locations of all noise sources and noise generating equipment. The information should be of such detail to enable the Class engineer to verify compliance with the criteria stipulated by Class. The test plan should also include the acoustic insulation plan and relevant supporting documentation for review and approval. The test plan must be submitted to Class Engineering for their review and approval.
- *Test personnel.* The test plan should provide information about the testing specialist who will be conducting the test and their approval and certification in accordance with the Class specific Procedural Requirements for Recognised Ambient Environmental Testing Specialists or equivalent.
- *Test conditions.* The test plan should detail the conditions under which the tests will be performed. Information must be submitted to Class Engineering within the test plan with respect to the geographical area of operation for an installation. Further details about test conditions are provided in the section under 'Test conditions'.
- *Measurement locations.* The test plan should document, in detail, on appropriate drawings, all spaces where measurements will be taken. In addition, transducer placement positions should be indicated. Details on selecting measurement locations and determining transducer placement positions are provided in the section 'Measurement locations'.
- *Data acquisition and instruments.* The test plan should provide information regarding the methods and instrumentation to be used for measurement and data collection. Instrumentation specification details should include the type of instruments to be used, the accuracy, calibration and sensitivity. Further details on data acquisition and instruments are provided in the section 'Data acquisition and instruments'.
- *Data* analysis. The test plan should provide information regarding the methods, software, and instrumentation to be used for data analysis.
- *Test schedule.* The test plan should provide information regarding the proposed test schedule. The test schedule should include information on the approximate date and duration of testing (including a summary of the scheduled measurements per day).
- *Data collection sheets.* The test plan should provide sample data sheets (in tabulated format) that will be used for reporting the measurement data.
- *Conflict of interest declaration.* In the case that the testing specialist has, in any capacity or to any extent, provided input into the design of the installation, their participation must be indicated in the test plan including details of the testing

specialist's involvement. Class will then review any potential conflicts of interest and determine if further action is necessary.

Test Requirements

In general, the noise measurements should be carried out in accordance with the requirements of IMO Resolution MSC.337(91) Code on Noise Levels Onboard Ships. However, when the requirements or procedures proscribed by Class deviate from those mentioned in IMO Resolution MSC.337(91), then the requirements or procedures proscribed by Class will take precedence unless otherwise stipulated. *Data acquisition and instruments.* The integrating-averaging sound level meter must meet the requirements for a Type 1 instrument as specified in IEC 61672-1. For each location sampled, a measurement must be taken in accordance with the requirements outlined in the next section, 'Test conditions'. *Test conditions.* The test conditions required for the noise measurements should be in accordance with each of the following sub paragraphs, based on ISO 2923. (1) *Location.* The installation should be on location, properly moored or fixed, and in normal operating mode. (2) *Machinery and equipment operation.* All machinery essential for installation operation should operate under normal conditions throughout the measurement period. HVAC systems are to run as for normal conditions during the noise measurements. For the galley/scullery, all equipment that runs for 20 min or more (e.g., ventilation hoods) should continue to run during the noise measurement. Equipment that is used only intermittently (e.g., blenders/mixers/meat cutters) do not need to run during the noise measurements. Any equipment that is running during the test should be recorded in the noise test report. Due to the nature of offshore operations, a log should be maintained with respect to equipment and facility operations during testing periods (i.e., flaring, compressors starts, automatic starts and stops of equipment, etc.). All machinery essential for operation (fluid, turbo generators, and additional machinery) should operate under normal conditions throughout the measurement period. (3) *Heading control.* Thrusters used to maintain or assist the heading of an installation should be operating as required under normal operating environmental conditions. (4) *Test interference.* During the noise measurements, noise arising from every kind of unnecessary human activity must be avoided. For this reason, only the personnel needed for the normal operation of the equipment in the space and those carrying out the measurements should be present in the space being tested. Doors and windows must be closed, except where they are normally left open (such as the door on the lee side of the navigation bridge, which may normally be open). Any open doors or windows should be noted in the noise test report. Spaces should be furnished with all usual equipment, fixtures and fittings normally found in the space. Equipment should be configured to operate in its normal operating mode.

Measurement locations. Selection of spaces where measurements are to be conducted. The aim when selecting noise measurement locations is to obtain a representative sample of data that reflects the actual conditions in manned spaces. For practical reasons, it is important to select the locations such that an appropriate amount of sample data can be collected during the testing phase. The measurement locations should be selected in accordance with the following criteria.

- Select potential worst-case locations based on their proximity to noise emitting sources or where noise is likely to be transmitted to manned spaces, personnel accommodation areas and recreation areas via the installation's structure. Measurements should be taken in all identified worst-case locations.
- Where a single instance of one type of manned space exists within the installation (e.g., the bridge, mess room, gymnasium, library, etc.), that location should be selected for measurement.
- Select a representative sample of personnel cabins throughout the installation. For installations with less than 20 cabins, 50% of cabins on each deck should be selected. For installations with greater than 20 cabins, 30% of cabins on each deck should be selected. The worst-case locations are to be considered part as the representative sample for personnel cabins, where applicable.

Walkthrough verification inspection locations. All normally manned spaces are to be subject to a walkthrough inspection by the Class approved surveyor. The number and locations of the walkthrough inspections are generally determined by the surveyor. The purpose of the walkthrough verification is to subjectively assess the noise qualities. At the discretion of the surveyor, additional measurements may be required. *Measurement positions.* The measurement positions described below are taken or adapted from ISO 2923 and IMO Resolution MSC.337(91).

> Measure at positions where persons will be seated or standing. The microphone shall be at a height of approximately 1,200 mm (47 in) from the deck to represent seated persons and approximately 1,600 mm (63 in) from the deck to represent standing persons, as appropriate for the measurement position.
>
> In personnel accommodation spaces, measure in the middle of the space.
>
> For all measurements, the microphone shall not be closer than 500 mm (20 in) from the boundary surface (e.g., bulkhead) of a space. The measurement time shall be at least 15 seconds and shall be long enough to enable the measurement of the equivalent continuous A-weighted sound pressure level for any specified time interval within the stated limits of overall measurement uncertainty.
>
> If practicable, do not measure closer than 1,000 mm (39.5 in) from operating machinery, air inlets, or from decks, bulkheads or other large surfaces. Where this is not possible, measure midway between the machinery and an adjacent reflecting surface.

Measurement procedures and recorded results. Persons present during measurements. When testing specialist personnel are conducting noise level measurements in any space, only personnel necessary for the operation of equipment or machinery in that space should be present. *Sampling duration.* Equivalent continuous A-weighted sound pressure levels (L_{Aeq}) should be reported for each measurement location. The

L_{Aeq} sampling duration should be sufficient to achieve a stable reading. Sampling time should be 15 s or longer. *Cyclic noise.* If the noise within a space is cyclic, the L_{Aeq} sampling duration should be sufficient to capture an integer number of complete cycles. If a long-duration sample is judged impractical, an L_{Aeq} value should be determined and reported for the high-noise portion of the cycle. *Intermittent noise.* If the noise within a space is present intermittently, an L_{Aeq} value should be determined and reported for a period of high-level noise. HVAC related noise. If HVAC system-related noise is a large contributor to the noise level in the space, a noise measurement should be made approximately 300 mm (12 in.) from the vent, measured in line with the direction of airflow, and recorded in the noise test report.

Test Report

As stated in Chap. 43 under the section 'Test reports', a test report must be submitted to the Class approved surveyor to determine whether the noise levels are at or below the limits and whether this part of the notation requirement has been met. The details listed below must therefore be provided in the noise test report. *Test details.* The following details are required be provided for each period of testing.

- Installation location including latitude and longitude coordinates and appropriate block identification.
- Machinery and equipment operated during the test.
- Number of personnel and total number of persons on board during tests.
- Any indication of abnormal activities during the test that might skew results.

Measurement positions. For transit and dynamic positioning measurements, actual measurement location positions must be indicated on appropriate drawings. *Measurement equipment details.* Details of measurement and analysis equipment (e.g., manufacturer, type and serial number, accuracy, sampling frequency and resolution) must be provided. Copies of the relevant instrumentation reference calibration certificates, together with the results of field setup and calibration checks before and after the field tests, should also be provided with the report. *Results.* The following results per measuring location, should be provided in a tabulated format.

- Measurement position (i.e., space and location within space).
- Measurement conditions (transit or dynamic positioning, power output and pitch angles, sea states).
- Number of people present in the space at time of measurement.
- Measurement period, if different from requirement.
- Equivalent continuous A-weighted sound pressure level (L_{Aeq}).
- Note any open doors and windows.
- Note equipment operating in proximity to the measurement position.
- Note observed direct sources of noise (such as ventilation devices) and any additional measurement data collected.
- Indication of pass/fail.

For all the remaining spaces that were checked through walk through verification inspection and spot check measurements, the following information should also be provided.

- Name and number of spaces.
- Walk through inspection observations.
- Measurement results if necessary.

Deviations. All deviations from the approved test plan should be reported. Where there are noted deviations from criteria and measurements were taken during Installation and Commissioning, those deviations can be retested during operation of the unit to determine if compliance is met in the operational setting. *Surveyor witnessing documentation*. The equipment field calibration and data collection process of vibration tests conducted at sea must be witnessed by a Class approved surveyor. The surveyor is required to sign or initial each page of the test report and prepare a witnessing document stating whether all steps of the noise testing were completed to their satisfaction. A copy of the witnessing document is then given to the person conducting the onboard testing, for insertion into the final noise test report. The original is usually retained by the surveyor and submitted to Class for inclusion in the installation's Record.

Results

On completion of the test, the noise test report will be reviewed by the Class approved surveyor against the appropriate Class Habitation notation requirements. Where the requirements have been satisfactorily met, the installation will be assigned the appropriate Class Habitation notation.

Chapter 47
Indoor Climate on Offshore Installations

General

Thermal comfort is defined in ISO 7730 as "… that condition of mind which expresses satisfaction with the thermal environment". The sensation of thermal comfort is therefore largely subjective and will vary from person to person. Due to differences in metabolism and expectations, there are distinct individual differences among people's perception of comfort as a function of temperature, humidity, and other atmospheric characteristics. Acclimatisation, habits, and expectations influence perceived comfort. These individual differences make it difficult to specify a single thermal environment that will be satisfactory to everyone. A thermal environment is therefore typically defined to be acceptable to at least 80% of the occupants of an interior space. Individually, the perception of thermal comfort is largely determined by the interaction of thermal environmental factors such as air temperature, air velocity, relative humidity, and factors related to activity and clothing. The thermal control or HVAC systems on an offshore installation should be designed to effectively control the indoor thermal environmental parameters to within acceptable limits to facilitate the thermal comfort of the occupants. To that end, this chapter provides the assessment criteria, verification and measurement methodology for indoor climate relating to habitability on offshore installations. The criteria are based on currently available standards and were selected to provide an index of personnel thermal comfort. The thermal environmental variables covered in this chapter include the ambient qualities of air temperature in personnel cabins and the average air temperature within the space serves as an indication of potential thermal comfort or discomfort. The thermal environmental criteria provided in this chapter are for persons wearing typical indoor clothing occupied with light, primarily sedentary activity and resulting in a thermal environment acceptable to at least 80% of the occupants. This chapter applies to enclosed manned spaces such as personnel accommodation areas, manned control rooms, hospital/medical facilities, and indoor workspaces. Compliance with the provisions outlined in this chapter are generally required to meet

A. A. Olsen, *Applying Physical Ergonomics to Modern Ship Design*,
https://doi.org/10.1007/978-3-031-57974-5_47

the Class Habitability (e.g., offshore installation Habitability (**HAB(OW)**), offshore installation Habitability Plus (**HAB+(OS)**), and offshore installation Habitability Plus Plus (**HAB++(OS)**) notations for ABS classed installations. Other classification societies will assign their own specific Class notations.

Terminology

Air supply quantity: The total amount of air supplied to a specific space consisting of a percentage of re-circulated air and a percentage of fresh air supply quantity, measured in litres per second (l/s).

Air temperature: The temperature of the air surrounding a person, measured with a standard thermometer.

Air velocity or movement: The rate of displacement of ambient air in a specific direction in metres-per-second (m/s) or feet-per-second (ft/s).

Comfort zone: That range of environmental conditions in which at least 80% of personnel experience thermal comfort.

Fresh air supply quantity: The amount of fresh/outdoor air supplied to a specific space, expressed in litres per second (l/s).

Horizontal gradient: The difference between the inside surface temperatures of the bulkheads bounding the berth and the average air temperature within the cabin. This is used as an indication of potential thermal discomfort due to radiant thermal sources.

HVAC zone: A space or group of spaces that is (are) independently controlled for temperature, humidity, and air distribution. A zone usually comprises common duct work fed from an air handler.

Reference calibration: Calibration of a measuring instrument, conducted at least every two years (or more frequently if specified by the manufacturer) by an accredited Testing and Calibration Laboratory accredited according to ISO 17025:2005, as amended, with traceability to a national or international standard.

Relative humidity (RH): The ratio of the amount of vapour contained in the air (absolute humidity) to the maximum amount of vapour the air can hold at a given temperature before precipitation (condensation) occurs.

Thermal comfort: Subjective index of "that condition of mind which expresses satisfaction with the thermal environment".

Ventilation: Ventilation is the process of supplying air to and removing air from any space by natural or mechanical means. From the standpoint of comfort and health, ventilation issues involve both quantity and quality.

Vertical gradient: The vertical air temperature difference within an enclosed space. The vertical gradient is used as an indication of potential local discomfort at the head and feet.

Associated Documentation

The following documents provide details about test plan preparation, test measurement procedures and/or test reporting.

- ANSI/ASHRAE 55-2013, Thermal Environmental Conditions for Human Occupancy.
- ISO 15138: 2007(E), Petroleum and Natural Gas Industries—Offshore Production Installations—Heating, Ventilation and Air-Conditioning.
- ISO 7547:2002, Ships and Marine Technology—Air-Conditioning and Ventilation of Accommodation Spaces—Design Conditions and Basis of Calculations.
- ISO 7726:1998, Ergonomics of the Thermal Environment—Instruments for Measuring Physical Quantities.
- NEBB:2005, Procedural Standards for Testing, Adjusting, Balancing of Environmental Systems.

Criteria

The indoor climate criteria for offshore installation Habitability notations (e.g., **HAB(OS), HAB+(OS),** and **HAB++(OS)**) are provided in Table 47.1, *Summary of indoor climate requirements.* Indoor climate measurements should be taken only in indoor manned spaces. Specific measurement locations are discussed in the section 'Measurement locations'. The thermal environmental comfort ranges and conditions should be achievable, under the test conditions specified in the section on 'Test conditions', for normal operating conditions, in all indoor manned spaces. For installations with lower and median notations (e.g., **HAB(OS)** and **HAB+(OS)**), the criteria provide for a range of air temperatures maintained by a temperature controller for each zone and are primarily aimed at HVAC systems that do not make provision for individual adjustment to suit personal preferences and activities within a specific space. Installations with an advanced notation (e.g., **HAB++(OS)**) should aim at enhancing personnel comfort by making provisions for the personnel to adjust indoor climate conditions, with regard to air temperature to suit personal needs within a particular space. *Air temperature.* For installations with a **HAB(OS)** or **HAB+(OS)** notation (or equivalent), the HVAC system should be capable of providing an air temperature within the range of 20–25 °C (68–77°F) during winter months and 22–27 °C (71.5–80°F) during summer months to an HVAC zone. This temperature should be maintained by a temperature controller. Each zone should also have a thermostat for reheat and dehumidification purposes. Alternatively, for installations with a **HAB++(OS)** notation (or equivalent), the HVAC system should be capable of sustaining an adjustable range of air temperatures between 20 and 25 °C (68–77°F) during winter months and 22–27 °C (71.5–80°F) during summer months in indoor manned personnel accommodation spaces. This temperature should be maintained by a temperature controller. Each indoor manned personnel accommodation space should have its own individual controller for temperature regulation. *Relative humidity.* The HVAC system should be capable of providing and maintaining a relative humidity within a range from 30% minimum to 70% maximum. *Enclosed space vertical gradient.* The difference in temperature at 100 mm (4 in.) above the deck and 1700 mm (67 in.) above the deck should be maintained within 3 °C (6°F). *Air velocity.* Air velocities should not exceed 30 m-per-minute or 100 feet-per-minute (0.5 m/s or 1.7 ft/s) at the measurement position in the space. *Personnel cabin area horizontal gradient.* In personnel cabin areas, the difference between the inside bulkhead surface temperature adjacent to personnel cabins and the average air temperature within the space should be less than 10 °C (18°F). *Air supply quantity.* The minimum quantity of fresh/outdoor air supply should be not less than 40% of the total air supplied to a specific space. The fresh/outdoor air supply quantity should not be less than 8 l/s per the number of person(s) for which the specific space is designed for (e.g., for cabins is the number of beds, for mess rooms is the number of seats, for workspaces is the number of workstations, etc.). *Summary.* A summary of the indoor climate requirements is presented in Table 47.1, Summary of indoor climate requirements.

Table 47.1 Summary of indoor climate requirements

Item	Requirement or criterion	
	HAB9OS) and HAB+(OS) or equivalent	HAB++(OS) or equivalent
Adjustability	Non-adjustable air temperature	Adjustable range of air temperatures
Air temperature	Winter: 20–25 °C (68–77°F) Summer: 23–28 °C (73.5–82.5°F)	
Relative humidity	A range from 30% minimum to 70% maximum	
Vertical gradient	The acceptable range is 0–3 °C (0–6°F)	
Air velocity	Not exceed 30 m-per-minute or 100 feet-per-minute (0.5 m/s or 1.7 ft/s)	
Horizontal gradient (crew cabin areas)	The horizontal temperature gradient in crew cabin areas should be < 10 °C (18°F)	
Air supply quantity	The minimum quantity of fresh/outdoor air supply should be not less than 8 l/s per person and in addition will need to be at least 40% of the total air supplied to a specific space	

Test Plan

As stated in the section 'Test plans' in Chap. 43, a test plan should be developed to serve as the principal means for verifying the measurements to be performed to verify compliance with the indoor climate criteria. This test plan should typically include the following.

- *Documentation.* The test plan should include appropriate drawings indicating the location of all indoor climate systems and equipment including the HVAC system design specifications and all schematics/layout drawings of the HVAC system.
- *Test personnel.* The test plan should provide information about the testing specialist who will be conducting the test and their approval and certification in accordance with the Class specific Procedural Requirements for Recognised Ambient Environmental Testing Specialists or equivalent.
- *Test conditions.* The test plan should detail the conditions under which the tests will be performed. Details about test conditions are provided in the section on 'Test conditions'.
- *Measurement locations.* The test plan should document, in detail, on appropriate drawings, all spaces where measurements will be taken. In addition, transducer placement positions should be indicated. Details on selecting measurement locations and determining transducer placement positions are provided in the section 'Measurement locations'.
- *Data acquisition and instruments.* The test plan should provide information regarding the methods and instrumentation to be used for measurement and data collection. Instrumentation specification details should include type of instruments to be used, accuracy, calibration, sensitivity. More details on data acquisition and instruments are provided in the section 'Data acquisition and instruments'.
- *Data analysis.* The test plan should provide information regarding the methods, software, and instrumentation to be used for data analysis.
- *Test schedule.* The test plan should provide information regarding the proposed test schedule. The test schedule should include information on the approximate date and duration of testing (including a summary of the scheduled measurements per day).
- *Data collection sheets.* The test plan should provide sample data sheets (in table format) that will be used for reporting the measurement data.
- *Conflict of interest declaration.* In the case that the testing specialist has, in any capacity or to any extent, provided input into the design of the installation, their participation should be indicated in the test plan including details of the testing specialist's involvement. Class may review any potential conflicts of interest and determine if any further action is necessary.

Test Requirements

Indoor climate measurements should be carried out in accordance with the procedures described in ANSI/ASHRAE 55-2013. When the requirements or procedures proscribed by Class deviate from those mentioned in ANSI/ASHRAE 55-2013 then the requirements or procedures proscribed by Class will take precedence unless otherwise stipulated. Data acquisition and instruments. The thermal measurement instrumentation used must meet or exceed the minimum characteristics of instruments for measuring physical quantities characterising an environment specified in ISO 7726. The following quantities should be measured in each of the spaces or zones identified in the test plan and the results noted in the indoor climate test report.

- Air temperature.
- Relative humidity.
- Air velocity.
- Vertical gradient.
- Horizontal gradient (in personnel cabin spaces only).

The air temperature and humidity measurements should be made at least every five minutes for a minimum period of one hour. The minimum, maximum, and average values for the one hour period should be reported for each space measured. For the **HAB++(OS)** notation, the air temperature measurements should be made at least every five minutes for a minimum period of one hour if the system has the capacity to reach within that hour the **HAB++(OS)** requirements for lower and upper temperatures. If not, the air temperature measurements should be made at least every 10 min for a minimum period of two hours. ***Note***: *data loggers are a type of device that can be left unattended to capture data. These type devices have been proven effective in gathering temperature and humidity values.* The measuring period for determining the average air velocity at any location should be three minutes. Surface temperatures for all wall surfaces that are adjacent to the head of the bed in cabins should be measured for determining the horizontal gradient in personnel cabin spaces. The horizontal gradient is to be calculated and recorded as the difference between the wall temperature and the average air temperature at 1100 mm (43 in.) above the deck personnel cabin areas. The test plan should include supporting documentation and calculations relating to the air supply for each of the spaces or zones identified in the test plan. Test conditions. In order to determine the effectiveness of the HVAC system at providing the environmental conditions specified in this chapter, measurements should be made under the following conditions.

- *Testing.* Testing of identified spaces can be performed in port and at sea, provided the required test conditions stated here are in compliance at the time the measurements are made and recorded. In the case where some testing is performed in port, confirmatory testing should be performed at sea with all normally functioning equipment in its operational mode. This confirmatory testing must usually consist of a sample of the manned spaces selected as follows.

- Spaces that were identified as "worst case" locations.
- Spaces which were out of compliance or close to non-compliance (refer to Table 47.1, Summary of indoor climate requirements).
- Spaces that required some type of mitigation.
- Additional testing may be required at the Class approved surveyor's discretion.

Equipment operation. The HVAC system should be operating in the normal operation or mode. All machinery essential for operation (fluid, turbo generators, and additional machinery) should operate under normal conditions throughout the measurement period. A written confirmation issued by the builder confirming that the HVAC system has been tested, adjusted and balanced should be provided to the Class approved surveyor before any measurement commences. The total system TAB should be conducted in accordance with the NEBB standard NEBB Procedural Standards for Testing, Adjusting, Balancing of Environmental Systems or other similar certification standard. *Doors and windows.* The space doors (including louvres) and windows should be closed during the evaluation period, except for routine entry and exit. Any open doors or windows should be noted in the indoor climate test report. *Equipment and furnishings.* Spaces should be furnished with all usual equipment, fixtures and fittings normally found in the space. Equipment must be configured to operate in its normal operating mode. *Weather and climatic conditions.* When thermal conditions in the manned space have a high sensitivity to time of day and weather conditions (e.g., spaces adjacent to exterior bulkheads), the measurement should be made such that the high and low extremes of the thermal parameters are determined (e.g., measurements could therefore be taken during the day and night in the same space). If possible, measurements should be taken with little or no cloud cover. *Test interference.* During the indoor climate measurements, any activity that might affect the indoor climatic variables in the space must be avoided. For this reason, only the personnel needed for the normal operation of the equipment in the space and those carrying out the measurements should be present in the space being tested. Doors and windows should be closed, except where they are normally left open (such as the door on the lee side of the navigation bridge, which may normally be open). Any open doors or windows should be noted in the indoor climate test report.

Measurement locations. Selection of spaces where measurements are to be conducted. The aim when selecting indoor climate measurement locations is to obtain a representative sample of data that reflects the actual conditions in manned spaces. For practical reasons, it is important to select the locations such that an appropriate amount of data can be collected during the testing phase. This means the measurement locations should be selected in accordance with the following criteria.

- Select potential problem areas where the influence of internal conditions or factors may adversely impact the quality of the indoor climate in enclosed manned spaces. Internal conditions include space proximity to equipment that radiates or absorbs heat (e.g., engine exhaust trunks, freezer spaces, galley, scullery, etc.) and surfaces with thermal differentials in excess of 10 °C (18°F) from the ambient temperature in the space. Living and working areas at the ends of HVAC ductwork or piping runs (for heating or cooling) should be selected as potential problem locations. Measurements should be taken in all identified potential problem areas.

- Select potential problem areas where the influence of external ambient environmental conditions (e.g., sun, wind, precipitation, etc.) may adversely impact the quality of the indoor climate. These areas include manned spaces, accommodation areas and recreation areas which may be outboard or adjacent to the installation's hull. Measurements should be taken in all identified problem areas.
- Where a single instance of one type of manned space exists within the installation (e.g., the control rooms, mess room, gymnasium, library, etc.), that location should be selected for measurement.
- Where multiple instances of one type of space exist (e.g., cabins or recreation areas), a representative sample of at least 15% of each type should be selected for measurement. The worst-case locations are to be considered part of the representative sample, where applicable.

Walk through verification inspection locations. All normally manned spaces are to be subject to a walkthrough inspection by the Class approved surveyor. The number and locations of the walkthrough inspections are generally determined by the surveyor. The purpose of the walkthrough verification is to subjectively assess the indoor climate qualities. At the discretion of the surveyor, additional measurements may be required. Transducer placement positions. For each space identified in the test plan, the transducer locations should be standardised as follows.

- Air temperature and relative humidity measuring instrumentation should be set up approximately in the middle of the space to measure general space temperature and humidity levels. Air temperature should be measured at approximately 100 mm (4 in.), 1100 mm (43 in.) and 1700 mm (67 in.) above the deck. Relative humidity should be measured at a height of approximately 1700 mm (67 in.) above the deck.
- Air velocity should be measured at approximately 100 mm (4 in.), 1100 mm (43 in.) and 1700 mm (67 in.) above the deck in manned spaces, as applicable (to assure air velocity is not excessive). This measurement should be taken in the centre of the space. Air velocity only needs to be measured once per testing location.
- For personnel cabin spaces, inside wall surface temperatures for determining the horizontal gradient should be captured at approximately 300 mm (12 in.) above the mattress.

Test Report

As stated in the section on 'Test reports' in Chap. 43, a test report must be submitted to the Class approved surveyor to determine whether the indoor climate levels meet the criteria and whether this part of the Class notation requirement has been met. The details listed in the following paragraphs are required to be provided in the indoor climate test report. Test details. The following details are required to be recorded for each period of testing.

- Installation's geographical location and appropriate block identification.
- Weather conditions and meteorological data (i.e., wind speed and direction, ambient outdoor air temperature, outdoor humidity, barometric pressure) at the onset of every data collection period and at intervals of every four hours during any data collection period. Weather conditions should also be reported at the end of each data collection period.
- Any indications of abnormal activities or conditions during the test that might skew the results.

Transducer placement positions. The actual transducer placement positions must be indicated on appropriate drawings. *Measurement equipment details.* Details of measurement and analysis equipment (e.g., manufacturer, type and serial number, accuracy and resolution must be provided. Copies of the relevant instrumentation reference calibration certificates, together with the results of field setup and calibration checks before and after the field tests, should also be provided with the report. *Results.* The following results per measuring location, should be provided in a tabulated format.

- Measurement position.
- Number of people present in the space at time of measurement.
- Measurement period if different from requirements.
- Air temperature (minimum, maximum, and average) at 100 mm (4 in.) above deck.
- Air temperature (minimum, maximum, and average) at 1100 mm (43 in.) above deck.
- Air temperature (minimum, maximum, and average) at 1700 mm (67 in.) above deck.
- Relative humidity (minimum, maximum, and average) at 1700 mm (67 in.) above deck.
- Air velocity at 100, 1100, and 1700 mm (4, 43, and 67 in.) above deck at measurement positions.
- Vertical gradient (average air temperature at 1700 mm (67 in.) minus average air temperature at 100 mm (4 in.) above deck).
- Horizontal gradient (side wall surface temperature in personnel cabin spaces minus average air temperature at 1100 mm (43 in.) above personnel cabin areas).
- Indication of pass/fail.

For all the remaining spaces that were checked through walk through verification inspection and spot check measurements, the following information should be provided.

- Name and number of spaces.
- Walk through inspection observations.
- Measurement results if necessary.

Deviations. All deviations from the approved test plan must be reported. Where there are noted deviations from criteria and measurements were taken during Installation and Commissioning, those deviations can be retested during operation of the unit to determine if compliance is met in the operational setting. *Surveyor witnessing documentation.* The equipment field calibration and data collection process of indoor climate tests must be witnessed by a Class approved surveyor. The surveyor is required to sign or initial each page of the test report and prepare a witnessing document stating whether all steps of the indoor climate testing were completed to their satisfaction. A copy of the witnessing document is then given to the person conducting the onboard testing, for insertion into the final indoor climate test report. The original is usually retained by the surveyor and submitted to Class for inclusion in the installation's Record.

Results

On completion of the test, the indoor climate test report will be reviewed by the Class approved surveyor against the appropriate Class Habitation notation requirements. Where the requirements have been satisfactorily met, the installation will be assigned the appropriate Class Habitation notation.

Chapter 48
Lighting and Illumination on Offshore Installations

General

The lighting of personnel spaces should facilitate visual task performance, facilitate movement in the space and aid in the creation of an appropriate visual environment. Lighting design involves integrating these aspects to provide adequate illumination for the safety and wellbeing of the personnel as well as for the various tasks performed on board offshore installations. The selection of appropriate illumination levels for specific tasks and seafarer spaces is an important consideration in the design of lighting systems. There is a difference of opinion as to what levels of light may be considered best for visual tasks. Since illumination recommendations are generally consensus values, for any task, a range of illumination levels may apply. Since visual tasks performed within habitable spaces on board an offshore installation are generally similar to tasks encountered ashore, requirements for illumination on offshore installations generally correspond to those tasks performed in living, working, and recreation areas on shore. To that end, this chapter outlines criteria for assessing the illumination levels of general lighting and task lighting on offshore installations. The main objective of the assessment is to determine whether the various lighting systems comply with minimum standards to accommodate personnel visual task performance and facilitate personnel movements and wellbeing and safety on board offshore installations. Lighting criteria are based on currently available objective standards and research data. Compliance with the criteria outlined in this chapter is typically a prerequisite for attaining a Class Habitability (e.g., **HAB(OS)**, **HAB+(OS)**, and **HAB++(OS)** or equivalent) notation.

Terminology

Disability glare: Glare which reduces the ability to perform a visual task.

A. A. Olsen, *Applying Physical Ergonomics to Modern Ship Design*,
https://doi.org/10.1007/978-3-031-57974-5_48

Discomfort glare: Glare which produces viewer discomfort, but which does not significantly interfere with visual task performance or visibility.

General lighting: Lighting designed to provide a substantially uniform level of illumination throughout an area, exclusive of any provision for special, localised tasks. Such lighting should be provided by fixed luminaires.

Glare: The discomfort or impairment of vision experienced when parts of the visual field are excessively bright in relation to the general surroundings.

Illumination: The luminous flux density at a surface (or the amount of light falling on an object or surface), i.e., the luminous flux incident-per-unit area. Illumination is measured in units of Lux (lm/m^2) or footcandles (fc; lm/ft^2). One foot-candle equals 10.76 Lux.

Lumen: The SI of luminous flux, used in describing a quantity of light emitted by a source or received by a surface.

Luminaire: A complete lighting unit consisting of a lamp(s) together with the parts designed to distribute the light, to position and protect the lamp and to connect the lamp to the power supply.

Luminance: The photometric brightness of an illuminated surface (or the amount of light emitted or reflected from the surface). The SI unit of luminance is candela-per-square metre (cd/m^2).

Luminous flux: The light emitted by a source, or received by a surface and indicates the intensity of a source. Flux is expressed in lumens.

Lux: a unit of illumination, equivalent to 0.0929 foot-candle and equal to the illumination produced by luminous flux of one lumen falling perpendicularly on a surface one metre square. Also called a metre candle.

Red or low-level white illumination: Lighting provided to accommodate efficient dark adaptation in areas where seeing tasks are performed during night-time operations and in areas where people need to move from a lit interior into a dark environment and maintain good vision.

Reference calibration: Calibration of a measuring instrument, conducted at least every two years (or more frequently if specified by the manufacturer) by an accredited Testing and Calibration Laboratory accredited according to ISO 17025:2005, as amended, with traceability to a national or international standard.

Task lighting: Lighting provided to meet the illumination requirements of a specific task. Task lighting refers to the total illumination requirement that may be obtained by supplementary lighting provided in addition to the general illumination. Such lighting may be provided by fixed luminaires or via floor lamps or table lamps.

Task plane: The horizontal, vertical, or inclined plane in which the visual task lies. If no information is available, the task plane may be considered to be the horizontal

and at 750 mm (29.5 in.) above the deck for seated tasks and 1000 mm (39.5 in.) for standing tasks.

Veiling reflection (or reflected glare): A reflection of incident light that partially or totally obscures the details to be seen on a surface by reducing the contrast.

Associated Documentation

The following documents provide details about test plan preparation, test measurement procedures and/or test reporting:

- ISO 8995-1: 2002, Lighting of Workplaces—Part 1: Indoor.
- DIN EN 12464-1:2011, Light and Lighting—Lighting of Workplaces—Part 1: Indoor Workplaces.
- DIN 5032-7:1985, Photometry; Classification of Illuminance Meters and Luminance Meters.
- CIE S 023/E:2013 (ISO/CIE 19476:2014), Characterisation of the Performance of Illuminance Meters and Luminance Meters.
- BS 667:2005, Illuminance meters. Requirements and test methods.

Criteria

The lighting and illumination criteria for Class Habitability notations are provided in:

- Table 48.1, Lighting criteria for personnel accommodation spaces.
- Table 48.2, Lighting criteria for navigation and control spaces.
- Table 48.3, Lighting criteria for service space.
- Table 48.4, Lighting criteria for operating and maintenance spaces/areas.
- Table 48.5, Lighting criteria for red or low-level white illumination.

For lighting, the criteria for **HAB(OS)**, **HAB+(OS)**, and **HAB++(OS)** notations are the same. In this chapter, general lighting and task lighting requirements are provided for personnel tasks and spaces normally encountered on offshore installations. The lighting levels provided in the tables are for new lamps. Emergency lighting is covered in SOLAS and IMO Resolutions and was not considered in the selection of the lighting levels provided in this chapter. The criteria in this chapter are applicable to the entire installation and not just manned spaces. *General and task lighting.* The minimum maintained illumination levels stated in the Lighting Criteria Tables 48.1 through 48.4 may be achieved under the test conditions specified in the section on 'Test conditions' and measured with task lighting turned on, where provided, but with external light sources (e.g., daylight, moonlight, shore light, etc.) excluded. These levels were selected from a variety of sources that are listed in the

Table 48.1 Lighting criteria for crew accommodation spaces

Space	Illumination level in Lux	Space	Illumination level in Lux
Cabins and sanitary spaces			
General lighting	150		
Reading and writing		Sanitary spaces	
• Desk	500	• Lavatory/toilet	200
• Bunk light	200	• Bath/shower area	200
Changing room	200		
Light during sleep periods[1]	<30		
Dining spaces			
Mess room and cafeteria	300		
Snack or coffee area	150		
Recreation spaces			
Recreation room/ lounges	200	Gymnasium	300
Library		Bulletin boards/ display areas	150
• General lighting	150	All other recreation spaces (e.g., game rooms)	200
• Reading area	500		
Multimedia resource centre (e.g., computer room)			
Medical, dental and first aid centre			
General lighting/ medical waiting areas	150	Dispensary	
Medical and dental treatment/examination room	500	• General lighting	150

(continued)

[1] If there is any opportunity for light to enter cabins at the times of day or night when people sleep (e.g., portlights, transoms, etc.), the maximum lighting level should be 30 Lux.

Table 48.1 (continued)

Space	Illumination level in Lux	Space	Illumination level in Lux
Reading and writing (desk)	500	• Over the dispensary table	500
Laboratories	500	Other medical spaces	300

Table 48.2 Lighting criteria for entrances and passageways

Space	Illumination level in Lux	Space	Illumination level in Lux
Interior walkways, passageways, corridors, stairways and access ways	100	Exterior walkways, passageways, and access ways (night)	50
Elevators	100	Exterior stairways	100

Table 48.3 Lighting criteria for navigation and control spaces

Space	Illumination level in Lux	Space	Illumination level in Lux
Wheelhouse, pilothouse, bridge		Offices	
• Day	300	• General lighting	300
• Night when underway	<30	• Reading and writing (desk)	500
Chart room		• Meeting, training rooms	500
• General lighting	150	Control rooms and control stations[2]	
• On chart table	500	• General lighting	300
Radar/gyro room[2]	300	• Control consoles and boards, panels, instruments, etc.	300
Radio room		• Reading and writing (desk)	500

References part of this book. *Red or low-level white illumination.* In workspaces where red or low-level white illumination is provided to facilitate dark adaptation, the maintained illumination levels in Table 48.5, Lighting criteria for red or low-level white illumination, are provided for guidance only and are not generally required for a Class Habitation notation.

[2] At night, the wheelhouse must be kept as dark as possible to permit visibility of objects on the water, hazards to navigation, and navigational lights.

Table 48.4 Lighting criteria for service spaces

Space	Illumination level in Lux	Space	Illumination level in Lux
Food preparation		Laundries	300
• Galley	500	Chemical storage (e.g., paint store, etc.)	300
• Pantry	300	Storerooms	
• Butcher shop	500	• Large parts	200
• Thaw room	300	• Small parts	300
• Working surfaces, food preparation counter and range tops	500		
• Scullery (dishwashing) food serving lines package handling/cutting	300		
Food storage			
• Non-refrigerated	200		
• Refrigerated	100		
Mail sorting	300		

Test Plan

As stated in the section 'Test plans' in Chap. 43, a test plan should be developed to serve as the principal means for verifying the measurements to be performed to verify compliance with the lighting and illumination criteria. This test plan should typically include the following:

- *Documentation.* The test plan should include appropriate design information and layout drawings showing the hull/installation outline, bulkheads, access routes, location of luminaires, outlines of major furniture and equipment, and the space name and number. The drawings should be to a scale and sized to permit the scaling of measurement points (as required in the section on 'Measurement locations') and lighting equipment and the recording of luminance and other relevant data.
- *Test personnel.* The test plan should provide information about the testing specialist who will be conducting the test and their approval and certification in accordance with the Class specific Procedural Requirements for Recognised Ambient Environmental Testing Specialists or equivalent.
- *Test conditions.* The test plan should detail the conditions under which the tests will be performed. Details about test conditions are provided in the section on 'Test conditions'.

Table 48.5 Lighting criteria for operating and maintenance spaces/areas

Space	Illumination level in Lux	Space	Illumination level in Lux
Machinery spaces	200	Cargo holds (portable lighting)	
Engine room (main propulsion machinery)	300	General lighting[2]	30
Generator, switchboard and transformer room emergency generator room	200	During cargo handling[3]	200
Fan room/HVAC room	200	Passageways and trunks	80
Bilge/void spaces (portable lighting)	75	Storerooms	200
Pump room	200	Inspection and repair tasks	
Fire pump room	200	• Rough	300
Shaft alley	100	• Medium	500
Steering gear room	200	• Fine	750
Windlass room	200	• Extra fine	1000
Battery room	200	Workshops	
Boiler rooms	100	• General lighting	300
		• On work bench top/ machine paint shop	500
Muster/embarkation area	200	Escape trunks	50
Cargo handling[3]	200		
Weather decks	200	Crane cabin	400
Lay down area	200		
General processing area	200	Hand signalling areas between crane shack and ship deck	300
Utility area	200		
Loading ramps and bays	200		
Cargo storage and manoeuvring areas	200		

- *Measurement locations.* The test plan should document, in detail, on appropriate drawings, all spaces where measurements will be taken. In addition, transducer placement positions should be indicated. Details on selecting measurement locations and determining transducer placement positions are provided in the section 'Measurement locations'.
- *Data acquisition and instruments.* The test plan should provide information regarding the methods and instrumentation to be used for measurement and data collection. Instrumentation specification details should include type of instruments to be used, accuracy, calibration, sensitivity. More details on data acquisition and instruments are provided in the section 'Data acquisition and instruments'.
- *Data analysis.* The test plan should provide information regarding the methods, software, and instrumentation to be used for data analysis.
- *Test schedule.* The test plan should provide information regarding the proposed test schedule. The test schedule should include information on the approximate date and duration of testing (including a summary of the scheduled measurements per day).
- *Data collection sheets.* The test plan should provide sample data sheets (in table format) that will be used for reporting the measurement data.
- *Conflict of interest declaration.* In the case that the testing specialist has, in any capacity or to any extent, provided input into the design of the installation, their participation should be indicated in the test plan including details of the testing specialist's involvement. Class may review any potential conflicts of interest and determine if any further action is necessary.

Test Requirements

Visual tasks encountered on installations vary and the lighting provided can influence the ability to see and perform those tasks. Some vision and lighting considerations include task duration, visual fatigue, task criticality, veiling reflections, shadows, and abilities of the observer. The presence of glare is a concern and is often difficult to identify, measure, and assess. Glare is often transient (based on factors such as the direction of the sun to illuminated components or from the placement of lighting fixtures) and therefore difficult to anticipate. As part of lighting data collection, glare should be subjectively evaluated jointly by the testing specialist and the Class approved surveyor witnessing the collection of lighting data. These assessments should be made as part of the activities discussed in the section 'Walk through verification inspection locations'. Measurement of ambient lighting need only be performed once during sea trials or at pier side when all external light sources can be blocked out. Data acquisition and instruments. The illuminance meter (light meter) must conform to any of the International Standards specified below (or equivalent):

- DIN 5032-7:1985 (Class A).
- CIE S 023/E:2013 (ISO/CIE 19476:2014).
- BS 667:2005 (Type F).

Note: *the measurement of red and/or low-level white illumination is not required for installation Habitability notations. Test conditions.* The test conditions required for lighting and illumination measurements must be in accordance with the following:

- *Location.* Lighting measurements may be taken in port, at sea, or both, since the measurements are not dependent on installation operation. If testing is done in port, the installation's power—and not shore power—should be used.
- *Spaces with windows/portlights.* In spaces with windows or portlights where the minimum lighting level is intended to be provided by artificial light sources only, lighting measurements should be taken after dark, or during the day provided all natural lighting is blocked out.
- *Spaces without windows/portlights.* Interior spaces with no windows or portlights can be measured during daylight hours.
- *Stray light.* Stray light (e.g., deck lighting and moonlight) should be capable of being blacked out as far as practicable. Where it is not possible, measurements of stray light, at appropriate positions, with all lighting turned off, should be obtained. These readings must then be deducted from readings taken at the same positions, with the lighting turned on, to determine the illumination from the lighting.
- *Light for sleep.* Lighting measurements should be taken in cabins with all cabin lights turned off and curtains, shutters, etc., closed.
- *Test interference.* During the lighting measurements, shadows on the light meter caused by any kind of human activity should be avoided. For this reason, only the personnel needed to operate any equipment during testing and those carrying out the measurements may be present in the space. Doors and windows must be closed, except where they are normally left open (such as the door on the lee side of the navigation bridge, which may normally be open). Any open doors or windows should be noted in the lighting test report. Spaces should be furnished with all usual equipment, fixtures and fittings normally found in the space. *Measurement locations.* Selection of spaces where measurements are to be conducted. The aim when selecting lighting and illumination measurement locations is to obtain a representative sample of data that represents the actual conditions in the manned personnel spaces listed in the Lighting Criteria Tables 48.1, 48.2, 48.3, 48.4, 48.5 and 48.6, inclusive. For practical reasons, it is important to select the locations such that an appropriate amount of sample data can be collected during the testing phase. The measurement locations must be selected in accordance with the following criteria:

  - Select problem areas based on the potential for excessive external illumination (daylight) into the space (e.g., bridge). Areas at the end of cable runs should be selected as potential problem locations. Measurements should be taken in all identified problem areas.
  - Select potential problem areas based on the positioning of luminaires in the space as indicated on the drawings (e.g., uneven spacing of luminaires resulting in non-uniform illumination levels and dimly lit areas). Measurements must be taken in all problem areas.

Table 48.6 Lighting criteria for red or low-level white illumination[3]

Space	Illumination level in Lux
Where seeing is essential for charts and instruments	1–20
Interiors or spaces	5–20
Bridge areas (including chart tables, obstacles and adjacent corridors and spaces)	0–20 (continuously variable)
Stairways	5–20
Corridors	5–20
Repair work (with smaller to larger size detail)	5–55

- Where a single instance of one type of space exists within the installation (e.g., control rooms, mess room, gymnasium, library, etc.), the location should be selected for measurement.
- Where multiple instances of one type of space exist (e.g., recreation areas) with the exception of cabins, a representative sample of at least 25% of each type should be selected for measurement. The problem area locations are to be considered part of the representative sample, if applicable.
- Where a number of personnel cabin spaces are identical in configuration in terms of lighting systems, surface treatments, geometry, furnishings and equipment layout, only two of the spaces on each deck should be selected to determine whether the lighting requirements are met. Where cabins are not identical in configuration in terms of lighting system, surface treatments, geometry, furnishing, and equipment at least 25% should be selected on each deck for measurement.
- Where external night-time measurements are taken, spaces should be selected according to human presence in the space or task location, including exterior passageways and access ways, operating consoles, hand signalling locations, anchor handling locations, mooring locations, personnel embarkations and debarkation areas, and cargo handling locations—particularly with regard to attaching and detaching cargo from crane hooks.

Walk through verification inspection locations. All normally manned spaces are to be subject to a walkthrough inspection by the Class approved surveyor. The number and locations of the walkthrough inspections are generally determined by the surveyor. The purpose of the walkthrough verification is to subjectively assess lighting and illumination qualities. The presence of discomfort or disability glare should be specifically and subjectively assessed by the Class approved surveyor and the testing specialist. Upon completion of the assessment, the result of "no glare present", "some glare present", "discomfort glare present", or "disability glare present" should be noted for each lighting test location. At the discretion of the surveyor, additional measurements may be required. *General illumination measurement positions.* General lighting levels should be measured with all lights turned on

[3] When the space is separate from the wheelhouse.

except supplementary task lighting, such as desk lights and berth lights. Daylight should be excluded during the measurements. Measurements should be taken on a horizontal plane approximately 750 mm (29.5 in.) above the deck. For traffic areas, readings should be taken on deck. For larger spaces (e.g., public rooms, messes, recreation areas, etc.), it may be necessary to take measurements at a number of locations in order to obtain a representative sample of the illumination for that space. Measurement locations should always be evenly distributed throughout the space. For a specific room size, the minimum number of measurement locations will be as indicated in Table 48.7, Distribution of measurement positions within spaces. The arithmetic mean of the measurement should be used for the measure of illumination of the general area. In the case of corridor lighting, measurements should be taken approximately every 3000 mm (10 ft) of corridor length. The arithmetic mean should not be used to determine compliance with corridor lighting requirements.

Task lighting measurement positions. For task area lighting (such as consoles, control stations, workbenches, computer workstations, desktops and meeting tables, medical stations, galley areas, and maintenance workstations), a representative set of readings should be taken over the task surface with the worker in the normal working position. It is best to assess the task lighting with both general and supplementary task lighting turned on. As much as possible, daylight should be excluded during the measurements. Measurements of task lighting must be made on the surface or in the plane of the task (horizontal, vertical or at an angle). For small task surfaces, i.e., smaller than 0.5 m^2 (5 ft^2), a single measurement is best taken at the centre of the task surface. For larger task surfaces (0.5 m^2 (5 ft^2) or larger), the illumination is best measured by dividing the task surface into a grid no larger than 0.5 m^2 (5 ft^2) and averaging the measurements taken at the grid intersections. Measurements for mirrors should be taken approximately 400 mm (16 in.) away from the surface of the mirror. For berths and bunks, measurements may be taken at a point approximately 300 mm (12 in.) above the top of the mattress and 600 mm (24 in.) from the bunk light fixture. *Red and low-level white illumination.* Guidance for the measurement of red and low-level white is provided in B.8.4 in IESNA RP-12-97. These measurements are not required for the Habitability notation.

Table 48.7 Distribution of measurement positions within spaces

Space	Minimum number of measurement positions in room
Less than 20 m^2 (215 ft^2)	1
> 20–40 m^2 (215–431 ft^2)	2
> 40–80 m^2 (431–861 ft^2)	3
> 80–120 m^2 (861–1291 ft^2)	4
> 120–200 m^2 (1291–2150 ft^2)	5
Greater than 200 m^2 (2150 ft^2)	6

Test Report

As stated in the section on 'Test reports' in Chap. 43, a test report must be submitted to the Class approved surveyor to determine whether the lighting and illumination levels meet the criteria and whether this part of the Class notation requirement has been met. The details listed in the following paragraphs are required to be provided in the lighting test report. *Test details.* The following details are required to be recorded for each period of testing:

- Time of day.
- External lighting conditions (e.g., were measurements taken in port or at sea and were measurements taken during day light hours or at night?).

Measurement positions. Actual measurement locations should be indicated on appropriate drawings. *Measurement equipment details.* Details of the measuring equipment used (e.g., manufacturer, type and serial number, accuracy and resolution) should be provided. Copies of the relevant instrumentation reference calibration certificates, together with the results of field setup and calibration checks before and after the field tests, should be included with the lighting test report. *Results.* The following results, per space, should be provided in a tabulated format:

- Name and number of space(s).
- Task areas (if any) in space.
- Average lighting level for general lighting.
- Average lighting level for task blighting on each task surface.
- Corridor lighting measurements.
- Lighting levels in cabins with lights turned off and all curtains, shutters, deadlights, etc. closed.
- Presence of glare at specific workstations within the space and any follow-up activity requirements.
- Indication of pass/fail.

For all the remaining spaces that were checked through walk through verification inspection and spot check measurements, the following information should be provided:

- Name and number of spaces.
- Walk through inspection observations.
- Measurement results if necessary.

Deviations. All deviations from the approved test plan must be reported. *Surveyor witnessing documentation.* The equipment field calibration and data collection process of indoor climate tests must be witnessed by a Class approved surveyor. The surveyor is required to sign or initial each page of the test report and prepare a witnessing document stating whether all steps of the indoor climate testing were completed to their satisfaction. A copy of the witnessing document is then given to the person conducting the onboard testing, for insertion into the final indoor climate test report. The original is usually retained by the surveyor and submitted to Class for inclusion in the installation's Record.

Results

On completion of the test, the lighting test report will be reviewed by the Class approved surveyor against the appropriate Class Habitation notation requirements. Where the requirements have been satisfactorily met, the installation will be assigned the appropriate Class Habitation notation.

Part VI
Living and Working in Low Temperature Marine Environments

Chapter 49
Weather Conditions of Interest

General

The dominant factor for operations in polar and sub-polar regions is the occurrence of extremely low temperatures and the associated formation of ice. In low temperatures, any precipitation will be in the form of snow, or at closer to the freezing point as freezing rain, sleet or ice pellets. Visibility in any of these conditions can be very limited and ice build-up can produce a range of hazards, as described earlier. Ice accumulation due to spray is most likely in air temperatures below 2 °C (36 °F), and wind speeds of above 20 knots (10 m/s). It will worsen as wind speeds increase beyond this, and in higher sea states. In very low temperatures, sea ice can form quite rapidly once the water temperature itself falls below 0 °C (32 °F). Vessels with little or no ice capability can find themselves at risk if caught in these conditions, which are most likely to occur towards the onset of winter. More generally, most vessels can be put at risk by ice movement, which can occur with considerable rapidity under conditions of high wind or currents. Conditions reported on ice charts or by remote imagery can change fast, particularly the reported positions of the ice edge and the location of leads through the pack. It is important for mariners to be able to recognise the conditions in which such changes can occur, and signs of the proximity of ice. These can include:

- "Ice blink"—a reflection of ice from the underside of cloud cover; most apparent when there is snow cover on the ice.
- The onset of fog, which is often present near the ice edge.
- The appearance of small ice floes, which can indicate that larger amounts are present nearby.
- Rapid moderation of waves, when approaching the ice edge from downwind.

A. A. Olsen, *Applying Physical Ergonomics to Modern Ship Design*,
https://doi.org/10.1007/978-3-031-57974-5_49

Ice Types

For vessels intended to operate in ice, it is important to be able to distinguish between the different ice types that may be encountered, both for efficiency and or safety. The two most fundamental properties of ice cover are thickness and age, both of which are reported on standard ice charts using World Meteorological Organisation (WMO) terminology and symbols, as outlined in Table 49.1, Sea ice stages of development. Other terminology is also used, particularly for the initial freezing stages, including frazil, grease, shuga, and slush ice. These are all thin forms of ice cover that will normally be reported as new ice.

Ice services use a number of techniques to derive the thickness of ice cover, but all of these are approximate. The symbols on ice charts should always be treated with some caution. Old ice is ice that has survived one or more melt seasons. It encompasses both second-year and multi-year ice, but the term multi-year is frequently applied to either old ice form. Multi-year ice becomes much stronger than first year ice, due in part to its reduced salinity. Floes also tend to have much more variable thickness than younger ice, as they incorporate weathered ridges and other features. This and other features help experienced ice navigators to distinguish between firstyear and multi-year ice. Ice "of land origin" is generally glacial ice, formed over thousands of years by the accumulation and re-crystallisation of packed snow. Ice islands and icebergs enter the sea from glaciers and ice sheets, and may in turn 'calve' smaller bergy bits and growlers as they degrade. Glacial ice is very hard, and represents a major hazard for vessels with even the highest level of ice

Table 49.1 Sea ice stages of development

Description	Thickness	WMO code
New ice	< 10 cm	1
Nilas; ice rind	0 – 10 cm	2
Young ice	10 – 30 cm	3
Grey ice	10 – 15 cm	4
Grey-white ice	15 – 30 cm	5
First-year ice	30 – 200 cm	6
Thin first-year ice	30 – 70 cm	7
Thin first-year ice first stage	30 – 50 cm	8
Thin first-year ice second stage	50 – 70 cm	9
Medium first-year ice	70 – 120 cm	1
Thick first-year ice	120 – 200 cm	4
Old ice	7	
Second-year ice	8	
Multi-year ice	9	
Ice of land origin	.	
Undetermined or unknown	x	

capability. Growlers and bergy bits have small freeboards and can be very difficult to detect either when part of the general ice cover or in open water with moderate sea states. Due to their origin, they are usually found in proximity to icebergs, whose own presence is a good indicator of the potential risk of encountering these smaller fragments. Ice cover is very rarely uniform or homogeneous. Near the coast, ice may be ‘land fast’, anchored in place by the shoreline and possibly by grounded pressure ridges. Land fast ice tends to have relatively consistent properties but may still include ridges and rubble piles. At the edge of the land fast ice, shear zones may occur where the free-floating pack and land fast ice collide. The shear zone can be a chaotic amalgamation of ridging and rubbling. It can be both difficult and dangerous to transit, especially if the pack is in motion at the time. Even the most powerful ice breakers have become trapped, and less capable vessels have suffered damage or been sunk by pressure events in shear zones. Shear zones should be transited, where necessary, with extreme caution. The general ice pack is typically a mix of ice types, thickness and floe sizes at various total ice concentrations. Patches or stretches of open water can be found even in the winter polar pack as floes move relative to each other. In some areas, more or less permanent polynyas of open water exist due to water upwelling. When ice floes move against each other, they may raft, increasing local thickness, form rubble, or generate ridging. All of these increase the difficulty of ice transit. Ridges may have sail and keel heights totalling in the tens of metres which can only be penetrated by repeated ramming. Ice charts consolidate all available information on ice cover using the “ice egg”, which in most sea areas will be developed using WMO principles and terminology. An example of how the ice egg is developed is shown in Fig. 49.1, The egg code. Each of the fields is filled out using codes, of which the ‘stage of development’ code given in Table 49.1, Sea ice stages of development is one example. More complete explanations can be found in sources such as the Canadian government’s MANICE, available online at: http://ice-glaces.ec.gc.ca/content_contenu/ice_codes/manice

Air Temperature

Traditionally, many operations in polar regions have been conducted in the summer and autumn seasons when air temperatures are not normally as low as may be encountered in mid-winter. In sub-polar areas such as the Baltic and Gulf of St. Lawrence, all-year operations have become the norm. For all of these services, air temperatures below −20 °C (−4 °F) are relatively rare; though they can be experienced. Midwinter operations are becoming increasingly common on some Arctic routes, and here, temperatures can be considerably lower, with −30 °C (−22 °F) to −35 °C (−31 °F) not uncommon. Year-round temperature data can be obtained for coastal sites throughout the Arctic and for most sub-polar regions. Temperatures at sea (in the ice) are on average slightly above those on the adjacent land, therefore, port temperature data will normally represent the worst case for equipment functionality and crew comfort. Isotherms for the surface air temperature for latitudes north of 60°

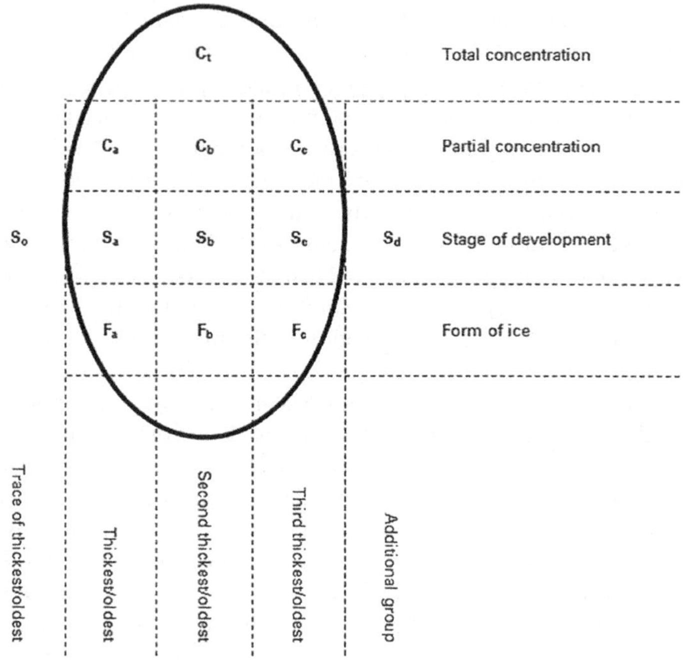

Fig. 49.1 The egg code

are presently being developed as this Guide goes to press. A Corrigendum will be issued when this information is available. This data is provided for guidance only in determining the design service temperature for the vessel.

Humidity

As noted earlier, at extremely cold temperatures, the relative humidity in the air drops to well below that found typically elsewhere. Unless moderated by the vessel's HVAC system, dryness may cause crew discomfort. It may also increase the probability of static electricity build up and discharge, which can damage sensitive equipment and increase fire and explosion risk.

White Outs, Blizzards and Fog

Fog is commonly observed in areas of sea ice. Advection fog is a common occurrence in summer months when warm air masses pass over cold-water areas and sea ice, and the air mass cools below the vapour dew point. Evaporation fog or steam fog is formed when water vapour is added to air which is much colder than the vapour's source. This most commonly occurs when very cold air drifts across relatively warm water. Blowing snow has a significant impact on visibility. Depending on grain size and compaction, visibility in snow at wind speeds over 35 knots become extremely limited, to a few metres in most cases.

Hours of Darkness

Although not strictly a weather condition, the extended hours of darkness at the higher latitudes during the winter months should not be ignored. Extra lighting capacity for deck operations and additional searchlight capability for safe navigation results from this limiting visibility. Additional deck watches for areas where sea ice and glacial ice may be encountered can cause crew fatigue and should be considered when operating in any ice-covered waters. An opposite problem is the effect of almost continuous daylight during the summer months. Crew fatigue needs to be considered when operating for extended periods in ice conditions. Additional blinds for crewing quarters will provide proper resting periods. Ice blindness may be a risk for lookouts and bridge crew, who should be provided with appropriate eye wear. Navigation lights should be of sufficient brightness to be visible in Arctic lighting conditions.

Chapter 50
Crew Safety in Low Temperature Marine Environments

General

This chapter relates to vessels operating in low temperature environments. This chapter contains recommendations only and specific advice should be sought wherever appropriate. The vessel's crew will need to make regular checks of safety–critical systems and equipment so that they are functioning or capable of functioning as intended. The vessel's owner/operator is responsible for developing relevant and appropriate operational guidelines and for keeping these procedures up to date. This chapter is organised into sections: deck (hull), machinery and safety systems. Within each section there is a list of operational issues that should be considered when developing the vessel's operations manual.

Deck Recommendations

Ice/snow removal. The deck and associated piping should be cleaned up periodically, particularly due to spray and bad weather. Ice can quickly build up on the vessel, affecting safety of the crew's movement and vessel stability. *Crew access. Access routes.* Access ways should be checked each day when snow or under-cooled rain occurs. De-ice chemical mix (glycol mix) should be applied to deck plating when freezing occurs. Deck drains should be opened using steam, pressure air and/or de-ice chemical mix. Snow and ice should be removed by use of shovels, hardwood clubs, steam or equal. Heated walkways are activated at 10 °C (50 °F) with increasing heat as the temperature drops. Where necessary, additional safety lines should be run and crew members instructed on tethering themselves. *Escape routes.* Escape routes should be clearly marked and checked at least once a day or more often when required. Escape routes should be cleared of debris, snow and ice. *Emergency exits.* Gaskets, hinges, dowels and locks should be checked and treated with approved de-ice

A. A. Olsen, *Applying Physical Ergonomics to Modern Ship Design*,
https://doi.org/10.1007/978-3-031-57974-5_50

chemicals before the temperature drops below 0 °C (32 °F). Exits should be checked once a day or more often when required. Snow and ice should be continuously removed using shovels, hardwood clubs, steam and/or compressed air. Pre-treatment with de-ice chemicals should be applied at least each month or when required. The Derrick man's escape (Geronimo Chutes) should be checked and treated with de-ice chemicals before the temperature drops below 0 °C (32 °F). The chute should be function tested every day when the temperature is zero or below. *External emergency lights.* External emergency lights should be operative at all times and kept free from snow and ice. The lights should be checked daily when the temperature is below 0 °C (32 °F).

Vessel Systems and Machinery Recommendations

Firefighting equipment readiness. Door gaskets should be treated with de-ice treatments at least each month or when required. All snow and ice accumulation on equipment should be removed by using steam, compressed air or equal. Fire water hoses that have been used should be drained and dried immediately after use or stored at a frost-free location. Fire mains should be drained until needed when the temperature is 0 °C (32 °F) or below. When temperature drops below 0 °C (32 °F), all external fire equipment should be checked daily or more often when required. Portable fire extinguishers at anchor stations should be kept in operator's cabins when the temperature is −15 °C (5 °F) or below. All the fire dampers directly exposed to the weather are to be checked and function tested every day when the temperature is 0 °C (32 °F) or below. Exposed deck machinery should be checked frequently to verify correct operation. *Tanks.* Any tank, its associated piping system and venting condition should be checked prior to loading or discharging operations. *Exterior alarms.* All external alarm equipment should be inspected daily when temperature is below 0 °C (32 °F). *Helicopter deck.* Foam monitors should be properly drained. De-ice chemical mix should be applied to the helicopter deck. Deck border lights and floodlights should be treated with de-ice cure. Portable fire extinguishers should be kept in a heated area close to the helicopter deck. Drains should be checked daily to verify heat tracing is working. When operating in snow and freezing rain, in addition to the items above, the following procedures should be performed: snow should be removed from helicopter deck and access/escape ways, when required, by using shovels, steam and/or pressure air; de-ice chemical mix should be applied to deck when required; and ice should be removed from deck border lights and floodlights. Steam and compressed air can be used. De-ice cure should be applied at least every month or when required. *Procedures before landing and take-off.* The helicopter deck should be checked before landing. Snow and ice should be removed. Special attention should be paid to landing area of helicopter deck and to walkways/escape routes. Ice should be removed from deck border lights and floodlights. Personnel should be kept in protected spaces until departure to avoid cold exposure.

Safety Systems

Evacuation systems. Launching/lifeboat stations. Steam hoses should be rigged and ready for use at stations. All hinges, lashes and gaskets on life-vest and survival suit boxes should be checked and treated with de-ice treatments. Means to de-ice should be applied to deck plating where required to prevent icing. Lifeboat stations should be inspected two times per day or more often when required. Ice and snow should be removed from deck, grating, handrails, rope escape ladder, life vest and survival suit boxes using shovels, hardwood clubs, steam or compressed air. Glycol/water mix or salt should be applied on deck plating as required. Lashings on life jacket and survival suit boxes should be checked. Ice can be removed by steam and compressed air. De-ice treatments should be applied at least each month or when required. *Lifeboats.* Lifeboat heaters in all lifeboats should be switched on. Hooks, latches and hinges should be checked and treated with de-ice treatments. Radio equipment with batteries should be thoroughly checked every second day. Lifeboats should be checked two times a day or more often if required. Engine heaters, lifeboat heating fans and battery chargers should be checked each day. Snow and ice should be removed. Hooks, lashes and hinges should be checked. Ice should be removed using steam and compressed air. De-ice treatments should be applied at least each month or when required. Lifeboat engine should be run each day until it reaches operation temperature. Lifeboat engine fuel should be checked daily for clouding or waxing. Lifeboat engine lubricating and hydraulic (starting) oils should have correct viscosity at design service temperature without the use of heaters. *Launching arrangements.* Hooks, latches and hydrostatic release couplings should be checked and treated with de-ice treatments. Brake guide wire and sheaves should be checked and properly greased, if necessary. Lifeboat launching arrangements should be inspected two times a day or more often when required. Snow and ice should be removed from winches. Ice on hooks, latches and hydrostatics release couplings should be removed using steam and compressed air. De-ice treatments should be applied at least every month or when required. Winches should be run each day. Free-fall lifeboats cannot be released onto ice-covered waters without risking the safety of occupants due to impacting the ice and/or surfacing under ice floes. Standard lifeboats have almost no ice-transit capability, and so they should be launched into the vessel's track to maximise their ability to get clear of a sinking or burning vessel.

Life rafts. Latches/hydrostatic release couplings should be checked and treated with de-ice cure. Launching arrangements should be checked two times per day or more often if required. Snow and ice should be removed. Ice on lashes/hydrostatic release couplings should be removed using steam and compressed air. De-ice cure should be applied at least every month or when required. *Life raft launching arrangements.* Winch should be checked and greased according to lubrication chart if required. Brake guide wire and sheaves should be checked and properly greased if required. Life raft launching arrangements should be checked two times per day or more often if required. Snow and ice should be removed using steam and compressed air. Ice should be removed from release hooks using steam and pressure air and should be treated with de-ice cure at least each month or when required. *Rescue boats.* Radio equipment with batteries should be thoroughly checked every second day. Rescue boat should be inspected two times a day or more often if required. Engine heaters and battery charger should be checked each day. Ice and snow should be removed using shovel, steam gun or pressure air. De-ice means such as glycol/water mix should be used if necessary. Engine should be run each day until operating temperature is reached. Rescue boat engine fuel should be checked daily for clouding or waxing. Rescue boat engine lubricating oil should have correct viscosity at design service temperature without the use of heaters. *Rescue boat launching arrangement.* Brake guide wire and sheaves should be checked and properly greased if necessary. Quick release hook should be checked and treated with de-ice cure. Rescue boat launching arrangement should be test run each week or more often if required. Brake guide wire should be thoroughly checked two times a day or more often if required. Snow and ice should be removed from winches. Ice on quick release hook should be removed using steam and compressed air. De-ice cure should be applied at least every month or when required. Launching arrangement should be checked three times a day or more often when required. *Escape systems. Escape chutes and life raft stations.* Steam hoses should be rigged and ready for use on both stations. De-ice chemical mix should be applied to deck plating to prevent icing. Life raft station should be checked two times per day or more often if required. Ice and snow should be removed from deck handrails, raft and launching arrangements using shovel, steam hardwood clubs or equal. De-ice chemical mix should be applied to deck if required. *Personal safety equipment. Immersion suits and life jackets.* If immersion suits are stored at stations, they should be checked for moisture and dried if necessary. Personal immersion suits should be stored in cabins to avoid moisture accumulating in the suits if stored at lifeboat stations. *Life Jackets.* Life jackets should be free from moisture when returned back into boxes after use.

Chapter 51
Crew Welfare in Low Temperature Marine Environments

General

Working in cold weather environments has significant implications on human capabilities and unless proper precautions are made can be hazardous to a person's health. In recognition of these implications on human health and performance due to working in cold climes, this chapter provides some basic guidance and information in relation to.

- Human performance and health hazards.
- Guidance for design or selection of clothing.
- Design of equipment to be operated in cold conditions.
- Information that can be used to help generate cold weather operations' safety and operating procedures.
- Information that can be used to preserve the health of persons working in cold environments.

The guidance that follows is simply guidance and carries no weight as Rules or requirements leading to Notations or Certifications. The information is provided for those vessel owners, designers, or operators that would like information of the sort provided as a reference to consider in the course of ship design and operation. The material in this chapter is applicable to extreme cold weather working conditions. Some of the hazards discussed here are a concern under exceptional circumstances which force personnel to stay in cold temperature for extended periods.

Terminology

Atrium: A body cavity or chamber, especially either of the upper chambers of the heart that receives blood from the veins and forces it into a ventricle.

A. A. Olsen, *Applying Physical Ergonomics to Modern Ship Design*,
https://doi.org/10.1007/978-3-031-57974-5_51

Chilblains: An inflammation followed by itchy irritation on the hands, feet, or ears, resulting from exposure to moist cold.

Clo-units: The rate of a person's heat loss as measured in watts per square meter of skin area per kelvin of temperature difference across the clothing; the value of insulation is measured by the reciprocal of this rate, in square meter kelvins per watt (m^2 K/W).

Ergonomics: The applied science of equipment design, as for the workplace, intended to maximise productivity by reducing operator fatigue and discomfort.

Fibrillation: Rapid uncoordinated twitching movements that replace the normal rhythmic contraction of the heart and may cause a lack of circulation and pulse.

Hypothermia: Abnormally low body temperature.

Melatonin: A hormone that plays a role in sleep.

Physiologic: Being in accord with or characteristic of the normal functioning of a living organism Raynaud's syndrome: A bluish discolouration of the extremities; can occur when a spasm of the blood vessels is caused by exposure to cold or by strong emotion.

Triglyceride: A naturally occurring ester of three fatty acids and glycerol that is the chief constituent of fats and oils.

Ventricle: The chamber on the right side of the heart that receives venous blood from the right atrium and forces it into the pulmonary artery; any of the interconnecting cavities of the brain.

Wind Chill: A still-air temperature that would have the same cooling effect on exposed human flesh as a given combination of temperature and wind speed; called also chill factor, wind chill factor, wind chill index.

Human Response to Cold and Arctic Exposure

The core (trunk) of the human body should remain within a small temperature range for healthy function. Excessive cooling or excessive heating will result in abnormal cardiovascular and neurological function. The skin is the organ through which a person regulates body temperature. With an average skin temperature of 33°C (91.4°F), conductive heat loss occurs at temperatures below this value. Therefore, it is easy to see how cold weather performance can significantly influence normal body function. As a person cools.

- Metabolism is increased to generate more body heat—as cooling continues a person will begin to "shiver"—a visible sign that body cooling has progressed beyond a comfortable level. Increased metabolism will reduce the amount of time a person can sustain work.

- Safe manual materials handling tasks require the use of tactile senses, hand dexterity, strength, and coordination. Decreases in the ability to produce force, exhibit fine control over objects, and sustain muscular workloads occur in cold working environment.
- Work in cold environments is related to an increased risk for musculoskeletal injury.
- Motor function impairments of the arms and hands will occur long before cognitive or hypothermic related disabilities occur. Impaired cognitive performance will lead to poor decision-making and increased risk for accident.
- Persons suffering from arthritis or rheumatism will generally experience increased levels of pain during cold weather operations.

Decreases in cognitive/reasoning ability due to cold exposure. Extremely cold conditions [i.e., below 10°C (50°F)] adversely influence mental skills and coordination. As operational temperatures decrease, the frequency of cognitive error will increase. Tasks requiring vigilance may be hampered after prolonged exposure to cold. Decision verification procedures should be implemented. Cold weather operations, coupled with other physical distracters, such as noise or motion environments, will influence the quality of perception, memory and reasoning and compound the risk of decision-making error. *Health hazards related to cold exposure.* The list of potential injuries and issues for occupational work in cold environments is lengthy. Personnel should have adequate training to enhance preparation for work in cold environments. Proper planning and precaution can deter the potential risks of cold work. *Non-freezing tissue injury.* With non-freezing tissue injuries, parts of the skin are chilled but not frozen.

- *Frostnip* is a cold injury in which the chilled areas of skin become numb, swollen and red. The only treatment needed is warming the area for a few minutes. During warming, the area may hurt or itch intensely. No permanent damage results, although sometimes the area is particularly sensitive to cold for months or years afterward.
- *Immersion foot (trench foot)* is a cold injury that develops when a foot is kept in wet, cold socks and boots for several days. The foot is pale, clammy and cold. After warming, the foot becomes red and painful to the touch. Sometimes blisters develop, which may become infected. Rarely, this type of injury occurs in the hands. Treatment consists primarily of gently warming, drying and cleaning the foot; elevating it; and keeping it dry and warm. Medical doctors (aboard or ashore) should be consulted regarding need for treatment with antibiotics to prevent infection or whether a tetanus booster is recommended if the person's tetanus vaccination is not current. Trench foot can often be prevented by changing socks and drying the feet at least daily.
- *Chilblains (pernio)* is an uncommon reaction that may occur with repeated exposure to cold. Symptoms include itching, pain and, in rare cases, discoloured areas or blisters on the affected area (usually the leg). The condition is uncomfortable and recurrent. Preventing exposure to cold is the best treatment. Treatment

includes re-warming, using blankets, lukewarm baths or heating pads positioned at low heat setting, which will allow blood flow to return to the peripheral tissues.

Freezing tissue injury. Damage to tissue during cold exposure is commonly caused by freezing of the tissue and surrounding area. Tissue freezing is more commonly referred to as frostbite and is a common occurrence in colder climates. Frostbite occurs most often at ambient temperature below −20°C (−2°F). Frostbite occurs as a result of limbs and facial areas being exposed to low ambient temperatures and wind. Generally, frostbite is accompanied with discolouration of the skin, along with burning and/or tingling sensations, partial or complete numbness, and possibly intense pain. Hands, feet, noses and ears are more commonly affected by frostbite. If the nerves and blood vessels have been severely damaged, gangrene may follow, and amputation may eventually be required. If left untreated, frostbitten skin gradually darkens after a few hours. Skin destroyed by frostbite is completely black and looks loose and flayed, as if burnt. The crucial factor in the development of frostbite is not necessarily the ambient temperature and wind, but whether the tissue reaches a temperature of −4.8–−7.8°C (23.4–18°F). Cellular structural damage occurs when cells freeze. There are four levels of frostbite:

- *First-degree* frostbite results in vesicles (blisters) that damage only the outer skin.
- *Second-degree* frostbite is associated with full thickness skin loss.
- *Third-degree* frostbite involves the skin and underlying tissue.
- *Fourth-degree* frostbite is the worst and results in freezing to the bone. This level of frostbite typically results in amputation of the affected tissue.

Diagnosis of whether superficial or deep tissue was damaged involves analysis of blisters; superficial injury will have clear blisters, while deeper tissue injury will have blisters filled with blood as a result of damage to vascular tissues. Frostbite requires emergency medical care. Procedures should be in place to provide first aid for casualties of frostbite. Means to get medical advice (from physicians either aboard or ashore) should be in place as part of these procedures. A person who has frostbite should be covered with a warm blanket and given a hot beverage because people with frostbite may also have hypothermia. Rubbing the area (particularly with snow) leads to further tissue damage. Because the area has no sensation, it should not be warmed in front of a fire or with a heating pad or electric blanket. The frostbitten area becomes extremely painful on warming and medical attention likely will be urgently needed for pain management. Blisters should not be broken. If blisters break, they should be covered with antibiotic ointment. It is common for individuals who experience severe frostbite to later experience such problems as: hypersensitivity to cold, cold feet, burning/prickling/itching or tingling sensations, chronic pain, loss of sensation of touch, white fingers, excessive sweating, pain when walking, and other such problems causing a hindrance to work performance. Causes are generally thought to be related to permanent nerve and tissue damage, and in severe frostbite there can even be degeneration of bone and onset of arthritis (Figs. 51.1 and 51.2).

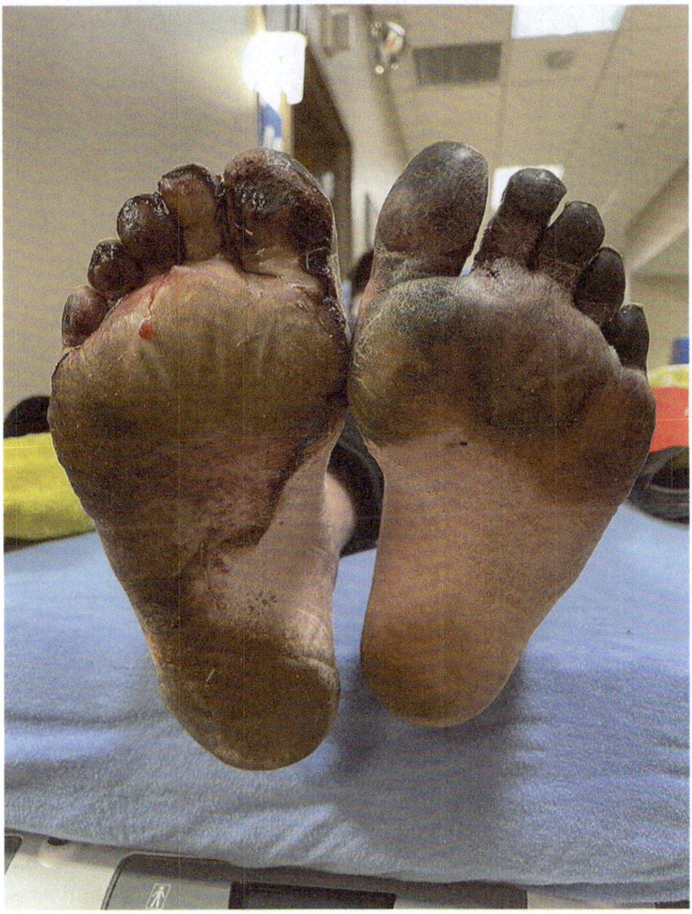

Fig. 51.1 Example of frostbite (feet)

Raynaud's syndrome (white finger). Raynaud's syndrome represents a disease and relates to the body's hypersensitivity to cold, causing the arterial musculature to spasm and constrict peripheral blood flow at a much higher level than typical for a given ambient temperature. Less severe cases of Raynaud's syndrome, referred to as primary Raynaud's syndrome, are idiopathic (a medical condition of unknown origin), while secondary cases occur as a result of an underlying physiologic condition or environmental exposure. Secondary Raynaud's syndrome often occurs as a result of thickening in underlying tissue and vascular beds through mechanisms such as previous severe frostbite or exposure to vibrating tools. The term most often associated with vibration-induced impairment in hand-arm circulation is termed vibration-induced white-finger. White-finger describes the loss of colour, increased numbness, and decreased tactile sense associated with circulation impairment for individuals exposed to vibration and/or cold. At the end of a white-finger attack, blood flow

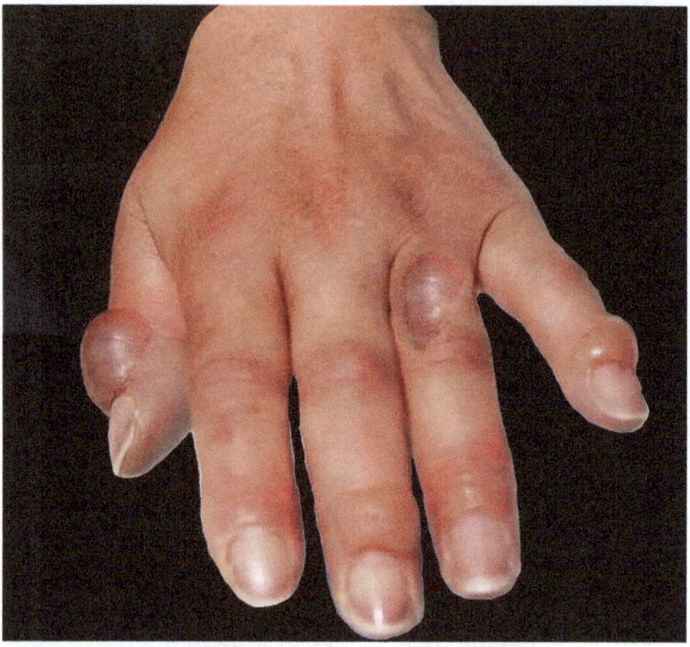

Fig. 51.2 Example of frostbite (hand)

returns to the area causing redness, swelling and pain. In severe cases, the symptoms can cause debilitation of performance. The condition may take a few months or several years to develop, depending on the individual physiologic tolerance and the level of cold. With the decreased ability for blood to be delivered to the peripheral areas, there are often symptoms of burning/throbbing pain, numbness, stiffness, diminished sensations and swelling from fluid pooling. These symptoms, even at mild levels, are likely to make manual performance uncomfortable and unreliable. There is no known cure for Raynaud's syndrome. However, there are known intervention techniques to alleviate symptoms. Exercise and using warming devices to increase blood flow to affected areas has been successful (Fig. 51.3).

Hypothermia. Hypothermia is a rapid, progressive mental and physical collapse due to the body's warming mechanisms failing to maintain normal body temperatures. While hypothermia is often associated with immersion in cold water, it can also occur in air when suitable cold weather protection is not employed. Conditions of extremely low dry ambient temperature or mildly cold ambient temperatures with wind and dampness can lead to a general cooling effect on the body. If metabolic heat production is less than the gradient of heat loss to the environment, hypothermia becomes an issue. As a result of loss of core temperature, damage to more central organs and systems presents a problem. As the temperature of the core body decreases, so does the heart and associated tissue, which ultimately results in a reduction in heart

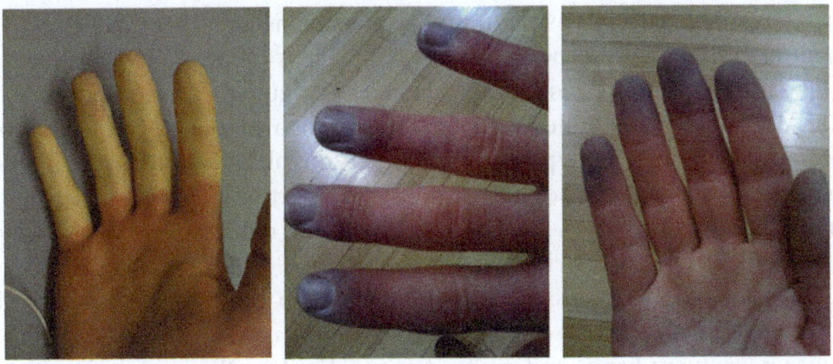

Fig. 51.3 Examples of Reynaud's Syndrome

rate and therefore blood output to the body's tissues and organs. A natural protective mechanism occurs where the brain progressively shuts down cerebral (brain) areas. As core temperature drops, more areas of the brain shut down, leading to reports of amnesia during hypothermic situations, and eventual loss of consciousness in severe hypothermia. At core temperatures less than 34°C (93°F), cognitive function begins to become impaired and at temperatures below 28°C (82.4°F), there is a risk of ventricular fibrillation ("twitching" of the heart and loss of pulse) and cardiac arrest. People with diabetes, injuries, kidney problems, epilepsy and arthritis are at a higher risk of hypothermia in comparison to healthy people. Classification of hypothermia ranges from mild to severe. Mild hypothermia involves core body temperatures between 32 and 35°C (89.6 and 95°F), moderate involves temperatures between 28 and 32°C (82.4 and 89.6°F), and severe hypothermia involves core temperatures below 28°C (82°F). Table 51.1, Symptoms of hypothermia, presents the symptoms of hypothermia.

Table 51.1 Symptoms of hypothermia

Mild hypothermia	Moderate hypothermia	Severe hypothermia
Increased heart rate, resulting from venous return increase Shivering Excessive discharge of urine—resulting from central pooling of blood flow and resulting in rapid delivery to the kidneys Increased muscular tone Decreased nerve conduction	Impaired respiration Decreased heart rate, blood pressure Blue-grey lips, nail beds or skin colour Muscle spasms Loss of feeling in or use of arms and legs Loss of muscle function Slurred speech Blurred vision Impaired cognition Confusion Shivering stops (resulting of used energy in prior stages)	Limited or no cognitive ability Twitching of the heart muscles (atrial and ventricular fibrillation) Possible cardiac arrest Unconsciousness Death

Procedures devised in consultation with physicians should be in place to guide treatment of a hypothermic person. Means to access medical advice should be in place (aboard or ashore access to physicians). The guidance contained in MSC.1/ Circular 1185, "Guide for Cold Water Survival" is also applicable to the treatment of hypothermia. Treating the hypothermic casualty involves quick action. Immediate treatment is much more important than any later action in reviving casualties of immersion hypothermia. Get the casualty out of the cold, wind, rain or water. Be gentle with the casualty, restricting their movements. Do not allow the casualty to walk unless absolutely necessary. Take precautions to warm the casualty up slowly. ***Note:*** *rapidly applying heat to a hypothermia casualty can send the person into shock and cause permanent brain and organ damage.* Replace the casualty's wet clothes with dry clothes. If the casualty is conscious and alert, you can allow him or her to drink warm liquids that do not contain alcohol or caffeine. Rewarm the casualty by covering them with a dry blanket or any other material that will help heat retention. If possible, arrange for a hot bath (the temperature no higher than the immersed hand will tolerate) as this is the most effective method of achieving rewarming. As the only non-invasive hospital treatment suitable for active core rewarming at sea, inhalation rewarming donates heat directly to the head, neck, and thoracic core (the critical core) through inhalation of warm, water-saturated air at 43–45°C (107–122°F). This method also warms the hypothalamus, the temperature regulation centre, the respiratory centre, and the cardiac centre at the base of the brainstem. In more severe cases of hypothermia, non-clinical attempts to rewarm the body should revolve around passive re-warming—covering the body to prevent any additional heat loss and letting the body's metabolism gradually raise internal temperature. Rapid rewarming can cause immediate changes in peripheral blood distribution, cardio-pulmonary function, perfusion ratio, and nervous control of circulation; with the impaired cardiac output associated with moderate and severe hypothermia, sudden changes in peripheral blood flow can force the individual into cardiovascular shock. If the casualty is semiconscious, keep the person awake; if the casualty is unconscious and shows no signs of life, then begin CPR immediately. Severely hypothermic individuals have a greater chance of survival if greater travel time is taken to reach a facility with appropriate treatment equipment, rather than settling for the closest facility and relying on more primitive re-warming methods.

Cardiovascular function and cold exposure. Sudden changes in cardiovascular function occur every day when individuals are exposed to cold weather after previous exposure to a thermoneutral climate. These changes cause a sudden increase in the workload placed on the heart and may cause subsequent cardiovascular problems. Systolic blood pressure can increase by about 30 mm Hg, and diastolic blood pressure can increase about 20–22 mm Hg in older subjects and increases by 23 mm Hg in systolic and 12 mm Hg in younger subjects. In both populations there is a significant increase in blood pressure, and for individuals who are hypertensive, the strain on the heart may result in a cardiovascular incident. Blood pressure can be easily monitored and should be recommended for persons with existing coronary and respiratory disorders. Cardiovascular problems associated with cold exposure rarely occur upon the initial exposure; usually a resultant case of heart attack or stroke will occur 24–54 hours after severe cold exposure. *Respiratory function and cold exposure.* Cold air has an influence on airway resistance and pulmonary function. This response can cause significant problems for individuals with asthma. There is evidence of long-term occupational exposure to cold causing chronic obstructive pulmonary disorders. The increased constriction of the bronchial air passages causes an increase in airway resistance and thus greater effort in breathing. These effects are not seen in thermoneutral air, and even when present, they are not likely to cause performance limitations, but the increased constriction may further complicate symptoms of asthmatic individuals.

Physiologic changes in Arctic environments unrelated to cold. Seasonal changes in circadian rhythms are common and result in changes in melatonin release. Melatonin (a hormone that plays a role in sleep) is released in response to decreased light exposure, and release is depressed with greater light. A change in melatonin release can cause disruption in the physiologic sleep-wake cycle, and this effect can reach extremes in Arctic and Antarctic environments where there are prolonged periods of light or darkness. There is also evidence of work schedules and daily routine having an influence on sleep-wake cycles; in Arctic environments where light conditions persist several months followed by months of darkness, the melatonin-mediated control of circadian rhythm can clash with the social and work-mediated sleep-wake cycle. Interruptions in sleep patterns are common in Arctic regions for this reason, often leading to increased fatigue and sleep deprivation. To counteract the physiologic reaction to decreased light, a strategy of matching lighting conditions to work patterns, using blinds and artificial lighting, has shown success. Releasing melatonin by natural or artificial conditions to match work schedules will allow for matching circadian rhythms to social demands and an overall improved sleep pattern. Shift work on ships is suspected to cause both physiological and psychological problems with crew, and these problems are expected to be amplified in Arctic conditions. Suntan beds have been employed in vessels operating at near-polar locations to counteract light deprivation periods of the year. Sleeping areas should be capable of being well "blacked out" to reduce problems in falling asleep due to excess light in the sleeping quarters. In snow-covered environments, the snow can act as a reflective surface for sunlight without absorbing the light in the same manner as pure-water reflective surfaces. Exposure to direct sunlight for more than 10 minutes requires use of protective eye wear. Using glass lenses for protective eye equipment to filter UV and IR light from entering the retina and using brimmed hats to reduce sun exposure is recommended.

Monitoring environmental conditions. Working in cold environments requires an understanding of the interaction between ambient temperature, wind speed, relative humidity, personnel protective equipment and task being performed. In order to limit the risk during operational activities due to cold stress and further prevent local cold injuries and general freezing, specific preventative measures should be evaluated and introduced during the planning and execution of the daily work activities. A plan for monitoring exposure to cold should be devised and should take account of

Table 51.2 Relationship between wind chill and exposure danger

		Ambient temperature (°C)								
		4	-1	-7	-12	-18	-23	-29	-34	-40
Wind km/h	Velocity mph	Equivalent wind chill temperature (°C)								
Calm										
0	0	4	-1	-7	-12	-18	-23	-29	-34	-40
8	5	3	-3	-9	-14	-21	-26	-32	-38	-44
16	10	-2	-9	-16	-23	-30	-35	-43	-50	-57
24	15	-6	-13	-20	-28	-36	-43	-50	-58	-65
32	20	-8	-16	-23	-32	-39	-47	-55	-63	-71
40	25	-9	-18	-26	-34	-42	-51	-59	-67	-76
48	30	-16	-19	-22	-36	-44	-53	-62	-70	-78
56	35	-11	-20	-29	-37	-46	-55	-63	-72	-81
64	40	-12	-21	-29	-38	-47	-56	-65	-73	-82
		Little danger in less than one hour exposure of dry skin			DANGER – exposed flesh freezes within one minute			EXTREME DANGER – flesh may freeze within 30 seconds		

Table 51.3 Maximum work times for different levels of exposure to cold[1]

Equivalent temperature	Consequence/action
Below −30°C	No outdoor work performed unless deemed critical from a safety or operational perspective
Below −21°C	Available outdoor working time is below 50% of working hour
Below −12°C	Available outdoor working time is below 75% of working hour
Below −6°C	Available outdoor working time is below 90% of working hour
Above −6°C	Normally 100 % Available working time

variations in thermal conditions. Ship managers should, if practicable, eliminate the need for extended work in cold conditions (for example, by rescheduling work to be performed in a warmer season, or by moving the work from outdoors to indoors, or separating the cold parts of a process from the workers, as far as practicable). If elimination of such work is impracticable, other measures to reduce risk from cold conditions should be devised. Employers should ensure that workers are not positioned near very cold surfaces or, if this cannot be avoided, that the workers are protected by radiation shields. For standing tasks, the floor should, where practicable, have an insulating surface. Climatic metrics such as temperature, wind speed and humidity should be regularly monitored in the locations where outside work is to be performed. Of primary importance is a regular reporting of the wind chill or equivalent temperature. Regular communications should be maintained regarding allowable time to work outside. Indoor personnel should regularly monitor outside workers so that best work-to-rest/ warming schedules are maintained (Tables 51.2 and 51.3).

Clothing and personal protective equipment. For appropriate protection/isolation against cold climate conditions, adequate clothing should be selected and used on

[1] Canadian Centre for Occupational Health and Safety, 1999.

board during cold periods. Such optimal clothing should be able to mitigate water and humidity during work and at the same time insulate sufficiently to maintain thermal comfort during rest. The insulating effect of the clothing is influenced by different factors including temperature, wind and humidity. The measure of clothing insulation is determined by clo-units. The rate of a person's heat loss is measured in watts per square meter of skin area per kelvin of temperature difference across the clothing; the value of insulation is measured by the reciprocal of this rate, in square metre kelvins per watt (m^2-K/W). One clo is equal to 0.155 m^2-K/W attire, which allows an individual to remain comfortable at an ambient temperature of 21°C (70°F). Garments are to be labelled with exposure protection information (for example, hours of protections at a particular temperature) (Tables 51.4 and 51.5).

General recommendations for clothing. In order to decide on the appropriate dress, it is important that the clothing be as comfortable as possible. It needs to provide optimal transportation of humidity from the body to the environment, good insulation and sufficient dexterity. Each garment in the clothing ensemble is important and all pieces should be considered part of the protective system. For temperatures between:

−8°C (17.6°F) and −25°C (−13°F) an average of 3.67 clo is recommended
−20°C (−4°F) to −48.2°C (−44°F) an average of 4.26 clo is recommended

With the addition of wind and humidity to low ambient temperature, greater clo units are required. Wind has an effect of compressing clothing, decreasing the overall amount of trapped air that creates an insulating layer. General body protection from cold should be provided. For optimal protection, the following three clothing protection layers are recommended for outdoor work activities on board (for an effective insulation of 2.6 clo).

1. *Inner layer* of two-layer underwear effectively absorbing and transporting perspiration: "super" underwear or similar products against the skin, and/or wool as second layer.
2. *Middle layer* which provides an isolating effect: fibre/fleece (pants and sweater/jacket), wool sweater, woollen socks.
3. *Outer layer* which protects against the environment and should be waterproof.

Sufficient amount of appropriate skin protective cold cream should be available at all times on board, located at the assigned heating huts, common rooms and sickbay. Someone should be designated to be responsible for ensuring that cold creams are available as required on board. Dispensers should be used instead of jars containing cold cream. Clothing should have adjustable neck and sleeves to influence air exchange when internal temperature becomes unfavourable and to prevent sweating/moisture build-up under the clothing. Cotton or polypropylene long underwear is recommended for all-over warmth. Multiple layers of light, loose-fitting clothes are recommended. Use suspenders instead of belts (tight belts can constrict blood circulation). *Hand protection.* Do not wear gloves or scarves that can get caught in moving parts of machinery. While glove use is always recommended for outside work, the use of cold-protective gloves likely impairs the ability to carry out fine

Table 51.4 Threshold limit values for work/warm-up schedule for four-hour shifts/watches[2]

Air temperature sunny sky		No noticeable wind		5 mph wind		10 mph wind		15 mph wind		20 mph wind	
°C	°F	Max work period	Nos breaks	Max work period	Nos breaks	Max work period	Nos breaks	Max work period	Nos breaks	Max work period	Nos breaks
−26° to −28°	−15° to −19°	(Norm breaks) 1		(Norm breaks) 1		75 min	2	55 min	3	40 min	4
−29° to −31°	−20° to −24°	(Norm breaks) 1		75 min.	2	55 min	3	40 min	4	30 min	5
−32° to −34°	−25° to −29°	75 min	2	55 min	3	40 min	4	30 min	5	↓ Non-emergency work should cease ↓	
−35° to −37°	−30° to −34°	55 min	3	40 min	4	30 min	5	↓ Non-emergency work should cease ↓			
−38° to −39°	−35° to −39°	40 min	4	30 min	5	↓ Non-emergency work should cease ↓					
−40° to −42°	−40° to −44°	30 min	5	↓ Non-emergency work should cease ↓							
−43° and below	−45° and below	Non-emergency work should cease									

[2] Adapted from Threshold Limit Values (TLV) and Biological Exposure Indices (BEI) booklet: published by ACGIH, Cincinnati, Ohio, 2000.

Table 51.5 Protective and functional properties for outdoor work garments[3,4]

Parameter	Limit	Basis	Method
Thermal insulation ($T < -20°C$)		BS EN ISO 11079 BS EN ISO 11079 Frostbites	BS DD ENV 342 BS DD ENV 343 BS EN 511
• Short-term cold exposure	≥3 clo		
• Gloves	≥ 2.5 clo		
• Long-term cold exposure	2.5…3.0 clo		
Air permeability		Preventing of cooling Evaporation	BS EN 31092 BS EN 31092
• In wind	< 20 l/m^2-s		
• Rest (100 W/m^2)	20…150 l/m^2-s		
• Heavy work (> 300 Wm^2)			
Ventilation			
• Heavy work (> 300 Wm^2)	≥ 300 l/min		
Water vapour permeability			
• Cold weather clothing	≤ 13 m^2 Pa/W	Evaporation	BS EN 31092
• Cold/foul weather clothing	≤ 20 m^2 Pa/W	Evaporation	BS EN 31092
Resistance to water penetration			
• Light rain	≥ 2200 Pa	Protection against moisture	BS EN 20811
• Heavy rain	≥ 33,000 Pa	Protection against moisture	BS EN 20811

motor skills, while gross motor skills are less affected. Mittens or gloves provide sufficient insulation to prevent frostbite. Mittens will retain heat for a longer period of time compared to wearing gloves. However, mittens can interfere with a person's

[3] Adapted from a collaboration of information by Biem, Koehncke, Classen, & Dosman, 2003; Giesbrecht, 2000; and Golden, & Tipton, 2002.

[4] List of Standards noted in the "Method" column of Table 50.5:

- British Standard EN ISO 11079: Ergonomics of the Thermal Environment – Determination and Interpretation of Cold Stress when Using Required Clothing Insulation (IREQ) and Local Cooling Effects.
- British Standard DD ENV 342: Protective Clothing - Ensembles for Protection against Cold.
- British Standard DD ENV 343: Protective Clothing - Protection against Foul Weather British Standard EN 511: Specification for Protective Gloves against Cold.
- British Standard EN 31092: Textiles - Determination Of Physiological Properties – Measurement of Thermal And Water-Vapour Resistance Under Steady-State Conditions (Sweating Guarded-Hotplate Test).

ability to grasp and manipulate objects. *Head and eye protection.* As much as 40% of body heat can be lost from an uncovered head. Use an appropriate hard hat liner to reduce heat loss when wearing protective headwear. It is often noted that the choice of safety hat takes priority over the choice of its insulating properties. This is not a desirable principle procurement criterion for cold weather operations. Wear a warm hat with ear protection to prevent heat loss from the head. Protect vision from UV rays by wearing appropriate sunglasses while working in snow and ice on a bright day. *Foot protection.* The surface contact area of the sole is important for frictional force on ice. It is suggested that shoes/boots with rounded heels increase the contact with the surface and reduce the risk of slipping. When there is a layer of water along icy surfaces, the contact with the sole of the shoe is reduced. Therefore, the addition of sharp cleats to shoes is suggested. However, wearing shoes/boots with steel cleats on steel decks can be quite dangerous because of the loss of frictional force. Boot selection should involve procuring a boot with proper insulation for the duration of the activity, required protection and the intensity of the work. Woollen, polypropylene, or other thermal socks should be worn to protect ankles and feet. Keep snow and water out of footwear and replace damp socks with a dry pair. *Maintenance of personnel protective equipment.* Dirty and oily clothing loses much of its insulation value and wearing of dirty/oily clothing should be avoided. Do not keep cold weather clothing stored or compressed in a duffel bag for long periods of time. Fluff waterfowl down garments when removing them from the duffel bag following initial boarding of vessel. Keep garments dry; brush off snow and frost before entering warm buildings or vehicles. Always follow the manufacturer's guidelines for cleaning clothing. Failure to do so may reduce the protective capacity of the clothing. Do not leave rips or tears in disrepair. Temporary repairs can be made with electrical or duct tape. *Immersion suit protection.* In work where there is a risk of accidental immersion or abandonment into cold-water, specific protective equipment is recommended.

- Dry-suits and marine anti-exposure suits are commonly worn during marine operations and have been found not to impair working ability. These provide both protection and buoyancy in the water.
- A fully insulated dry suit has a clo value of approximately 0.7, leading to survival time up to 15 hours of immersion at 12°C (53.6°F), compared to 1–6 hours without the suit.
- Protective suits should be checked regularly for wear and accidental rips. In some cases, replacement of the suit is recommended.

There are marine immersion suits available that protect the wearer from wind and cold in a dry environment and have isolative properties and buoyancy in water as well. While on land, these suits also have the ability to increase ventilation by opening collars and sleeves as needed, thereby accounting for individual variation in thermal comfort.

Nutrition considerations in cold climates. The added weight of protective clothing and the limitations in mobility created by protective equipment will increase the mobility demands of the operator, thus increasing the metabolic needs for a given task. A mixed diet of protein, carbohydrates and fats combining for a caloric intake of 4500 kCal for an average man is recommended for demanding labour in the cold. Normal physiologic water loss is unavoidable in cold environments. There is a loss during physiologic activities such as metabolism, from sweat production inside protective equipment, and from increased urination related to cold-induced excessive discharge of urine. It is recommended that 3–6 litres of daily intake of water be maintained. For warming purposes, hot non-alcoholic beverages or soup are suggested. Caffeinated drinks such as coffee should be limited because it increases urine production and contributes to dehydration. Generally, tobacco prevents the periodic blood vessel dilation needed for preventing tissue injury. The risk of frostbite and peripheral injury may increase with tobacco use. Alcohol use should be avoided. The use of alcohol increases the flow of blood to the skin, and can lead to rapid loss of body temperature, increasing the likelihood of hypothermia. Frequent, small meals will allow a constantly high metabolism to help maintain body temperature, whereas large meals served only 2–3 times per day will result in a decreased metabolic rate after digestion and decrease the "heat production to heat loss" ratio. In the event of ship abandonment, a food stores carry-off kit should be provided, containing water and non-perishable foods. The carry-off kits should be stored in areas not subject to freezing temperatures. The carry-off kits should be stored in immediately accessible locations.

Workstation design and operational considerations. The analysis of outdoor work situations should be performed early in design/layout development and should be updated when design changes are made that will influence personnel's exposure to cold stress. Outdoor operations analyses (an examination of the tasks to be carried out in cold conditions) should be carried out for open work areas and semi-open work areas. The objective of these analyses is to identify and remedy task performance issues due to overall exposure to temperature, wind, icing and precipitation, including investigation of the weather protection necessary to comply with exposure limits. Environmental measurements and assessments (air temperature, humidity and wind speed and direction) should be performed on a routine basis and at regular intervals throughout the day during cold periods with temperatures below zero degrees. All personnel working in exposed locations should be monitored at regular intervals. A person is usually unable to recognise their own signs and symptoms of hypothermia. As a precaution, use of the "buddy system" to detect signs of cold injury in colleagues is recommended. Crew should be allowed sufficient time to acclimatise to an extremely hot or cold environment, including major changes in climatic conditions. Monitoring of thermal conditions should be performed and should take account of:

- All stages of work cycles and the range of temperature and humidity under which tasks are performed.
- The range of clothing worn during the tasks.
- Major changes in physical activity level (metabolic heat production).
- Occasional tasks such as cleaning and maintenance of hot equipment and cold areas, and renewal of hot or cold insulation.

A log that is to be maintained by the onboard Safety Officer during operations in cold periods to keep records of the specific areas and work activities where administrative measures are introduced is recommended. Work areas where administrative measures are introduced on a regular basis should be identified by periodical review of such records. Workers in the cold will often need to urinate more frequently, and employers should ensure that suitable arrangements are available, where feasible, and that the design of protective clothing allows easy urination. Suitable protection should be given to the hands and fingers, particularly where dexterity is needed, as well as other exposed parts of the body. Employers should provide:

- Facilities for warming the hands, for example, by warm air, where appropriate.
- Tools with insulated handles, especially in temperatures below the freezing point.
- Measures to ensure that the bare hand does not touch surfaces below −7°C (19.4°F) (workplace design or protective clothing).
- Measures to ensure that bare skin does not touch liquids below 4°C (39.2°F).
- Appropriate measures to be taken in the event of insulating clothing getting wet.
- Face and eye protection, as appropriate, for outdoor work and working in snow (e.g., safety goggles against glare).
- Adequate facilities for changing.
- Arrangements for cleaning such clothing and drying clothing and footwear between shifts.

Slips and falls are much more common in winter seasons:

- Walk slowly on slippery, icy surfaces to prevent danger of slipping. Stop occasionally to break the momentum.
- Avoid carrying heavy loads or loads that obstruct vision during transportation.
- Use approved snow removal compounds.
- Use mats at entrances to prevent slippery conditions indoors caused by melting snow or ice.
- Stairs, ladders and walking surfaces should be provided with non-skid surfaces.
- Hand railings should be provided with nonslip surfaces or should be heated to prevent formation of ice.
- Exterior stairs should be made of grated material to aid in snow and ice removal.

Post visible signage emphasising the risk for cold-related injuries at all outdoor workstation locations. Knowledge of the work area at risk for slippery conditions, augmented with proper hazard signage, is important. If individuals are aware of the cold weather conditions, they will rely on additional muscle-based reflexes (i.e., reaction time), postural control, and muscular strength to adjust to the environment and prevent slipping and falling. Consider providing enclosed bridge wings. Consider providing a temperature transition area (such as a heat vestibule) at exit/entry points to soften the often-harsh transition from bitter cold to warmth, and vice versa). Assess possibility for temporary local shielding around working area(s) if permanent shielding is not possible. Clearly post locations for nearest re-warming or break areas. Power tools and other equipment generally require specific maintenance schedules for usage in cold weather. Cold tool handles reduce grip force. Therefore, tools with larger handles may be required for cold weather usage to accommodate the protective hand wear and reduced grip capacity. Cold exposure aggravates the effects of mechanical vibration, making manual work more difficult and painful. Effort should be made to minimise exposure to vibration at its source. Hospital/medical kits should be stocked with materials appropriate to treat cold weather illnesses and maladies. Cold weather illnesses training, and procedures should be provided with emphasis on both symptom identification and treatment of the cold-related illness. *Accommodations and environmental control.* All personnel accommodations should be designed and arranged to protect the occupants from unfavourable environmental conditions and minimise risk of injury during normal (including ice transiting or ice breaking) operations and emergency conditions. Table 50.6, Exposure standards and action-limits for vibration, and 49.7, Vibration effects on function and performance, provide guidance on exposure standards for vibration and vibration effects on various personnel functions. All personnel accommodations, public spaces and the equipment installed in them should be designed so that each person making proper use of them will not suffer injury during normal open water operations and emergency manoeuvring conditions. All facilities should include non-slip decking, three rigid sides, handholds and insulation from exposed hot water pipes. Personal quarters should have sufficient room to dry and store cold weather protective equipment. Cabin heating systems should be thermostat controlled and have shielded elements, if appropriate.

- Protective clothing should not be dried directly upon heating system.
- Protective skin cream dispenser should be supplied in each room.

Consideration should be given for providing a sauna for crew recreational use and for re-warming cold persons (Tables 51.6 and 51.7).

Table 51.6 Exposure standards and action-limits for vibration[5]

	Hand-transmitted vibration		Whole-body vibration			
	Exposure action value	Exposure limit value	Exposure action value		Exposure limit value	
Exposure duration	RMS method	VDV method	RMS method	VDV method	RMS method	VDV method
1 s	424.26	848.53	84.85	6.51	195.16	14.98
10 s	134.16	268.33	26.83	3.66	61.72	8.42
1 m	54.77	109.54	10.95	2.34	25.20	5.38
10 m	17.32	34.64	3.46	1.32	7.97	3.03
1 h	7.07	14.14	1.41	0.84	3.25	1.93
2 h	5.00	10.00	1.00	0.71	2.30	1.63
4 h	3.54	7.07	0.71	0.59	1.63	1.37
8 h	2.50	5.00	0.50	0.50	1.15	1.15
12 h	2.04	4.08	0.41	0.45	0.94	1.04
16 h	1.77	3.54	0.35	0.42	0.81	0.97
24 h	1.44	2.89	0.29	0.38	0.66	0.87

Table 51.7 Vibration effects on function and performance

Vibration induced effect	Frequency (Hz) at which there is occurrence
Decreased tactile sensation	0.1–1,000,000
Motion sickness	0.1–1
Vibrating hand tools diseases	5–10
Respiration difficulties	15–15
Abdominal pain	5–15
Increased muscle tone	10–20
Lower back pain	7–20
Head sensations	10–20
Disturbance of speech	1–100
Urge to defecate and urinate	10–20
Speech difficulties	7–20

[5] Adapted from Griffin M. J. (1990). Handbook of human vibration. London: Academic Press.

Chapter 52
Crew Training for Low Temperature Conditions

General

Training and manning are both important considerations for all vessels. However, this is especially true on those vessels which operate in cold climate areas. Operations in ice require special skills if they are to be accomplished safely and efficiently; and even the most capable ice breakers cannot be piloted without the exercise of due caution and respect for the environment. The importance of having an experienced ice master or ice navigator has been recognised for centuries, on voyages of discovery and for commercial operations. Ice breakers and other vessels that spend much of their time in ice-congested waters may have masters and officers with sufficient expertise to provide the ice navigator function. For many other vessels, it is necessary to supplement the regular crew with supernumerary ice advisors or ice navigators, either to provide a basic capability or to comply with national regulations for local experience. Where such personnel are used, the number carried should be matched to the nature of the service and the anticipated nature of the operation. The terminology used to describe personnel with ice-related expertise is not standardised, and so its usage may lead to some confusion. In this chapter, the following definitions apply:

- Ice advisor: an individual with expertise in the interpretation of ice information and in forecasting future ice conditions.
- Ice navigator: a mariner (usually a master mariner) with extensive experience in the operation of vessels in ice who can provide guidance to the vessel's master and other watchkeepers.
- Ice pilot: a mariner individual with similar qualifications and experience to an ice navigator who may be required in some areas and jurisdictions to undertake pilotage duties.
- Ice master: a vessel's master with sufficient ice-going expertise to undertake the functions of ice navigator.

A. A. Olsen, *Applying Physical Ergonomics to Modern Ship Design*,
https://doi.org/10.1007/978-3-031-57974-5_52

While these personnel require extensive specialised training, some level of training should be given to all officers and crew of vessels involved in ice operations. It should also be recognised that operations in remote regions may have very limited support from shore-based infrastructure, and sufficient skills should be available on board to cover both normal and emergency situations. General guidance on training and crewing is offered in the IMO Guidelines; MSC Circular 1056, MEPC Circular 399. National administrations, including those of Canada and Russia, have specific requirements that must be respected in waters under their jurisdiction. This chapter aims to provide additional useful information that vessel owners, operators and vessels' crews can utilise in the planning and conduct of cold climate operations.

Training

Seafarers exposed to cold and their supervisors should be trained:

- To recognise symptoms which may lead to hypothermia in themselves or others, and the steps to be taken to prevent onset and/or emergencies.
- In the use of rescue and first-aid measures, ice control equipment and procedures, personal protective equipment, and other devices and procedures intended to counter the effects or working in cold conditions.
- About action to be taken in the event of increased risks of accidents because of low temperatures.

Seafarers should be advised of:

- The importance of physical fitness for work in cold environments.
- The importance of drinking sufficient quantities of liquid and the dietary requirements providing intake of salt and potassium and other elements that are depleted due to sweating.
- Effects of drugs which can reduce their tolerance to thermal extremes.

Seafarers normally use classroom training, self-study, and operating experience to develop competencies and achieve certifications. Until recently, there have been very limited formal training resources for developing ice-operational knowledge. Now, several training institutes around the world can offer some level of course for ice advisors, ice navigators and/or ice masters. There are also some simulator facilities that offer training in ice recognition and other aspects of ice operations. However, these still have very limited fidelity in representing ship-ice interaction. On board training remains the most valuable means of developing expertise, and minimum durations of operation and watchkeeping experience are specified by most national administrations before personnel can be qualified as ice navigators. There are no formally mandated requirements for the content of training courses in ice navigation. However, any comprehensive training program needs to address the demands of Strategic, Tactical and Close Quarters navigation. Strategic navigation is concerned with the general routing of a vessel in the geographic region in which a vessel will operate, for example, the Baltic, Northern Sea Route, Northwest Passage, etc. At the tactical level, considerations include the selection of way points according to current and forecast ice conditions. At the close quarter level, active navigation evaluates the local ice conditions and the manoeuvring of the vessel around and through ice floes. Strategic level (voyage planning) decision-making includes:

- Obtaining relevant ice information from ice maps, satellite imagery and medium-term forecasts.
- Evaluating ice conditions and environmental forecasts to verify the feasibility of the voyage and its probable duration.
- Optimising route to minimise the time that will be spent in the ice and to avoid potentially dangerous ice features.

Tactical level (transiting) decision-making includes:

- Obtaining more local ice information from enhanced satellite imagery, airborne radar images and visual reports, reports from other vessels in the area, etc.
- Optimising the route to minimise encounters with ice that will impede vessel's progress or cause structural damage.

Close quarters level (manoeuvring) decision-making includes:

- Using radar and visual data to assess the properties of the ice cover.
- Selecting safe speeds to reduce the risk of structural or mechanical damage.
- Manoeuvring in a manner that minimises impacts against vulnerable regions of the hull, steering gear, etc.

As can be seen from the above, an essential area of expertise—and therefore, training—is the ability to interpret various ice information ranging from charts and imagery to direct visual observation. Several good pictorial guides to ice recognition have been developed, examples of which are included in the reference list. Simulators with high quality graphics can also be valuable training aids. Ice advisors will have received training in these areas, but not necessarily in navigational skills. A key element of expertise for an ice navigator, pilot or master is the ability to relate the ice conditions to the capabilities of a vessel, based on its ice class, powering and other features. This can be taught in general terms by training institutes, though all vessels have their own specific capabilities that will influence performance and safety. Amongst issues where not only the ice navigator but all deck officers should have some level of understanding is the relationship of the vessel's ice-strengthening class to the limiting safe ice conditions. For the new IMO/IACS Polar Classes, the general definitions provided in Table 51.1, IMO/IACS polar classes, offer some guidance (see this chapter for additional information on ice types and conditions). Vessels operating in many Russian waters are provided with an Ice Certificate, which provides more detailed information on safe speeds in different ice conditions. However, it is also necessary for operators to understand how the overall ice strengthening of any vessel can vary around the hull; and, for example, how operating in ballast conditions may expose weaker areas of the hull or appendages to ice impacts. The sides and stern of most vessels are considerably less strong than the bow area, and vessels should not be operated in ways that expose these areas to significant ice impact loads.

The manoeuvrability of most vessels is reduced dramatically by the presence of any ice, and as ice cover thickens, it may become almost impossible to alter course by conventional manoeuvres. This is particularly true of vessels with long parallel mid bodies and no side flare (i.e., most commercial vessels). However, even ice breakers may find it necessary to change course by backing into a broken track in order to generate a change in heading. This type of manoeuvre is known as the Captain's Manoeuvre, or Star Turn, due to the appearance of the track. Another situation that may require a vessel to reverse course is the need to make repeated rams to traverse thick ice or pressure ridges. Only vessels with reasonably high levels of ice capability should attempt to ram, and great caution is required during the backing phase of any manoeuvre to avoid damage to propellers. It is particularly dangerous for propellers to hit ice when the shaft is stopped, which can exert very high bending loads on the blades. Ramming speeds should be selected to avoid beaching the vessel in an attitude that creates stability problems, or that will lead to problems of extrication. Ice navigators should be able to recognise the onset of pressure events, which can cause a vessel to become beset and may pose both direct and indirect safety hazards. This pressure can be sufficient to damage the hull, and general ice drift may carry the vessel into shallow or dangerous waters. Pressure can be managed to some degree by ice breaker intervention, or if time permits, the vessel can be oriented into rather than across the direction of compression. As pressure events are usually of reasonable short duration, stronger vessels may find it best to wait them out. The ice navigator should therefore also be sensitive to signs of reducing pressure.

A significant number of high ice class vessels incorporate unusual design features ranging from azimuthing propulsors to bubbler, water wash and heeling systems. Ice navigators and other personnel should understand why such systems are used and how they can be operated for maximum effectiveness. Training for deck officers should address ice breaking procedures and the conduct of escorted operations. While these differ somewhat in different national waters, many general principles apply. These include ensuring that communications are established, and protocols understood. Potential problem areas for escort include manoeuvring (if the ice breaker makes too close a turn); stopping (if the ice breaker is brought to a halt by heavy ice); the risk of damage due to impact with track sides for a wide vessel escorted by a narrower ice breaker; and the risk of damage due to bow or bottom impact with large ice pieces submerged below the ice breaker. Masters should understand that an escort does not take over responsibility for the safety of the escorted vessel. The points covered above are not intended to represent a complete listing of subject areas for training courses, but rather an illustration of the types of knowledge that training should aim to provide. They are also amongst the type of material that should be covered in a vessel's onboard documentation, as discussed further in Table 52.1.

Table 52.1 IMO/IACS polar classes

Polar class	Ice description (based on WMO sea ice nomenclature)
PC 1	Year-round operation in all Polar waters
PC 2	Year-round operation in moderate multi-year ice conditions
PC 3	Year-round operation in second-year ice which may include multi-year ice inclusions
PC 4	Year-round operation in thick first-year ice which may include old ice inclusions
PC 5	Year-round operation in medium first-year ice which may include old ice inclusions
PC 6	Summer/autumn operation in medium first-year ice which may include old ice inclusions
PC 7	Summer/autumn operation in thin first-year ice which may include old ice inclusions

Documentation

Under the IMO Guidelines, "all ships operating in Arctic ice-covered waters should carry on board at all times an operating and training manual for all Ice Navigators on board the ship." The IMO Guidelines provide a basic overview of the type of content that each manual should cover. Manuals and other documentation should conform to the intent of other requirements, such as those of the ISM Code, Contingency Planning for Shipboard Emergencies (IMO A852(20)), and other international and national standards. As such, they need to address the information needs of all crew members and procedures for making any passengers aware of specific requirements and challenges. *Operating manual.* This manual should contain relevant information directly related to operations in ice-covered waters including contingency planning in the event that the vessel suffers damage. The typical format of any operating manual should include the following sections, tailored to operations in ice and in cold temperature conditions: Normal Operation, Risk Management, and Emergency. *Instructions. Normal operation.* This should include measures to be taken when planning a voyage into cold or ice-infested waters, including:

- Contacting maritime administrations well in advance to confirm if special requirements, crewing or personnel are required on board.
- Arranging for the availability of ice and weather data and forecasts.
- Identifying any loading restrictions that may be required for safety or pollution prevention reasons.
- Ensuring that the vessel is adequately outfitted and supplied for the voyage.

As the vessel approaches the cold region, additional checks should be undertaken, such as:

- Contact the vessel traffic control services for the area.
- Verify operation of VHF radio including confirming channel used by ice breakers operating in the area.

- Take measures to prevent freezing damage; (e.g., piping on deck, sounding pipes and air vents, etc., are to be drained of water).
- Check drafts and trim to verify that the ice belt covers the waterline area, and that propeller and rudders are adequately submerged.
- Verify the operation of searchlights and signalling devices.
- Ready anchors and tow lines.
- Position protective covers on exposed equipment.
- Configure HVAC and other systems (where possible) for cold weather operations.
- Implement any other measures required for safe navigation.

The completion of check lists should be recorded. Once the vessel is operating in cold and ice, other issues that should be covered by the operations manual (as applicable) may include:

- Use of special features and systems installed on the vessel, such as water wash and bubblers, etc.
- Cold start and run-up of systems, including preheating of lubricating and hydraulic oils.
- Procedures for clearing ice and snow accumulations without damage to equipment.
- Recording and reporting local conditions to assist in forecasting and to comply with regulatory requirements.

Risk management. Risk Management for cold regions should cover all hazards that may result from or be accentuated by the cold temperature environment, and by operations in ice-infested waters. This includes ensuring the operability and checking the functionality of all operational and safety systems, including firefighting and lifesaving, as discussed in previous chapters. Other issues that should be addressed include monitoring of:

- Hull integrity, by sounding tanks and voids at regular intervals.
- Water temperatures in tanks and systems, to provide warning of freeze-up problems.
- Performance of pollution prevention systems, to prevent the discharge of wastes.

In general, risk management requires an understanding that many systems and components will be operating at or near their design limits, and their performance may degrade rapidly. Monitoring and testing should be conducted more frequently than is the case in normal operations. *Emergency instructions.* Emergency instructions should address the unique aspects of certain types of incidents in cold and/or ice-infested regions. These range from communicating with support and search and rescue organisations to abandoning ship in the presence of ice. Emergency instructions cannot take the place of proper training in emergency response management but can be extremely valuable in reminding crew members of the procedures to be followed. As with all other elements of the operating manual, the instructions must be tailored to the specifics of the vessel and its modes of operation. They should also be integrated with the drills and exercises covered by the Training Manual and onboard training procedures. *Training manual.* A training manual should be provided on board all vessels. The contents of a manual for vessels operating in cold regions should be an informational and educational tool for personnel and a guide for the types of training exercise that should form part of risk management for cold regions operation. As such, it should cover at least the following subject matter areas:

- Ice recognition.
- Safe navigation in ice.
- Conduct of escorted operations.
- Instructions for drills and emergency response.

Drills are covered in some detail by the IMO Arctic Guidelines, which addresses lifesaving, firefighting, damage control and other exercises. These should be carried out under as realistic conditions as possible consistent with maintaining the safety of those on board. Crew members should all be given the opportunity to undertake drills while wearing cold weather survival gear to gain a proper appreciation of how this will affect their mobility and dexterity. Where specialised equipment is carried for any type of emergency response, an adequate number of crew members should be trained in advance (and certified as necessary) in its use. Some examples may include firearms training, if firearms are carried as part of the evacuation kit, and any specialised pollution clean-up systems.

Correction to: Evaluating Bridge Design

Correction to:
Chapter 29 in: A. A. Olsen, ***Applying Physical Ergonomics to Modern Ship Design*****,**
https://doi.org/10.1007/978-3-031-57974-5_29

In the original version of the book, Table 29.1 contains overlap in text with previously published content [1] that was inadvertently omitted. The author failed to attribute the reference [1]. The author has now obtained permission to re-use this content from CRC Press. Where [1] is Ross Jonathan M 2009 Human Factors for Naval Marine Vehicle Design and Operation, CRC Press, London. The book has been updated with the changes.

The updated version of this chapter can be found at
https://doi.org/10.1007/978-3-031-57974-5_29

A. A. Olsen, *Applying Physical Ergonomics to Modern Ship Design*,
https://doi.org/10.1007/978-3-031-57974-5_53

Annex 1

List of Maritime Administrations and Contact Information

Various national Administrations have additional regulations for vessels operating in their territorial waters beyond those listed in this book. It is recommended that vessel designers, owners and operators contact these Administrations directly regarding these regulations. The table below provides a list of Administrations in areas subject to low temperatures (correct as of February 2024).

Country	Address	Contact/telephone	Web site address
Argentina, Republic of Prefectura Naval Argentina	Av. Eduardo Madero, 235 (C1106 CCA) Buenos Aires, Rep. of Argentina	Director Policia Seguridad de la Navegacion Jefe Departmento Reglamentacion de la Navegacion: (54-11) 4318-7475 ext 2411 or (54-11) 4318-7467	http://www.prefecturanaval.gov.ar
Australia, Commonwealth of Australian Maritime Safety Authority	GPO Box 2181 Canberra City ACT 2601 Commonwealth of Australia	Principal Adviser Technical: (61-2) 6279 5049 or (61-2) 6279 5966	http://www.amsa.gov.au/
Canada Transport Canada Safety and Security	Tower "C", Place De Ville 11th Floor 330 Sparks Street Ottawa, Ontario K1A 0N8, Canada	Director of Ships and Operations Standards: (1-613) 991-3131 or (1-613) 993-8196	http://www.tc.gc.ca/en/menu.htm

(continued)

A. A. Olsen, *Applying Physical Ergonomics to Modern Ship Design*,
https://doi.org/10.1007/978-3-031-57974-5

(continued)

Country	Address	Contact/telephone	Web site address
Chile Direccion General del Territorio Maritimo y de Marina Mercante Armada de Chile	Avenida Errazuriz 537 Valparaiso, Chile	Director General: (56-32) 2208102 or (56-32) 2208097 (fax)	http://www.directemar.cl/
Denmark Danish Maritime Authority	Skibsregistret 38C, Vermundsgade P. O. Box 2605 DK-2100 Copenhagen O, Denmark	Chief Ship Surveyor (45-39) 17 44 00 or (45-39) 17 44 01 (fax)	http://www.dma.dk/
Estonia Estonian Maritime Administration	Valge 4 11413 Tallinn, Estonia	Director general (372) 6205 700 or (372) 6205 706 (fax)	http://www.vta.ee/atp/?lang=en
Finland Finnish Transport Agency Maritime Department	Finnish Transport Agency P. O. Box 33 FIN-00521 Helsinki, Finland www.fma.fi	Technical Approval (358-20) 448 4249 or (358-20) 448 4336 (358-20) 63 7373 (Operator)	http://portal.fma.fi/sivu/www/fma_fi_en/services/winter_navigation
Greenland Danish Maritime Authority	Skibsregistret 38C, Vermundsgade P. O. Box 2605 DK-2100 Copenhagen O, Denmark	Chief Ship Surveyor (45-39) 17 44 00 or (45-39) 17 44 01 (fax)	http://www.dma.dk/
Iceland Icelandic Maritime Administration	Vesturvor 2 P. O. Box 120 202 Kopavogur, Iceland	Head of Technical Regulations Section: (354) 560 0000 or (354) 560 0060 (fax)	http://www.sigling.is/English
Latvia Ministry of Transport of Latvia, Maritime Administration	Trijadibas iela 5 Riga, LV-1048, Latvia	Director: (371) 67062101 or (371) 67860082 (fax)	http://www.jurasadministracija.lv/index.php?pid=211
Lithuania Ministry of Transport and Communications of the Republic of Lithuania	Gedimino Av. 17 Vilnius 2679, Republic of Lithuania	Director: 370 46 469657	http://www.msa.lt/index.php/en/27717
Netherlands Inspectorate Transport and Water Management	Netherlands Shipping Inspectorate Gebouw Prinsenpoort 's-Gravenweg 665 (PO Box 8634, 3009 AP) Rotterdam, The Netherlands	Director: (31-10) 266 8500 or (31-10) 202 3424	http://www.ivw.nl/en/water/koopvaardij/

(continued)

(continued)

Country	Address	Contact/telephone	Web site address
NEW Zealand Maritime New Zealand	Level 8, gen-i Tower 109 Featherston Street (P. O. Box 27006) Wellington, New Zealand	General Manager, Maritime Operations: (64-4) 4730111 or (64-4) 4941263	http://www.maritimenz.govt.nz/
Norway, Kingdom of Norwegian Maritime Directorate	P.O. Box 2222 N-5509 Haugesund, Norway	Vessels and Seafarers 47 52 74 50 00 47 52 74 50 01	http://www.sjofartsdir.no/en/
Poland Ministry of Transport and Maritime Economy	Department of Maritime and Inland Waters Administration U1. Chalubinskiego 4/6 00-928 Warsaw, Poland	Ships inspection—Head Office (48-22) 621 1448 or (48-22) 621 9437	http://www.en.mi.gov.pl/2-482076faaa403.htm
Russian Federation Ministry of Transport of the Russian Federation Russian Northern Sea Route Administration	Maritime Administration ul. Rozhdestvenka, 1/4 103759 Moscow, Russia	Head of Maritime Administration (7-495) 926 10 00 or (7-495) 978 91 70	Ministry of Transport http://www.mintrans.ru/ Northern Sea Route http://www.morflot.ru/
Sweden, Kingdom of **Swedish Maritime Administration**	Maritime Safety Inspectorate Slottsgatan 82, SE-601 78 Norrkoping Kingdom of Sweden	Director of Maritime Safety: (46-11) 19 10 00 or (46-11) 10 19 49	http://www.sjofartsverket.se/
Ukraine **Ministry of Transport of Ukraine**	State Department of Maritime and Inland Water Transport 14, prospect Peremogy Kyiv 01135, Ukraine	Director General: (380 44) 461 5320 or (380 44) 461 5304 (Emergency Officer)	http://www.mintrans.gov.ua/# http://www.kmu.gov.ua E-mail: security@morrechflot.gol.ua
United States OF America **U. S. Coast Guard**	Commandant (G-MOC) 2100s Street, SW Washington, DC 20593-0001 Federal Communications Commission Wireless Telecommunications Bureau 445 12th Street SW Washington DC 20554	Chief, Office of Vessel Activities (202) 372-1210 (202) 372-1918 (fax) Public Safety and Private Wireless Division (202) 418-2771 (202) 418-2643	http://www.uscg.mil

Annex 2

Meteorological Organisation Contact Information

Various government organisations provide readily accessible climatic, environmental and meteorological data sets and historic atlases. The table below provides a list of these organisations along with their contact information.

Country	Address	Contact/telephone	Web site address
All Russian Research Institute- of Hydrometeorological Information—World Data Centre	RIHMI-WDC, 6, Korolev St., Obninsk, Kaluga Reg., Russia	Tel: (7-495) 255-21-94 Fax: (7-495) 255-22-25 E-mail: wdcb@meteo.ru	http://www.meteo.ru/
Arctic and Antarctic Research Institute	38 Bering str., St.Petersburg, Russian Federation, 199397	Tel: (7-812) 352 1520 Fax: (7-812) 352 2688 E-mail: aaricoop@aari.nw.ru	http://www.aari.nw.ru/default_en.asp
Argentine Navy Meteorological Service (SMARA) of the Naval Hydrographic Service	Av.Montes de Ocal 2124, 1217 Buenos Aires, Argentina	E-mail: cnhsmara@rina.ara.mil.ar	
Australian Bureau of Meteorology	GPO Box 1289 Melbourne VIC 3001 (700 Collins Street Docklands) Australia	Tel: (61-(0)3) 9669 4000 Fax: (61-(0)3) 9669 4699 E-mail: webclim@bom.gov.au	http://www.bom.gov.au/

(continued)

A. A. Olsen, *Applying Physical Ergonomics to Modern Ship Design*,
https://doi.org/10.1007/978-3-031-57974-5

(continued)

Country	Address	Contact/telephone	Web site address
British Antarctic Survey;	High Cross, Madingley Road Cambridge CB3 0ET United Kingdom	Tel: (44 (0)12) 2322 1400 Fax: (44 (0)12) 2336 2616 Email: basweb@bas.ac.uk	http://www.antarctica.ac.uk/
Canadian Ice Service	373 Sussex Drive Block E, Third Floor Ottawa, Ontario K1A 0H3 Canada	Tel: (1-613) 996-1550 Fax: (1-613) 947-9160 Email: cis-scg.client@ec.gc.ca	http://www.ice-glaces.ec.gc.ca
Climatic Research Unit, University of East Anglia	School of Environmental Sciences University of East Anglia Norwich NR4 7TJ, United Kingdom	Tel: (44-1603) 592722 Fax: (44-1603) 507784 Email: cru@uea.ac.uk	http://www.cru.uea.ac.uk/
Cold Regions Research and Engineering Laboratory- US Army	72 Lyme Road, Hanover New Hampshire, 03755-1290 USA	Tel: (1-603) 646-4582 Email: contactcrrel@crrel.usace.army.mil	http://www.crrel.usace.army.mil/
Danish Meteorological Institute	DMI Lyngbyvej 100, 2100 København Ø Denmark	Tel: (45-39) 15 75 00 Fax: (45-39) 27 10 80 Email: vmarked@dmi.dk	http://www.dmi.dk/dmi/index/
Environment Canada	Meteorological Service of Canada Atmospheric and Climate Science Directorate (ACSD) 4905 Dufferin Street Toronto, Ontario, M3H 5T4 Canada	Tel: (1-416) 739-4239 Email: firstname.lastname@ec.gc.ca	http://www.climate.weatheroffice.ec.gc.ca/climateData/canada_e.html
European Climate Support Network	KNMI PO Box 201 3730 AE De Bilt The Netherlands	Tel: 31 302 206-522 Fax: ECSN Programme Manager	http://www.eumetnet.eu/ECSN_home.htm
Icelandic Meteorological Office	Bustadavegur 9 150 Reykjavík Iceland	Tel: (354) 522 6000 Fax: (354) 522 6001 Email: climate@vedur.is	http://www.vedur.is/english/
International Oceanographic Data and Information Exchange (IODE)	IOC Project Office for IODE Wandelaarkaai 7 B-8400 Oostende Belgium	Tel: (32-59) 34 01 58 Fax: (32-59) 34 01 52 IODE Programme Coordinator	http://www.iode.org/

(continued)

(continued)

Country	Address	Contact/telephone	Web site address
International Polarview Organisation		Tel: (1-709) 737 3735 Email: info@ polarview.org	http://polarview.org/ project/index.htm
National Snow and Ice Data Center, University of Colorado	449 UCB, University of Colorado Boulder, CO 80309-0449 USA	Tel: (1-303) 492 6199 Fax: (1-303) 492 2468 Email: nsidc@nsidc. org	http://nsidc.org/data/ ewg/index.html
Network of European Meteorological Services	*Refer to European Climate Support Network*		http://www.eum etnet.eu.org/
Norwegian Meteorological Institute (DNMI)	PO Box 43 Blindern 0313 Oslo Norway	Phone: (47-22) 96 30 00 Fax: (47-22) 96 30 50 Email: met.inst@met. no	http://www.met.no/ english/index.html
Norwegian Polar Institute	Norwegian Polar Institute, Polar Environmental Centre N-9296 Tromsø, Norway (Norsk Polarinstitutt, Polarmiljøsenteret 9296 Tromsø)	Tel.: (47-77) 75 05 00 Fax. (47-77) 75 05 01 Email: postmottak@ npolar.no	http://npiweb.npo lar.no/

Annex 3

IMO Resolutions

IACS Rec. 95. Recommendation for the Application of SOLAS Regulation V/15, Bridge Design, Equipment Arrangement and Procedures (BDEAP).
IACS UI SC 235. Navigation Bridge Visibility.
IMO Res. A.342(IX). Recommendation on Performance Standards for Automatic Pilots [see also Res. MSC.64(17)].
IMO Res. A.382(X). Magnetic Compass Carriage and Performance Standards.
IMO Res. A.384(X). Performance Standards for Radar Reflector.
IMO Res. A.424(XI). Performance Standards for Gyro-compass.
IMO Res. A.477(XII). Performance Standards for Radar Equipment [see also Res. MSC.64(67), Annex 4].
IMO Res. A.526(13). Performance Standards for Rate-of-Turn Indicators.
IMO Res. A.601(15). Provision and Display of Manoeuvring Information Onboard Ships.
IMO Res. A.617(15) Implementation of the NAVTEX System as a Component of the Worldwide Navigational Warning Service.
IMO Res. A.665(16). Performance Standards for Radio Direction-finding Systems [revokes Res. A223 (VII)].
IMO Res. A.694(17). General Requirements for Shipborne Radio Equipment Forming Part of the Global Maritime Distress and Safety System (GMDSS) and for Electronic Navigational Aids [revokes Res.A574(14)].
IMO Res. A.816(19). Recommendation on Performance Standards for Shipborne Decca Navigator Receivers.
IMO Res. A.817(19). Recommendation on Performance Standards for Electronic Chart Display and Information Systems (ECDIS) [see also Res. MSC.64(67), Annex 5].
IMO Res. A.818(19). Performance Standards for Shipborne Loran-C and Chayka Receivers.

A. A. Olsen, *Applying Physical Ergonomics to Modern Ship Design*,
https://doi.org/10.1007/978-3-031-57974-5

IMO Res. A.823(19). Recommendation on Performance Standards for Automatic Radar Plotting Aids (ARPA) [for installations before 1/1/97, see Res. A.422(IX)].
IMO Res. MSC.452(99). Amendments to revised Performance Standards for Integrated Navigation Systems (INS).
IMO Res. MSC.96(72). Performance Standards for Devices to indicate Speed and Distance [for installations before 1/1/97, see Res. A.478(XII)].
IMO Res. MSC, 252(83). Adoption of the Revised Performance Standards for Integrated Navigation Systems (INS).
IMO Res.MSC.113(73). Performance Standards for Shipborne GLONASS Receiver Equipment.
IMO Res.MSC.128(75). Performance Standard for a Bridge Navigational Watch Alarm System (BNWAS).
IMO Res.MSC.137(76). Standards For Ship Manoeuvrability.
IMO Res.MSC.191(79). Performance Standards for the Presentation of Navigation Related Information on Shipborne Navigational Displays.
IMO Res.MSC.192(79). Revised Performance Standards for Radar Equipment.
IMO Res.MSC.232(82). Adoption of the Revised Performance Standards for Electronic Chart Display and Information System (ECDIS).
IMO Res.MSC.64(67), Annex 3. Recommendation on Performance Standards for Automatic Pilots.
IMO Res.MSC.64(67), Annex 3. Recommendation on Performance Standards for Heading Control Systems.
IMO Res.MSC.74(69), Annex 2. Recommendation on Performance Standards for Track Control Systems.
IMO Res.MSC.74(69), Annex 3. Performance Standards for Automatic Identification System (AIS).
IMO Res.MSC.74(69), Annex 4. Recommendation on Performance Standards for Echo-sounding Equipment.
IMO Res.MSC.96(72), Annex 14. Performance Standards for Devices to Measure and Indicate Speed and Distance.
MSC. Circular 53(66). Performance Standards for Shipborne GLONASS Receiver Equipment.
MSC. Circular 64(67). Adoption of New and Amended Performance Standards [revokes Res.A.574; amends Res.A342(IX), Res. A477(XII) and Res. A.817(19)].

Glossary

Access aid A device that makes operation of, or access to, operating or maintenance components possible, or more immediate. For example, a reach rod that extends access to a valve actuator below a deck grating, or a ladder of an access platform.

Accessibility The ability for personnel to access equipment that requires maintenance, inspection, removal, or replacement while wearing the appropriate clothing, including personal protective equipment, and the ability to use all necessary tools and test equipment.

Accessible valves Readily or Immediately Accessible Valves are located in a space entered without requiring tools or keys; clear of obstructions, moving equipment and hot surfaces; within operator's reach without requiring use of extenders, ladders, or other access aids.

Accommodations Areas where the primary purpose is to rest or recreate. Accommodations spaces include cabins and staterooms, medical facilities, offices, and public and recreational rooms. Accommodations also include service spaces, such as mess rooms, laundry, storerooms, and workshops.

Achromatic Designating colour perceived to have zero saturation and therefore no hue, such as neutral greys, white or black.

Acquisition The selection of those target vessels requiring a tracking procedure and the initiation of their tracking.

Active protection A safety design or device that actively (or directly) requires a person to take specific actions before a potential loss, for example, donning a fall arrestor fitted to both the ladder and the climber.

Alarm priority A predicted assessment of the potential consequence of a condition or situation and the resulting urgency of mitigating responses required of personnel. Generally, the more severe the potential consequence, the higher the alarm priority.

Alarm A visual and/or audible signal of a condition, or a predetermined out-of-tolerance condition, for machinery, equipment, components, or systems that require attention.

Ambient environment Ambient environmental conditions that personnel are exposed to during periods of work, leisure, or rest. Ambient environmental

A. A. Olsen, *Applying Physical Ergonomics to Modern Ship Design*,
https://doi.org/10.1007/978-3-031-57974-5

conditions include human whole-body vibration, noise, indoor climate, and lighting.

Ambient temperature The air temperature in the vicinity of the vessel.

Annunciation An audible signal indicating a condition or system state (usually of an emergency or off-normal nature) via audible signals.

Anthropometric percentile Given the range of variability of [human] bodily dimensions, anthropometric data are typically expressed as percentile statistics, such as 5th or 95th percentile. A percentile statistic defines the anthropometric point at which a percentage of a population falls above or below that value. For example, the seated eye height of a 95th percentile North American male is 853 mm (33.5 in), so by definition, 5% of North American males will have a seated eye height of greater that this figure and 95% will have a lesser seated eye height.

Anthropometrics Data relating to physical body dimensions. It includes body characteristics such as size, breadth and the distances between anatomical landmarks, such as elbow to fingertip.

ARPA Automatic radar plotting aid.

Ascenders/descenders The point size of a given font is measured from the top of the font's ascenders to the bottom of its descenders. The ascender is the measurement from the top of the lowercase letter to the top of the uppercase letter. The descender is measured from the bottom of the uppercase letter to the bottom of the lowercase letter.

Aspect ratio The ratio of the font size (measurement from the baseline to the top of the uppercase letters) to the x-height (measurement from the baseline to the top of the lowercase letters).

Assisted lifting The use of devices such as cranes, hoists, counter-balancing mechanisms, trolleys, monorails, come-alongs, padeyes, or A-frames by personnel to perform materials handling tasks.

Audible alarm An alarm comprised of tones, verbal messages, or verbal messages preceded by tones. Not all audible alarms are associated with visual alarms. Audible through pitch, frequency and modulation.

Auditory display A device that provides readings, status or condition of machinery, equipment, or systems operating parameters through the use of sound signals or spoken messages.

Back-up navigator Any individual, generally an officer, designated by the vessel master to be on call if assistance is needed on the navigation bridge.

BNWAS Bridge navigational watch alarm system.

Bridge wing workstation Workstation from which the vessel can be manoeuvred, and operated during unmooring and mooring, lock passage, taking or dropping the pilot, etc.

Bridge wings The parts of the bridge extending from both sides of the wheelhouse from which the ship's sides are visible. Ship docking is normally conducted from workstations on the bridge wings.

Bridge Area from which the navigation and control of the vessel is exercised, including the wheelhouse and bridge wings.

Case The part of equipment that encloses and protects the equipment from its surroundings and protects the surroundings–including personnel–from the equipment.

Catwalk Extension of a deck that is wide enough to allow the passage of a person.

Chart area Part of the wheelhouse situated and equipped for the performance of voyage planning and plotting activities.

CIE Commission Internationale de l'Eclairage is an international organisation that focuses on lighting.

Closest point of approach (CPA) The shortest target vessel-own vessel calculated distance that will occur if there is no change in course and speed data.

Coding Coding refers to a system of sensory stimuli used to represent specific information. There are several mechanisms for coding information. These include, but are not limited to.

Cognition The mental process of knowing, including aspects such as awareness, perception, reasoning, and judgment.

Commanding view View without obstructions which would interfere with the ability to perform immediate navigation tasks.

Communications workstation Workstation for operation and control of the equipment for distress/safety communications and public correspondence communications.

Communications Design for clear, reliable and understandable internal to vessel, vessel-to-vessel and vessel-to-shore voice communications.

Conning position Place on the bridge with a commanding view and which is used by navigators and pilots, when monitoring, manoeuvring and controlling a vessel.

Console A group of controls and displays associated with one or more individual pieces of equipment or systems mounted together on a structure dedicated to the control and monitoring of those individual equipment of systems. Consoles may be free standing units, and include angled and vertical surfaces.

Continuous control A continuous control is an actuator that operates at any point or value along a continuous scale (e.g., engine throttle).

Control A device an operator or maintainer uses to input a signal or uses to change the operating status of equipment or systems. Examples include switches, knobs, cranks, thumb wheels, levers, keyboards, and foot pedals.

Course The direction in which a vessel is steered or intended to be steered, expressed as an angular distance from North over the water.

Crew habitability The acceptability of vessel or offshore installation conditions in terms of vibration, noise, indoor climate, and lighting as well as physical and spatial characteristics, according to prevailing research and standards for human efficiency and comfort.

Crew member Any person on board a vessel, including the Master, who is not a passenger. This term is used interchangeably throughout this document with "seafarer".

Cross track alarm Comparison of the vessel's position with the track (control error).

Cross-coupling The movement of the follower (e.g., cursor or tracking symbol) in the orthogonal direction.

Dark adaptation Dark adaptation is the process by which the human eye adjusts to lower levels of illumination through increased sensitivity to light. Visual function in very dim light has the following characteristics.

dB(A) Sound level in decibels, measured using A-weighting. The use of A-weighting causes the frequency response of sound level measurement devices to mimic that of the human ear, where response is maximum at about 2KHertz, less at very low or very high frequencies. A-weighted measurements correlate well with measures of speech interference and judgments of loudness.

Deck machinery Any machinery or equipment that operates as designed at the ambient temperature. Some examples include winches, windlasses, cranes or other lifting appliances, hatch covers, vehicle ramps, boat davits and mooring fittings.

Design load The maximum intended load, being the total of all loads including the weight of the personnel, materials, and equipment, including the means of access structure.

Design service temperature The design service temperature is to be taken as the lowest mean daily average air temperature in the area of operation where:

Mean. Statistical mean over observation period (at least 20 years).

Average. Average during one day and night.

Lowest. Lowest during year. For seasonally restricted service the lowest value within the period of operation applies.

Detent controls A type of discrete control, characterised by the control locking into each position setting until the operator exerts extra force to move the control out of the setting. These types of controls are preferable for machinery equipment or system operation requiring control in discrete steps or different modes.

Discrete control A discrete control is an actuator that allows for the selection between two or more mutually exclusive operating functions or points along a scale (e.g., switching a machine ON or OFF, or selecting one of three pumps to run).

Display Means by which a device presents visual information to the navigator, including conventional instrumentation.

Effective temperature An index that combines into a single value the effect of temperature, humidity and air movement on the sensation of warmth or cold. The numerical value is that of the temperature of still, saturated air that would induce an identical sensation to the human body.

Electronic chart display and information system (ECDIS) A system which displays hydrographic information and the vessel's position along a pre-planned route.

Emergency A serious, unexpected, often dangerous event requiring immediate action.

Enclosed space The interior portions of a vessel where people work and live.

EPIRB Emergency Position Indicating Radio Beacon Equipment and room/space identification label. A type of label used to identify individual equipment, (e.g., valves, gauges, junction boxes, filters, pumps, transmitters, controls, pressure vessels), or spaces (e.g., rooms, buildings, etc.).

Ergonomics The study and design of working environments and their components, work practices, and work procedures for the benefit of the worker's productivity, health, comfort, and safety. Application of the human factor in the analysis and design of equipment and working environment.

Error control Exercising control over the probability, frequency, criticality and recoverability of human errors that may occur.

Essential Systems or equipment necessary to maintain vessel propulsion, manoeuvring, electrical services, firefighting (along with bilge system), ballasting or personnel safety.

Expectation A belief about or mental picture of the future. In ergonomic contexts, an expectation is a belief concerning the effect of an action, especially a control action.

Field of vision (FOV) Angular size of a scene that can be observed from a given position.

Function The appropriate action of any special components or part of a system to meet some defined purpose or objective; specific occupation or role; an assigned duty or activity.

General emergency alarms An alarm given in the case of an emergency to all persons on a vessel or offshore installation. Often, its purpose is to summon personnel to muster stations. These alarms typically sound throughout a vessel or offshore installation and are intended to be heard by all personnel. General emergency alarms relate to conditions of a serious nature such as announcing a fire or flooding, demanding evacuation of an area, or demanding abandonment of a vessel or offshore installation.

Glare Glare is the sensation produced by luminance within the visual field that is sufficiently greater than the luminance to which the eyes are adapted causing annoyance, discomfort or loss of visual performance and visibility. Glare is experienced if windows or light sources, seen directly or by reflection in shiny surfaces, are too bright compared to the general brightness within the interior of the bridge.

GLONASS Global Orbiting Navigating Satellite Systems.

GMDSS Global Maritime Distress and Safety System.

Grab-rails Hand grabs mounted to bridge surfaces that are used to help negotiate awkward accesses.

Graphic label A type of label used to present information through line schematics, diagrams, charts, tables, pictures, etc.

Guardrail (or safety rail) Device for protection against accidental fall or accidental access to a hazardous area, with which stairs, step ladders or landings, platforms and walkways, or deck edges and other fall points, may be equipped.

Habituation The reduction of a task performance likelihood following repeated false exposures to the stimulus (e.g., an oft repeated false alarm will reduce its ability to command a timely response).

Handle or hand grab Bent round bar of various sizes for use above access openings, on miscellaneous access opening covers or on hatch covers.

Handrail Top element designed to be grasped by the hand for body support which can be used individually or as the upper part of a rail. Rails mounted along passageways, ladder ways, etc., that are used to safely negotiate spaces and work areas.

Hazard identification sign A type of sign used to identify and provide information about situations that may be hazardous to personnel, equipment, or the environment. There are three types of hazards, DANGER, WARNING, and CAUTION.

Heading monitor Comparison of the heading source in use with a second independent heading sensor.

Heading The horizontal direction in which the longitudinal axis of a ship actually points or heads at any instant, expressed in angular units from a referenced direction.

Helmsman Person who steers the vessel.

Heuristic A speculative formula or conceptual model that serves as a guide in the investigation or solution of a problem; a "rule of thumb".

Human task A piece of work performed as part of one's duties; a function performed; an objective.

Human whole-body vibration Mechanical vibration (or shock) transmitted to the human body as a whole. Human whole-body vibration is often due to the vibration of a surface supporting the body.

Icon A graphical representation of an object or function. A graphic symbol.

Illumination The amount of light falling on an object or surface. Illumination is measured using units of Lux (lm/m^2) or foot-candles (fc; lm/ft^2). One foot-candle equals 10.76 lx.

In the notch An operational mode in which a trailing vessel is moored in a specially constructed notch in the stern of an ice breaker.

Information label or placard A type of label or placard used to present non-procedural information of a general nature.

Instruction label A type of label used to present instructions for accomplishing a specific task or tasks, along with any hazard or safety information related to performing the task(s).

Integrated bridge system (IBS) A combination of interconnected systems allowing centralised access to sensor information or command/control from workstations, with the aim of increasing safe and efficient vessel's management. For the purpose of this book, the integrated bridge system pertains only to aspects dealing with navigational, monitoring/alarming and communication functions as covered in this book.

Labels Within the context of this book, the terms "labels" and "labelling" refer to the descriptors used to identify controls, displays and other components on the navigation bridge and not to the internal components of displays such as meter scales or fields in a computer display.

Layout Physical arrangement of the parts and components that make up a module or a unit of equipment.

Lookout Activity carried out by sight and hearing as well as by all available means appropriate in the prevailing circumstances and conditions so to make a full appraisal of the situation and of the risk of collision. Also, the person performing these tasks.

Low temperature environment Areas of vessel operations with ambient temperatures less than or equal to −10 °C (14°F).

Luminance The amount of light emitted from a source (including that reflected from a surface). Luminance is measured in candela-per-square metre (cd/m^2).

Machinery Equipment and systems subject to Class design requirements.

Main workstation See "navigation and traffic surveillance/manoeuvring workstation".

Maintainability The ability to carry out maintenance tasks, such as testing, servicing, calibrating, adjusting, removing, replacing or repairing rapidly and effectively in order to allow equipment and systems to achieve a specified level of performance.

Maintenance item, critical Includes those maintenance actions which are system and safety critical, meaning a system critical to the safe operation of the vessel cannot function without that aspect being functional, and moderate to frequent maintenance actions are required. These are referred to as Category 1 maintenance items for maintenance access criticality categorisation purposes.

Maintenance item, infrequently used non-critical Includes those maintenance actions which are considered to be non-critical to system status and safety, and that do not require frequent access. These are referred to as Category 3 maintenance items for maintenance access criticality categorisation purposes.

Maintenance item, non-critical Includes those maintenance actions which are not system critical as defined above but are maintenance actions that are required frequently for the proper operation of the vessel. These are referred to as Category 2 maintenance items for maintenance access criticality categorisation purposes.

Maintenance of situation awareness Persons on watch duty understanding of the condition of equipment, hull performance, sea and weather conditions, hazards to navigation, crew capabilities and presence and intentions of other occupants of the waterways

Maintenance All activities necessary to keep equipment in or restore it to a specified level of performance.

Manoeuvring mode The lowest vessel speed necessary to maintain manoeuvrability.

Manual materials handling Actions taken by personnel to physically (manually) lift, lower, push, pull, hold, or carry loads.

Manual steering workstation Workstation from which the vessel can be steered by a helmsman.

Manual valve A valve that is manually opened by a human operator using a handwheel, lever, chain, extender, or other hand-operated device.

MCR The maximum continuous rating of the prime mover.

Minimum anticipated temperature The minimum anticipated temperature is determined as the Design Service Temperature minus 20 °C (36°F) for exposed machinery or plus 20 °C (36°F) for machinery in unheated spaces. In no case is the minimum anticipated temperature to be greater than 0 °C (−32°F).

Monitoring workstation Workstation from where equipment and environment can be checked constantly; when several persons are working on the bridge it serves for relieving the navigator at the navigation and traffic surveillance/manoeuvring workstation and/or for carrying out advisory functions by the master or pilot.

Monitoring Act of constantly checking the status of equipment and environment.

Motor task Human motion in moving, posturing, and manipulating equipment.

Nanometre One billionth of a metre.

Navigation and manoeuvring workstation Workstation where vessel's speed and course are considered and controlled.

Navigation and traffic surveillance/manoeuvring workstation Main workstation at which the vessel's course, speed and position in relation to the waters and traffic can be controlled and monitored, and where communication relevant to navigation can be performed. It is generally conceived for working in seated or standing position with optimum visibility and with an integrated presentation of information and operating equipment. It is to be possible from this location, particularly when a fast action is required.

Navigation All tasks relevant for deciding, executing and maintaining course and speed in relation to waters and traffic.

Navigator Person navigating, operating bridge equipment and manoeuvring the vessel.

NAVTEX A narrow-band direct printing telegraph providing navigational and meteorological warnings and urgent information to vessels.

Newton SI unit of force. One Newton is equal to the amount of net force required to accelerate a mass of one kilogram at a rate of one meter per second squared.

Newton-metres (n-m) SI unit of torque. One Newton-meter is equal to the torque resulting from a force of one Newton applied perpendicularly to a moment arm which is one meter long.

Non-saturated colours Saturation in colour usage refers to the degree that a hue differs from a grey of the same lightness. Therefore, a non-saturated colour is closer to grey and appears washed out.

Normal sailing conditions A condition wherein systems and equipment related to navigation operate within design limits, and environmental conditions such as weather and traffic do not cause excessive workload to the officer of the watch.

Occupational safety The science of preventing individuals from being injured or contracting diseases in the course of their work.

Off heading alarm An alarm that monitors the comparison of heading and pre-set heading (control error).

Officer of the watch (OOW) The person responsible for safe navigating, operation of bridge equipment and manoeuvring of the vessel.

Operating alarms Alarms presented within specific operating spaces that are intended to prompt a response by personnel in those spaces.

Operating torque The force an operator or maintainer exerts by pushing or pulling on a lever or turning a handwheel.

Operating/technical manual Manuals or operational instructions for equipment/systems installed on the bridge for the use of bridge personnel.

Panel Any surface where controls, displays, and/or alarms relating to equipment or system conditions are placed. Panels typically are flat vertical surfaces. Panels are sometimes referred to as control boards.

Parallax An apparent change in the direction of an object, caused by a change in observational position that provides a new line of sight.

Passage planning workstation Workstation at which vessel's operations are planned (i.e., route planning, deck log, etc.) and where determining and documenting all vessel's operation takes place.

Passive protection A safety design or device that requires a person to take no specific action prior to a potential loss, for example, a safety cage permanently fitted to a ladder. See also "Active protection" above.

Percentile Given the range of variability of [human] bodily dimensions, anthropometric data are typically expressed as percentile statistics, such as 5th or 95th percentile. A percentile statistic defines the anthropometric point at which a percentage of a population falls above or below that value. For example, the seated eye height of a 95th percentile North American male is 853 mm (33.5 in), so by definition, 5% of North American males will have a seated eye height of greater that this figure, and 95% will have a lesser seated eye height.

Perception The process of becoming aware of the immediate environment through any of the senses.

Pipe marker labels Labels with coloured markings (e.g., bands of colour), text, and flow arrows. These are placed on pipes to identify pipe contents and flow direction.

Position monitor Comparison of the position source in use with a second independent position sensor or source.

Pound force One pound force is equal to the gravitational force exerted on a mass of one pound on the surface of the Earth.

Primary bridge navigational equipment/systems For the purpose of this book, equipment/systems essential for the performance of primary bridge navigational functions such as gyro compass, radar, position-fixing system and electronic chart system.

Public rooms Those accommodations areas used for halls, dining rooms, lounges and similar permanently enclosed spaces. For the purpose of this book, the library, gymnasium, vessel's office and conference rooms are not considered.

Radar plotting The process of target detection, tracking calculation, calculation of relative and true motion, course, speed and display of information.

Reach rod A rod for operating a remote valve that would otherwise not be accessible (for example, to operate a valve below a deck grating).

Reflectance The amount of light that is reflected (bounced back) from a surface.

Saturation The extent to which a chromatic colour differs from a grey of the same brightness. It is a measure on an arbitrary scale from 0% (grey) to 100%.

Seafarer Any person on board a vessel, including the Master, who is not a passenger. This term is used interchangeably throughout this document with "crew member".

Sea-going vessel Any vessel navigating on the high seas, i.e., areas along coasts and from coast to coast.

Shall Expresses a provision that is mandatory.

Should Expresses a provision that is recommended or preferred practice.

Situation awareness The degree of accuracy by which one's perception of the current environment mirrors reality.

Staffing levels The number of crew who will staff the bridge during operations. This addresses bridge workload, both immediate (number of tasks that must be performed in a short period of time) and ongoing (work done over the course of a voyage). This also addresses bridge personnel fatigue.

Stair One of a flight or series of steps for going from one level to another, as opposed to a ladder offering two vertical or nearly vertical risers between which rungs forming a means of climbing up or down.

Static lift The physical act of a person lifting a load using muscle contraction but no motion. A static lift is also known as an isometric lift.

Stroke width The width of the line that defines the font.

Synthesised speech Machine-generated speech that is produced by a voice synthesiser.

Task-based design Design of interfaces based on the complexity, sequence, importance and error tolerance of tasks performed on the bridge.

TCPA Time to closest point of approach.

Toe board A barrier placed along the edge of a walking surface to prevent personnel from placing their foot over the edge of an elevated walking surface. It should also prevent objects from sliding or rolling over the edge onto personnel below.

Track The intended or desired direction of travel of a vessel.

Tracking The process of observing the sequential changes in the position of a target, to establish its motion.

Valve, critical Valves that require rapid and/or frequent access during normal operations or emergency conditions. These are referred to as Category 1 Valves for valve criticality categorisation purposes.

Valve, infrequently used non-critical Valves that are not critical for operations or routine maintenance and that are infrequently used for particular tasks like commissioning, start-up, shut down, or rarely performed maintenance tasks. These are referred to as Category 3 Valves for valve criticality categorisation purposes.

Valve, non-critical Valves that are not critical for operations but that are required for routine maintenance. These are referred to as Category 2 Valves for valve criticality categorisation purposes.

Visual alarms Alarms presented visually, and carry visually loaded information, such as coding priority by colour, or the presenting of text information.

Visual display A device that provides readings, status, or condition of machinery, equipment, or systems operating parameters. Examples of visual displays include. indicators, annunciators, sight glasses, gauges, digital counters, computer monitors, or video display units.

Warning A cautionary or deterrent indication of an impending abnormal situation.

Way-point Any of the various intermediate points on a route.

Wheelhouse The enclosed area of the bridge.

Whole-body vibration Mechanical vibration (or shock) transmitted to the human body as a whole. Whole body vibration is often due to the vibration of a surface supporting the body.

Work schedules The number of continuous hours and the work shift patterns performed by bridge persons on watch duty.

Work schedules The number of continuous hours and the work shift patterns performed by personnel.

Workspaces Areas allocated for work. Categories of workspaces include navigation spaces, service spaces (galley, laundry) and machinery spaces.

Workstation Position at which one or several tasks constituting a particular activity are carried out. This position includes a combination of job-related items necessary to fulfil job tasks such as the console with all devices, equipment and furniture.

References

Part 1

ADULTDATA: the handbook of adult anthropometric and strength measurements—Data for design safety. Institute for Occupational Ergonomics, University of Nottingham, UK, 1998

Advanced Technology to Optimise Manpower Onboard Ships (1994) Task 2309 design requirements for ship control centre (Bridge)

American Bureau of Shipping (2000) Guide for bridge design and navigational equipment/systems. Houston, TX

American Bureau of Shipping (2001a) Guide for crew habitability on ships. Houston, TX

American Bureau of Shipping (2001b) Rules for building and classing steel vessels under 90 meters (295 feet) in length. Houston, TX

American Bureau of Shipping (2002a) Guide for crew habitability on offshore installations. Houston, TX

American Bureau of Shipping (2002b) Rules for building and classing steel vessels. Houston, TX

American Bureau of Shipping (2003) Guidance notes on the application of ergonomics to marine systems. Houston, TX

American National Standard Institute (1991) Safety color code—Environmental facility safety signs—Criteria for safety symbols—Product safety sign and labels and accident prevention tags New York, NY

American National Stardards Institute (1997) American national standard for product safety signs and labels (ANSI Z535.4-1997). Author, New York, NY

American National Standards Institute (2002) Criteria for safety symbols requirements (ANSI A14.3). Author, New York, NY

American Society for Testing and Materials (2007) Standard practice for human engineering design for marine systems, equipment and facilities (ASTM F 1166-2007). West Conshohocken, PA

American Society for Testing and Materials (2010) Standard practice for human engineering program requirements for ships and marine systems, equipment, and facilities (ASTM F 1337-2010). West Conshohocken, PA

American Society of Testing and Materials. F 1166 (2000) Standard practices for human engineering design for marine systems, equipment and facilities. West Conshohocken, PA

Baker CC (1997) Advanced manning/human systems integration (HSI) for sealift ship technology development program (SSTDP). technical report prepared for John J. McMullen Associates, Inc and the naval sea systems command

Boff K (ed) (1986) Handbook of perception and human performance, vol I. John Wiley and Sons Inc, New York

A. A. Olsen, *Applying Physical Ergonomics to Modern Ship Design*,
https://doi.org/10.1007/978-3-031-57974-5

British Standard EN ISO 8468 (1995) Ship's bridge layout and associated equipment—Requirements and guidelines. London
British Standards Institution (2001) Ergonomics of the thermal environment—Principles and application of relevant international standards (BS ISO 11399: 1995). London
British Standards Institution (2007) Graphical symbols and signs—Safety signs Part 1–11 (BS5499). London
British Standards Institution. Recommendations for the design of VDT tasks—Parts 1–3 (BS7179). London
Carey M (2001) Proposed framework for addressing human factors in IEC 61508. HSE Books, Sudbury, Suffolk, England
Courtney AJ (1986) Chinese population stereotypes: color association. Hum Factors 28:97–99
Dunnette MD (ed) (1983) Handbook of industrial and organisational psychology. John Wiley and Sons, New York
Electric Power Research Institute (1984) Human factors guide for nuclear plant control room development (EPRI NP-3659). Palo Alto, CA
European Committee for Standardisation (2007) Safety of machinery—Human body measurement—Part 1: principles for determining the dimensions required for openings for whole body access into machinery (EN547-1). Brussels
Garg A, Chaffin D, Herrin G (1978) Prediction of metabolic rates for manual materials handling jobs. Am Ind Hyg J 38:661–674
Grandjean E (1988) Fitting the task to the man: a textbook of occupational ergonomics. Taylor and Francis, London
Griffin MJ (1990) Handbook of human vibration. Academic Press, London
Hammer W (1993) Product safety management and engineering, 2nd edn. American Society of Safety Engineers, Des Plaines, IL
Hendrikse EJ et al (2002) Effectively including human factors in the design of new facilities. In: Proceedings of the 2nd international workshop on human factors in offshore operations: demystifying human factors—Practical solutions to reduce incidents and improve safety, quality and reliability. Houston, Texas
Human Factors Society, Inc (2007) Human factors engineering of computer workstations (ANSI/HFS Standard No.100-2007). Santa Monica, CA
Hutchinson RD (1980) New horizons in human factors. McGraw-Hill Inc., New York
IEC 61209 (1999) Maritime navigation and radio communication equipment and systems—Integrated bridge systems (IBS)—Operational and performance requirements, methods of testing and required test results
International Association of Classification Societies (1992) Requirements concerning navigation. London
International Association of Classification Societies (2000) Recommendation 71: guide for the development of technical manuals. London
International Association of Classification Societies (2001) Requirements concerning machinery installations. London
International Association of Classifications Societies (2002) Guidelines for assessing the application of ergonomics to the development process of shipboard complex systems. Guidance No.16
International Electrotechnical Commission (1983) Code for designation of colours. (international standard IEC 60757). IEC Central Office, Geneva
International Electrotechnical Commission (2004) Electroacoustics—Sound level meters—Part 1: specifications (international standard IEC 61672-1). IEC Central Office, Geneva
International Ergonomics Association (2012) What is ergonomics. Retrieved 25 November 2023 from International Ergonomics Association (Home Page): http://www.iea.cc
International Labour Organisation (1990) International data on anthropometry. Occupational safety and health series: No 65. Geneva
International Labour Organisation (1996a) Accident prevention on board ship at sea and in port. Geneva

International Labour Organisation (1996b) Accident prevention on board ship at sea and in port
International Labour Organisation (2006) Maritime labour convention. Geneva
International Maritime Organisation A.343(IX) Recommendation on methods of measuring noise levels at listening posts. London
International Maritime Organisation Resolution A.468(XII) Code on noise levels on board ships. London
International Maritime Organisation Resolution A.694(17) General requirements for shipborne radio equipment forming part of the global maritime distress and safety system (GMDSS) and for electronic navigational aids. London
International Maritime Organisation (1972, as amended) International regulations for preventing collisions at sea (Annex III), rules 6, 7, and 8. London
International Maritime Organisation (1974) International convention for the safety of life at sea (SOLAS), as amended. London
International Maritime Organisation (1981a) Code on noise levels on board ships (IMO resolution A.468(XII)). London
International Maritime Organisation (1981b) International convention on load lines, 1966. London
International Maritime Organisation (1995) Code on alarms and indicators. Resolution A.830(19). London
International Maritime Organisation (1997) Symbols related to life-saving appliances and arrangements (IMO Resolution A.760 (18)). London
International Maritime Organisation (1998) Guidelines for safe access to tanker bows (MSC/Circ.62 (67)). London
International Maritime Organisation (2000) MSC/Circular 982 guidelines on ergonomic criteria for bridge equipment and layout. London
International Maritime Organisation (2002) Resolution 918(22) standard marine communication phrases. London
International Maritime Organisation (2006) Graphic symbols for fire control plans (IMO resolution A.654 (16)). London
International Maritime Organisation (2009) The code for the construction and equipment of mobile offshore drilling units (MODU code)
International Maritime Organisation (2010a) Code on alerts and indicators. London
International Maritime Organisation (2010b) International management code for the safe operation of ships and for pollution prevention (ISM Code). London
International Maritime Organisation (2010c) International convention on standards of training, certification, and watchkeeping for seafarers. London
International Maritime Organisation (2010d) International life-saving appliance code. London
International Maritime Organisation (2010e) International maritime dangerous goods code. London
International Maritime Organisation (2011) International convention of the safety of life at sea (SOLAS). London
International Organisation for Standardisation, ISO 2412 (1982) Shipbuilding—Colours of indicator lights. Geneva
International Organisation for Standardisation, ISO 3434 (1992) Ship building and marine structures—Heated glass panes for ships' rectangular windows. Geneva
International Organisation for Standardisation, ISO 3904 (1990) Ship building and marine structures—Clear-view screens. Geneva
International Organisation for Standardisation, ISO 8864 (1987) Air conditioning and ventilation of wheelhouse on-board ships—Design conditions and basis of calculations. Author, Geneva
International Organisation for Standardisation (1987a) Air conditioning and ventilation of machinery control rooms on board ships—Design conditions and basis of calculations (ISO 8862: 1987(E)). Geneva
International Organisation for Standardisation (1987b) Air conditioning and ventilation of wheelhouse on board ships—Design conditions and basis of calculations (ISO 8864: 1987(E)). Geneva

International Organisation for Standardisation (1999) Human-centred design processes for interactive systems (ISO 13407–1999). Author, Geneva
International Organisation for Standardisation (2001) Mechanical vibration—Guidelines for the measurement, reporting and evaluation of vibration with regard to habitability on passenger and merchant ships. (ISO 6954:2000). Geneva
International Organisation for Standardisation (2003) Mechanical vibration and shock—Evaluation of human exposure to whole body vibration—Part 2, vibration in buildings (ISO 2631-2:2003). Geneva
International Organisation for Standardisation (2005) Lighting of indoor workplaces (ISO 8995)
International Organisation for Standardisation (2008a) Mechanical vibration on ships—Part 2: measurement of structural vibration (ISO 20283-2:2008). Geneva
International Organisation for Standardisation (2008b) Air conditioning and ventilation of accommodations spaces on board ships—Design conditions and basis of calculations (ISO 7547: 2002 (E)). Author, Geneva
International Organisation for Standardisation (2009) Ship and marine technology—Identification colours for the contents of piping systems—Part 1: main colours and media (ISO 14726-1: 2002). Geneva
International Organisation for Standardisation. Ergonomic requirements for work with visual display terminals—Parts 1–16. (VDT's) (ISO/BN/EN 29241). Geneva
Karwolski W, Marras WS (eds) (1998) Occupational ergonomics handbook. CRC Press, Boca Raton
Kearney D (2003) Ergonomic analysis: the fifty factors, job smart, McMurray, Pa
Kirwan B, Ainsworth LK (1992) A guide to task analysis. Taylor and Francis, Ltd, London
Kryter KD (1994) The handbook of hearing and the effects of noise: physiology, psychology and public health. Academic Press, San Diego
Marine Accident Investigation Branch (MAIB) Annual Report 1999. Retrieved 30 November, 2023 from http://www.maib.detr.gov.uk/ar1999/04.htm
Meister D (1985) Behaviour analysis and measurement methods. John Wiley and Sons, New York
Mital A, Nicholson A, Ayoub M (1993) A guide to manual materials handling. Taylor and Francis, Washington, DC
National Aeronautical and Space Administration (1995) Man systems integration standards (NASA-STD-3000, Vol. 1, Revision B). Houston, TX
National Fire Protection Association (NFPA) (2012) Life safety code 101. Quincy, MA
Norwegian Oil Industry Association and the Federation of Norwegian Engineering Industries (NORSOK) (1996) Common requirements, architectural components and equipment (C-CR-002). Oslo
Norwegian Oil Industry Association and the Federation of Norwegian Engineering Industries (NORSOK) (2004) Working environment (S-002). Oslo
O'Neil WA (2001) The human element in shipping. In: Keynote address, biennial symposium of the seafarers international research centre, Cardiff, Wales
Panero J, Zelnik M (1979) Human dimension and interior space: a source book of design reference standards. Whitney Library of Design, New York
Pheasant S (1987) Ergonomics: standards and guidelines for designers. British Standards Institution, Great Britain
Pheasant S (1998) Body space: anthropometry, ergonomics, and the design of work, 2nd edn. Taylor & Francis Ltd., London
Rea MS (ed) (1999) IESNA lighting handbook, 9th edn. Illuminating Engineering Society of North America, New York
Salvendy G (ed) (1997) Handbook of human factors and ergonomics, 2nd edn. John Wiley and Sons Inc., New York
Sanders MS, McCormick EJ (1993) Human factors in engineering and design, 7th edn. McGraw-Hill Inc., New York
Snook SH, Ciriello VM (1991) The design of manual handling tasks: revised tables of maximum acceptable weights and forces. Ergonomics 34:1197

Templer JA (1994) The staircase: studies of hazards, falls and safer design. The MIT Press, Cambridge
Thomas J et al (2002) Application of human factors engineering in reducing human error in existing offshore systems. In: Proceedings of the 2nd international workshop on human factors in offshore operations: demystifying human factors—Practical solutions to reduce incidents and improve safety, quality and reliability. Houston, Texas
UK Department of Trade (1998) Adultdata, the handbook of adult anthropometric and strength measurements—Data for design safety. Department of Trade, Nottingham
US Coast Guard, Office of Marine Safety (1990) Load line technical manual. Washington, DC
US Department of Defence (1995) Handbook for human engineering design guidelines: human engineering (MIL-HDBK-759C). US Government Printing Office, Washington, DC
US Department of Defence (1997) Design criteria standard: noise limits (MIL-STD-1474D). US Government Printing Office, Washington, DC
US Department of Defence (2012) Design criteria standard: human engineering (MIL-STD-1472G). US Government Printing Office, Washington, DC
US Department of Labour (2000a) Code of Federal Regulation, 29 CFR 1910. 23. Subpart D—Walking-working surfaces—Guarding floor and wall openings and holes. Washington, DC
US Department of Labour (2000b) Code of federal regulation, 29 CFR 1910. 24. Subpart D—Walking-working surfaces—Fixed industrial stairs. Washington, DC
Van Cott HP, Kinkade RG (eds) (1992) Human engineering guide to equipment design. McGraw-Hill Book Company, New York, NY
Waters T, Putz-Anderson V, Garg A (1994) Applications manual for the revised NIOSH lifting equation. US Department of Health and Human Services (NIOSH) Publication No. 94-110
Wieringa D, Moore C, Barnes V (1990) Procedure writing: principles and practices. Battelle Press, Columbus Ohio
Woodson WE, Tillman B, Tillman P (1992) Human factors design handbook: information and guidelines for the design of systems, facilities, equipment and products for human use, 2nd edn. McGraw-Hill Inc., New York
Wu S-P (1999) Psychophysically determined infrequent lifting capacity of Chinese participants. Ergonomics 42(7)

Part 2

Distress and Safety System (GMDSS) and for electronic navigational aids [revokes Res.A574(14)]
IACS Rec. 95 Recommendation for the application of SOLAS regulation V/15, bridge design, equipment arrangement and procedures (BDEAP)
IACS UI SC 235 navigation bridge visibility
IMO Res. A.342(IX) Recommendation on performance standards for automatic pilots [see also Res. MSC.64(17)]
IMO Res. A.382(X) Magnetic compasses carriage and performance standards
IMO Res. A.384(X) Performance standards for radar reflector
IMO Res. A.424(XI) Performance standards for gyro-compass
IMO Res. A.477(XII) Performance standards for radar equipment [see also Res. MSC.64(67), Annex 4]
IMO Res. A.526(13) Performance standards for rate-of-turn indicators
IMO Res. A.601(15) Provision and display of manoeuvring information on board ships
IMO Res. A.617(15) Implementation of the NAVTEX system as a component of the worldwide navigational warning service
IMO Res. A.665(16) Performance standards for radio direction-finding systems [revokes Res. A223 (VII)]

IMO Res. A.694(17) General requirements for shipborne radio equipment forming part of the global maritime
IMO Res. A.816(19) Recommendation on performance standards for shipborne decca navigator receivers
IMO Res. A.817(19) Recommendation on performance standards for electronic chart display and information system (ECDIS) [see also Res. MSC.64(67), Annex 5]
IMO Res. A.818(19) Performance standards for Shipborne Loran-C and Chayka receivers
IMO Res. A.823(19) Recommendation on performance standards for automatic radar plotting aids (ARPAS) [for installations before 1/1/97, see Res. A.422(IX)]
IMO Res. MSC.452(99) Amendments to revised performance standards for integrated navigation systems (INS)
IMO Res. MSC.96(72) Performance standards for devices to indicate speed and distance [for installations before 1/1/97, see Res. A.478(XII)]
IMO Res.MSC,252(83) Adoption of the revised performance standards for integrated navigation systems (INS)
IMO Res.MSC.113(73) Performance standards for shipborne GLONASS receiver equipment
IMO Res.MSC.128(75) Performance standard for a bridge navigational watch alarm system (BNWAS)
IMO Res.MSC.137(76) Standards for ship manoeuvrability
IMO Res.MSC.191(79) Performance standards for the presentation of navigation related information on shipborne navigational displays
IMO Res.MSC.192(79) Revised performance standards for radar equipment
IMO Res.MSC.232(82) Adoption of the revised performance standards for electronic chart display and information system (ECDIS)
IMO Res.MSC.64(67), Annex 3 recommendation on performance standards for heading control systems
IMO Res.MSC.64(67), Annex 3. recommendation on performance standards for automatic pilots
IMO Res.MSC.74(69), Annex 2 recommendation on performance standards for track control systems
IMO Res.MSC.74(69), Annex 3 performance standards for automatic identification system
IMO Res.MSC.74(69), Annex 4. recommendation on performance standards for echo-sounding equipment
IMO Res.MSC.96(72), Annex 14 performance standards for devices to measure and indicate speed and distance
MSC. Circular 53(66) Performance standards for shipborne GLONASS receiver equipment
MSC. Circular 64(67) Adoption of new an amended performance standards [revokes Res. A.574; amends Res.A342(IX), Res. A477(XII) and Res. A.817(19)]

Part 3

General

American Bureau of Shipping. Guide for crew habitability on workboats. Houston, TX
American Bureau of Shipping. Rules for building and classing marine vessels. Houston, TX
American Bureau of Shipping. Rules for building and classing steel vessels for service on rivers and intracoastal waterways. Houston, TX
American Bureau of Shipping. Rules for materials and welding: aluminium and fibre reinforced plastics. Houston, TX

Accommodation Area

Accommodations for officers and crew, (code of federal regulations, 46CFR32.40). US Government Printing Office, Washington, DC, pp 398–400
American Society for Testing and Materials (2013a) Standard practice for human engineering design for marine systems, equipment and facilities (ASTM F 1166 –2000). West Conshohocken, PA
International Labour Office (1949) Accommodation of crews convention (revised) (convention C92). Geneva
International Labour Office (1970) Accommodation of crews (supplementary provisions convention (convention C133). Geneva
International Labour Office (1998) Maritime labour conventions and recommendations (4th ed). Geneva
International Maritime Organisation (1974) International convention for the safety of life at sea 600 human factors in ship design and marine operations (SOLAS), 1974: means of escape (Chapter II-2, regulation 13). London
International Maritime Organisation (2006b) Maritime labour convention. Geneva
International Organisation of Masters, Mates, and Pilots. Contract requirements for both existing and new ship construction. Linthicum Heights, MD
National Fire Protection Association (NFPA) (2015) Life safety code 101. Quincy, MA
Norwegian Oil Industry Association and The Federation of Norwegian Engineering Industries (NORSOK) (2004) Working environment (S-002). Oslo
Norwegian Oil Industry Association and The Federation of Norwegian Engineering Industries (NORSOK) (2006) Architectural components and equipment (C-CR-002). Oslo
Panero J, Zelnik M (1979) Human dimension and interior space: a source book of design reference standards. Whitney Library of Design Sailor's Union of the Pacific, New York; (1996) Agreement between Sailor's Union of the Pacific and American
President Lines, Ltd. In the offshore and intercoastal trades. San Francisco, CA
Seafarers International Union. Contract requirements for unlicensed seamen. Camp Springs, MD
US Coast Guard, Department of Transportation (1998a) Coast guard regulation 32. Subpart 32:40
US Coast Guard, Department of Transportation (1998b) Coast guard regulation 92, subparts 92.10 to 92.25, 92.10-1 to 92.25-90 Construction and arrangement, (code of federal regulations, 46CFR92.10-92.25). US Government Printing Office, Washington, DC, pp 51–56
US Coast Guard, Department of Transportation (2008) Coast guard regulation 177, subpart I, 177.900-177.970 construction and arrangement, (code of federal regulations, 46CFR177.900 177.970). US Government Printing Office, Washington, DC, pp 214–216
US Coast Guard, Department of Transportation (2011a) Coast guard regulation 116, subparts D to I, 116.400-116.970 construction and arrangement, (code of federal regulations, 46CFR116.400-116.970). US Government Printing Office, Washington, DC, pp 339–357
US Coast Guard, Department of Transportation (2011b) Coast guard regulation 116, subpart k,116.1110-116.1160 construction and arrangement, (code of federal regulations, 46CFR116.1110-116.1160, p. 358). US Government Printing Office, Washington, DC
US Coast Guard, Office of Marine Safety (2006) Load line technical manual, report number USCG-M-1-90. Washington, DC
US Department of Defence (1999) Design criteria standard: human engineering (MIL-STD-1472F). US Government Printing Office, Washington, DC
US Department of Health and Human Service: Centres for Disease Control and Prevention (2001) Recommended shipbuilding construction guidelines for cruise vessels destined to call on US Ports. Atlanta, GA
US Department of Health, Education and Welfare: Food and Drug Administration (1996) Handbook on sanitation of vessel construction: standards of sanitation and ratproofing for the construction of vessels. Washington, DC
US Department of Labour (2000a) Code of federal regulation, 29 CFR 1910. 23. Subpart D—Walking-working surfaces—Guarding floor and wall openings and holes. Washington, DC

US Department of Labour (2000b) Code of federal regulation, 29 CFR 1910. 24. Subpart D—Walking-working surfaces—Fixed industrial stairs. Washington, DC
US Maritime Administration: National Maritime Resource and Education Centre (1995) Guideline specification for merchant ship construction, prepared by the Office of Ship Design. US Department of Transport, Washington, DC
US Navy (1995a) General specifications for ships of the United States Navy (section 612—Liferails and lifelines). Naval Sea Systems Command, Arlington, VA
US Navy (1995b) General specifications for ships of the united states navy (section 622—Ladders, handrails, floor plates, staging, gratings and service platforms). Naval Sea Systems Command, Arlington, VA
US Navy (1995c) General specifications for ships of the United States Navy (section 640—General requirements for habitability spaces). Naval Sea Systems Command, Arlington, VA
US Navy (1995d) General specifications for ships of the United States Navy (section 644—Plumbing fixtures and fittings). Naval Sea Systems Command, Arlington, VA
US Navy (1995e) General specifications for ships of the United States Navy (section 645—Leisure and community facilities). Naval Sea Systems Command, Arlington, VA
US Navy (1995f) General specifications for ships of the United States Navy (section 652—Medical and dental spaces). Naval Sea Systems Command, Arlington, VA
US Navy (1995g) General specifications for ships of the United States Navy (section 655—Laundry and dry cleaning facilities). Naval Sea Systems Command, Arlington, VA
US Navy (1995h) Shipboard habitability design criteria manual (T9640-AB-DDT- 010/HAB). Naval Sea Systems Command, Arlington, VA
US Navy (2013) Shipboard habitability design criteria and practices manual (surface ships) for new ship designs and modernisation (T9640-AA-PRO-010/HAB). Naval Sea Systems Command, Arlington, VA
Woodson WE, Tillman B, Tillman P (1992) Human factors design handbook: information and guidelines for the design of systems, facilities, equipment and products for human use, 2nd edn. McGraw-Hill Inc., New York

Whole-Body Vibration

Griffin MJ (1990) Handbook of human vibration. Academic Press, London
International Organisation for Standardisation (1998) Mechanical mounting of accelerometers for measuring mechanical vibration and shock (ISO 5348: 1998(E)). Geneva
International Organisation for Standardisation (2000) Mechanical vibration and shock—Guidelines for the measurement, reporting and evaluation of vibration with regard to habitability on passenger and merchant ships. (ISO 6954:2000). Geneva
International Organisation for Standardisation (2003) Mechanical vibration and shock—Evaluation of human exposure to whole body vibration—Part 2, vibration in buildings (ISO 2631-2:2003). Geneva
International Organisation for Standardisation (2005) Human response to vibration—Measuring instrumentation amendment 1 (ISO 8041: 2005). Geneva
International Organisation for Standardisation (2008) Mechanical vibration on ships—Part 2: measurement of structural vibration (ISO 20283-2:2008). Geneva
World Meteorological Organisation (WMO) Manual on codes, No. 306, Part A, alphanumerical codes (2011 Edition, Updated in 2014)

Noise

International Electrotechnical Commission (2013) Electroacoustics—Sound level meters—Part 1: specifications (international standard IEC 61672–1). IEC Central Office, Geneva

International Maritime Organisation (2012) Code on noise levels on board ships (IMO resolution MSC.337(91). London

International Organisation for Standardisation (1996) Acoustics—Measurement of noise onboard vessels (ISO 2923: 1996). Geneva

ISO 10140-2: 2010, Acoustics—Laboratory measurement of sound insulation of building elements—Part 2: measurements of airborne sound insulation

ISO 717-1: 2013, Acoustics—Rating of sound insulation in buildings and of building elements—Part 1: airborne Sound Insulation

Kryter KD (1994) The handbook of hearing and the effects of noise: physiology, psychology and public health. Academic Press, San Diego

World Meteorological Organisation (WHO) Manual on codes, No. 306, Part A, alphanumerical codes (2011 edition, updated in 2014)

Indoor Climate

American National Standards Institute (2013) Thermal environmental conditions for human occupancy (ANSI/ASHRAE 55-2013). Atlanta, GA

International Organisation for Standardisation (1998) Ergonomics of the thermal environment—Instruments for measuring physical quantities (ISO 7726: 1998 (E)). Geneva

International Organisation for Standardisation (2002) Ships and marine technology—Air-conditioning and ventilation of accommodation spaces—Design conditions and basis of calculations (ISO 7547: 2002 (E)). Geneva

National Environmental Balancing Bureau (2005) NEBB procedural standards for testing, adjusting, balancing of environmental systems. Gaithersburg, MD

Lighting and Illumination

BS 667:2005, Illuminance meters. Requirements and test methods

CIE S 023/E:2013 (ISO/CIE 19476:2014), Characterisation of the performance of illuminance meters and luminance meters

DIN 5032-7:1985, Photometry; classification of illuminance meters and luminance meters

DIN EN 12464-1:2011, Light and lighting—Lighting of workplaces—Part 1: indoor workplaces international organisation for standardisation. (2002). lighting of indoor workplaces—Part 1: indoor (ISO 8995:2002). Geneva

Part 4

General

American Bureau of Shipping. Guide for crew habitability on ships. Houston, TX

American Bureau of Shipping. Rules for building and classing steel vessels. Houston, TX

American Bureau of Shipping. Rules for building and classing steel vessels for service on rivers and intracoastal waterways. Houston, TX

American Bureau of Shipping. Rules for building and classing steel vessels under 90 meters (295 feet) in length. Houston, TX

American Bureau of Shipping. Rules for materials and welding: aluminium and fibre reinforced plastics. Houston, TX

Accommodation Area

American Society for Testing and Materials (2006) Standard practice for human engineering design for marine systems, equipment and facilities (ASTM F 1166). West Conshohocken, PA

International Labour Office (1949) Accommodation of crews convention (revised) (convention C92). Geneva

International Labour Office (1970) Accommodation of crews (supplementary provisions convention (convention C92). Geneva

International Labour Office (1998) Maritime labour conventions and recommendations, 4th ed. Geneva

International Labour Organisation (2006) Maritime labour convention. Geneva

International Maritime Organisation (1997) International convention of the safety of life at sea: means of escape (Chapter II-2, regulation 28). London

International Organisation of Masters, Mates, and Pilots. Contract requirements for both existing and new ship construction. Linthicum Heights, MD

National Fire Protection Association (NFPA) (1997) Life safety code 101. Quincy, MA

Norwegian Oil Industry Association and The Federation of Norwegian Engineering Industries (NORSOK) (1996) Common requirements, architectural components and equipment (C-CR-002). Oslo

Panero J, Zelnik M (1979c) Human dimension and interior space: a source book of design reference standards. Whitney Library of Design Sailor's Union of the Pacific, New York; (1996) Agreement between Sailor's Union of the Pacific and American

President Lines, Ltd. In the offshore and intercoastal trades. San Francisco, CA

Seafarers International Union. Contract requirements for unlicensed seamen. Camp Springs, MD

US Coast Guard, Department of Transportation (1998a) Coast guard regulation 32, subpart 32.40, accommodations for officers and crew, (code of federal regulations, 46CFR32.40). US Government Printing Office, Washington, DC, pp 398–400

US Coast Guard, Department of Transportation (1998b) Coast guard regulation 92, subparts 92.10 to 92.25, 92.10-1 to 92.25-90 construction and arrangement, (code of federal regulations, 46CFR92.10-92.25). US Government Printing Office, Washington, DC, pp 51–56

US Coast Guard, Department of Transportation (1998c) Coast guard regulation 116, subparts d to I, 116.400-116.970 construction and arrangement, (code of federal regulations, 46CFR116.400-116.970). US Government Printing Office, Washington, DC, pp 339–357

US Coast Guard, Department of Transportation (1998d) coast guard regulation 116, subpart K, 116.1110-116.1160 construction and arrangement, (code of federal regulations, 46CFR116.1110-116.1160, p 358). US Government Printing Office, Washington, DC

US Coast Guard, Department of Transportation (1998e) Coast guard regulation 177, subpart I, 177.900-177.970 construction and arrangement, (code of federal regulations, 46CFR177.900-177.970). US Government Printing Office, Washington, DC, pp 214–216

US Coast Guard, Office of Marine Safety (1990) Load line technical manual, report number USCG-M-1-90. Washington, DC

US Department of Defence (1998) Design criteria standard: human engineering (MIL-STD-1472F). US Government Printing Office, Washington, DC

US Department of Health and Human Service: Centres for Disease Control and Prevention (1999) Recommended shipbuilding construction guidelines for cruise vessels destined to call on US Ports. Atlanta, GA

US Department of Health, Education and Welfare: Food and Drug Administration (1996) Handbook on sanitation of vessel construction: Standards of sanitation and ratproofing for the construction of vessels. Washington, DC
US Department of Labour (2000a) Code of federal regulation, 29 CFR 1910. 23. Subpart D—Walking-working surfaces—Guarding floor and wall openings and holes. Washington, DC
US Department of Labour (2000b) Code of federal regulation, 29 CFR 1910. 24. Subpart D—US Maritime Administration: National Maritime Resource and Education Centre. (1995) Guideline specification for merchant ship construction, prepared by the office of ship design. US Department of Transport, Washington, DC
US Navy (1994) Shipboard habitability design practices manual (T9640-AA-PRO-010/HAB). Naval Sea Systems Command, Arlington, VA
US Navy (1995a) General specifications for ships of the United States Navy (Section 612—Liferails and lifelines). Naval Sea Systems Command, Arlington, VA
US Navy (1995b) General specifications for ships of the United States Navy (Section 622—Ladders, handrails, floor plates, staging, gratings and service platforms). Naval Sea Systems Command, Arlington, VA
US Navy (1995c) General specifications for ships of the United States Navy (Section 640—General requirements for habitability spaces). Naval Sea Systems Command, Arlington, VA
US Navy (1995d) General specifications for ships of the United States Navy (Section 644—Plumbing fixtures and fittings). Naval Sea Systems Command, Arlington, VA
US Navy (1995e) General specifications for ships of the United States Navy (Section 645—Leisure and community facilities). Naval Sea Systems Command, Arlington, VA
US Navy (1995f) General specifications for ships of the United States Navy (Section 652—Medical and dental spaces). Naval Sea Systems Command, Arlington, VA
US Navy (1995g) General specifications for ships of the United States Navy (Section 655—Laundry and dry cleaning facilities). Naval Sea Systems Command, Arlington, VA
US Navy (1995h) Shipboard habitability design criteria manual (T9640-AB-DDT- 010/HAB). Naval Sea Systems Command, Arlington, VA
Walking-Working Surfaces—Fixed industrial stairs. Washington, DC
Woodson WE, Tillman B, Tillman P (1992) Human factors design handbook: information and guidelines for the design of systems, facilities, equipment and products for human use, 2nd edn. McGraw-Hill Inc., New York

Whole-Body Vibration

Griffin MJ (1990) Handbook of human vibration. Academic Press, London
International Organisation for Standardisation (1996) Acoustics—Measurement of noise shipboard vessels (ISO 2923: 1996). Geneva
International Organisation for Standardisation (1998) Mechanical mounting of accelerometers for measuring mechanical vibration and shock (ISO 5348: 1998(E)). Geneva
International Organisation for Standardisation (2000) Mechanical vibration and shock—Guidelines for the measurement, reporting and evaluation of vibration with regard to habitability on Passenger and Merchant Ships. (ISO 6954:2000). Geneva
International Organisation for Standardisation (2003) Mechanical vibration and shock—Evaluation of human exposure to whole body vibration—Part 2, vibration in buildings. (ISO 2631-2:2003). Geneva
International Organisation for Standardisation (2005) Human response to vibration—Measuring instrumentation (ISO 8041: 2005). Geneva
International Organisation for Standardisation (2008) Mechanical vibration on ships—Part 2: measurement of structural vibration (ISO 20283-2:2008). Geneva
World Meteorological Organisation (1995) International codes vol I.1 Part A alphanumeric codes. American Meteorological Society, Boston, MA

Noise

International Electrotechnical Commission (2000) Integrating-averaging sound level meters
International Electrotechnical Commission (2004) Electroacoustics—Sound level meters—Part 1: specifications (international standard IEC 61672–1). IEC Central Office, Geneva
International Maritime Organisation (1981) Code on noise levels on-board ships (IMO resolution A.468(XII)). London
International Organisation for Standardisation (1996) Acoustics—Measurement of noise onboard vessels (ISO 2923: 1996). Geneva
International Standard IEC 60804(2000-10). IEC Central Office, Geneva
Kryter KD (1994) The handbook of hearing and the effects of noise: physiology, psychology and public health. Academic Press, San Diego
World Meteorological Organisation (1995) International codes Vol I.1 Part A alphanumeric codes. American Meteorological Society, Boston, MA

Indoor Climate

American National Standards Institute (2010a) Practices for measuring, testing, adjusting, and balancing shipboard HVAC&R systems. (ANSI/ASHRAE 15-2010). The American Society of Heating, Refrigerating and Air Conditioning Engineers, Inc., Atlanta, GA
American National Standards Institute (2010b) Thermal environmental conditions for human occupancy (ANSI/ASHRAE 55a-2010). The American Society of Heating, Refrigerating and Air Conditioning Engineers, Inc., Atlanta, GA
International Organisation for Standardisation (1998) Ergonomics of the thermal environment—Instruments for measuring physical quantities (ISO 7726: 1998 (E)). Geneva
National Environmental Balancing Bureau (2005) NEBB procedural standards for testing, adjusting, balancing of environmental systems. Gaithersburg, MD

Lighting and Illumination

Illuminating Engineering Society of North America (1997) Recommended practice for marine lighting (IESNA RP-12-97). New York
International Organisation for Standardisation (2000) Lighting of indoor workplaces (ISO 8995:2000). Geneva

Part 5

General

American Bureau of Shipping. Guide for crew habitability on ships. Houston, TX
American Bureau of Shipping. Rules for building and classing facilities on offshore installations. Houston, TX
American Bureau of Shipping. Rules for building and classing floating production installations. Houston, TX
American Bureau of Shipping. Rules for building and classing offshore installations. Houston, TX
American Bureau of Shipping. Rules for building and classing steel vessels. Houston, TX

Accommodation Area

American Society for Testing and Materials (2013) Standard practice for human engineering design for marine systems, equipment and facilities (ASTM F 1166). West Conshohocken, PA

International Labour Office (1949) Accommodation of crews convention (revised) (convention C92). Geneva

International Labour Office (1970) Accommodation of crews (supplementary provisions convention (convention C92). Geneva

International Labour Office (1998) Maritime labour conventions and recommendations (4th ed). Geneva international labour organisation. maritime labour convention, 2006. Geneva

International Maritime Organisation (1974) International convention for the safety of life at sea (SOLAS), 1974: means of escape (Chapter II-2, Regulation 13). London

International Organisation of Masters, Mates, and Pilots. Contract requirements for both existing and new ship construction. Linthicum Heights, MD

National Fire Protection Association (NFPA) (2015) Life safety code 101. Quincy, MA

Norwegian Oil Industry Association and The Federation of Norwegian Engineering Industries (NORSOK) (2004) Working environment (S-002). Oslo

Norwegian Oil Industry Association and The Federation of Norwegian Engineering Industries (NORSOK) (2006) Architectural components and equipment (C-002). Oslo

Panero J, Zelnik M (1979) Human dimension and interior space: a source book of design reference standards. Whitney Library of Design, New York

Sailor's Union of the Pacific (1996) Agreement between Sailor's Union of the Pacific and American President Lines, Ltd. In the offshore and intercoastal trades. San Francisco, CA

Seafarers International Union. Contract requirements for unlicensed seamen. Camp Springs, MD

US Coast Guard, Department of Transportation (1998a) Coast guard regulation 32, subpart 32.40, accommodation for officers and crew, (code of federal regulations, 46CFR32.40). US Government Printing Office, Washington, DC, pp 398–400

US Coast Guard, Department of Transportation (1998b) Coast guard regulation 92, subparts 92.10 to 92.25, 92.10-1 to 92.25-90 construction and arrangement, (code of federal regulations, 46CFR92.10-92.25). US Government Printing Office, Washington, DC, pp 51–56

US Coast Guard, Department of Transportation (2008) Coast guard regulation 177, subpart I, 177.900-177.970 construction and arrangement, (code of federal regulations, 46CFR177.900-177.970). US Government Printing Office, Washington, DC, pp 214–216

US Coast Guard, Department of Transportation (2011a) Coast guard regulation 116, subparts D to I, 116.400-116.970 construction and arrangement, (code of federal regulations, 46CFR116.400 116.970). US Government Printing Office, Washington, DC, pp 339–357

US Coast Guard, Department of Transportation (2011b) Coast guard regulation 116, subpart K, 116.1110-116.1160 construction and arrangement, (code of federal regulations, 46CFR116.1110-116.1160). US Government Printing Office, Washington, DC, p 358

US Coast Guard, Office of Marine Safety (2006) Load line technical manual, report number USCG-M-1-90. Washington, DC

US Department of Defence (1999) Design criteria standard: human engineering (MIL-STD-1472F). US Government Printing Office, Washington, DC

UK Department of Energy (1996) Offshore installations: guidance on design, construction and certification, 4th edn. HMSO, London

US Department of Health and Human Service: Centres for Disease Control and Prevention (2001) Recommended shipbuilding construction guidelines for cruise vessels destined to call on US Ports. Atlanta, GA

US Department of Health, Education and Welfare: Food and Drug Administration (1996) Handbook on sanitation of vessel construction: standards of sanitation and ratproofing for the construction of vessels. Washington, DC

US Department of Labour (2000a) Code of federal regulation, 29 CFR 1910. 23. Subpart D—Walking-working surfaces—Guarding floor and wall openings and holes. Washington, DC

US Department of Labour (2000b) Code of federal regulation, 29 CFR 1910. 24. Subpart D—Walking-working surfaces—Fixed industrial stairs. Washington, DC

US Maritime Administration: National Maritime Resource and Education Centre (1995) Guideline specification for merchant ship construction, prepared by the office of ship design. US Department of Transport, Washington, DC

US Navy (1995a) General specifications for ships of the United States Navy (Section 612—Liferails and lifelines). Naval Sea Systems Command, Arlington, VA

US Navy (1995b) General specifications for ships of the United States Navy (section 622—Ladders, handrails, floor plates, staging, gratings and service platforms). Naval Sea Systems Command, Arlington, VA

US Navy (1995c) General specifications for ships of the United States Navy (section 640—General requirements for habitability spaces). Naval Sea Systems Command, Arlington, VA

US Navy (1995d) General specifications for ships of the United States Navy (Section 644—Plumbing fixtures and fittings). Naval Sea Systems Command, Arlington, VA

US Navy (1995e) General specifications for ships of the United States Navy (Section 645—Leisure and community facilities). Naval Sea Systems Command, Arlington, VA

US Navy (1995f) General specifications for ships of the United States Navy (Section 652—Medical and dental spaces). Naval Sea Systems Command, Arlington, VA

US Navy (1995g) General specifications for ships of the United States Navy (Section 655—Laundry and dry cleaning facilities). Naval Sea Systems Command, Arlington, VA

US Navy (1995h) Shipboard habitability design criteria manual (T9640-AB-DDT- 010/HAB). Naval Sea Systems Command, Arlington, VA

US Navy (2013) Shipboard habitability design criteria and practices manual (surface ships) for new ship designs and modernization (T9640-AA-PRO-010/HAB). Naval Sea Systems Command, Arlington, VA

Woodson WE, Tillman B, Tillman P (1992) Human factors design handbook: information and guidelines for the design of systems, facilities, equipment and products for human use, 2nd edn. McGraw-Hill Inc., New York

Whole-Body Vibration

Griffin MJ (1990) Handbook of human vibration. Academic Press, London

International Organisation for Standardisation (1998) Mechanical mounting of accelerometers for measuring mechanical vibration and shock (ISO 5348: 1998(E)). Geneva

International Organisation for Standardisation (2000) Mechanical vibration and shock—Guidelines for the measurement, reporting and evaluation of vibration with regard to habitability on passenger and merchant ships. (ISO 6954:2000). Geneva

International Organisation for Standardisation (2003) Mechanical vibration and shock—Evaluation of human exposure to whole body vibration—Part 2, vibration in buildings. (ISO 2631-2:2003). Geneva

International Organisation for Standardisation (2005) Human response to vibration—Measuring instrumentation (ISO 8041:2005). Geneva

International Organisation for Standardisation (2008) Mechanical vibration on ships—Part 2: measurement of structural vibration (ISO 20283-2:2008). Geneva

World Meteorological Organisation (WMO) (2011) Manual on codes, No 306, Part A, alphanumerical codes (2011 Edition, Updated in 2014)

Noise

International Electrotechnical Commission (2013) Electroacoustics—Sound level meters—Part 1: specifications (international standard IEC 61672-1). IEC Central Office, Geneva

International Maritime Organisation (2012) Code on noise levels on-board ships (IMO resolution MSC.337(91) (2012). London

International Organisation for Standardisation (1996) Acoustics—Measurement of noise onboard vessels (ISO 2923: 1996). Geneva

ISO 10140-2: 2010, Acoustics—Laboratory measurement of sound insulation of building elements—Part 2: measurements of airborne sound insulation

ISO 717-1: 2013, Acoustics—Rating of sound insulation in buildings and of building elements—Part 1: airbourne sound insulation

Kryter KD (1994) The handbook of hearing and the effects of noise: physiology, psychology and public health. Academic Press, San Diego

World Meteorological Organisation (WMO) (2011) Manual on codes, No 306, Part A, alphanumerical codes (2011 edition, updated in 2014)

Indoor Climate

American National Standards Institute (2013) Thermal environmental conditions for human occupancy (ANSI/ASHRAE 55-2013). Atlanta, GA

International Organisation for Standardisation (1998) Ergonomics of the thermal environment—Instruments for measuring physical quantities (ISO 7726: 1998 (E)). Geneva

International Organisation for Standardisation (2002) Ships and marine technology—Air-conditioning and ventilation of accommodation spaces—Design conditions and basis of calculations (ISO 7547: 2002 (E)). Geneva

International Organisation for Standardisation (2007) Petroleum and natural gas industries—Offshore production installations—Heating, ventilation and air-conditioning (ISO 15138:2007 (E)). Geneva

National Environmental Balancing Bureau (2005) NEBB procedural standards for testing, adjusting, balancing of environmental systems. Gaithersburg, MD

Lighting and Illumination

BS 667:2005, Illuminance meters. requirements and test methods

CIE S 023/E:2013 (ISO/CIE 19476:2014), characterization of the performance of illuminance meters and luminance meters

DIN 5032-7:1985, Photometry; classification of illuminance meters and luminance meters

DIN EN 12464-1:2011, light and lighting—Lighting of workplaces—Part 1: indoor workplaces

International Organisation for Standardisation (2002) Lighting of workplaces—Part 1: indoor (ISO 8995:2002). Geneva

Part 6

Bercha FG (2006) Recent developments in arctic EER (escape, evacuation and rescue). The Bercha Group

BIMCO (2005) Ice handbook. Bjorn Tomsen A/S. Denmark
Bridges R (2005) Cold climate navigation, design and operation considerations. Lloyd's Register Technical Notes
Canadian Coast Guard (1999) Ice navigation in canadian waters
Finnish Maritime Administration (2005) Winter navigation 2004–2005
Harrington RL (ed) (1992) Marine engineering. Society of Naval Architects and Marine Engineers, Jersey City, NJ
IACS (2006) UR I1—Polar class descriptions and application; UR I2—Structural requirements for polar class ships; UR I3—Machinery requirements for polar class ships (effective for vessels contracted for construction after 1 July 2007)
International Labour Organisation (2001) Ambient factors in the workplace. Geneva. IMO DE 48 Guidelines
IMO MEPC 54/2/5—Draft practical guidelines for a ballast water exchange management strategy in the antarctic treaty area
IMO MSC Circular 1056/MEPC Circular 399 (2002) Guidelines for ships operating in Arctic ice-covered waters
IMO MSC Circular 1175 (2005) Guidance on shipboard towing and mooring equipment
IMO MSC Circular 1185 (2006) Guide for cold water survival
IMO Life Saving Appliances (2003) (from the following two codes)
IMO Resolution A.749(18) (1993) and updates (2002), code on intact stability for all types of ships covered by IMO instruments
IMO Resolution MSC.81(70) (1998) revised recommendation on testing of life-saving appliances
IMO Resolution MSC.48(66) (1996) life saving appliance code
IMO (2002) MSC.152(78), adoption of amendments to the international convention for the safety of life at sea, 1974, as amended. 5-20-04 (lifeboats, EPIRB's, immersion suits)
Kitagawa H et al (2000) The northern sea route—The shortest sea route linking East Asia and Europe. Ship and Ocean Foundation, Japan
Lamb T (ed) (2004) Ship design and construction, Volume II, Chapter 40, ice-capable ships, society of naval architects and marine engineers. Jersey City, NJ
MacKinnon S (2006) Operational and human factors in cold maritime environments, memorial university
Magelssen W, Grubisic R (2004) Operation of ships in cold climate with emphasis on tankers and new requirements. Brodo Gradnja
Mahmood M, Revenga A (2006) Design aspects of winterised and Arctic LNG carriers—A classification perspective. American Bureau of Shipping
Ministry of Transport of the Russian Federation (1996) Guide to navigation though the northern sea route
Russian Maritime Register of Shipping (2003) Rules for the classification and construction of sea-going ships. St. Petersburg, Russia
Santos-Pedro VM (2005) Ice regime system, exporting Canada's experience. Transport Canada
Swedish Maritime Administration (2005) Winter navigation 2004–2005
Transport Canada (2005) Winter navigation on the river and Gulf of St. Lawrence: practical notebook for marine engineers and deck officers
Tsoy LG (2002) Report on current regulations and practices and impact of IMO regulations). Central Marine Research and Design Institute
Tunik A (1998) Major factors in tanker design for oil transportation from Arctic Regions. American Bureau of Shipping
Tustin R (2005) Recent developments in LNG and sice-class tanker design and the potential application to future Arctic LNG ships. Lloyd's Register

The manufacturer's authorised representative in the EU is Springer Nature Customer Service Centre GmbH, Europaplatz 3, 69115 Heidelberg, Germany. If you have any concerns regarding our products, please contact ProductSafety@springernature.com

Printed and bound by CPI Group (UK) Ltd, Croydon, CR0 4YY
28/09/2026
02224065-0020